# 3rd International Conference on Science and Engineering in Mathematics, Chemistry and Physics 2015



# (ScieTech 2015)

Journal of Physics: Conference Series Volume 622

## Bali, Indonesia
## 31 January – 1 February 2015

**Editor:**

**Ford Lumban Gaol**

ISBN: 978-1-5108-0802-7
ISSN: 1742-6588

**Printed from e-media with permission by:**

Curran Associates, Inc.
57 Morehouse Lane
Red Hook, NY 12571

CURRAN ASSOCIATES INC. proceedings.com

**Some format issues inherent in the e-media version may also appear in this print version.**



Printed by Curran Associates, Inc. (2015)



Institute of Physics
Dirac House, Temple Back
Bristol BS1 6BE UK

Phone: 44 1 17 929 7481
Fax: 44 1 17 920 0979

techtracking@iop.org

**Additional copies of this publication are available from:**

Curran Associates, Inc.
57 Morehouse Lane
Red Hook, NY 12571 USA
Phone: 845-758-0400
Fax: 845-758-2634
Email: curran@proceedings.com
Web: www.proceedings.com

## *Table of contents*

## Volume 622

**3rd International Conference on Science & Engineering in Mathematics, Chemistry and Physics 2015 (ScieTech 2015)**
**31 January to 1 February 2015, Bali, Indonesia**

**Accepted papers received: 15 May 2015**
**Published online: 22 June 2015**

### Preface



### Papers

The 3rd International Conference on Science & Engineering in Mathematics, Chemistry and Physics 2015 (ScieTech 2015), was held at The Westin Resort Nusa Dua, Bali on 31 January - 1 February 2015.

The ScieTech 2015 conference is aimed to bring together researchers, engineers and scientists from around the world. ScieTech 2015 is placed on promoting interaction between the theoretical, experimental, and applied communities, so that a high level exchange is achieved in new and emerging areas within mathematics, chemistry and physics.

As we already know that science and technology have brought tremendous benefits for human civilization. People are becoming healthier, wealthier, better educated, more peaceful, increasingly connected, and living longer. Of course, science and technology provide many answers to global challenges, but we will face more complex problems in the next decade due to increasing world population, limitation of energy, and climate change. Therefore, researchers should be more active in conducting research that enables collaboration between one and the others.

Interdisciplinary cooperation is absolutely necessary in order to create a smart system for solving the global problems. We need a global and general long-term view of the future with long-range goals for solving complex problems in next decade. Therefore the conference was held to be a forum for researchers from different disciplines to start collaborating and conducting research that provides a solution to the global issues.

The theme of ScieTech 2015 was "The interdisciplinary Application between Mathematics, Chemistry and Physics to enhance the Quality of Life".

We would like to express our sincere gratitude to all in the Technical Program Committee who have reviewed the papers and developed a very interesting conference program as well as the invited and plenary speakers.

This year, we received 197 papers and after rigorous review, 59 papers were accepted. The participants came from 19 countries, and there were six paralell sessions and four keynote speakers.

It is an honour to present this volume of *Journal of Physics: Conference Series* (JPCS) and we deeply thank the authors for their enthusiastic and high-grade contributions. Finally, we would like to thank the conference chairmen, members of the steering committee, the organizing committee, the organizing secretariat and the financial support from the conference sponsors that allowed the success of ScieTech 2015.

Editor of the ScieTech 2015

Dr. Ford Lumban Gaol
Bina Nusantara University

Journal of Physics: Conference Series **622** (2015) 011002 doi:10.1088/1742-6596/622/1/011002

# The 3rd International Conference on Science & Engineering in Mathematics, Chemistry and Physics 2015 (ScieTech 2015)

## The Westin Resort Nusa Dua, Bali

## 31 January - 1 February 2015

Journal of Physics: Conference Series **622** (2015) 011002 doi:10.1088/1742-6596/622/1/011002

Journal of Physics: Conference Series **622** (2015) 011002 doi:10.1088/1742-6596/622/1/011002

# The 3rd International Conference on Science & Engineering in Mathematics, Chemistry and Physics 2015 (ScieTech 2015), The Westin Resort Nusa Dua, Bali

## 31 January - 1 February 2015

## Keynote Speakers

**Professor Frank Stephan, National University of Singapore**

**Title:** Automatic Structures - Recent Results and Open Questions

**Abstract:**

Regular languages are languages recognised by finite automata; automatic structures are a generalisation where one looks also at automatic relations which are relations recognised by synchronous finite automata and at automatic functions which are functions whose graph is an automatic relation. Functions and relations first-order definable from other automatic functions and relations are again automatic. Automatic functions coincide with the functions computed by position-faithful one-tape Turing machines in linear time. This talk gives an overview on recent results on automatic structures and provides some concurrent research questions.

**ShortBio**

Frank Stephan received his doctorate from the University of Karlsruhe (T.H.) in the year 1990 in mathematics. He since then worked at the University of Karlsruhe (T.H.) until 1995, then at the Univerisity of Heidelberg until 2003, at the National ICT Australia located at the University of New South Wales until 2004 and since then at the National University of Singapore. His working areas are in mathematical logic and theoretical computer science; in particular he works in algorithmic randomness, automata theory, inductive inference and recursion theory.

Journal of Physics: Conference Series **622** (2015) 011003 doi:10.1088/1742-6596/622/1/011003

**Professor Belal E. Baaquie, National University of Singapore, Singapore**

**Title:** Quantum Mathematics and Financial Modeling

**Abstract:**

The mathematics of quantum mechanics is a natural and efficient formalism for modeling financial instruments. A brief review is made of the modeling of option theory using the Feynman path integral. The modeling of forward interest rates is carried out using the framework of a two dimensional quantum field theory and the Hamiltonian is used for determining the martingale condition. An empirical analysis shows that the quantum modeling of forward interest rates is well supported by market data.

**ShortBio:**

Belal E Baaquie is Professor in Department of Physics National University of Singapore, Singapore. He received a BS, Physics (1972) from California Institute of Technology and a PhD, Theoretical Physics from Cornell University in 1976. He was Research Associate at Stanford University, 1976-78; Member, Institute for Advanced Study, Princeton 1991; Physics Department, Harvard University 1998; Physics Department, National University of Singapore, 1984-2000; University Scholars Programme, Vice-Dean of Physics Dept NUS, 1998-2002.

He has published these books: Path Integrals and Hamiltonians: Principles and Methods (Cambridge University Press), The Theoretical Foundations of Quantum Mechanics (Springer Publications), Exploring Integrated Science (CRC Press), Interest Rates and Coupon Bonds in Quantum Finance (Cambridge University Press), Quantum Finance: Path Integrals and Hamiltonians for Options and Interest Rates (Cambridge University Press). His research interests are: Quantum field theory and Financial modeling based on techniques of quantum theory.

Journal of Physics: Conference Series **622** (2015) 011003 doi:10.1088/1742-6596/622/1/011003

**Professor Dr. Jamil Akhtar, CSIR-CEERI, Pilani-333 031, India**

**Title:** Development of Diamond Detectors for Radiotherapy; Material & Technology

**Abstract**:
Diamond is a wide band gap semiconductor with atomic number close to biological tissues and cells and therefore a potential candidate among Tissue Equivalent Materials (TEM). Synthetic diamond technology has been geared up using Chemical Vapour Deposition (CVD) route to produce poly crystal and single crystal diamond sheets. Whereas sc-Diamond is more useful as dosimeter, pc-Diamond has further applications as UV and high energy particle detector. With the advancement in CVD processes for synthetic diamond substrate, understanding of material properties and related technologies has taken up momentum. The polar surface of synthetic diamond due to CVD process is often Hydrogen terminated. The oxygen termination is therefore required to suppress leakage currents in sensor applications. These issues will be addressed in the talk while explaining technology development for diamond detector in Metal-Insulator-Metal (MIM) configuration. Some experimental results shall also be shared covering I-V at different temperatures.

**ShortBio:**

Jamil Akhtar was born in Ghaziabad, India, in 1959. He received B.Sc (Hons) and M.Sc degrees in Physics with specialization in Electronics in 1977 and 1980 respectively. He joined BEL Ghaziabad as Planner B during 1979-80. From 1980 to 1983, he worked at CSIR-CEERI, Pilani, firstly as CSIR-JRF and then CSIR-SRF for his Ph.D. thesis work on “Study of Two-Dimensional Breakdown Phenomena in Semiconductor Devices”.

Since 1983, he has been associated with the Semiconductor Devices Area of CEERI, Pilani, as Scientist. At present he holds the grade of Chief Scientist at CSIR-CEERI and heads Sensors and Nano-Technology Group. He has visited Technical University of Munich, Germany, in the year 1991-92, under DAAD fellowship program. From 1998 to 2001, he stayed at School of Physical Sciences at J.N.U., New Delhi as research fellow and received Ph.D in 2008 on Mev Ion induced re-ordering in single crystalline silicon.

He has been involved in a number of projects sponsored by DRDO, DAE, ISRO and CSIR and completed successfully with working prototypes. He has been instrumental in imparting Hands-on-Training in MEMS under NPMASS program to faculties from Engineering Institutions including IITs and central universities from all over India. His passion is to develop working prototypes.

His research interest includes; Technology for silicon based millimeterwave IMPATTs and BARITTs, Numerical techniques for semiconductor device simulation, Silicon Microstrip detectors, MEMS micro sensors, Digital Microfluidics, Silicon Carbide detectors, Soft magnetic nano-composite materials, Nanostructures and Nanotechnology, Vibrational energy localization in discrete systems, RF MEMS and Diamond technology for radiation detection in health care and high energy applications.

He is associated with AcSIR (Academy of Scientific and Innovative Research, India) as Professor and coordinates courses in Advanced Semiconductor Electronics and teaches semiconductor Device Physics. He has guided a number of Ph.D, M.Tech and B.Tech students for their thesis work registered in different universities including AcSIR. He holds four patents to his credit and more than sixty papers in international journals and more than hundred papers in international/national conferences besides a number of invited talks, internal research/technical reports and three chapters in Books. He is coeditor of a book by Springer. Dr Akhtar has been awarded Bharat Jyoti and Bharat Shiksha Ratan in 2013 and 2014 respectively. He is a life member of Indian Physics Association, Indian Nuclear Society, Instrument Society of India and IETE, India. He is a member of IEEE, USA. He has been nominated as honorary member of Chemical Society of Georgia in May 2014.

**P. N. Gajjar, Department of Physics, University School of Sciences, Gujarat University,**

**Title:** Engineering New Materials for Thermal Devices: Simulation experiments

**Abstract:**
Scientists have made sufficient progress in developing devices based electronics, spintronics, and photonics. But the progresses to design and develop the devices that can utilize the flow of phonons are still rare. If flow of heat in solids could be controlled as electric current in semiconductor circuits, very large numbers of innovations could happen in thermal engineering.

Journal of Physics: Conference Series **622** (2015) 011003 doi:10.1088/1742-6596/622/1/011003

We know, the thermal radiation heat-ups the surface exposed by it and the efforts are always made to cool the surface. We have not made sufficient efforts to store this heat for its fruitful use at later. The first-most requirement to utilize the heat is to have a material which allows us to control the flow of heat in desire direction and/or stores the heat for longer time. At nanoscale, to study heat transport thorough a material is not an easy task because of thermal contacts to be made on nanoscale devices, heat baths and temperature sensors are to be connected at nanoscale. Not only this, the carrier of heat – Phonons is not point particle with definite properties but bundles of energy that have no mass or charge which remains unaffected by electromagnetic field. These are main causes why the devices that control the flow of heat in a desire manner are still not available for practical use.

Computer simulation helps in overcoming such problems. Because of some simulation studies, temperature profile, heat flux, thermal conductivity, negative differential thermal resistance, interface thermal resistance, thermal rectification are now known at nanoscale. Very recently, we have engineered exponential mass graded material which works as a better option for thermal rectifier as it produces 70-75% rectification of heat flow. We are also in the process of engineering new materials which may be useful in developing thermal devices. Our results of heat transport in monoatomic, diatomic chain, linear mass graded, exponential mass graded, mass defected chain, thermal diode, thermal transistor and thermal logic gates will be presented during the talk. Such study will help in understanding heat transport mechanism in conducting and non-conducting polymers, RNA/DNA chains, nanowires, nanotubes, etc.. Such study is useful in designing molecular/nano-heat pumps and nano-thermal devices.

**Shortbio:**

Prof. P. N. Gajjar born on September 15, 1966. He had his M. Sc., M. Phil. and Ph. D. in Physics. His field of specialization is Condensed Matter Theory / Computational Materials Science / Non linear dynamics. He is Professor and Head, Department of Physics, Gujarat University, Ahmedabad, Gujarat, India. He has visited Italy, Vietnam, Sri Lanka, France, Poland, Singapore, Malaysia, Hong Kong, Indonesia for the presentation of research work and invited talks. He is a member of many academic committees and Fellow of Gujarat Science Academy. He had successfully completed many research projects and guided doctorate students. He has authored/co-authored more than 160 scientific papers in refereed journals of repute. He is recipients of Bharat Jyoti Award. He has credit of receiving best research papers award for five times.

# Peer review statement

All papers published in this volume of *Journal of Physics: Conference Series* have been peer reviewed through processes administered by the proceedings Editors. Reviews were conducted by expert referees to the professional and scientific standards expected of a proceedings journal published by IOP Publishing.

# Comparison of solar panel cooling system by using dc brushless fan and dc water

Y.M.Irwan[1,a], W.Z.Leow[1,b], M.Irwanto[1], Fareq.M[1], S.I.S,Hassan[1], I.Safwati[2] and A.R.Amelia[1]
[1] Centre of Excellence for Renewable Energy, School of Electrical System Engineering, University Malaysia Perlis, (UniMAP), Malaysia.
[2]Institute of Engineering Mathematics, University Malaysia Perlis, (UniMAP), Malaysia.

E-mail: [a]irwanyusoff@unimap.edu.my, [b]wzjohnnyleow_90@hotmail.com

**Abstract.** The purpose of this article is to discuss comparison of solar panel cooling system by using DC brushless fan and DC water pump. Solar photovoltaic (PV) power generation is an interesting technique to reduce non-renewable energy consumption and as a renewable energy. The temperature of PV modules increases when it absorbs solar radiation, causing a decrease in efficiency. A solar cooling system is design, construct and experimentally researched within this work. To make an effort to cool the PV module, Direct Current (DC) brushless fan and DC water pump with inlet/outlet manifold are designed for constant air movement and water flow circulation at the back side and front side of PV module representatively. Temperature sensors were installed on the PV module to detect temperature of PV. PIC microcontroller was used to control the DC brushless fan and water pump for switch ON or OFF depend on the temperature of PV module automatically. The performance with and without cooling system are shown in this experiment. The PV module with DC water pump cooling system increase 3.52 %, 36.27 %, 38.98 % in term of output voltage, output current, output power respectively. It decrease 6.36 °C compare than to PV module without DC water pump cooling system. While DC brushless fan cooling system increase 3.47 %, 29.55 %, 32.23 % in term of output voltage, output current, and output power respectively. It decrease 6.1 °C compare than to PV module without DC brushless fan cooling system. The efficiency of PV module with cooling system was increasing compared to PV module without cooling system; this is because the ambient temperature dropped significantly. The higher efficiency of PV cell, the payback period of the system can be shorted and the lifespan of PV module can also be longer.

## 1. Introduction

With worldwide, energy represents an essential part within human modern society. The energy will be separating in two kinds that are non-renewable energy and renewable energy. Just as the effect of planet global economic development and also growing requirement regarding energy, the non-renewable energy (coal, oil and also natural gas) will be quickly raising. Other than that, the consumption of non-renewable energy raise and lead to the level of atmospheric carbon dioxide to increase that result in temperature of planet to increase. Renewable energy often known as non-conventional energy just likes tidal wave energy, solar energy, geothermal energy, hydropower, and so on. Once the non-conventional energy demand can be first individual make use of its decision, the conventional energy demand will likely be reduce. The actual offered involving green energy to produce an alternative in order to resolve the situation involving worldwide warming.

PV energy is a renewable energy which it is can be a process which will be made of sun ray. Cellular of PV makes use of semi-conducting products, to make certain in case the sunlight hit, which is often became electrical power. Solar cells are made regarding various layers associated with thin silicon. PV is usually desirable electrical energy; it can be abundant, hushed in addition to environmentally friendly, and it can certainly apply in numerous applications. In general, three major sorts of technology are utilized from the manufacturing of photovoltaic cells: monocrystalline; polycrystalline; and also amorphous silicon [1]. PV technology is usually any one of a very useful solution for renewable energy, seeing that it could be natural resources and also pollute-free. Moreover PV technology can certainly to decrease varieties of greenhouse gas emission. The PV cell necessary substance is actually silicon semiconductor (Si).

The specific ambient temperatures and also sun irradiance participate in significant aspect intended for productivity of the PV module. When the ambient temperature of PV module is usually raise, the actual efficiency of PV module is decrease and vice versa. Whenever sun irradiance is usually increased, the output power of PV furthermore increases. More efficiency of PV module, investment payback period of the system can certainly shorten as well the lifespan of PV module will also be extended. This standard rule is certainly the decline through the open cell voltage and also short circuit current drop is determined. Consequently, an efficient PV performance circumstance required cooling. Although PV module operating outside; PV modules have a high irradiance condition upon attaining high temperatures.

J.K. Tonui, [2] researches the improvement of the low cost air typed PV/T collector with the problem of less material consumption and low running costs. Using the ways of add thin flat metal plate in the centre or just add fins in the flow channel, It will achieve high heat energy output, and decrease the temperature of PV, in order to improve the efficiency of electric. J. K. Tonui [2] described air cooling is preferred than the other cooling arrangements due to minimal use of material and low operating cost despite its poor thermo physical properties. Kozak et al. [3] discussed a study of temperature variation of the water cooled PV solar template where the system included the heat exchanger mathematical module has been established and the variation of template temperature with the amount of cooling water has been calculated and analyzed under the different power and the inlet water temperature. In solar PV water-cooling type template system, temperature of PV wall reduced with increasing the cooling water flow, but reduced the amount minimized [3].

Arab, A. [4] reported the water spraying is atomized by control system and spraying unit. The control system includes temperature sensor and microcontroller circuit. Odeh S. [5] carried out the heat energy generated by the modules due to high temperature sunlight will be absorbed by the water particles, allowing the temperature of the module not to rise very high. Gang et al. [6] experimentally studied the performance of a novel heat pipe photovoltaic/ thermal system and validated the model output with measured data. The experimental results showed an improvement in the system efficiency with cooling with water circulation. Recently, Tina et al. [7] investigated the effect of submerging a photovoltaic system in water on the efficiency of the system under high irradiance and ambient temperature conditions.

Moharram et al. [8] developed a cooling system based on water spraying of PV panels. A mathematical model has been used to determine when to start cooling of the PV panels as the temperature of the panels reaches the maximum allowable temperature (MAT). Dan M. J. Doble [9] mentioned when comparing the efficiency reduction of the module operating at high temperature and the module with water layer reducing the solar irradiation towards the module, the performance of the module with the water layer on surface top has better efficiency improvement, thus the effects of refraction on sunlight due to existence of water is negligible, as the efficiency decrease due to this

Journal of Physics: Conference Series **622** (2015) 012001 doi:10.1088/1742-6596/622/1/012001

reason is relatively small. Chinamhora et al [10] used a water cooling system on the front and back of the PV module and the found that the cooling system could improve the efficiency of PV module during clear days, while it had disadvantages during cloudy days. Asachi [11] presents a combined photovoltaic and thermal Solar Panels in order to reduce the heat produced by PV system and enhance the output energy of PV and thermal collector.

Sharp Solar Module ND-130T1J has been chosen to analysis PV modules performance in this investigation. The primary concentration is compared with parameter performances of the PV module with and without cooling system. A report consists of the key parameters and the provisional result through the component to the user.

## 2. Methodology

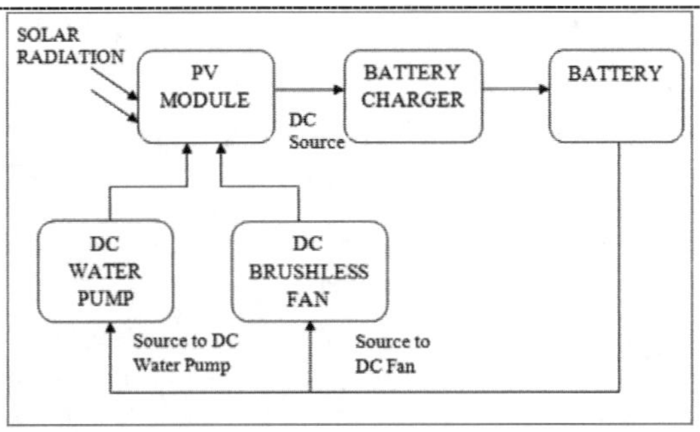


**Figure1.** Block diagram overview of the investigation.

Figure 1 displays the block diagram summary of the investigation. As a way to satisfy the requirement of investigation, the solar energy is chosen as a main supply of this design. The PV module produces electrical energy and provides DC supply to battery charger. The output power of PV module is used to charge the 12 $V_{DC}$ batteries by using battery chargers. It continually charges the battery until it shows at the sign of complete condition on which will cut-off charging process. The utilization of battery is used to store electrical energy that generate by PV module. Battery provide DC source to DC cooling system that is places at back side (DC brushless fan) or front side (DC water pump) of PV module. DC cooling system to reduces temperature for improving efficiency output power of PV module. Two various methods of DC cooling systems have been constructed that is DC brushless fan cooling system and DC water pump cooling system.

**Figure 2.** DC brushless fan cooling system.

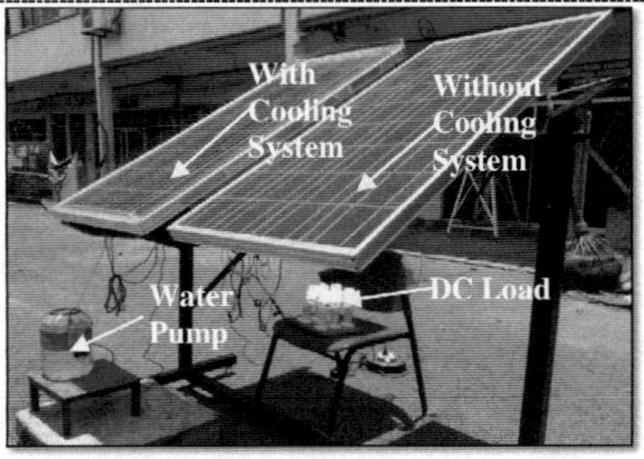


**Figure 3.** DC water pump cooling system.

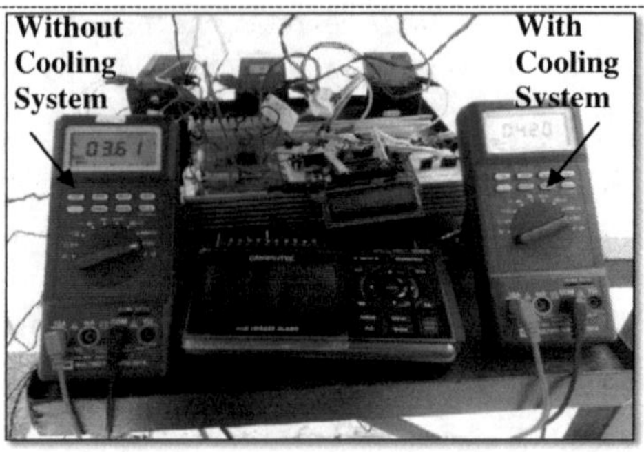


**Figure 4.** Setup experiment with measurement devices.

Within this investigation, two Sharp Solar Module ND-130TIJ polycrystalline solar modules were utilized to convert solar energy into electrical energy. A Sharp Solar Module ND-130T1J was used with 130 W peak power, 22.0V $V_{oc}$, 8.09A $I_{sc}$, module efficiency is 13 % under STC. A PV module set up DC brushless fan cooling system whiles another PV module without DC brushless fan cooling system. Both PV modules were constructed connected with DC lamps (84 W).

Besides that, the DC cooling system is controlled by control system which is microcontroller circuit and temperature sensor. LM 35 as temperature sensor is placed on the back side of the PV module. Four temperatures sensors installed at every PV module which a couple of temperature sensors at the top side along with another a couple of temperature sensors at the top bottom of PV module. The data of temperature ranges were assessed and recorded by using Midi Logger GL220 in most ten moments. Function involving PIC 18F4550 microcontroller is to switch ON / OFF the DC cooling method automatically. When the temperature of PV module reach at or more than 35 °C that detected by using LM 35, the PIC 18F4550 is switched ON the DC cooling system and the other way around. After switch on the cooling system, the brushless fans were blowing at the backside or water pump was spray water at front side of PV module to cool down the temperatures of PV module. This controller system is an intelligent system because it will run the cooling system automatically when the temperature of PV module reaches setting level that detected by temperature sensors and avoid waste electrical energy.

Midi Logger GL220 was using to measure and collect output voltage of both PV modules. In each ten minutes, the output current of both PV modules were measured and recorded by using Digital Multimeter. The daily ambient temperature and solar irradiance were determined by using Davis Vantage PRO2 Weather.

## 3. Results and discussion

### 3.1 The Result of DC Brushless Fan Cooling System

Two various methods of cooling system were conducted at Centre of Excellence Renewable Energy (CERE) on 31 March 2014 from 9:00 a.m until 5:00 p.m. The DC cooling systems were testing at the outdoor CERE.

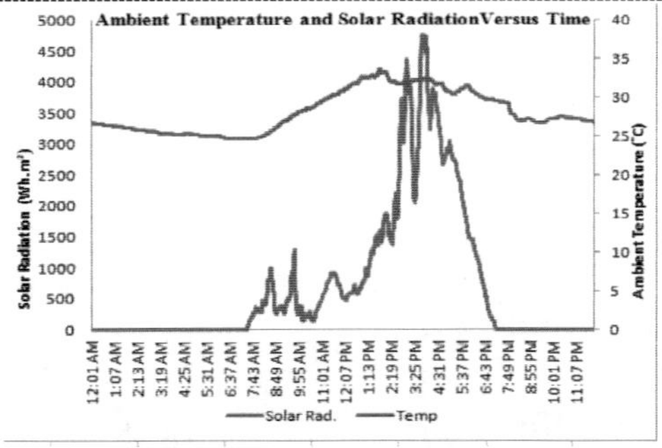


**Figure 5.** Solar radiation and ambient temperature on 31 March 2014 for testing DC cooling system.

Figure 5 displays solar radiation and ambient temperature on 31 March 2014. The maximum solar radiation occurs at 15:48 p.m which was 4752 Wh/m² and the maximum ambient temperature was 33.6 °C which occur at 13:47 p.m.

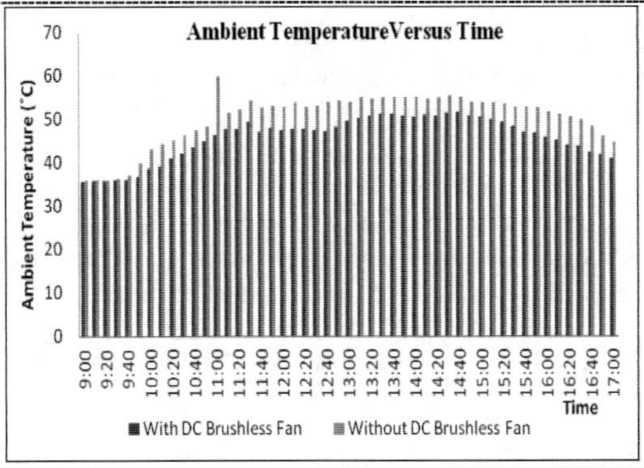


**Figure 6.** Ambient temperature versus time for PV module with and without DC brushless fan.

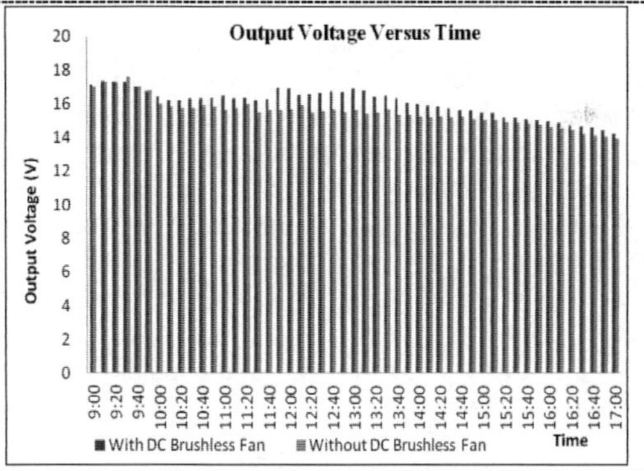


**Figure 7.** Output voltage versus time for PV module with and without DC brushless fan.

Figure 6 displays ambient temperature versus time for PV module with brushless fan cooling system and PV module without brushless fan cooling system. The average temperature of PV module along with cooling system can be 44.77 °C while the average temperature of PV without cooling system seems to be 50.87 °C. The temperature variant of PV module without cooling system seemed to be raising 6.1 °C reviews to be able PV module with cooling system. Efficiency of PV module can certainly be affected by ambient temperature.

Figure 7 shows output voltage versus time for PV module with and without DC brushless fan. The figure displays the maximum output voltage of PV module with DC brushless fan cooling system was

17.31 V while minimum output voltage was 14.21 V. The average output voltage of PV module with DC brushless fan cooling system was 16.05 V. In the same time, the maximum output voltage of PV module without DC brushless fan cooling system was 17.61 V while minimum output voltage was 13.93 V. The average of output voltage which produced by PV module without DC brushless fan cooling system has been 15.49 V. Solar radiation was in fragile condition and some of solar cells cannot be found worked well that afflicted the productivity voltage reduction in this situation. Compared between these kinds of the two systems, the output voltage increased 3.47 % whenever using cooling system. This kind of is caused by the ambient temperature of PV module surface with cooling system decrease; the output voltage of PV module with cooling system has been increasing.

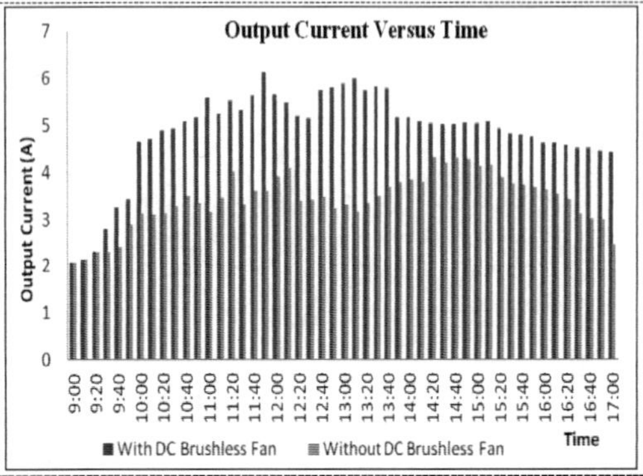


**Figure 8.** Output current versus time for PV module with and without DC brushless fan.

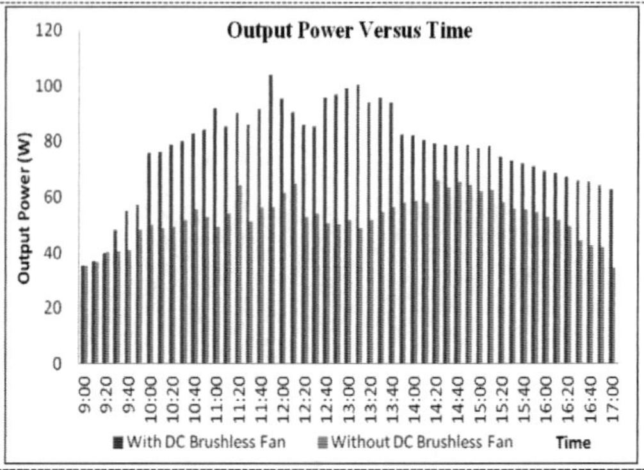


**Figure 9.** Output power versus time for PV module with and without DC brushless fan.

Figure 8 displays examine concerning output current for PV module with cooling and without cooling situation. It could be discovered maximum output current that is generated by PV module with cooling system was 6.13 A while the minimum output current seemed to be 2.07 A. The average output current seemed to be 4.84 A. In the same time, the absolute maximum output current of PV module without cooling system achieved seemed to be 4.32 A while the minimum output current shown on 2.08 A. 3.41 A seem to be the average output current of PV module without cooling system. Through in comparison concerning these kinds of each system, the output current improved 29.55 % when using cooling system. From the end result, it could be discovered that whenever temperature along with solar irradiance increase, the output current likewise increased. Consequently, the output current increased with major percentage in comparison to output current when using DC brushless fan cooling system.

Figure 9 shows output power for PV module with DC brushless fan cooling system and PV module without DC brushless fan cooling system. That figure displays PV module with DC brushless fan cooling system, the output power ended up being a bit improved through 9:00 a.m as well as attained with maximum level at 11:50 p.m after which lowered until finally 5:00 p.m. The adjusting regarding solar irradiance as well as ambient temperature which effect on if which output power unable to keep firm. This condition happen due to the impression regarding winds and clouds the act as barriers within reflected the intensity of light. The maximum output power of PV with DC brushless fan cooling system was measured and calculated at 104.09 W while the minimum output power was 35.44 W. The average output power of PV module with DC brushless fan cooling system was 77.63 W.

In the other sides, the output power of PV module without cooling system was below than 70.0 W. That Figure 9 shows that it can be determined output power starting move increased and reached maximum output power at 65.92 W which is at 14:20 p.m then moved decrease until 5:00 p.m. While the minimum output power of PV module without DC brushless fan cooling system was 34.41 W. and the average output power of PV module without DC brushless fan cooling system was 52.61 W. Through in comparison concerning these kinds of each system, the output power improved 32.23 % when using DC brushless fan cooling system.

### 3.2 The Result of DC Water Pump Cooling System

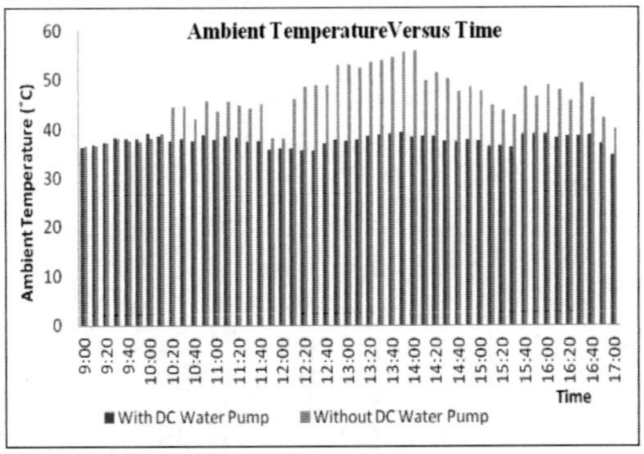


**Figure 10.** Ambient temperature versus time for PV module with and without DC water pump cooling system.

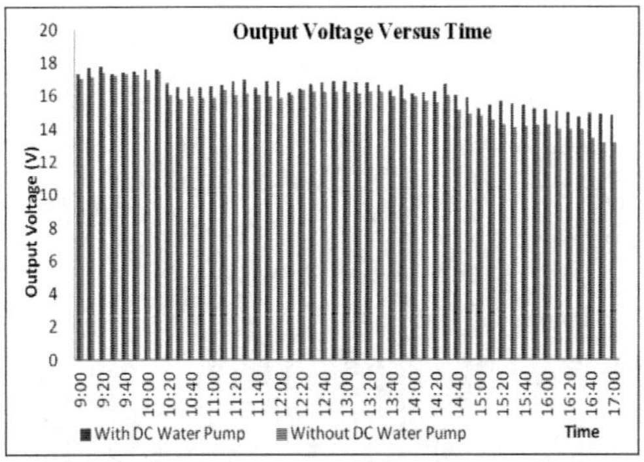


**Figure 11.** Output voltage versus time for PV module with and without DC water pump cooling system.

Figure 10 demonstrates ambient temperature versus time under PV module with water pump cooling system and PV module without water pump cooling system. The average temperature of PV module with cooling system is 37.65 °C while the average temperature of PV without cooling system was 45.74 °C. The temperature variation of PV module without cooling system was increase 6.36 °C compare to PV module with cooling system. Performance of PV module can be affected by ambient temperature. Figure 11 displays the average output voltage of PV module with DC water pump cooling system was 16.453 V while the average of output voltage that produced by PV module without DC water pump cooling system was 15.87 V. Compared between these both systems, the output voltage increased 3.52 % when using cooling system.

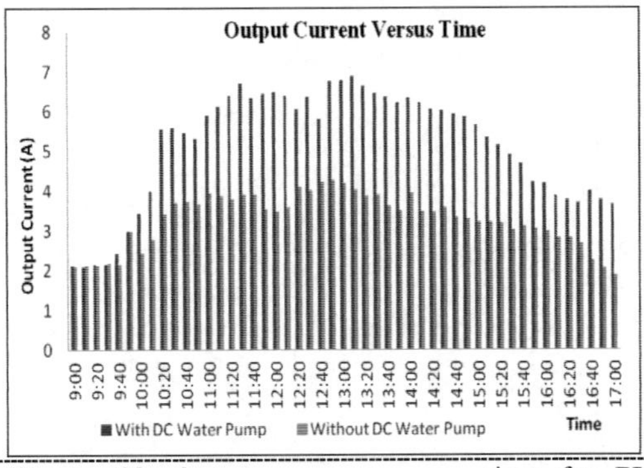


**Figure 12.** Output current versus time for PV module with and without DC water pump cooling system.

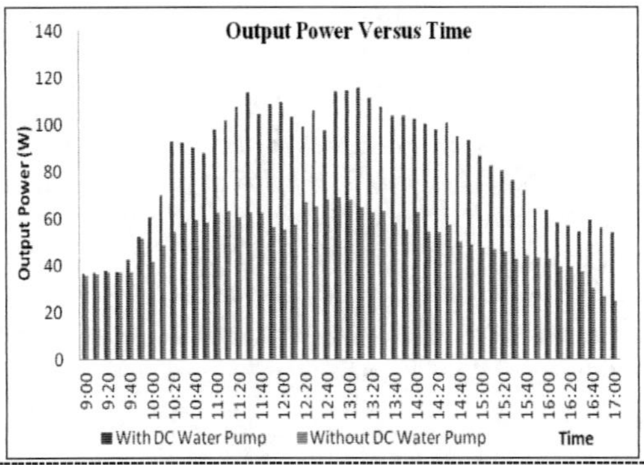


**Figure 13.** Output power of DC water pump under cooling and without DC water pump cooling system.

Figure 12 shows the average output current that produced by PV module with DC water pump cooling system was 5.13 A while 3.27 A was the average output current of PV module without DC water pump cooling system. By compared between these both systems, the output current increased 36.27 % when using DC water pump cooling system.

Figure 13 displays output power for PV module with DC water pump cooling system and PV module without cooling system. The average output power of PV module with cooling system seemed to be 83.79 W whereas the average output power of PV module without cooling system seemed to be 51.13 W. The increase percentage in the output power in unit of electricity is calculated to be 38.98 %.

### 3.3 The Result of the Result of Comparison of Solar Panel Cooling System by Using DC Brushless Fan and DC Water Pump

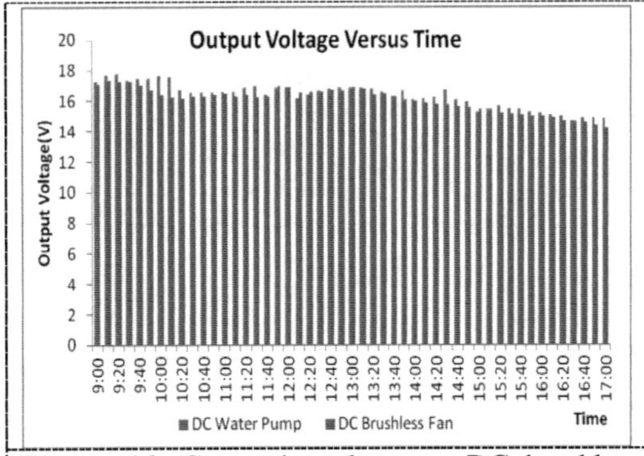


**Figure 14.** Comparison between DC brushless fan and DC water pump in term of output voltage.

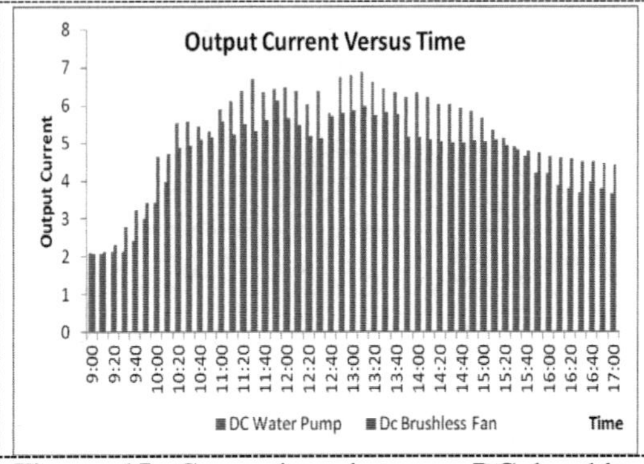


**Figure 15.** Comparison between DC brushless fan and DC water pump in term of output current

Figure 14 presents comparison of output voltage between DC brushless fan and DC water pump. The average output voltage of PV module with DC brushless fan cooling system was 16.05 V while the average output voltage of PV module with DC water pump cooling system was 16.45 V. The output voltage of PV module with DC brushless fan cooling system increase 3.47 % compared to PV module without DC brushless fan cooling system. While output voltage of PV module with DC water pump cooling system increase 3.52 % compare to PV module without DC water pump cooling system. PV module with DC water pump cooling system will generate more 0. 05 % output voltage compared to PV module with DC brushless fan cooling system.

Figure 15 demonstrates comparison of output current between DC brushless fan and DC water pump. The average output current of PV module with DC brushless fan cooling system was 4.84 A while the average output current of PV module with DC water pump cooling system was 3.27 A. The output current of PV module with DC brushless fan cooling system increase 29.55 % compared to PV module without DC brushless fan cooling system. While output current of PV module with DC water pump cooling system increase 36.27 % compare to PV module without DC water pump cooling system. PV module with DC water pump cooling system will generate more 6.72 % output current compared to PV module with DC brushless fan cooling system.

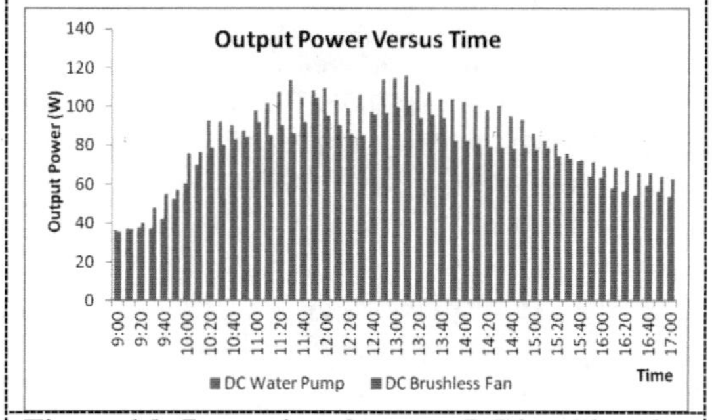


**Figure 16.** Comparison between DC brushless fan and DC water pump in term of output power.

Figure 16 illustrates comparison of output power between DC brushless fan and DC water pump. The average output power of PV module with DC brushless fan cooling system was 77.63 W while the average output power of PV module with DC water pump cooling system was 51.13 W. The output power of PV module with DC brushless fan cooling system increase 32.23 % compared to PV module without DC brushless fan cooling system. While output power of PV module with DC water pump cooling system increase 38.98% compare to PV module without DC water pump cooling system. PV module with DC water pump cooling system will generate more 6.75 % output power compared to PV module with DC brushless fan cooling system.

**Table 1.** Percentage in compare between with cooling system and without cooling system in term of ambient temperature, output voltage, output current and output power.

| Method / Parameters | Brushless Fan | Water Pump |
|---|---|---|
| Ambient Temperature | 6.1 °C | 6.36 °C |
| Output Voltage | 3.47 % | 3.52 % |
| Output Current | 29.55 % | 36.27 % |
| Output Power | 32.23 % | 38.98 % |

## 4. Conclusion

This paper has presented various type of DC cooling system with this PV module cooling system. PIC 18F4550 microcontroller to control the particular cooling system that will diagnose by LM 35 either switches ON or perhaps switches OFF cooling system automatically. Solar irradiance and ambient temperature could be the have effect on the efficiency of PV module. Any time temperature increase, output current is going to be increase but output voltage and power will probably reduce and the other way around. When the solar irradiance increase, output current and power will increase with linear and output voltage will increase with marginal and vice versa. With compare between both systems of various types DC cooling system, the PV module with DC water pump cooling system is most efficient compare to DC brushless fan cooling system. The PV module with DC water pump cooling system increase 3.52 %, 36.27 %, 38.98 % in term of output voltage, output current and output power. It decrease 6.36 °C compare than to PV module without DC water pump cooling system. The

DC brushless fan cooling system increase 3.47 %, 29.55 %, 32.23 % in term of output voltage, output current and output power. It decrease 6.1 °C compare than to PV module without DC brushless fan cooling system. The efficiency of PV module with cooling system was be improving as compared to PV module without cooling system, the reason being that the ambient temperature decreased considerably. More efficiency associated with PV module, investment payback period of the system can minimize and also the lifespan associated with PV module can also be prolonged.

## 5. Acknowledgment

The authors gratefully acknowledge to Kementerian Pengajian Tinggi Malaysia supported under Knowledge Transfer Program (KTP) and the contributions and cooperation from member Centre of Excellence for Renewable Energy (CERE), University Malaysia Perlis (UniMAP) for their work on the original version of this document.

## 6. References

[1] AC Moreira Soares, *et al.* "Simulation of a photovoltaic model using bisection method". 2011, pp. 807-811.

[2] J.K. Tonui, Y. Tripanagnostopoulos Air-cooled PV/T solar collectors with low cost performance improvements. *Solar Energy*, 2007, pp.498–511.

[3] T. Kozak, W. Maranda, A. Napieralski, "Influence of Ambient Temperature on the Amount of Electric Energy Produced by Solar Modules." MIXDES 2009, *16th International Conference Mixed Design of Integrated Circuits and Systems*, June 25-27, 2009.

[4] Arab, A., F. Chenlo and M. Benghanem, Loss-of-load probability of photovoltaic water pumping systems. *Solar Energy,* 76,2004: 713-23.

[5] Odeh. S., Behnia. M., Improving Photovoltaic Module Efficiency using Water Cooling, *Taylor and Francis Ltd, Heat Transfer Engineering*, Volume 30, 2009, pp. 499 – 505.

[6] Gang P, Huide F, Huijuan Z, Jie J. Performance study and parametric analysis of a novel heat pipe PV/T system. *Energy* 2012;37(1):384e95.

[7] Tina GM, Rosa-Clot M, Rosa-Clot P, Scandura PF. Optical and thermal behavior of submerged photovoltaic solar panel: SP2. *Energy* 2012;39(1):17e26.

[8] K.A. Moharram, M.S. Abd-Elhady, H.A. Kandil, H. El-Sherif, Enhancing the per-formance of photovoltaic panels by water cooling, *Ain Shams Engineering Journal 4* (4) (2013) 869–877.

[9] El-Shobokshy M. S., H. F. Effect of dust with different physical properties on the performance of photovoltaic cells. *Solar Energy* 51, 2008, pp. 505-511.

[10] H.G. Doble. D. M. J., Minimization of Reflected Light in Photovoltaic Module, *Renewable Energy World*, Article 03, 2013.

[11] T. Chinamhora, G. Cheng, Y. Tham and W. Irshad: PV Panel Cooling System for Malaysian Climate Conditions, International Conference on Energy and Sustainability, April 27, 2013, Karachi, Pakistan.

[12] Gheorghe Asachi: Combined Photovoltaic and Thermal Solar Panels -Enhanced Energy Conversion and Heat Transfer, *Termotehnica* (2013):135-140.

[13] H.G.Teo, P.S. Lee, M.N.A. Hawlader, "An Active Cooling System for PV Modules," *Applied Energy*, Vol.90, pp. 309-315, 2012.

[14]T. H. Guan, "PV Thermal (PV/T) System: Effect of Active Cooling," Thesis (B. Sc Eng (NCKU)), National University of Singapore, Singapore, 2010.

[15] Brinkworth, B. J., Cross, B. M., Marshall, R. H., and Yang, H., 1997, "Thermal Regulation of Photovoltaic Cladding," *Sol. Energy*, 61, pp. 169–178.

Journal of Physics: Conference Series **622** (2015) 012002 doi:10.1088/1742-6596/622/1/012002

# AN OPTIMIZATION MODEL TO AGROINDUSTRIAL SECTOR IN ANTIOQUIA (COLOMBIA, SOUTH AMERICA)

**J Fernandez**[1]
Titular Professor, *University Pontificia Bolivariana*,

E-mail: javier.fernandez@upb.edu.co

**Abstract.** This paper develops a proposal of a general optimization model for the flower industry, which is defined by using discrete simulation and nonlinear optimization, whose mathematical models have been solved by using ProModel simulation tools and Gams optimization. It defines the operations that constitute the production and marketing of the sector, statistically validated data taken directly from each operation through field work, the discrete simulation model of the operations and the linear optimization model of the entire industry chain are raised. The model is solved with the tools described above and presents the results validated in a case study.

## 1. Introduction

Today the floriculture sector is one of the main economic sectors in Colombia, exporting more than a billion dollars annually. Generating not only economic impact in the country but also a social impact, as it is the largest employer in the agribusiness sector as [1]. However, although Colombia is one of the largest producers of flowers in the world, the Colombian flower industry is going through a difficult economic situation, marked by a cyclical revaluation of the peso against the dollar, by increased production costs, including the labor [2], while sale prices remain the same [1]. Given this scenario it is required the use of various tools to maximize the use in this sector of human capital, which represents between 50% and 60% of total production costs. However, progress in the field largely involves standardizing working methods, determination of capacities, volumes and profitability indicators among other things that in this work are referred to the three phases in which the model has been divided: harvesting, post-harvest and marketing.

In this context, we have developed an optimization model for the flower industry, which has been applied to a real case study with discrete simulation techniques and linear optimization with the help of the tools Gams and Promodel respectively.

This article is organized as follows. Section 2 discusses some related work. Section 3 provides an analysis of the sector. Section 4 presents the model and discusses the results. Finally, we present the conclusions and following work in Section 5.

## 2. Related works

[1] Industrial Engineering Faculty, Circ. 1 N-70-01, B 11 Of. 234, Medellin, Colombia

Some of the related work is found in proposals such as [3] in which is developed a model of optimization and improvement of a process, using time study in a company in the floriculture sector. In this proposal were identified the activities that presented problems and subsequently were defined improvement actions. Diagrams tours technique was implemented.

Meanwhile the work of [4] is based on characterizing the system costs of some floral species whose aim was to provide a tool for the industry to develop their business competitively. The ABC costing technique was used.

However, previous studies did not demonstrate the use of combined discrete simulation techniques and linear optimization for the flower industry, which is the added value of this work.

## 3. Sector situation

### *3.1. International situation*

The global floriculture is characterized by a large number of producing countries, competing in a globalized free market, trying to meet the diverse global markets, which are mainly in developed countries. According to [5] cut flower cultivation is widely spread throughout the world, is included in statistics of 145 countries, according to the International Horticultural Society (ISHS, for its acronym in English), the global surface designed to flowers cut is 60,000 hectares.

It is therefore, that in 2008 global production of flowers per area was distributed as follows: Asia with 75%, Europe 10%, Central and South America 9% 4% North America, Middle East 1% and Africa 1%. As for exports, six (6) countries cover 75% of world exports. The main flower exporting country is the Netherlands with 42%, followed by Colombia, which exports 13%, the other major exporters in the world are: Kenya, Ecuador, Israel and China, with 7%, 5%, 4% and 3% respectively. Of the total supply, 60% is demanded by Germany, United Kingdom, United States, Netherlands, France and Japan. [6]

It is then, that the European continent is the largest consumer of fresh flowers, demanding nearly two-thirds of world exports (65%) in which Germany imports about the same amount that is exported by Colombia (1.043 billion dollars). The second continent that import the most, is the American continent, with 16% share in global consumption of imports of flowers, where the main buyer is the U.S., which imports 90% of total demand for the continent, then followed by Canada and Chile [7].

Undoubtedly the biggest competitor for Colombia in the continent is Ecuador, who although nowadays exported $ 507 million, representing a little more than half of Colombia's exports, this is tending to grow in the coming years. [7]

### *3.2. National situation*

The flower industry in Colombia is one of the main export sectors of the country, which generated in 2011 more than 180,000 direct and indirect jobs in more than 7,000 hectares in 48 municipalities in the country and by 2012 it is estimated that the total of jobs created was 219,323 jobs, of which 120.640 were direct and 68,683 were indirect jobs. [1] [8] becoming a local development actor and a major player in the global market, according to [10] the flower industry is, after coffee, the largest foreign exchange earner of agricultural goods. 95% of production is exported: 84% goes to the U.S. and 9.8% to the European Union [5]

The flower industry is today the largest exporter of non-traditional agriculture, with sales exceeding one billion dollars, representing 25% of female rural labor, is the largest employer in the area of agribusiness in the country (15 employees per hectare). [1] [7]. In addition, the flower industry for the national economy is one of its main sources of income to represent 6.6% of the Colombian agricultural GDP.

The country has three regions that produce the most flowers. These regions are: Cundinamarca that produces 79% of the national flowers, Antioquia 17% and the rest of the country mainly in the Coffee-

Growers Axis and Valle del Cauca 4%. [8]. in 2009, the flower industry had 7,290 hectares cultivated and offered fresh cut flowers for export. The 79% of the cultivated area is located in the Bogotá savannah, 17% in Antioquia and 4% in other departments, which include Valle del Cauca and the Coffee-Growers Axis [8]. Likewise, Colombia produces more than 50 species and 1500 varieties of flowers, making it the largest exporter of flowers to the United States and the second largest exporter in the world after The Netherlands. [8] [9].

The most produced flower species in Colombia today are: roses, carnations, chrysanthemums, alstroemeria and pompons, being the largest producer of carnations. Whose behavior at the production level for 2012, according to [1] was the following:

**Table 1.** Percentages of export per of type of flower.

| Types | % |
|---|---|
| Roses | 32 |
| Carnations | 14 |
| Mini Carnations | 7 |
| chrysanthemums and pompons | 8 |
| Alstromelies | 5 |
| Others | 33 |
| The rest | 1 |

On the other hand, Colombia exports 95% of its production of flowers (cut flowers), mainly roses and carnations. A large percentage of these exports is done in an atomized way, i.e., participate in them many agents and / or brokers, which does not allow competitive pricing in the market, and limit the direct involvement of Colombian flower growers in international distribution channels. Although, currently they are getting higher profits by selling the flowers packaged as "bouquets" (flower arranging) and not in boxes, to marketing centers.

Furthermore, the sector is characterized by flower production made mainly by SMEs that export in an atomized way. The 90% of total exports in 2004 (U.S. $ 703 m) was traded by 228 companies, each of which, did not export on average more than $ 3 m. In this sense, the production of flowers is basically for external market and sales have a seasonal pattern. This condition causes the domestic market takes surplus production and non-exportable flowers. It is also characterized by high working capital requirements (cash flow) for the continuous recruitment of labor, with an imported component in low production elements such as agrochemicals and cuttings.

Likewise, the industry benefits from advantages like the FLORVERDE ® seal, with which certify the quality of their products and the efficiency in production processes worldwide. It has robust social responsibility programs, along with the support of national and international entities; tariff preferences to enter the U.S. market granted by the ATPDEA system, facilitate the competitiveness of Colombian flowers in that market.

By having these advantages and features, the Colombian flower industry is in the ability to export 95% of total flower production, specifically cut flowers (roses and carnations mostly) relatively cheap labor and low costs of transport and storage by proximity to the ports of embarkation. However it is also characterized by low utilization of technical change and little research and development of new varieties and production techniques at national level; high costs associated with the specialized transportation of flowers and the introduction to the U.S. market (Cold management systems, hiring phitosanitary control inspectors and controls to prevent drug trafficking, among others). These costs are borne by exporters, as well as the high exposure to uncontrollable factors such as climate change (hail, frost, excessive rain, and wind), diseases and pests.

## 4. Proposed model

It has been proposed a model based on the floriculture production chain, specifically in the Environmental Guide for floriculture proposed by [11], in which the development process that allows the production of flower comprises four stages: Propagation (plants mothers), Propagation (Banks rooting), production and Post-harvest and support stages of the production process: Construction and maintenance of infrastructure and activities. And the production scheme referenced in [12] called Colombian Florist cluster anatomy, which evidence the elements that feed the system and provide inputs for both harvest and post-harvest also shows the stages of production and presentation of the finished product. Then there are the distribution systems, transportation and marketing, which are mostly in foreign markets, and finally, the cluster support institutions.

Then, the proposed model has been defined, based on three sub-models: the cutting sub-model, post-harvest sub-model and marketing model.

### *4.1. Cutting submodel*

In this study we proceeded to the respective time-making for the case study and their respective statistical validation using Statgraphics tool and statistical module Input Analyzer Arena simulation software and the establishment of model assumptions, the definition of model variables such as: the processing time cutting stages, initial resource number (flower), number of locations, time between cutting cycles, number of branches and branches average time for the simulated model entities of 5, 7 and 9 stems.

Then with the help of the tool Promodel, we proceeded to define the elements of the model as the locations (Consolidation Bouquet, Measurement and Cut, Place rubber and cap, Shaking and place in bucket, receipt and deposit), entities (Flower type (1,2 and 3), consolidation of stems, branches type (1,2 and 3)), arrivals (arrivals were configured like flowers (1,2 and 3) receiving the beds by the number of stems), networks (from receiving to the warehouse), variables (defined quantities and times for each operation of the process, the cycle time for each bed, ie for each number of stems, the total number of stems of each bed and the total number of branches) and renders of the model. As shown in the following figure:

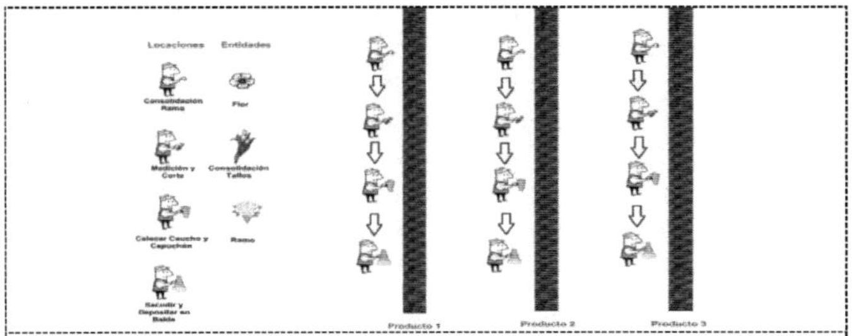


**Figure 1.** Cutting Simulation Model

Finally, we proceeded to set the simulation model, with the aim of determining the standard operation time in the whole process.

**Table 2.** Cutting times.

| Operation | Real average | Simulated average |
|---|---|---|
| Consolidation Bouquet | 0,26 | 0,26 |
| Measuring and cutting | 0,11 | 0,11 |
| Place rubber and cap | 0,23 | 0,21 |
| Shake and Placing | 0,04 | 0,05 |

Where it was determined that the current average cutting time for a bouquet is 0.63 minutes versus an average simulated time of 0.62 minutes, so that an skilled operator in normal conditions, cut 483 stems per hour on average.

It was observed also that the most critical operations of the cutting process are: consolidation stems at an occupancy rate of 28.21% of the total cycle time and place rubber cap with a percentage of 23.26%. Data was statistically validated by applying the central limit theorem creating confidence intervals for each of the 30 replicas simulated operations.

### *4.2. Postharvest Submodel*

We proceeded to the respective time-making for the case study and their respective statistical validation using the Statgraphics tool, as well as the establishment of model assumptions, the definition of model variables as the amounts of meshes in each transaction processed.

Then, with the help of the tool Promodel, we proceeded to define the elements of the model as the locations (Cold Storage, classification, Boncheo, Court, Labeling, and Packaging Rinse), entities (meshes, stems, packages and boxes), arrivals (meshes arrivals were set), nets (from the classification to the storage), resources (operators defined for each operation of the process), attributes (att_meshes and att_general), variables (number of ordered meshes defined, number of branches in boncheo, number of sectors in cutting, the number of branches in labeling and rinsing, the number of packed boxes and number of stored boxes) and renders of the model as shown in the following figure:

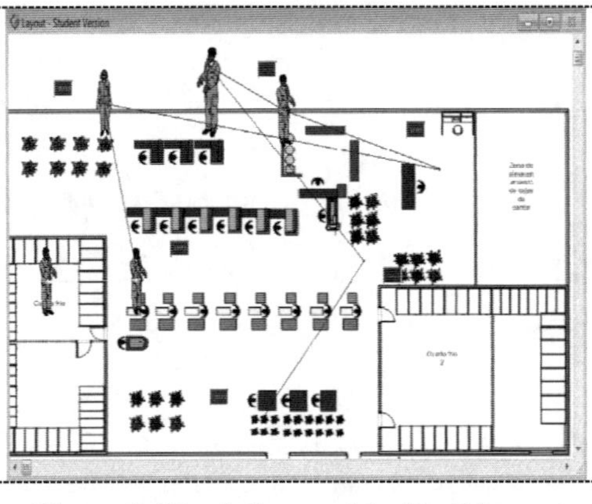

**Figure 2.** Simulation model of post-harvest

Finally, we proceeded to set the simulation model, with the aim of determining the standard operation time in the whole process.

**Table 3.** Times post-harvest

| Operation | Real average | Simulated average |
|---|---|---|
| Classification | 4,52 | 3,66 |
| Boncheo | 1,37 | 1,11 |
| Cutting | 0,19 | 0,15 |
| Labeled | 0,11 | 0,09 |
| Rinse | 0,24 | 0,19 |
| Packing | 2,39 | 1,94 |

From this, it was determined that the real average time of a post-harvest to a mesh that is transformed in a bouquet, and then in box is 8.82 minutes, compared to an simulated time of 7.14 minutes.

*4.3. Marketing Submodel*

For the Marketing submodel, has been taken as reference a linear optimization, whose role was raised around throughput, term used in the theory of constraints (TOC) to refer to operating profit, expressing utility in terms of sales minus unitary costs.

We grouped the varieties of flowers according to the classification proposed by the catalog of flowers [13] and the classification made by the [14]. We selected only those varieties that in addition to high demand in Europe are also produced in Colombia. Defining high demand in Europe as those varieties of flowers that had been, according to Market News Center [14], among the top 15 varieties market for the years 2011 and 2012. Both demand and selling price in Europe were obtained from weekly reports Market News Service [14], taking the years 2011 and 2012.

The behavior of price and demand was separately analyzed in monthly series to identify seasonality. The cost of transport was obtained from Proexport (Colombia) through the Fee Routes tool; the rate provided by this tool was in terms of the weight of the goods, so it was necessary to convert the average weight for a stem of different types of flowers.

The proposed optimization model is of the form:

$$Max\ \sum((Dij * Pij * Xij) - ((Ccij + Ctij) * Xij)) * \mathit{Throughput}$$

S.A:

$$Xijk \leq Dijk: \text{(Restriction of demand)}$$

$$\sum(\sum(Xij) * Hi) \leq Hti: \text{(Restriction of capacity).}$$

Where:

*i: Variety of flower*
*j: Month of the year*
*Xij: The number of flower stalks types i, to market in month j.*
*Dij: The demand for the type of flower stalks units i in month j.*
*CCIJ: Cost of harvest and post-harvest flower per stem type i in month j*
*Ctij: Shipping Cost flower per stem type i in month j.*
*Pij: The selling price of the flower i in month j.*

Journal of Physics: Conference Series **622** (2015) 012002 doi:10.1088/1742-6596/622/1/012002

*Hti: The total capacity of Antioquia square meters for the European market*
*Hi: Area required annually to plant a flower stem type i.*

Taking data for the period 2011-2012, we preceded to the implementation of GAMS tool to solve the issue, so that:

```
Positive Variables
X(i,j) N° de tallos a comercializar de la flor tipo i en el preiodo j;
Variables
Z Función Objetivo
He(i) Hectarias Anuales por tipo de flor i

Equations
Demanda(i,j) Restricción de demanda
Hectareas Resticción de capacidad
FO Funcion Objetivo;

Demanda(i,j).. X(i,j) =L= D(i,j);
Hectareas.. sum(i,sum (j, X(i,j))*Ha(i)) =L= 403900;
Fo.. Z =E= Sum ((i,j), P(i,j)* X(i,j))+ sum(i,sum (j, X(i,j))*Ct(i))+ sum((i,j),Cc(i,j)*X(i,j));

model Ejemplo /All/;
solve Ejemplo MAXIMIZE Z USING LP;
```

**Figure 3.** Flowers marketing optimization model

The following are the results obtained by the proposed model for the analyzed varieties of flowers:

| [N° Stems ] | Ene | Feb | Mar | Abr | May | Jun |
|---|---|---|---|---|---|---|
| Rose Large | 185349 | 239030 | 217196 | 213480 | 229778 | 209979 |
| Cris. Spray | 56514 | 59444 | 61425 | 71361 | 86746 | 0 |
| Cris. S.Bud | 11189 | 11934 | 11444 | 11819 | 14543 | 13695 |
| Rosa M&P | 21592 | 25929 | 26357 | 23760 | 21847 | 21962 |
| Gerbera Min | 34794 | 37970 | 47667 | 47214 | 57542 | 46928 |
| Cris. Santini | 12074 | 13007 | 13040 | 16062 | 18378 | 0 |
| Gerbera Lar | 7665 | 8206 | 10333 | 10093 | 13274 | 10958 |
| Alstroemer | 10458 | 10844 | 11730 | 12967 | 17111 | 17961 |
| Rosa Spray | 7804 | 7598 | 8191 | 9577 | 10800 | 10138 |
| [N° Stems] | **Jul** | **Ago** | **Sep** | **Oct** | **Nov** | **Dic** |
| Rose Large | 186747 | 179219 | 179800 | 198825 | 199715 | 187571 |
| Cris. Spray | 0 | 0 | 77748 | 18353 | 64690 | 58143 |
| Cris. S.Bud | 13760 | 15185 | 16879 | 22277 | 21238 | 11645 |
| Rosa M&P | 18196 | 13640 | 14954 | 18798 | 18205 | 20511 |
| Gerbera Min | 44106 | 45624 | 49412 | 46267 | 42597 | 38325 |
| Cris. Santini | 0 | 0 | 0 | 16798 | 14903 | 11533 |
| Gerbera Lar | 9122 | 9288 | 10280 | 9500 | 8087 | 7112 |
| Alstroemer | 18244 | 17281 | 15828 | 13211 | 11935 | 10688 |
| Rose Spray | 9652 | 9034 | 9099 | 8995 | 9726 | 9089 |

**Figure 4.** Estimated flowers demand optimization model

## 5. Conclusions and future work

The proposed model for the flower industry has managed to develop from three sub-models: the cutting sub-model, the post-harvest sub-model and the marketing sub-model, the first two have been resolved with discrete simulation and the last one with linear programming, whose support tools for resolution have been adapted to the needs of the system.

Facing the model itself, it was established average standard times for cutting threads and post-harvest to ensure adequate amounts of supplies for marketing model, from this, it was concluded that the simulated time for cutting a stem is 0.62 minutes, the post-harvest time for a box is 7.14 minutes, so the time of producing a flower box for export under the proposed models is cutting 6.21 minutes and 7.14 minutes for post-harvest, for a total of 13.35 minutes per box.

Now, for a total volume estimated by the optimization model of 4,448,592 stems, i.e. 88 972 boxes, the cycle time to meet this demand will be 1,187,774 minutes, equivalent to 2200 workdays of 9 hours, which is a critical indicator in the sector to meet a demand of this kind, if we take that the sector directly employs 120.640 direct jobs and under the assumption that all employees are in this lawsuit, we would expect a productivity of 0.73 cases per direct employment against which must be set efficiency strategies to meet the new market approaches.

As future work, it is been carrying out research to optimize the dairy sector, to provide clear evidence against the new schemes and productivity models for the country.

## 6. References

[1]ASOCOLFLORES. Report of activities 2011. Available at: http://issuu.com/asocolflorespc/docs/informe_de_actividades_2011-12?mode=window&backgroundColor=%23222222. 2012.

[2] Agriculture in the Americas. Creativity and Quality Key to Floricultores and present and future of marketing. The Journal of the Agricultural Sector. Volume 36 Number 345. October 2005. Media and Media Publishers.

[3] Jurado, T. M. Application of a study of methods for improving processes snapdragon in the Cornfield Flowers LTDA. 2008.

[4] Duke, A. M., Trejos, E. A., & Duke, V. E. System characterization of chain costs of flowers Tropical (Heliconia) in the department of Risaralda. Scientia Et Technica, 331-336. 2007.

[5] GRINCOMEX. Characterization of International Distributors Flowers of Eastern Antioquia. Catholic University of East Magazine. Issue 27 January-July 2009.

[6] ICESI University, ASOCOLFLORES. Floriculture: A world-class export sector. Available at: http://www.icesi.edu.co/icecomex/images/stories/pdfs/asocolflores.pdf. 2009.

[7]Bolivian Foreign Trade Institute. Market Profile flowers. Available at: http://www.ibce.org.bo/documentos/perfil_mercado_flores.pdf. 2011.

[8] National Administrative Department of Statistics-DANE. Report of Results: Census Flores Production Farms. In 28 municipalities of Sabana de Bogotá and Cundinamarca 2009. Available:<http://www.dane.gov.co/files/investigaciones/boletines/flores/Informe_resultados_2009.pdf> Accessed: February 7, 2012. 2010.

[9] ASOCOLFLORES, Ministry of Agriculture and Rural Development. The finest quality Grown Colombian cut flowers in the world. Bogota: Editorial Design Books and Journals Ltd. 2008.

[10] Reina, M. P. Acosta Oviedo S. Florist Sector against Revaluation: Current Situation and Policy Alternatives. ASOCOLFLORES Project prepared by FEDESARROLLO. Available: Accessed: February 7, 2012. 2008.

[11] ASOCOLFLORES. Environmental Guide for floriculture ASOCOLFLORES. Publishing and Audiovisual Producers Produmedios Available:

http://www.minambiente.gov.co/documentos/floricultor.pdf> Accessed: January 28, 2012. 2002.
[12] Henao, O.L and Herrera A. The cluster model in floriculture: technifies and modernizes business strategies. Agribusiness and Proflora Acopaflor. 2011.
[13] ASOCOLFLORES. Catalog of flowers 2000. 2000.
[14] International Trade Center. Market News Service: Cut Flowers and Ornamental Plants. Disponible en: http://www.intracen.org/policy/cut-flowers/market-news/. 2013.

# Comparison of conformal Toda model quantisation approaches

**Alexander Zuevsky**
Institute of Mathematics, Academy of Sciences of the Czech Republic, Prague
E-mail: zuevsky@yahoo.com

**Abstract.** We make a comparison among quantization schemes for conformal Toda models.

## 1. Introduction

In this review we discuss various approaches to the construction of the Toda Heisenberg feld operator equatons and their solutions. In the case of the non-periodic boundary condition we start with the light-cone formalism which was applied in [12] to the quantization of the conformal Toda model. Then we remind results obtained in [8] in frames of the perturbative Yang–Feldman approach [13] and its coincidence with the approach of [9] quantum group based construction. For the periodic boundary condition case, we recall the quantizaion of the Liouville field operators in the sense of [3], [4]. Finally we compare results in the above mentioned constructions in the most simple case of conformal Toda model, i.e., the quantum Liouville equation.

## 2. Conformal Toda models: classical region

Let $M$ be the manifold $\mathbf{R}^2$ with the standard coordinates $z^{\pm}$. Let $G$ be a complex simple Lie group of rank $r$ with the Lie algebra [7] $\mathcal{G}$ endowed with the principal gradation. In this, in the decomposition $\mathcal{G} = \oplus_{m \in \mathbf{Z}} \mathcal{G}_m$ the subspace $\mathcal{G}_0$ is abelian. Denote by $G_0$ and $G_{\pm}$ the subspaces corresponding to $\mathcal{G}_0$ and $\oplus_{m>1}\mathcal{G}_{\pm m}$, respectively. Denote by $h_i$ and $x_{\pm i}$ the Cartan and Chevalley generators of $\mathcal{G}$, i.e., for its principal gradation, the elements of $\mathcal{G}_0$ and $\mathcal{G}_{\pm 1}$, respectively, satisfying the defining relations

$$[h_i, h_j] = 0,\ [h_i, x_{\pm j}] = \pm k_{ij} x_{\pm j}, \quad [x_{+i}, x_{-j}] = \delta_{ij} h_i, \quad 1 \le i, j \le r, \tag{1}$$

where $k$ is the Cartan matrix of $\mathcal{G}$. There are also the Serre relations.

Conformal abelian Toda fields $\phi = \sum_{i=1}^{r} h_i \phi_i$, satisfy the equations [12]

$$\partial_+ \partial_- \phi + \frac{4\eta^2}{\beta} \sum_{i=1}^{r} m_i \frac{\alpha_i}{\alpha_i^2} e^{\beta \alpha_i \cdot \phi} = 0, \tag{2}$$

with some constants $\beta$, $m_i$, $\mathcal{G}$-roots $\alpha_i$, and a length scale factor $\eta$. Their general solution written in the [10] in a holomorphically factorisable form given by

$$e^{-\beta\lambda_j \cdot \phi} =< \Lambda_j | \gamma_+^{-1} \mu_+^{-1} \mu_- \gamma_- | \Lambda_j >, \tag{3}$$

where $\gamma_\pm(z_\pm)$ and $\mu_\pm(z_\pm)$ are holomorphic and anti-holomorphic mappings $M \to G_0$ and $M \to G_\pm$, respectively; $|\Lambda_i>$ is the highest vector of the $i$-th fundamental representation of $\mathcal{G}$. Moreover, the mappings $\mu_\pm(z^\pm)$ satisfy the initial value problem $\partial_\pm\mu_\pm = \mu_\pm\kappa_\pm$, where $\kappa_\pm$ realize mappings $M \to \mathcal{G}_{\pm 1}$, i.e., $\kappa_\pm(z^\pm) = \sum_{i=1}^{r} \Psi^{(0)}_{\pm i}\cdot x_{\pm i}$, $\Psi^{(0)}_{\pm j} = m_j \exp\left(\mp\beta \sum_{i=1}^{r} k_{ji}\phi^{(0)}_{\pm i}\right)$, with $\phi^{(0)}_{\pm i}$ being solutions of the free field equation.

## 3. Light-cone quantization

The conformal Toda field theory was quantized in the light-cone formalism in [12], [5]. Introduce a set of $r$ scalar fields subject to the canonical commutation relations

$$[\phi_i(z^+,z^-),\phi_j(\tilde{z}^+,z^-)] = -1/4i\hbar\delta_{ij}\epsilon(z^+ - \tilde{z}^+),$$
$$[\phi_i(z^+,z^-),\phi_j(z^+,\tilde{z}^-)] = -1/4i\hbar\delta_{ij}\epsilon(z^- - \tilde{z}^-),$$
$$[\phi_i(z^+,z^-),\phi_i(\tilde{z}^+,\tilde{z}^-)] = 0,$$

when $(z^+ - \tilde{z}^+)(z^- - \tilde{z}^-) < 0$, ($\epsilon(z)$ is a standard sign function). Then one takes [12]

$$\partial_+\partial_-(\beta\Phi_i) + \alpha^2\sigma\sum_{j=1}^{r} k_{ij} : e^{\beta\Phi_i} := 0, \tag{4}$$

as the quantum Toda equation for the Heisenberg field operators. Here $\alpha$, $\beta$, $\sigma$ are constants, $\Phi_i = \sum_{s=1}^{r} M_{is}\phi_s$, and $: \ldots :$ denotes the normal ordering in the light-cone $I^\pm_\tau = \{z^\pm = \frac{\tau}{2}, z^\mp \geq \frac{\tau}{2}\}$, with respect to associated Fock space oscillator operators (see [12] for detailes) $a_j(p) = \int_{I^+_\tau} dz^- e^{i\vec{p}\cdot\vec{x}} i\hat{\partial}_-\phi_i + \int_{I^-_\tau} dz^+ e^{i\vec{p}\cdot\vec{x}} i\hat{\partial}_+\phi_i$, where $a\hat{\partial}_\pm b = a(\partial_\pm b) - (\partial_\pm a)b$. Then the exponential of the field $m^{-\frac{\hbar\beta^2}{2\pi}} : e^{\beta\Phi_i}:$ does not depend on the mass $m$. The equation (4) can be rewritten as the zero curvature condition

$$[\partial_+ + \omega_+, \partial_- + \omega_-] = 0, \tag{5}$$

on the quantum Lax pair operators containing both non-commuting Heisenberg field operators $\psi_i$ and Lie algebra generators $x_{\pm i}$, $h_i$ [12]

$$\omega_+ = \partial_+\psi\cdot h + \alpha\sum_{i=1}^{r} x_{+i} + \zeta\Omega, \quad \omega_- = -\alpha\sigma\sum_{i=1}^{r} : e^{\beta\Phi_i} : x_{-i}, \tag{6}$$

where $\psi\cdot h = \sum_{i=1}^{r}\psi_i h_i$, and $\psi_i = \beta\sum_{j=1}^{r} k^{-1}_{ij}\Phi_j$. To make the left hand side of (5) equal to zero we have to put $\zeta = -1/4i\hbar\beta^2\delta(0)$. Then one can construct a solution to (4) using properties of the highest weight representation of Lie algebras. In this formalism, the Heisenberg field operator solution is of the form

$$: e^{-\psi_j(z^+,z^-)} := \; <\lambda_i| \; g(z^+,0)\cdot g^{-1}(0,z^-) \; |\lambda_i> \; : e^{-\psi_j(z^+,0)} :,$$

where $P = (z^+,z^-)$ is an arbitrary point inside the light cone and $(z^+,0)$, $(0,z^-)$ are points on $I^\pm_\tau$ with $z_\pm$-coordinates same as at point $P$. Evaluating $g(z^+,0)$, $g^{-1}(0,z^-)$ elements one finally finds

$$: e^{-\psi_j(z^+,z^-)} : \; =: e^{-\psi_j(z^+,0)} :< \lambda_j| \left(Texp\int_{(0,0)}^{(z^+,0)} dz^+ B_+\right)\cdot u : e^{-\psi(0,0)\cdot h} :$$
$$\times\left(Texp\int_{(0,0)}^{(0,z^-)} dz^-\omega_-\right)|\lambda_j> \; : e^{-\psi_j(0,z^-)} : \cdot e^{\frac{1}{2}\hbar\beta^2\Delta(0)k^{-1}_{jj}}, \tag{7}$$

where $B_+ = -\alpha \sum\limits_{j=1}^{r} x_{+i} : e^{\beta\Phi_j} : e^{\hbar\beta^2\Delta(0)}$, and $\Delta(\vec{x}) = \int\limits_{-\infty}^{+\infty} \frac{dp}{4\pi\omega} e^{-i\vec{p}\cdot\vec{x}}$. A regularization $\Delta(\vec{x}) = \int\limits_{p_-}^{p_+} \frac{dp}{4\pi\omega} e^{-i\vec{p}\cdot\vec{x}}$ with some $p_-$, $p_+$ is implied in $B_+$. The $T$-exponents in the last formula denote $z^{\pm}$ ordering on the light-cone.

## 4. Yang–Feldman perturbative approach

It has been observed in [9] that the explicit expressions for the conformal Toda Heisenberg operators (satisfying the equations of the form (2)) obtained via of the Yang–Feldman perturbation procedure [8], [13] exactly, order by order, coincide with the expression for the general solution of the form (3) constructed by means of corresponding quantum group. We end up with the exact explicit expressions for the Heisenberg operators $\phi_i^{(q)}$ as finite series in terms of the free fields operators ${}_q\phi_i^{(0)}$ satisfying the canonical commutation relations. Here, in the same way as for the classical solutions, the number of terms in the series is precisely equal to the dimension of the $i$-th fundamental representation of corresponding Lis algebra $\mathcal{G}$.

In the Yang–Feldman formalism [11], [8] the expression for the exponential of the Heisenberg field operator in the $n$-th order is

$$\left(e^{-\phi_\alpha}\right)_{(n)} = \frac{1}{(i\hbar)^n} \int\limits_{-\infty}^{+\infty} \mathbf{dz}\theta(z - z_1)...\theta(z_{n-1} - z_n) \sum_{\substack{P(k_1,\\ ...,k_n)}} \left[e^{-\phi_\alpha^{(0)}}, V_{1k_1}, ..., V_{nk_n}\right],$$

where $\mathbf{dz} := dz_1...dz_n$, $\theta(z - z') = \theta(z^+ - z^{+'})\ \theta(z^- - z^{-'})$ is a usual step function, $V_{ij} = \sum\limits_{\alpha=1}^{r} 2w_e^{\psi_{+\alpha}^{(0)}(z^{+i})} e^{\psi_{-\alpha}^{(0)}(z^{-j})}$, $\psi_{\pm\alpha}^{(0)} = \sum\limits_{\beta=1}^{r} k_{\alpha\beta}\phi_{\pm\beta}^{(0)}$, $\phi_{\pm\alpha}^{(0)} = \sum\limits_{\beta=1}^{r} k_{\alpha\beta}^{-1}\psi_{\pm\beta}^{(0)}$, $[A, B, ..., C] = [[...[A, B], ...], C]$. Here $\phi_{\pm\alpha}^{(0)}$ are asymptotic fields satisfying canonical commutation relations

$$\left[\phi_{\pm i}^{(0)}(z^\pm), \phi_{\pm j}^{(0)}(z^{\pm'})\right] = \frac{\hbar}{4i} k_{ij}^{-1} w_j^{-1} \epsilon(z^\pm - z^{\pm'}), \quad \left[\phi_{\pm i}^{(0)}(z^\pm), \phi_{\mp j}^{(0)}(z^{\mp'})\right] = 0,$$

$$\left[\psi_{\pm i}^{(0)}(z^\pm), \psi_{\pm j}^{(0)}(z^{\pm'})\right] = \frac{\hbar}{4i} \hat{k}_{ij}^{-1} w_j^{-1} \epsilon(z^\pm - z^{\pm'}),$$

$$\left[\psi_{\pm i}^{(0)}(z^\pm), \phi_{\mp j}^{(0)}(z^{\mp'})\right] = \frac{\hbar\delta_{ij}}{4iw_j} \epsilon(z^\pm - z^{\mp'}), \quad \hat{k}_{ij} = k_{ij} w_j^{-1}.$$

The three first orders in the Yang–Feldman formula are

$$\left(e^{-\phi_\alpha}\right)_{(0)} = e^{-\phi_\alpha^{(0)}}, \quad \left(e^{-\phi_\alpha}\right)_{(1)} = e^{-\phi_\alpha^{(0)}} \frac{2w_\alpha}{i\hbar} \left[1 - e^{-\frac{i\hbar}{2w_\alpha}}\right] \Phi_\alpha^+ \Phi_\alpha^-, \tag{8}$$

$$\left(e^{-\phi_\alpha}\right)_{(2)} = e^{-\phi_\alpha^{(0)}} \frac{2w_\alpha}{i\hbar} \left[1 - e^{-\frac{i\hbar}{2w_\alpha}}\right] \sum_{\gamma\neq\alpha} \frac{2w_\alpha}{i\hbar} \left[1 - e^{\frac{i\hbar}{2}\hat{k}_{\alpha\gamma}}\right] \Phi_{\alpha\gamma}^+ \Phi_{\alpha\gamma}^-,$$

$$\Phi_{\alpha_1,...,\alpha_m}^\pm(z^\pm) = \int_{-\infty}^{z^\pm} dz_1^\pm e^{\psi_{\alpha_1}^{(0)}(z_1^\pm)} \int_{-\infty}^{z_1^\pm} dz_2^\pm e^{\psi_{\alpha_2}^{(0)}(z_2^\pm)} ... \int_{-\infty}^{z_{m-1}^\pm} e^{\psi_{\alpha_m}^{(0)}(z_m^\pm)}, \tag{9}$$

e.g.,

$$\Phi_\alpha^\pm(z^\pm) = \int_{-\infty}^{z^\pm} dz_1^\pm e^{\psi_{\alpha_1}^{(0)}(z_1^\pm)}, \tag{10}$$

$$\Phi_{\alpha_1,\beta}^\pm(z^\pm) = \int_{-\infty}^{z^\pm} dz_1^\pm e^{\psi_{\alpha_1}^{(0)}(z_1^\pm)} \int_{-\infty}^{z_1^\pm} dz_2^\pm e^{\psi_\beta^{(0)}(z_2^\pm)}. \tag{11}$$

## 5. Solution construction based on quantum group

Now take the general solution (3) to the conformal Toda equations (2) and substitute [9] the group elements and representation vectors in the right hand side by quantum group elements and representation vectors correspondingly. In particular, substitute the mappings $\kappa_\pm$ with ${}_q\kappa_\pm(z^\pm) = \sum_{i=1}^{r} {}_q\Psi^{(0)}_{\pm i} \cdot x^q_{\pm i}$. Here ${}_q\Psi^{(0)}_{\pm j} = m_j \exp\left(\mp\beta \sum_{i=1}^{r} k_{ji\,q}\phi^{(0)}_{\pm i}\right)$. Here as before ${}_q\phi^{(0)}$ are operators of the asymptotic fields, i.e., fields in absence of interactions. ${}_q\Psi^{(0)}_{\pm j}$ is contains elements $x^q_{\pm i}$ that belong to the quantized universal enveloping algebra $U_q(\mathcal{G})$ for a Lie algebra $\mathcal{G}$. Instead of (1) the quantum group $U_q(\mathcal{G})$ has the commutation relations (e.g., in Jimbo–Drinfeld form [6], [1])

$$[h^q_i, h^q_j] = 0, \quad [h^q_i, x^q_{\pm j}] = \pm k_{ij} x^q_{\pm j}, \quad [x^q_i, x^q_{-j}] = \delta_{ij} \frac{q_i^{h^q_i} - q_i^{-h^q_i}}{q_i - q_i^{-1}}, \tag{12}$$

where $q_i$ is defined as $q_i = e^{d_i \hbar}$ in terms of the Planck constant $\hbar$ and coprime integers $d_i$ such that $d\,k$ is a symmetric matrix. There are also analogues of the Serre relations.

In place of (3) we get a similar expression for the exponential of a Heisenberg field operator solution

$$e^{-\beta\lambda^{(q)}_j \cdot \phi^{(q)}} = {}_q < \Lambda_j | {}_q\gamma_+^{-1}{}_q \cdot \mu_+^{-1} \cdot {}_q\mu_- \cdot {}_q\gamma_- | \Lambda_j >_q, \tag{13}$$

As in the classical case the group elements ${}_q\mu_\pm$ satisfy $\partial_\pm {}_q\mu_\pm = {}_q\mu_\pm \cdot {}_q\kappa_\pm$. In this approach it is easy to find the first three order terms of (13):

$$\left(e^{-{}_q\phi_\alpha}\right)_{(0)} = {}_q < \alpha | e^{-\sum_{\beta=1}^{r} h^q_\beta \left({}_q\phi^{(0)}_{+\beta} + {}_q\phi^{(0)}_{-\beta}\right)} | \alpha >_q = e^{-\left({}_q\phi^{(0)}_{\alpha+} + {}_q\phi^{(0)}_{\alpha-}\right)} = e^{-{}_q\phi^{(0)}_\alpha}, \tag{14}$$

$$\begin{aligned} \left(e^{-{}_q\phi_\alpha}\right)_{(1)} &= -e^{-{}_q\phi^{(0)}_{+\alpha}} \sum_{\theta,p=1}^{r} \Phi^+_\theta \Phi^-_p \cdot {}_q < \alpha | x^q_{+\theta}\, x^q_{-p} | \alpha >_q \, e^{-{}_q\phi^{(0)}_{-\alpha}} \\ &= e^{-{}_q\phi^{(0)}_\alpha} \frac{2w_\alpha}{i\hbar}\left[1 - e^{-\frac{i\hbar}{2w_\alpha}}\right] \Phi^+_\alpha \Phi^-_\alpha, \end{aligned} \tag{15}$$

$$\begin{aligned} \left(e^{-{}_q\phi_\alpha}\right)_{(2)} &= -e^{-{}_q\phi^{(0)}_{+\alpha}} \sum_{\theta,\tau,p,q=1}^{r} {}_q < \alpha | x^q_{+\theta}\, x^q_{+\tau}\, x^q_{-p}\, x^q_{-q}\, | \alpha >_q \Phi^+_{\theta\tau} \tilde{\Phi}^-_{pq} \quad e^{-{}_q\phi^-_\alpha} \\ &= e^{-{}_q\phi^{(0)}_{+\alpha}} \sum_\theta e^{\frac{i\hbar \hat{k}_{\theta\alpha}}{4}} \frac{2w_\alpha}{i\hbar} \sinh\left(\frac{i\hbar}{4w_\alpha}\right) \frac{2w_\theta}{i\hbar} \sinh\left(\frac{i\hbar}{4}\left(2\delta_{\alpha\theta} - k_{\alpha\theta}\right)\right) \Phi^+_{\theta\alpha} \Phi^-_{\theta\alpha} e^{-{}_q\phi^{(0)}_{-\alpha}}, \end{aligned} \tag{16}$$

$$\tilde{\Phi}^-_{pq} = \int^{z^-} dz_1^- \int^{z^-} dz_2^- \theta(z_2^- - z_1^-) e^{{}_q\psi^-_p} e^{{}_q\psi^-_p} \ldots = e^{-\frac{\hbar}{4i}\hat{k}_{pq}} \Phi^-_{pq},$$

and $\Phi^\pm_\alpha$, $\Phi^\pm_{\alpha\beta}$ is the same as in (10), (11). One can see that (14)–(16) are the same as (8)–(9) correspondingly.

## 6. Solution construction for quantum conformal Toda model

In order to explain and compare formulae given in [9], one considers the quantum group construction of operator solutions to the conformal Toda equations. Recall the light-cone quantization of the conformal Toda field theory given in [12]. The way one derives solutions (3) in classical region for the conformal Toda model (2) can be modified to get solutions in the quantum group case. The zero-curvature condition (5) implies the gradient form of ${}_qA_\pm$. The pure gauge ${}_qA_\pm = {}_qg^{-1}\partial_\pm ({}_qg)$ therefore implies $\partial_\pm {}_qg^{-1} = -{}_qA_\pm {}_qg^{-1}$, $\partial_\pm {}_qg = {}_qg_qA_\pm$, (here

${}_qA_\pm = \eta \sum_{i=1}^{r} \phi^0_{\pm i}(x_{+i}\hat{\otimes}...)$ (i.e., the same ( in $x_{\pm i}$-part) as in [9]). Then the properties of the fundamental highest weight representation of a quantum group implies that

$$\partial_-\left({}_q < \Lambda_i|e^{-\beta\phi(z^+,\tilde{z}^-)\cdot h}{}_qg^{-1}\right) =< \Lambda_i|e^{-\beta\Lambda\cdot\phi(z^+,\tilde{z}^-)}\left(-{}_qA_{-q}g^{-1}\right) = 0,$$

and $\partial_+\left({}_qg|\Lambda_i>_q\right) = {}_qg_qA_+|\Lambda_i>_q = 0$. Here $|\Lambda_i>_q$ are vectors in the fundamental highest weight representation of the $U_q(\mathcal{G})$. Therefore we derive that

$${}_q < \Lambda_i|e^{-\beta\phi(z^+,\tilde{z}^-)\cdot h} \quad {}_qg^{-1}(z^+,\tilde{z}^-) \quad \cdot {}_qg(\tilde{z}^+,z^-)|\Lambda_i>_q,$$

does not depend on $\tilde{z}^\pm$ and resultes in $e^{-\beta\Lambda\cdot\phi(z^+,z^-)}$ when $\tilde{z}^\pm = z^\pm$. Thus for a point $p = (z^+,z^-)$ and two arbitrary points $(z^+,0)$, $(0,z^-)$ on the null rays, we have

$$e^{-\beta\Lambda\cdot\phi(z^+,z^-)} = {}_q < \Lambda_i|e^{-\beta\phi(z^+,0)\cdot h}{}_qg^{-1}(z^+,0)\cdot{}_qg(0,z^-)|\Lambda_i>_q . \tag{17}$$

The quantum Gauss decomposition [2] can be represented as ${}_qg(z^+,z^-) = {}_q\mu_+(z^+)_q\nu_-(z^-)_q\gamma_{0+}$, so that ${}_qg^{-1}(z^+,z^-) = {}_q\gamma^{-1}_{0+}{}_q\nu^{-1}_-(z^-)_q\mu^{-1}_+(z^+)$, and ${}_qg^{-1}(z^+,0) = {}_q\gamma^{-1}_{0+}{}_q\mu^{-1}_+(z^+)$, and elements ${}_qg(z^+,0)$ and ${}_qg^{-1}(z^-,0)$ are just ${}_q\mu^{-1}_+(z^+)$ , ${}_q\mu_-(z^-)$ elements of the modified Gauss decomposition [2] (same for ${}_qg(0,z^-)$ ). All that gives us a solution to the quantum conformal Toda equations (2). Note also that solutions of an inhomogeneous equation (line conformal Toda equations) with fixed $z^\mp$ dependence satisfy the homogeneous equation, namely $\partial_+\partial_-\phi(z^+,0) = \partial_+\partial_-\phi(0,z^-) = 0$, i. e., we retain the formula for the general solution to the quantized conformal Toda equation given in [8].

## 7. Quantization scheme for periodic boundary case

An alternative construction for the quantum Liouvillle exponentials was elaborated in the papers [3], [4]. The final expression for the quantum Liouvillle exponential was found in this approach in a form similar to the classical solution in [10], [11]

$$e^{-J\alpha_-\Phi(z_+,z_-)} =< J,J|\, E^{(J)}_{q+}(\eta_+(z^+),J_+)\, E^{(J)}_{q-}(\eta_-(z^-),J_-)\,|J,J>,$$

$$E^+_q = \sum_{J+M=0}^{\infty} e^{ih(J+M)}(-1)^{J+M}\eta^{(J)}_{+,-M}\frac{(J_+)^{J+M}}{[J+M]!},$$

$$E^-_q = \sum_{J+M=0}^{\infty} \eta^{(J)}_{-,M}\frac{(J_-)^{J+M}}{[J+M]!}, \quad \eta^{(J)}_{\mp,\mp M} = \xi^{(J)}_{\mp,\mp M}\begin{pmatrix} 2J \\ J\mp M\end{pmatrix}_q^{-\frac{1}{2}},$$

$$\begin{pmatrix} A \\ B\end{pmatrix}_q = \frac{[A]!}{[B]![A-B]!}, \quad [A]! = \prod_{k=1}^n [k], \quad [k] = \frac{\sin(hx)}{\sin(h)},$$

$$\xi^{(J)}_{\pm,M}(\sigma) = \sum_{-J\le m\le J} |J,\omega)^m_{\pm M}\psi^{(J)}_m(\sigma),$$

$$|J,\omega)^m_M = \left(e^{-ih\frac{a_2}{a_1}}\right)_q^{J+M} k_J\begin{pmatrix} 2J \\ J+M\end{pmatrix}_q^{\frac{1}{2}} e^{\frac{ihm}{2}}$$

$$\times \sum_{\frac{1}{2}(J_M+m-s)\in\mathbf{Z}} e^{ihs(\omega+m)}\begin{pmatrix} J-M \\ \frac{1}{2}(J-M+m-s)\end{pmatrix}_q\begin{pmatrix} J+M \\ \frac{1}{2}(J+M+m+s)\end{pmatrix}_q,$$

$$\psi^{(J)}_m = \psi^{J-m,J+m}, \quad \psi^{\mu,\nu} \sim \psi^\mu_1\psi^\nu_2 \sim N^{(1)}\left(e^{\mu\sqrt{\frac{h}{2\pi}}\phi_1}\right)N^{(2)}\left(e^{\nu\sqrt{\frac{h}{2\pi}}\phi_2}\right),$$

and $\psi_j$, $j = 1,2$ are two independent solutions of Schrodinger equation $-\psi_j''+T\psi_j = 0$, associated to the Liouvillle equation which implies $\partial_-\left(e^{-\frac{\phi}{2}}\partial_+ e^{\frac{\phi}{2}}\right) = 0$. Here the normal ordering is defined as

$$N^{(j)}(e^{\gamma\phi_j(z)}) = e^{\gamma q_0^{(j)}} e^{-iz(\pm\omega+\frac{1}{2})\frac{h}{2\pi}} exp\left(\gamma\sum_{n<0}\frac{p_n^{(j)}}{n}e^{-inz}\right) exp\left(\gamma\sum_{n>0}\frac{p_n^{(j)}}{n}e^{-inz}\right),$$

where $\phi_j(z) = q_0^{(j)} + p_0^{(j)} + i\sum\limits_{n<0}\frac{e^{-inz}}{n}p_n^{(j)}$. For the case $J = \frac{1}{2}$ the general formula gives $e^{-J\alpha_-\Phi(z^+,z^-)} = (a_1\kappa_1\kappa_2 - a_1\kappa_1\kappa_2)N^{(1)}N^{(2)}...$ where $\kappa_i$ are some infinite series containing $h$. We have then $\psi_j = d_j N^{(j)}(e^{\gamma\phi_j})$. Let's put $\phi_2 = 0$ and $\phi_1 = \phi_0^+$, $\phi_1 = \phi_0^-$. Then we have

$$e^{-J\alpha_-\Phi(z,\bar{z})} = \sigma_1 + \sigma_2 N^{(1)}(e^{\gamma}) + \sigma_3 N^{(1)}(e^{\gamma}) + \sigma_4 N^{(1)}(e^{\gamma})\,N^{(1)}(e^{\gamma}), \tag{18}$$

which is very close to the formulas for the quantum Liouville exponential in [9] and [12], [5] formulations in appropriate limits.

## 8. Comparison of approaches

In the subsection 3 we recalled the light-cone quantization procedure applied in [12] to the conformal Toda field theory. In this approach the Lax operators (6) corresponding to the quantum Heisenberg field operator equations (4) contain elements of ordinary Lie algebra, but, as it was shown in [5], this construction possesses an underlying quantum group structure. This case gives as an example of the Goursat problem sinse the exponent of the field operator depends on its value on the characteristics. Finally one obtains a solution (17) to the equation (4) which is of the form given in [10] in the classical case. At the same time this way of Lax pair construction for the quantum conformal Toda has some obvious drawbacks. One may think of the Lax pair (6) as a result of some gauge transformation sending the symmetric form of Lax operators $A_\pm = u_\pm h + f_\pm x_\pm$ (where $u_\pm$ , $f_\pm$ are operators ) to (6) with a constant operator near $\sum\limits_{i=1}^{r} x_{+i}$. Though infinite constant $\xi$ in (6) does not appear in the final expression for the solution (7) still it looks very discouraging in the operator form.

Now we would like to compare the solution in the most simple case of $\mathcal{G} = sl_2$ which corresponds to the quantum Liouville equation $\partial_+\partial_-(\beta\Phi)+2\alpha^2\sigma : e^{\beta\Phi} := 0$. The final expression for the exponent of the Liouville field in the construction of [12], [5] has the form

$$: e^{-\frac{\beta}{2}\Phi(z^+,z^-)} :=: e^{-\frac{\beta}{2}\Phi(z^+,0)} : \left( : e^{\frac{\beta}{2}\Phi(0,0)} : e^{\frac{\hbar\beta^2}{2}\Delta(0)} + \alpha^2\sigma\left(\int\limits_{(0,0)}^{(z^+,0)} : e^{\beta\Phi} : dz^+\right)\right.$$
$$\left.\times : e^{-\frac{\beta}{2}\Phi(0,0)} : \left(\int\limits_{(0,0)}^{(0,z^-)} : e^{\beta\Phi} : dz^-\right) e^{\frac{3\hbar\beta^2}{2}\Delta(0)}\right) : e^{-\frac{\beta}{2}\Phi(0,z^-)} : .$$

If we get rid of the normal ordering in the last formulausing the fact that $e^{\beta\Phi} =: e^{\beta\Phi} : e^{\hbar\beta^2\Delta(0)}$, then we get the expression

$$e^{-\frac{\beta}{2}\Phi(z^+,z^-)} = e^{-\frac{\beta}{2}\Phi(z^+,0)}\left(e^{-\frac{\beta}{2}\Phi(0,0)} + \alpha^2\sigma\left(\int\limits_{(0,0)}^{(z^+,0)} e^{\beta\Phi(\tilde{z}^+,0)}d\tilde{z}^+\right)\cdot e^{-\hbar\beta^2\Delta(0)}\right.$$
$$\left.\times e^{-\frac{\beta}{2}\Phi(0,0)}\cdot\left(\int\limits_{(0,0)}^{(0,z^-)} e^{\beta\Phi(0,\tilde{z}^-)}d\tilde{z}^-\right) e^{-\frac{\beta}{2}\Phi(0,z^-)}\right)$$
$$= e^{-\frac{\beta}{2}\Phi(z^+,0)}\cdot e^{-\frac{\beta}{2}\Phi(0,0)}\cdot e^{-\frac{\beta}{2}\Phi(0,z^-)}\cdot\left(1+\alpha^2\sigma e^{-\hbar\beta^2\delta(0)}\Phi^+\Phi^-\right).$$

Thus we see that the quantum Toda exponent is expressed through its values at points on characteristics. Therefore this is an example of the Goursat problem.

In the approach of [9] one deals with the quantum exponents depending on asymptotic fields which are in fact solutions fo the homogeneous operator equation $\partial_+\partial_-\phi = 0$. Then in the Liouville case using (14)–(15) we obtain

$$\left(e^{-\phi}\right)_{(0)} = e^{-{}_q\phi^{(0)}},\ \left(e^{-\phi}\right)_{(1)} = -e^{-{}_q\phi^{(0)}}\frac{2w}{i\hbar}\left[1 - e^{-\frac{i\hbar}{2w}}\right]\Phi^+\Phi^-,\ \left(e^{-\phi}\right)_{(2)} = 0.$$

Thus the series terminates after the first order. It is not a surprise that the formulas (8)–(9) and (14)–(16) coincide in [12] and [9] approaches. The reason is that the elements of the quantum group are used in (13) and give extra multipliers in the first order due to the non-linear right hand side of the commutation relation (12) of the $x_+^q$, $x_-^q$ generators of the quantum group.

One can also try to construct the Lax pair for the quantum conformal Toda model equation (4) in the same as in the approach of [12] but using generators of the quantized universal enveloping algebra. Indeed, consider the following pair of operators

$${}_q\omega_+ = \partial_+\psi \cdot h^q + \alpha\sum_{i=1}^{r} x_{+i}^q + \zeta\Omega_q, \quad {}_q\omega_- = -\alpha\sigma\sum_{i=1}^{r} :e^{\beta\Phi_i}: x_{-i}^q, \tag{19}$$

where $h_i^q$, $x_{\pm i}^q$ are generators of the quantum group. Here $\Omega_q \in U_q(\mathcal{G})$ is chosen in such a way that it would kill higher $h$-terms in the zero curvature condition (5). Thus the zero curvainteractions. ture condition would give us (4). The problem is to compute $\Omega_q$ precisely. Then one could find a solution to (4) using the Lax pair (19) which comply with the solution in [9] solution under an appropriate limit.

## References

[1] Drinfeld V G 1985 *Sov. Math. Dokl.***32** 254
[2] Faddeev L D , Reshetikhin N Yu and Takhtajan L A 1989 *Quantum groups. Proc. Yang C.N., Ge M.L. (eds.) Braid groups, knot theory and statistical mechanics. Adv. Ser. Math. Phys.* **9** 97(Singapore: World Scientific, Teaneck, NJ)
[3] Gervais J–L and Schnittger J 1994 *Nuclear Phys. B* **413** no 1-2 433
[4] Gervais J–L and Schnittger J 1994 *Nuclear Phys. B* **431** no 1-2 273
[5] Hollowood T, Mansfield P 1990 *Nucl.Phys.* B**330** 2-3 720
[6] Jimbo M 1985 *Lett. Math.Phys.***10** no 1 63
[7] Kac V G 1990 *Infinite dimensional Lie algebras, Third edition* (Cambridge university Press, Cambridge)
[8] Leznov A N and Fedosseev M A 1982 *Theor. and Math. Phys.* **53** 3 358
[9] Leznov A N and Mukhtarov M A 1987 *Theor. and Math.Phys.* **71** 1 46
[10] Leznov A N and Saveliev M V 1979 *Lett. Math. Phys.* **3** 489
[11] Leznov A N and Saveliev M V 1992 *Group–theoretical methods for integration of non-linear dynamical systems. Progress in Physics Series* **15** (Birkhauser-Verlag, Basel)
[12] Mansfield P 1983 *Nucl. Phys. B.* **222** 419
[13] Schweber S S 1955 *Nuovo Cimento* **10** 2 39

# Transport phenomena in chitosan synthetic membranes with emphasis on the effect of variations in the ratio of matrix/solvent

**N Nyoman Rupiasih**[1,2,3,4], **Yayuk Eka Puspita** [1] **and Made Sumadiyasa** [1]
[1] Department of Physics, Faculty of Mathematics and Natural Sciences, Udayana University, Kampus Bukit Jimbaran, Badung, Bali 80362, Indonesia
[2]Group Research Material Sciences and Technology-Polymer and Biomaterial

E-mail: rupiasih69@yahoo.com

**Abstract.** The object of this research was investigating the transport phenomena of chitosan synthetic membranes with emphasis on the effect of variations in the ratio of matrix/solvent. The study was focused on the effect of amount of chitosan as matrix and electrolytes solutions on the characteristics of current density of chitosan membranes. A series of chitosan membranes with various ratios of components was used such as 1%, 2%, 3% and 4%. The electrolytes solutions, NaCl and $CaCl_2$, with various concentrations, 0.1 mM, 1 mM, 10 mM, 100 mM and 1000 mM, were used. Ion transport processes were carried out in a cell membrane model which composed of two compartments named compartment 1 and 2, and the potential difference was measured using a pair of Activon AEP jnct Single 12 x 120 mm calomel electrodes. All the measurements were conducted at room temperature, 28.8 °C. The result showed that the current density increased with some parameters e.g. increased in the ratio of concentration of solution, C1:C2; increased in the amount of chitosan, 1%, 2%, 3% and 4%; and increased in the size of Stokes radii of the selected cations, $Na^+$ and $Ca^{+2}$.

**Keywords:** Chitosan synthetic membrane, electrolyte solution, current density, ion diffusion

## 1. Introduction

Chitosan is a biopolymer composed of glucosamine and N-acetyl glucosamine units linked by 1-4 glucosidic bonds. It is a cationic polysaccharide obtained by de-*N*-acetylation of chitin under alkaline conditions [1-3]. Recently, chitosan has attracted great interest due to its specific properties such as biodegradability, biocompatibility, non-toxic, antibacterial, physiological inert and bioactivity [4-5]. These biological characteristics provide chitosan a wide field of applications as a material part in bio-environments systems, such as pharmaceutical and biomedical engineering, food industry, agriculture and wastewater treatment. Such applications involve as drug carriers, wound-healing agents, chelating

[3]Author to whom any correspondence should be addressed: rupiasih69@yahoo.com; [4] Grant: RM Udayana University No: 104.41/UN14.2/PNL.01.03.00/2014 date 3 Maret 2014

agents, hemodialysis membranes and surgical dressing materials, biodegradable coating or film for food packaging, coating of seeds and leaves to improve plants' resistance to diseases, and coating of fertilizers and pesticides for their controlled release to soil [1, 5-10]. Chitosan is readily converted to fibers, membranes, coatings, powders and solutions, further enhancing its usefulness [1].

Chitosan membranes have been used for active transport of chloride ions in aqueous solution [1]. The other applications are as a carrier and/or a selective barrier controlling the transport rate of the substances involved, chitosan having a powerful chelating ability provides protection of the systems against destructive effects of heavy metal ions [11].

The development of synthetic membranes had always been inspired by the fact that the selective transport through biological membranes is enabled by highly specialized macromolecular and supramolecular assemblies based on and involved in molecular recognition [12], the use of membranes as ion exchanger. The ion exchange membrane has been used for the traditional applications, such as electrodialysis concentration or desalting of solutions, diffusion dialysis to recover acids, and electrolysis of sodium chloride solution. Recently it also used in various fields as a polymeric film having ionic groups [13] such as ion-exchange membranes for protein separation [14].

Most ion-exchange membranes are made of synthetic polymers; only a few are from natural polymers e.g. chitosan, cellulose and alginate [14]. Chitosan can be used as an anion-exchange membrane directly [15], cellulose derivatives such as cellulose phosphate [16], cellulose acetate [17] and alginate [18] are used as cation-exchange membranes.

In the present study, we prepared a series of chitosan membranes by varying the ratio of components e.g. matrix/solvent. The purpose of this work was to study of transport phenomena in those various synthetic membranes. The effects of concentration and electrolytes on their characteristics were also reported.

## 2. Materials and methods

### *2.1 Materials*
NaCl, $CaCl_2$ and other reagents used in this study were obtained from commercial sources.

### *2.2 Transport in chitosan membranes*
A series of chitosan membranes e.g. M1, M2, M3 and M4, have prepared by varying the ratios of components (matrix/solvent) such as 1%, 2%, 3% and 4% respectively. The electrolytes solutions used were NaCl and $CaCl_2$ with various concentrations, 0.1 mM, 1 mM, 10 mM, 100 mM and 1000 mM. Ion transport processes carried out in a cell membrane model which composed of two compartments named compartment 1 and 2, and the potential difference was measured using a pair of Activon AEP jnct Single 12 x 120 mm calomel electrodes. C1 and C2 is concentration of solution in compartment 1 and compartment 2, respectively. In the transport processes, C1 are varies as 0.1 mM, 1 mM, 10 mM, 100 mM and 1000 mM whereas C2 is kept constant 0.1 mM. All the measurements were conducted at room temperature, 28.8 °C.

## 3. Results and discussion

### *3.1 Characteristic of potential difference (V) vs Log (C1/C2)*
Figure 1a and 1b show the potential difference of membranes as function of Log (C1/C2) when the membranes in contact with NaCl and $CaCl_2$ solution, respectively. In general, Figure 1a and 1b show that the potential difference (V) increased as increasing the ratio of concentration, C1:C2. This result is in accordance to the Nernst equation (Hobbie, Russell K, 1978) [19], which illustrates that the potential difference is due to differences in concentration in compartment 1 and 2 e.g. proportional to Log ($C_1/C_2$). Both figures also show that the increased of potential difference in $CaCl_2$ solution was bigger compared with NaCl solution, the graph look sharper.

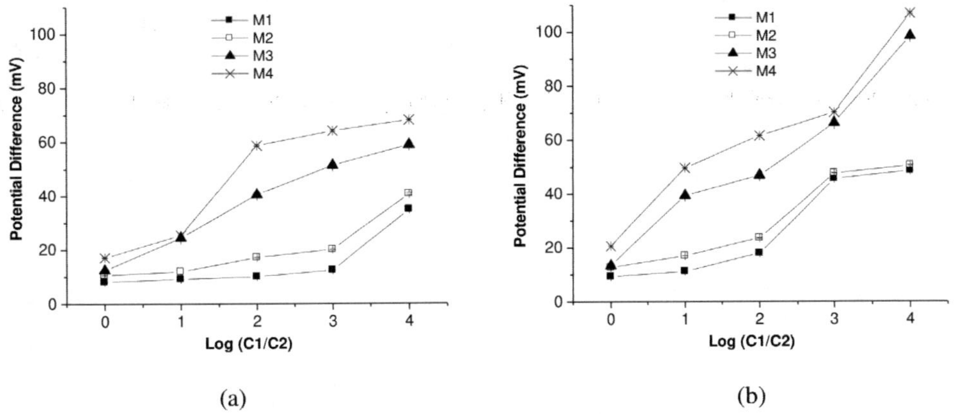


**Figure 1.** The potential difference (V) vs. Log (C1/C2) graphs of chitosan membranes in contact with a) NaCl solution and b) $CaCl_2$ solution.

Figure 2a and 2b show the potential difference in each membrane in contact with various concentrations (0.1, 1, 10, 100 and 1000 mM) of NaCl and $CaCl_2$ solution, respectively. It shows an increased in the potential difference with increased in the amount of matrix constituents of membranes from 1%, 2%, 3% and 4% which named as membrane M1, M2, M3 and M4, respectively. This explains that increased in the amount of matrix constituent result a more positive membranes formed, which affects the diffusion characteristics of the membranes itself.

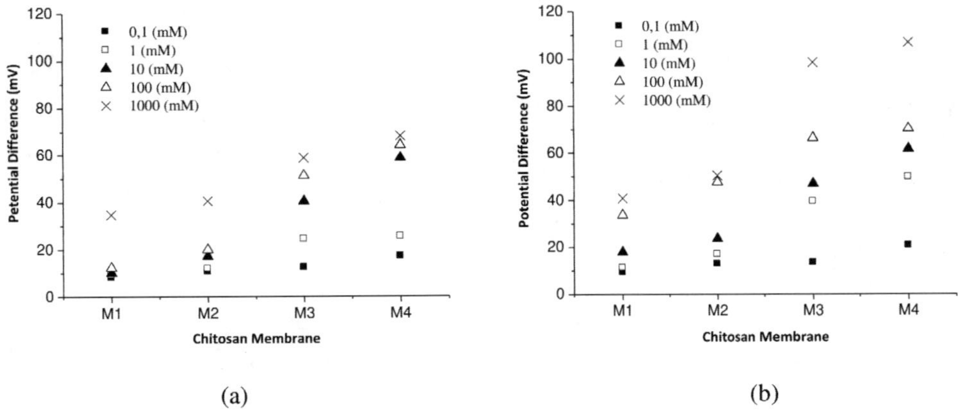


**Figure 2.** The potential difference (V) vs. chitosan membranes (M1, M2, M3 and M4) graphs in various concentrations e.g. 0.1, 1, 10, 100 and 1000 mM of a) NaCl solution and b) $CaCl_2$ solution.

*3.2 Characteristic of current density vs. Log (C1)*

Figure 3 and Figure 4 show that at small ratios (C1: C2 is small) e.g. 0.1 mM : 0.1 mM, 1 mM : 0.1 mM and 10 mM : 0.1 mM, the current density (J) increased gently. The increased is rising sharply at bigger ratios e.g. 100 mM : 0.1 mM and at the highest ratio e.g. 1000 mM : 0.1 mM, the increased is very sharp. The both figures also show that the increased of the current density is greater when the membranes in contact with $CaCl_2$ solution compared with NaCl solution. This result can be described that the numbers of $Cl^{-1}$ ions in $CaCl_2$ solution are double compared in NaCl solution, that is one $Cl^{-1}$ ion for one $Na^+$ ion and two $Cl^{-1}$ ions for one $Ca^{2+}$ ion. These mean that the driving force is double when the membranes in contact with $CaCl_2$ solution, the graphs in Figure 4 look sharper.

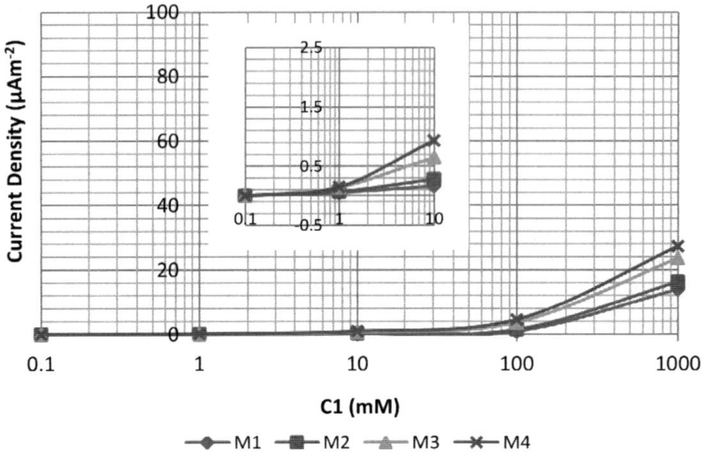


**Figure 3.** Current density vs. C1 (concentration of NaCl solution in compartment 1) semi log graphs.

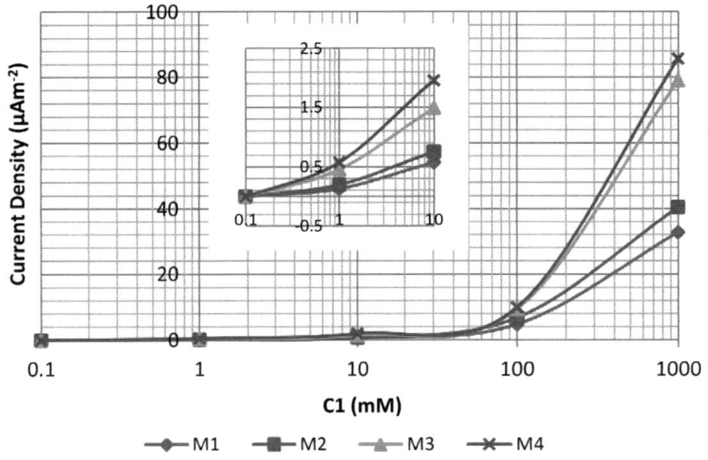


**Figure 4.** Current density vs. C1 (concentration of $CaCl_2$ solution in compartment 1) semi log graphs.

Increasing the amount of matrix (chitosan) also affects ion transport characteristics of the membranes as shown in Figure 5. It shows that the current density is increased with increasing the amount of matrix e.g. 1%, 2%, 3% and 4% which named as membrane M1, M2, M3 and M4 respectively. This result is consistent with the increased in the value of the potential difference measured.

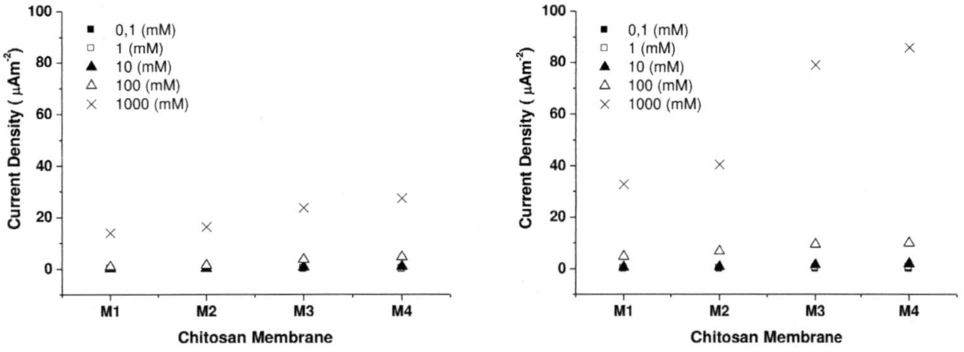


(a) (b)

**Figure 5.** Current density vs. chitosan membranes (M1, M2, M3 and M4) graphs in various concentrations (0.1, 1, 10, 100 and 1000 mM) of a) NaCl solution and b) $CaCl_2$ solution.

*3.3 Effect of electrolyte solutions on current density*

Figure 6 describe the effect of electrolyte solutions to the ion transport characteristics of chitosan membranes. The graphs show similar pattern for all membranes, M1, M2, M3 and M4, e.g. the graph is sharper as membranes in contact with $CaCl_2$ solution compared with NaCl solution. This is in accordance with the increased in the size of Stokes radii of the selected cations, $Na^+$ and $Ca^{+2}$. The numbers of $Cl^{-1}$ ions in $CaCl_2$ solution are double compared in NaCl solution, that is one $Cl^{-1}$ ion for one $Na^+$ ion and two $Cl^{-1}$ ions for one $Ca^{2+}$ ion.

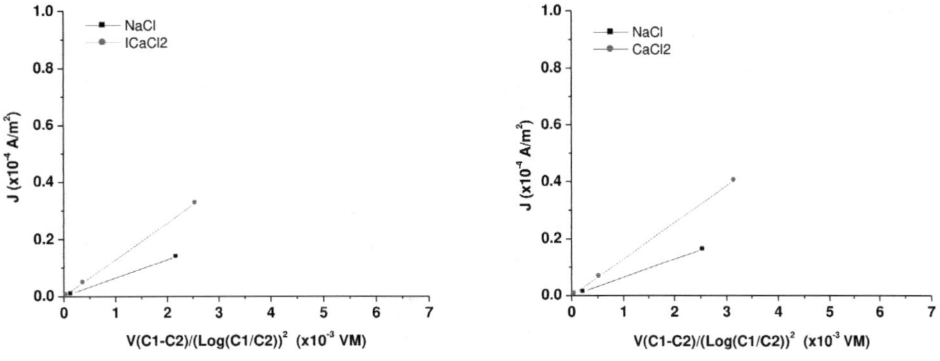


(a) (b)

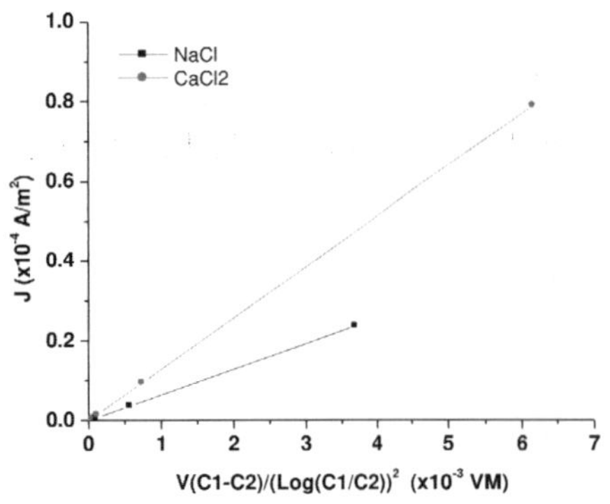


(c)

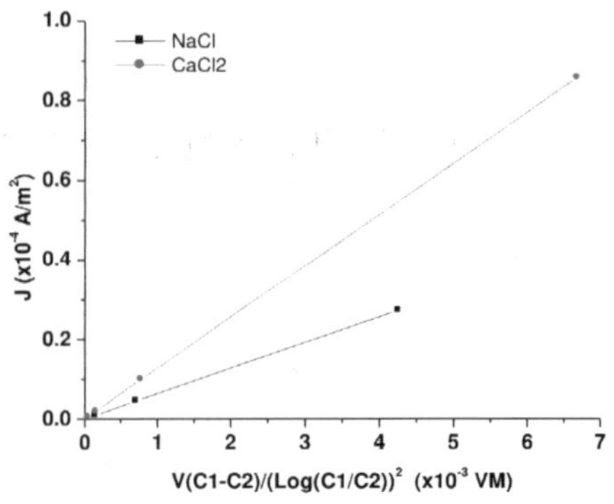


(d)

**Figure 6.** Current density vs. $\{V(C1\text{-}C2) / (Log(C1/C2))^2\}$, subsequently: a) M1, b) M2, c) M3 and d) M4.

## 4. Conclusion

The study about transport phenomena in chitosan synthetic membranes with emphasis on the effect of variations in the ratio of matrix/solvent has been described. The results showed that the current density is increased with some parameters such as increased in the ratio of concentration of solution, C1:C2; increased in the amount of matrix (chitosan), 1%, 2%, 3% and 4% which named as M1, M2, M3 and M4 respectively; and increased in the size of Stokes radii of the selected cations, $Na^+$ and $Ca^{+2}$.

***Acknowledgments***

This work was financially supported by the Ministry of National Education, Republic of Indonesia, under Fundamental Research Grant of Udayana University No: 023.04.2.415253/2014, 5 December 2013, MAK 2013.109.011.521119, Year Budget 2014 are highly acknowledged.

***References:***

[1] Ying Wan, Katherine A. M. Creber, Brant Peppley and V. Tam Bui, 2003, Synthesis, Characterization and Ionic Conductive Properties of Phosphorylated Chitosan Membranes. *Macromol. Chem. Phys.* **204**, pp 850-858.

[2] Jolanta Kumirska, Mirko X. Weinhold, Jorg Thöming and Piotr Stepnowski, 2011, Biomedical Activity of Chitin/Chitosan Based Materials-Influence of Physicochemical Properties Apart from Molecular Weight and Degree of *N*-Acetylation. *Polymers* **3**, pp 1875-1901.

[3] Srinivasa R. Popuri, Y. Vijaya, Veera M. Boddu and Krishnaiah Abburi, 2009, Adsorptive removal of copper and nickel ions from water using chitosan coated PVC beads. *Bioresource Technology* **100**, pp 194-199.

[4] Krajewska B, 2004, Application of chitin-and chitosan-based materials for enzyme immobilizations: a review. *Enzyme and Microbial Technology* **35**, pp 126-139.

[5] Sandford PA, 1989, Chitosan: Commercial Uses and Potential Applications. In: Skjak-Braek G, Anthonsen T, Sandford P. Chitin and Chitosan – Sources, Chemistry, Biochemistry, Physical Properties and Applications. London: Elsevier Applied Science, pp 51-69.

[6] Dutta PK, Rayikumar MNV and Dutta J, 2002, Chitin and chitosan for versatile applications. *Journal of Macromolecular Science*, **C42 (3)**, pp 307-354.

[7] Brine CJ, 1989, Controlled Release Pharmaceutical Applications of Chitosan. In: Skjak-Braek G, Anthonsen T, Sandford P (Eds.). Chitin and Chitosan – Sources, Chemistry, Biochemistry,

Physical Properties and Applications. London: Elsevier Applied Science, pp 679-691.
[8] Ueno H, Yamada H, Tanaka I, Kaba N, Matsuura M, Okumura M, Kadosawa T and Fujinaga T. 1999, Accelerating effects of chitosan for healing at early phase of experimental open wound in dogs. *Biomaterials* **20**, pp 1407-1414.
[9] Cho YW, Cho YN, Chung SH, Yoo G and Ko SW, 1999, Water-soluble chitin as a wound healing accelerator. *Biomaterials* **20**, pp 2139-2145.
[10] Hadwiger, LA, Fristensky B and Riggleman RC. Chitosan, 1984, A Natural Regulator in Plant-Fungal Pathogen Interactions, Increases Crop Yields. In: Zikakis JP (Ed.). Chitin, Chitosan, and Related Enzymes. Orlando: Academic Press, pp 291-302.
[11] Barbara Krajewska, 2001, Diffusion of metal ions through gel chitosan membranes, *Reactive & Functional Polymers* **47**, pp 37-47.
[12] Mathias Ulbricht, 2006, Advanced functional polymer membranes. *Polymer* **47**, pp 2217-2262.
[13] T. Sata, 1991, Ion Exchange Membranes and Separation Processes with Chemical Reactions. Reviews Of Applied Electrochemistry 26, *Journal Of Applied Electrochemistry* **21**, pp 283-294.
[14] Xin Chen, Jiahao Liu, Zhicheng Feng and Zhengzhong Shao, 2005, Macroporous Chitosan/Carboxymethylcellulose Blend Membranes and Their Application for Lysozyme Adsorption. *Journal of Applied Polymer Science* **96**, pp 1267-1274.
[15] Zeng, X. F. and Ruckenstein, E. 1998, *J Membr Sci* **148**, p. 195.
[16] Li, W., Zhao, H., Teasdale, R. R., John, R. and Zhang, S. 2002, *Anal Chim Acta* **464**, p. 331.
[17] Barragan, V. M., Rueda, C. and Ruizbauza, C. J 1995, *Colloid Interface Sci*, **172**, p. 361.
[18] Zhang, L. N., Zhou, J. R., Zhou, D. C. and Tang, Y. R. 1999, *J Membr Sci* **162**, p. 103.
[19] Hobbie, Russel K. 1978, *Intermediate Physic for Medicine and Biology*. Second Edition. John Willey and Son. Singapore, pp 45-105.

# On the Diophantine Equation $Ax^2 - KXY + Y^2 + Lx = 0$

**J. D. Urrutia , J. M. E. Arañas, J. A. C. L. Lara, D. L. P. Maceda**
*Polytechnic University of the Philippines*

E-mail: math_urrutia@yahoo.com.ph / jackieurrutia20@gmail.com /
jonaldmarvin_aranas@yahoo.com/ math_lara025@yahoo.com.ph /
holyprayer17@yahoo.com /

**Abstract.** For any positive integer $a \geq 2$, we have shown that there are infinitely many pairs of positive integer $l$ and $k$ so that the equation $Ax^2 - Kxy + y^2 + Lx = 0$ has infinitely many positive integer solutions. Moreover, we have shown the forms of $k$ and $l$ that may take. We also presented the tables of solution for $2 \leq a \leq 10$, for some integers $1 \leq n \leq 5$.
**Keywords:** Diophantine Equation, Pell equation, Continued fraction, Convergent

## 1. Introduction

In [1], Marlewski and Zarzycki introduce and prove that the Diophantine Equation

$$x^2 - kxy + y^2 + x = 0 \tag{1}$$

has infinitely many solution $(x, y)$ if and only if $k = 3$.

The preceded paper by Yuan and Hu [2] show that for any positive integer $l$, there exist finitely many positive integer $k$ so that the equation

$$x^2 - kxy + y^2 + lx = 0 \tag{2}$$

has infinitely positive integer solution $(x, y)$

The main purpose of the this paper is to determine the cardinality of pair of positive integer $l$ and $k$ for any positive integer $a \geq 2$ so that the equation

$$ax^2 - kxy + y^2 + lx = 0 \tag{3}$$

has positive integer solution $(x, y)$. Moreover, we want to know how many positive integer solutions it may have. The researchers also distinguished the possible forms of $l$ and $k$.

## 2. Methodology

This section discussed the concepts and ideas being used by the researchers to solve the problems.

**Def.** The simple continued fraction is any expression of the form

Journal of Physics: Conference Series **622** (2015) 012005 doi:10.1088/1742-6596/622/1/012005

$$a_1 + \cfrac{1}{a_2 + \cfrac{1}{a_3 + \cfrac{1}{a_4 + \ddots}}}$$

where the first term $a_1$ is usually a positive or negative integer, and where the terms $a_2, a_3, a_4, \ldots$ are positive integers.

**Lemma 1.** Let $d$ be a positive integer, which is not a square. Then the $(n+1)st$ complete quotient $\alpha_n$ of the continued fraction $\alpha = \sqrt{d}$ is of the form

$$\frac{\sqrt{d}+P_n}{Q_n}, {P_n}^2 \equiv d(mod\ Q_n),$$

**Lemma 2**. Let $d, P_n, and\ Q_n$ be as in Lemma (1). Then the quadratic equation

$$x^2 - dy^2 = (-1)^n Q_n$$

has a positive integer solution $(x, y)$. If $l \neq (-1)^n Q_n$ and $|l| < \sqrt{d}$, then the Diophantine equation

$$x^2 - dy^2 = l$$

has no integer solutions $(x, y)$.

***Proof.***

$$\sqrt{d} = \frac{\alpha_n p_{n-1} + P_{n-2}}{\alpha_n q_{n-1} + q_{n-2}}$$

$$= \frac{(P_n + \sqrt{d})p_{n-1} + p_{n-2}Q_n}{(P_n + \sqrt{d})q_{n-1} + q_{n-2}Q_n}$$

$$\Rightarrow \sqrt{d}((P_n + \sqrt{d}))q_{n-1} + q_{n-2}Q_n) = (P_n + \sqrt{d})p_{n-1} + p_{n-2}Q_n$$

$$\Rightarrow \sqrt{d}(P_n q_{n-1} + Q_n q_{n-2}\sqrt{d}q_{n-1}) = p_{n-1}\sqrt{d} + P_n p_{n-1} + p_{n-2}Q_n$$

$$\Rightarrow (P_n q_{n-1} + Q_n q_{n-2})\sqrt{d} + q_{n-1}d) = (p_{n-1}\sqrt{d} + P_n p_{n-1} + p_{n-2}Q_n$$

$$\Rightarrow p_{n-1} = P_n q_{n-1} + Q_n q_{n-2}$$

and

$$\Rightarrow dq_{n-1} = P_n p_{n-1} + p_{n-2} Q_n$$

Multiply the first by $p_{n-1}$ and the second by $q_{n-1}$, and obtain:

$$p_{n-1}{}^2 - dq_{n-1}{}^2 = Q_n(p_{n-1}q_{n-2} - p_{n-2}q_{n-1})$$

$$p_{n-1}{}^2 - dq_{n-1}{}^2 = (-1)^n Q_n$$

■

**Lemma 3**. Let $d > 1$ be a positive integer which is not a square and $c \neq 0$ a given integer. If the Diophantine equation

$$x^2 - dy^2 = c \qquad (*)$$

has a positive integer solution $(x, y)$ with $\gcd(x, y) = 1$, then the equation $(*)$ has infinite many positive integer solutions $(x, y)$ with $\gcd(x, y) = 1$.

The Lemma 1 and 3 are well known theorems so we preferred to omit their proofs.

## 3. **Results**

The following are the results of this study.

**Lemma 4.**

I. Let $a$ be positive even integer and $k = \dfrac{a(n+2)}{2}$, $n \in \mathbb{N}$, then

$$k - 1 < \sqrt{k^2 - a} < k.$$

II. Let $a$ be positive odd integer greater than or equal to 3 and $k = a(n+1)$, $n \in \mathbb{N}$, then

$$k - 1 < \sqrt{k^2 - a} < k.$$

Journal of Physics: Conference Series **622** (2015) 012005 doi:10.1088/1742-6596/622/1/012005

***Proof.***

We will start for the first case.

Let $a$ be positive even integer and $k = \dfrac{a(n+2)}{2}$. Note that for any $n \in \mathbb{N}, k > a$

implies that $2k > a$,

$\Rightarrow 2k - 1 > a$

$\Rightarrow -2k+1 < a$

$\Rightarrow k^2 - 2k + 1 < k^2 - a$

$\Rightarrow (k-1)^2 < k^2 - a$

$\Rightarrow k - 1 < \sqrt{k^2 - a}.$

Also, $k^2 - a < k^2 \Rightarrow \sqrt{k^2 - a} < k,$

Thus, $k - 1 < \sqrt{k^2 - a} < k.$

The proof of the second case is analogous to the first case.

■

**Remark**. The previous lemma state that if we arbitrary choose $a$, and $k$ is in either of those two forms, with respect to $a$, $\lfloor \sqrt{k^2 - a} \rfloor = k - 1$. This implication will be use to the next results introduce as lemmas.

**Lemma 5.**

**Case I.** For $a$ is even.

Let $a$ be a positive even integer and $k = \dfrac{a(n+2)}{2}, n \in \mathbb{N}$. The continued fraction of $\sqrt{k^2 - a}$ is of the form

$$\sqrt{k^2 - a} = \left[k - a; \overline{1, \tfrac{2(k-a)}{a}, 1, 2k - 2}\right] \tag{4}$$

with

$$Q_n = \begin{cases} a(n+1) - 1 & \text{if } Q_{4t+1} = Q_{4t+3} \\ a & \text{if } Q_{4t+2} \\ 1 & \text{if } Q_{4t+4} \end{cases}, \quad t \geq 0 \tag{5}$$

**Case II.** For $a$ is odd.

Let $a$ be $k = a(n+1), n \in \mathbb{N}$. The continued fraction of $\sqrt{k^2 - a}$ is of the form

$$\sqrt{k^2 - a} = \left[k - a; \overline{1, \tfrac{2(k-a)}{a}, 1, 2k - 2}\right] \tag{6}$$

Journal of Physics: Conference Series **622** (2015) 012005 doi:10.1088/1742-6596/622/1/012005

with

$$Q_n = \begin{cases} a(2n+1)-1 & \text{if } Q_{4t+1} = Q_{4t+3} \\ a & \text{if } Q_{4t+2} \\ 1 & \text{if } Q_{4t+4} \end{cases} \quad , t \geq 0 \tag{7}$$

*Proof.*

Let $a$ be positive even integer and $k = \dfrac{a(n+2)}{2}$. By Lemma 4,

$$\left\lfloor \sqrt{k^2 - a} \right\rfloor = k - 1$$

Take $\alpha = \sqrt{k^2 - a}$.

$\Rightarrow a_1 = k - 1$

$$\alpha_1 = \frac{1}{\sqrt{k^2 - a} - (k-1)} = \frac{\sqrt{k^2 - a} + (k-1)}{k^2 - a - (k-1)^2} = \frac{\sqrt{k^2 - a} + (k-1)}{2k - (a+1)}$$

$$\Rightarrow \alpha_1 = \frac{\sqrt{k^2 - a} + (k-1)}{2k - (a+1)}$$

$$a_2 = \left\lfloor \frac{\sqrt{k^2 - a} + (k-1)}{2k - (a+1)} \right\rfloor = \left\lfloor \frac{\left\lfloor \sqrt{k^2 - a} \right\rfloor + (k-1)}{2k - (a+1)} \right\rfloor = \left\lfloor \frac{2k-2}{2k - (a+1)} \right\rfloor = 1$$

$\Rightarrow a_2 = 1$

$$\alpha_2 = \frac{2k - (a+1)}{\sqrt{k^2 - a} - (k-a)} = \frac{\sqrt{k^2 - a} + (k-a)}{a}$$

$$\Rightarrow \alpha_2 = \frac{\sqrt{k^2 - a} + (k-a)}{a}$$

$$a_3 = \left\lfloor \frac{\sqrt{k^2 - a} + (k-a)}{a} \right\rfloor = \left\lfloor \frac{\left\lfloor \sqrt{k^2 - a} \right\rfloor + (k-a)}{a} \right\rfloor = \left\lfloor \frac{2k - (a+1)}{a} \right\rfloor$$

since $k = \dfrac{a(n+2)}{2}$, substitituting this, we obtain

$$a_3 = \left\lfloor (n+2) - \frac{(a+1)}{a} \right\rfloor$$

applying the property of floor function,

$$a_3 = (n+2) - \left\lceil \frac{a+1}{a} \right\rceil = (n+2) - 2 = \frac{2k}{a} - 2 = \frac{2(k-a)}{a}$$

$$\Rightarrow a_3 = \frac{2(k-a)}{a}$$

For $\alpha_3$ and $a_4$,

$$\alpha_3 = \frac{a}{\sqrt{k^2-a}-(k-a)} = \frac{\sqrt{k^2-a}+(k-a)}{2k-(a+1)}$$

$$\Rightarrow \alpha_3 = \frac{\sqrt{k^2-a}+(k-a)}{2k-(a+1)}$$

$$a_4 = \left\lfloor \frac{\sqrt{k^2-a}+(k-a)}{2k-(a+1)} \right\rfloor = \left\lfloor \frac{\lfloor\sqrt{k^2-a}\rfloor+(k-a)}{2k-(a+1)} \right\rfloor = \left\lfloor \frac{2k-(a+1)}{2k-(a+1)} \right\rfloor = 1$$

$$\Rightarrow a_4 = 1$$

To get $\alpha_4$,

$$\alpha_4 = \frac{2k-(a+1)}{\sqrt{k^2-a}-(k-1)} = \frac{\sqrt{k^2-a}+(k-1)}{1}$$

$$\Rightarrow \alpha_4 = \frac{\sqrt{k^2-a}+(k-1)}{1}$$

Finally,

$$a_5 = \left\lfloor \frac{\sqrt{k^2-a}+(k-1)}{1} \right\rfloor = \left\lfloor \frac{\lfloor\sqrt{k^2-a}\rfloor+(k-1)}{1} \right\rfloor = 2k-2$$

$$\Rightarrow a_5 = 2k-2$$

Therefore, the simple continued fraction of $\sqrt{k^2-a}$ is of the form

$$\sqrt{k^2-a} = \left[k-a; \overline{1, \frac{2(k-a)}{a}, 1, 2k-2}\right]$$

which proves the case I.

The proof of Case II for $a$ is odd is analogous to the first case.

**Theorem**.
For any positive integer $a \geq 2$, there exist infinitely many pairs of positive integers $l$ and $k$ such that $ax^2 - kxy + y^2 + lx = 0$ has infinitely many solutions. Moreover, for an even $a$, $l$ may take the form of $a(n+1)-1$ and $k$ may take the form $a(n+2)$ while for an odd $a$, that is $a \geq 3$, $l$ may take the form of $a(2n+1)-1$ and $k$ may take the form $2a(n+1)$.

Proof:

Consider the equation

$$au^2 - kuv + v^2 + l = 0 \qquad (8)$$

Note that the equation (*) may manipulate and transform into

$$\left(v - \frac{ku}{2}\right)^2 - \left(\frac{k^2}{4} - a\right)u^2 = -l \qquad (9)$$

From here, we ware going the divide the proof into two cases.

Case I: a is even

Take $k = 2k_1$ where $k_1 = \frac{a(n+2)}{2}$ $n \in \mathbb{N}$. Now substitute $k = 2k_1$ to equation (9) which make the equation

$$(v - k_1u)^2 - (k_1^2 - a)u^2 = -l \qquad (10)$$

which is a pell equation. By lemma (2), since we are looking for a positive integer $l$ let us take

$$-l = (-1)^n Q_n = (-1)^{4t+3}[a(n+1) - 1] = -[a(n+1) - 1]$$

thus,

$$l = a(n+1) - 1$$

So the equation

$$x^2 - (k^2 - a)y^2 = -l \qquad (11)$$

has positive integer solutions $(x, y)$.

Implies that from equation (10) and (11),

$$v - k_1u = x \text{ and } u = y$$

$$\Rightarrow v = k_1y \text{ and } u = y$$

Since $u$ is already define, let us multiply $u^2$ to equation (8),

$$u^2(u^2 - kuv + v^2 + l) = 0$$

$$(u^2)^2 - k(u^2)(uv) + (uv)^2 + lu^2 = 0$$

which is the original equation with solutions $x = u^2$ , $y = uv$. Moreover by Lemma (3), it has infinitely many positive integer solutions.

**Case II:** when $a$ is odd, $a \geq 3$.

Take $k = 2k_1$ where $k_1 = a(n+1), n \in \mathbb{N}$. Now substitute $k = 2k_1$ to equation (8) which make the equation

$$(v - k_1u)^2 - (k_1^2 - a)u^2 = -l \qquad (12)$$

which is a pell equation. By lemma (2), since we are looking for a positive integer $l$ let us take

$$-l = (-1)^n Q_n = (-1)^{4t+3}[a(2n+1) - 1] = -[a(2n+1) - 1]$$

thus,

$$l = a(2n+1) - 1$$

So the equation

$$x^2 - (k^2 - a)y^2 = -l \qquad (13)$$

has positive integer solutions $(x, y)$.
Implies that from equation (12) and (13),

$$v - k_1 u = x \text{ and } u = y$$
$$\Rightarrow v = k_1 y \text{ and } u = y$$

Since $u$ is already define, let us multiply $u^2$ to equation (8),

$$u^2(u^2 - kuv + v^2 + l) = 0$$

$$(u^2)^2 - k(u^2)(uv) + (uv)^2 + lu^2 = 0$$

which is the original equation with solutions $x = u^2$, $y = uv$. Moreover by Lemma (5), it has infinitely many positive integer solutions. This completes the proof.

The table below shows the value may take for the equation (3).

| $a$ | $n$ | $k$ | $l$ | $(x, y)$ |
|---|---|---|---|---|
| 2 | 1 | 6 | 3 | (1,5), (4,22), (196,1106) |
| | 2 | 8 | 5 | (1,7), (9, 69), (729, 5643) |
| | 3 | 10 | 7 | (1, 9), (16,156), (1936, 18964) |
| | 4 | 12 | 9 | (1,11), (25,295), (4225, 49955) |
| | 5 | 14 | 11 | (1,13), (36,498), (8100, 112230) |

| $a$ | $n$ | $k$ | $l$ | $(x, y)$ |
|---|---|---|---|---|
| 3 | 1 | 12 | 8 | (1,11), (9,105), (1849,21715) |
| | 2 | 18 | 14 | (1,17), (25, 445), (10201, 181901) |
| | 3 | 24 | 20 | (1, 23), (49,1169), (33489, 799527) |
| | 4 | 30 | 26 | (1,29), (81,2421), (83521, 2497249) |
| | 5 | 36 | 32 | (1,35), (121,4345), (175561, 6305531) |

| $a$ | $n$ | $k$ | $l$ | $(x, y)$ |
|---|---|---|---|---|
| 4 | 1 | 12 | 7 | (1,11), (4,46), (1024,11936) |
| | 2 | 16 | 11 | (1,15), (9, 141), (3481, 54811) |
| | 3 | 20 | 15 | (1, 19), (16,316), (8836, 174934) |
| | 4 | 24 | 19 | (1,23), (25,595), (18769, 447305) |
| | 5 | 28 | 23 | (1,27), (36,1002), (35344, 984556) |

| $a$ | $n$ | $k$ | $l$ | $(x, y)$ |
|---|---|---|---|---|
| 5 | 1 | 20 | 14 | (1,19), (9,177), (5625,111075) |
| | 2 | 30 | 24 | (1,29), (25, 745), (29929, 892853) |
| | 3 | 40 | 34 | (1, 39), (49,1953), (96721, 3856711) |
| | 4 | 50 | 44 | (1,49), (81,4041), (239121, 11932089) |
| | 5 | 60 | 54 | (1,59), (121,7249), (499849, 29949227) |

| $a$ | $n$ | $k$ | $l$ | $(x,y)$ |
|---|---|---|---|---|
| 6 | 1 | 18 | 11 | (1,17), (4,70), (2500,44150) |
| | 2 | 24 | 17 | (1,23), (9, 213), (8281, 196651) |
| | 3 | 30 | 23 | (1, 29), (16,476), (20736, 617904) |
| | 4 | 36 | 29 | (1,35), (25,895), (43681, 1565201) |
| | 5 | 42 | 35 | (1,59), (121,7249), (499849, 29949227) |

| $a$ | $n$ | $k$ | $l$ | $(x,y)$ |
|---|---|---|---|---|
| 7 | 1 | 28 | 20 | (1,27), (9,249), (11449,317683) |
| | 2 | 42 | 34 | (1,41), (25, 1045), (60025, 2511005) |
| | 3 | 56 | 48 | (1, 55), (49,2737), (192721, 10768231) |
| | 4 | 70 | 62 | (1,69), (81,5661), (474721, 47920195) |
| | 5 | 84 | 76 | (1,83), (121,10153), (990025, 83079515) |

| $a$ | $n$ | $k$ | $l$ | $(x,y)$ |
|---|---|---|---|---|
| 8 | 1 | 24 | 15 | (1,23), (4,94), (4624,109412) |
| | 2 | 32 | 23 | (1,31), (9, 285), (15129, 480315) |
| | 3 | 40 | 31 | (1, 39), (16,636), (37636, 1497874) |
| | 4 | 48 | 39 | (1,47), (25,1195), (78961, 3776921) |
| | 5 | 56 | 47 | (1,55), (36,2010), (147456, 8236416) |

| $a$ | $n$ | $k$ | $l$ | $(x,y)$ |
|---|---|---|---|---|
| 9 | 1 | 36 | 26 | (1,35), (9,321), (19321,690691) |
| | 2 | 54 | 44 | (1,53), (25, 1345), (100489, 5409605) |
| | 3 | 72 | 62 | (1, 71), (49,3521), (321489, 23106951) |
| | 4 | 90 | 80 | (1,89), (81,7281), (790321, 71049769) |
| | 5 | 108 | 98 | (1,107), (121,13057), (1646089, 177640331) |

| $a$ | $n$ | $k$ | $l$ | $(x,y)$ |
|---|---|---|---|---|
| 10 | 1 | 30 | 19 | (1,29), (4,118), (7396,219386) |
| | 2 | 40 | 29 | (1,39), (9, 357), (24025, 954955) |
| | 3 | 50 | 39 | (1, 49), (16, 796), (59536, 2964844) |
| | 4 | 60 | 49 | (1,59), (25,1495), (124609, 7455713) |
| | 5 | 70 | 59 | (1,69), (36,2514), (232324, 16229422) |

Journal of Physics: Conference Series **622** (2015) 012005 doi:10.1088/1742-6596/622/1/012005

**Reference:**


[1] A.Marlewski, P.Zarzycki, Infinitely many positive solutions of the Diophantine $x^2 - kxy - y^2 + x = 0$, Elsevier, international journal Computers and Mathematics with Applications 2002

[2] Li Feng, Pingzhi Yuan and Yongzhong Hu, ON THE DIOPHANTINE EQUATION X^2-KXY+Y^2+LX=0, Integers 13, 2013

[3] Richard A. Mollin, Fundamental Number Theory with Applications Second Edition 2007

[4] Frits Beukers, Elementary Number Theory, 2013

[5] Hua Loo Keng, Introduction to Number Theory, 1982

[6] Olds C.D., Continued Fractions New Mathematical Library 1963

# Study of The Vector Product using Three Dimensions Vector Card of Engineering in Pathumwan Institute of Technology

Wannapa Mueanploy

Assist. Prof., Department of Science and Mathematics
Faculty of Science and Technology, Pathumwan Institute of Technology
Bangkok, Thailand 10330
Email: bpuifai06@hotmail.com

**Abstract.** The objective of this research was to offer the way to improve engineering students in Physics topic of vector product. The sampling of this research was the engineering students at Pathumwan Institute of Technology during the first semester of academic year 2013. 1) Select 120 students by random sampling are asked to fill in a satisfaction questionnaire scale, to select size of three dimensions vector card in order to apply in the classroom. 2) Select 60 students by random sampling to do achievement test and take the test to be used in the classroom. The methods used in analysis of achievement test by the Kuder-Richardson Method (KR- 20). The results show that 12 items of achievement test are appropriate to be applied in the classroom. The achievement test gets Difficulty (P) = 0.40-0.67, Discrimination = 0.33-0.73 and Reliability (r) = 0.70.The experimental in the classroom. 3) Select 60 students by random sampling divide into two groups; group one (the controlled group) with 30 students was chosen to study in the vector product lesson by the regular teaching method. Group two (the experimental group) with 30 students was chosen to learn the vector product lesson with three dimensions vector card. 4) Analyzed data between the controlled group and the experimental group, the result showed that experimental group got higher achievement test than the controlled group significant at .01 level.

## 1. Introduction

### Objective

Physics are difficult for many engineering students; some of them feel that physics are not important. They may not find it interesting or useful because physics is very "abstract". Nguyen and Meltzer (2003) many students retained significant conceptual difficulties regarding vector methods that are heavily employed throughout the physics curriculum. Kustusch (2011) students in introductory physics struggle with vector algebra and with cross product direction in particular. Some have suggested that this may be due to misapplied right-hand rules but there are few studies that have had the resolve to explore this. Additionally previous research on students understanding has noted several kinds of representation-dependence of students' performance with vector algebra in both physics and non-physics. Cataloglu (2006) the effort on teaching vector algebra using free open source software (FOSS) recent studies showed that students have difficulties in learning basic physics concepts. Barniol and Zavala (2011) we can establish two general errors: 1) Misconception in the magnitude of the unit vector and in the *x*- and *y*-components of this vector, 2) Confusion between the unit vector in the direction of vector **A** with the two component vectors of the vector **A** written in the unit-vector notation.

In this study three dimensions vector cards assist students to understand in the Right Hand Rule and it is the aid for the direction of the vector product. Group of sampling were engineering students at Pathumwan Institute of Technology (PIT's students) by random sampling

Three dimensions vector cards assistant PIT's students to be able to learn and revise it all the time.

**Vector**

Vector in physics is a quantity that has both magnitude and direction. It is typically represent by an arrow whose direction is the same as that of the quantity and whose length is proportional to the quantity's magnitude. Vectors can be used to represent displacement, velocity, and acceleration. The length represents the magnitude and the direction of that quantity is the direction in which the vector is pointing. Because vectors are construct this way it is helpful to analyze physical quantities as vectors.

**Product of vector**

We will define two different kinds of product of vectors. The first called the scalar product and second called the vector product.

**The scalar product**

The scalar product of two vector $\boldsymbol{A}$ and $\boldsymbol{B}$ is donated by $\boldsymbol{A.B}$. The scalar product called the dot product. The scalar product of two vectors can be constructed by taking the component of one vector in the direction of the other and multiplying it times the magnitude of the other vector. This can be expressed in two ways: in terms of the magnitude of $\boldsymbol{A}$ and $\boldsymbol{B}$ and the angle $\boldsymbol{\theta}$ between the two vectors or in term of components of $\boldsymbol{A}$ and. The scalar product is commutative; for any two vector $\boldsymbol{A}$ and $\boldsymbol{B}$ $\boldsymbol{A.B} = \boldsymbol{B.A}$ the scalar product of two perpendicular vector is zero

$$\boldsymbol{A.B} = \boldsymbol{AB\cos\theta} = \|\boldsymbol{A}\|\,\|\boldsymbol{B}\|\boldsymbol{\cos\theta}$$

$$\boldsymbol{A.B} = \boldsymbol{A_xB_x + A_yB_y + A_zB_z}$$

**The vector product**

The cross product $\boldsymbol{AxB}$ is defined as a vector $\boldsymbol{C}$ that is perpendicular to both $\boldsymbol{A}$ and $\boldsymbol{B}$ with $\boldsymbol{A}$ direction given by the right-hand rule and a magnitude equal to the area of the parallelogram that the vectors span. The cross product is defined by the formula $\boldsymbol{A\,x\,B} = \|\boldsymbol{A}\|\,\|\boldsymbol{B}\|\boldsymbol{\sin\theta}$. Where $\theta$ is the angle between $\boldsymbol{A}$ and $\boldsymbol{B}$ in the plane, it is between 0° and 180°, $\|\boldsymbol{A}\|$ *and* $\|\boldsymbol{B}\|$ are the magnitudes of vectors $\boldsymbol{A}$ and $\boldsymbol{B}$ and in the direction given by the right-hand rule. The vector product is not commutative; for any two vectors $\boldsymbol{A}$ and $\boldsymbol{B}$.

$$\boldsymbol{A\,x\,B} = \boldsymbol{-B\,x\,A}$$

The vector product of two parallel the angle $\theta$ between them is either 0° or 180° by the above formula the cross product of $\boldsymbol{A}$ and $\boldsymbol{B}$ is zero.

$$C = AB \sin\theta$$

$$C_x = A_y B_z - A_z B_y$$

$$C_y = A_z B_x - A_x B_z$$

$$C_z = A_x B_y - A_y B_x$$

## 2. Methodology and Experimental

In experiments to determine the dimensions of the three dimensions vector card by choosing from three sizes and the most popular will be used to create a three dimensions vector card to use in the vector product classroom. The methodology and experimental was divided into four stages.

Stage 1 Make the Satisfaction questionnaires
Stage 2 Make the Achievement test
Stage 3 Make the three dimensions vector card and how to use the three dimensions vector card
Stage 4 Experimental

### 2.1 Satisfaction questionnaires

Made the satisfaction questionnaires to selection the paper size of three dimensions card by random sampling of 120 PIT's students choose the card that will be useful and applied. The results of this questionnaire will be used to make three dimensions vector card in classroom. The three dimensions, size card is:

1. Paper in 9 cm x 12 cm (double size of ID card)
2. Paper in 11.5 cm x 16 cm (half of note book)
3. Paper in 15 cm x 21 cm (half of A-4 paper size)

### 2.2 Achievement test

The steps make the achievement test.

1. Random sampling 60 PIT's students to do the achievement test (15 item)
2. Sort the answer sheet score from lowest to highest
3. Use 25% method to get 20 answer sheets in low group and 20 answer sheets in high group
4. Use KR-20 method to find Difficulty (P), Discrimination and Reliability (r)
5. Get only 12 items achievement test for pre-test and post- test

### 2.3The three dimensions vector card

The steps make three dimensions vector card.

1. Cut the paper form reuse paper (candy box, old book cover and other cardboard) in 11.5 cm x 16 cm (half of the notebook)

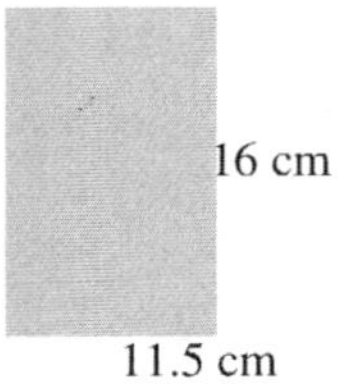


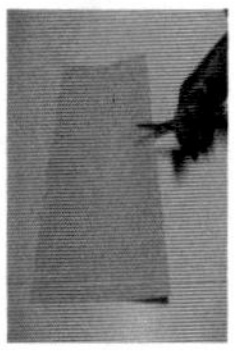

2. Half along it

3. Arrow are drawn to represent three perpendicular directions in space

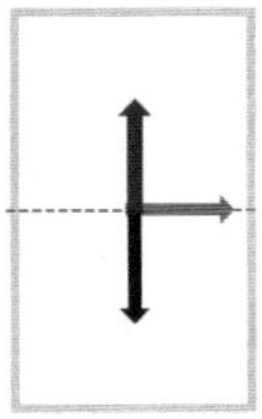

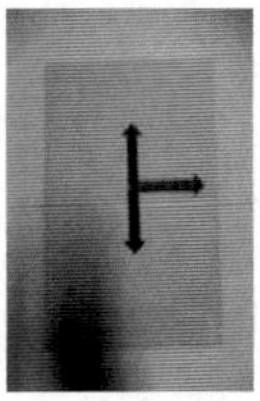

4.Three dimensions vector card for ***A B*** and ***C***

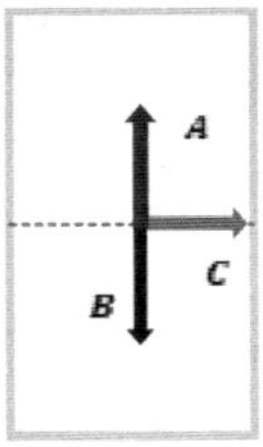


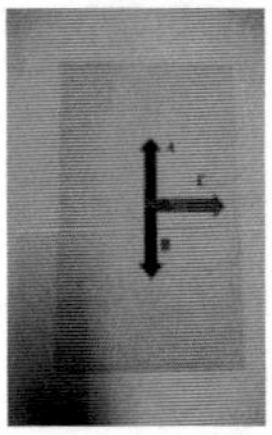

5. Two pink dot is placed on the card to assist in orienting the axis

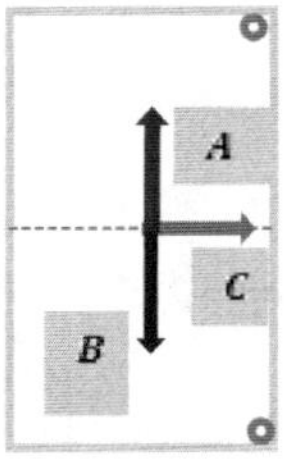


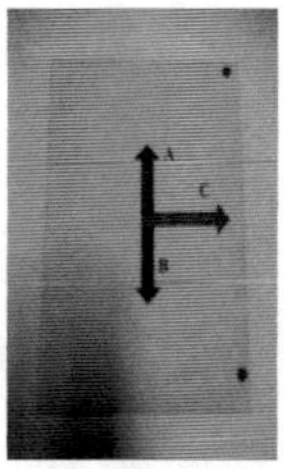

Journal of Physics: Conference Series **622** (2015) 012006 doi:10.1088/1742-6596/622/1/012006

**How to use the three dimensions vector card**

The students can fold the card in the direction from vector A (blue arrow) to vector B (black arrow) the result is the direction of the vector C (red arrow), can used in the angle $\theta$ between 0° to 90°. The pink dots are placed on the cards to assist students in orienting the axis. In the vector product lesson students can made the three dimensions vector card using recycled paper and took it on their own to find the direction of vector in three dimensions in classrooms.

**2.4 Experimental**

Studied in class

1. Random sampling 60 PIT's students (new group) to do the pre –test
2. The controlled group 30 PIT's students went to studied by the regular teaching method in the vector product lesson and post – test
3. The experimental group 30 PIT's students went to learn the vector product by three dimensions vector card
4. The experimental group 30 PIT's students made three dimensions vector card by themselves and used to find the direction in the classroom in the vector product lesson and post-test

## 3. Analyzed data

1. The satisfaction questionnaires analyzed data finding the average$(\bar{x})$. The simplest number used to characterize a sample is the *mean*, which for $N$ values $x_i$, $(i = 1,2,\ldots,N.)$ is defined by

$$\bar{X} = \frac{1}{N}\sum_{i=1}^{N} X_i$$

to select the size paper to make three dimensions vector card in the classroom

2. Analyzed data finding the average $(\bar{x})$ and the sample standard deviation (SD) is the positive square root of the sample variance i.e.

$$SD = \sqrt{\frac{1}{N}\sum_{i=1}^{N}(x_i - \bar{X}^2)}$$

We may therefore write the sample variance $SD^2$ as

$$SD^2 = \frac{1}{N}\sum_{i=1}^{N} x_i^2 - \left(\frac{1}{N}\sum_{i=1}^{N} x_i^2\right)^2$$

From which the sample standard deviation is found by taking the positive square root. Thurs., by evaluating the quantities $\sum_{i=1}^{N} X_i$ and $\sum_{i=1}^{N} X_i^2$ for our sample, we can calculate the sample mean and sample standard deviation at the same time.

Compare pre-test and post-test of physics achievement of PIT's students, studied the vector product by using three dimensions vector and PIT's students studied the vector product by the regular teaching method by t-test independent

## 4. The result of data analysis

1. The result of satisfaction questionnaire 120 students select the three dimensions vector size card:

**Table 1** The result of satisfaction questionnaire select size card

| Size card | GOOD=3 | OK=2 | POOR=1 | $\bar{x}$ |
|---|---|---|---|---|
| 9 cm x 12 cm | 11 | 62 | 47 | 1.7 |
| 11.5 cm x 16 cm | 91 | 22 | 7 | 2.7 |
| 15 cm x 21 cm | 45 | 59 | 16 | 2.24 |

The result form the satisfaction questionnaire show 11.5 cm x 16cm card good at $\bar{x} = 2.7$ this size is better useful and applied this size will be making in the classroom.

2. The results of analyzing pre-test and post-test physics achievement the vector product of PIT's students are as the following.

**Table 2** The result of physics achievement study, the vector product pre-test

| Statistics Sample | N | $\vec{X}$ | $SD$ | t-test |
|---|---|---|---|---|
| Controlled group | 30 | 5.93 | 1.08 | 1.03 |
| Experiment group | 30 | 5.63 | 1.18 | |

From table (1) Finding an average of the pre-test in physics achievement, the vector product the controlled group studied the vector product by the regular teaching method is 5.93, the experimental group studied the vector product by using three dimensions vector card is 5.63. Using t-test independent sample to compare achievement finds that the experimental group and controlled group achievement is 1.03 it mean different insignificantly.

**Table 3** The result of physics achievement study, the vector product post-test

| Statistics Sample | N | $\vec{X}$ | $SD$ | t-test |
|---|---|---|---|---|
| Controlled group | 30 | 7.66 | 1.02 | 2.68** |
| Experiment group | 30 | 8.33 | 0.97 | |

$^{**}p < .01$

From table (2) Finding the average of the post-test in physics achievement, the controlled group, studied the vector product by the regular teaching method is 7.66, the experimental group studied the vector product by using three dimensions vector card is 8.33 and an average of the post-test using t-test independent sample to compare achievement finds that is significantly higher at level of 0.01.

It means the achievement of PIT's students studying the vector product by using three dimensions vector card is higher than the achievement of PIT's students studying the vector product by the regular teaching method.

## 5. Conclusion

In this study intended to help improve teaching and learning in the physics of PIT's students in vector product lesson. The Right Hand Rule is an important mnemonic techniques to understand in three dimensions vector but difficult to use and difficult to alignment of their hand and fingers, unusual to watch and difficult to apply. In this experiment the three dimensions vector card that the student created can help the student understand in vector product lesson. The three dimensions vector size card with 11.5 cm x 16 cm is better useful and applied. This size is used to construct in the classroom. This size is half of the notebook can easily be flattened again to be stored in a pocket of the student's. The students can made the three dimensions vector card form the reused paper by themselves and took it in classroom and use to find the direction of the vector product.

In achievement test this studied used 12 items of the vector product to be analyzed by KR-20 program got Difficulty (P) = 0.40-0.67, Discrimination = 0.33-0.73 and Reliability (r) = 0.70. This test is used to measure the achievement of students .The student do the pre-test and after that the students have created a card and bring the card to use in the vector product lesson, and do the post- test. The achievement of PIT's students studied the vector product by using three dimensions vector card is higher than the achievement of PIT's students studied the vector product by the regular teaching method at a level of 0.01 which relates to the specified hypothesis.

This study was taken from the remaining material to create a tool to assist students understand the three dimensions vector better , in future could have to a reusable material used in other lessons of physics, such as wood cube or metal cube used to make three dimensions vector. The instructor uses repeats it in many times for the sake of the classroom environment and reduce global warming.

The three dimensions vector card is simply created, low cost and uses recycled materials, the accomplishment check showed a big distinction between students with and without three dimensions vector card. In this reason the instruction in physics and mathematics is used the three dimensions cards to assistant in the Right Hand Rule and vectors in three dimensions. The three dimensions vector card is applied to the study of three dimensions vector in the other, for example finding the direction of the magnetic field and force, the concepts of torque and angular momentum.

**Acknowledgment**

The author would like to thank Pathumwan Institute of Technology for the financial support and permission to experiment.

**References**

[1] Cataloglu Erdat 2006 Open Source Software In Teaching Physics: A Case Study On Vector Algebra And Visual Representations *The Turkish Online Journal of Educational Technology* **5** Issue 1 Article 8

[2] Fishbane, P.M. & et al. 2005 "*Physics for Scientists and Engineers with Modern Physics*" USA Prentice Hall

[3] Hugh D. Young and Roder A. Freedman 2004 "*University Physics with Modern Physics*" 11th Edition Pearson Education Inc.

[4] Mary Bridget Kustusch 2011 "Student Understanding of Cross Product Direction and Use of Right-Hand Rules: An Exploration of Representation and Context-Dependence" *Physics education research* March 23 p 236

[5] Ngoc-Loan Nguyen and David E. Meltzer 2005 "Visualization Tool for 3-D Relationships and the Right-Hand Rule" *The Physics Teacher* **43** p 155-157

[6] Pablo Barniol and Genaro Zavala 2014 "Test of understanding of vectors: A reliable multiple-choice vector concept test" *Physics Education Research* **10** 010121

[7] R. A. Serway 1999 Physics for Scientists & Engineers with Modern Physics 4th Ed. Philadelphia: Saunders College Pub USA

# Radiation from Secondary Planar Surfaces Sources in Quantum Field Theory

**A Maksuwan**[1*, 3] **P Viriyasrisuwattana** [2,3]
[1]Department of Science and Mathematics (Physics), Faculty of Science and Technology, Pathumwan Institute of Technology, Wangmai, Pathumwan, Bangkok, 10330, Thailand.
[2]Department of Physics, Faculty of Science, Ramkhamhaeng University, Huamark, Bangkapi, Bangkok, 10240,Thailand.
[3]Quantum Field Theory and High Energy Physics Group (QFTHEP) Ramkhamhaeng University, Huamark, Bangkapi, Bangkok, 10240, Thailand.

E-mail: iiibigiii@hotmail.com

**Abstract.** In quantum optics (optical coherence) theorem, many sources are employed in the laboratory are secondary planar sources. A source of this kind is usually an aperture in an opaque planar surface screen, illuminated either directly or via an optical system with primary sources. The expression following into the formulation of the radiation in radiant intensity which, analytical solution has not given the time evolution of amplitude. In this research, we consider this problem in quantum field theory (QFT) view point. By using the method of study based on a configuration space for explaining characteristic of the complete process, which begins and ends with the vacuum state so-called vacuum-to-vacuum transition amplitude between emitters and detectors. Propose in this research, we attend and explain the radiation for radiant intensity with time evolution process. To calculation amplitude transition of massive quantum particles propagator stimulated emission by secondary planar surface sources in space-time. Finally, we use the mathematical program for corresponding numerical evaluations between quantum optics and quantum field theory situation.

## 1. Introduction

In quantum optics (optical coherence) theorem, many sources are employed in the laboratory are secondary planar sources. A source of this kind is usually an aperture in an opaque planar surface screen, illuminated either directly or via an optical system with a primary source [3].

(a) (b) (c)

**Figure1**: Secondary planar surfaces sources on opaque screen (a) Circular aperture (b) Rectangular aperture (c) Crossed aperture

Representation of the source distribution as a correlation Green's function in the space-frequency domain $\tilde{G}(\bar{r}-\bar{r}',\upsilon)$, is given by

$$W(\bar{r},\upsilon)=\int J(\bar{r},\bar{r}')\tilde{G}(\bar{r}-\bar{r}',\upsilon)d^3r', \tag{1}$$

where
$$\tilde{G}(\bar{r}-\bar{r}',\upsilon)=\frac{k^2}{\pi c}\left(\frac{\sin k(\bar{r}-\bar{r}')}{k(\bar{r}-\bar{r}')}\right),\ k=\frac{2\pi\upsilon}{c}. \tag{2}$$

The expression following into the formulation of the radiation in radiant intensity which, analytical solution has not given the time evolution of amplitude [5]

Propose in this research focus on radiation from secondary planar surface sources of a crossed aperture. We attend and explain the radiation for radiant intensity together with Feynman's so-called spin-zero electron are technically called Klein-Gordon particles of finite mass [2, 12]. To calculation amplitude transition of the massive spin zero particle propagator stimulated emission by secondary planar surface sources of a crossed aperture in space-time. Finally, we use the mathematical program for corresponding numerical evaluations between quantum optics and quantum field theory situation.

## 2. Vacuum-to-Vacuum Transition Amplitude in The Presence of Source

We already obtained the transition amplitude

$$\langle q_f t_f | q_i t_i\rangle = N\int Dq\ \exp\left[\frac{i}{\hbar}\int_{t_i}^{t_f} dt\ L(q,\dot{q})\right] \tag{3}$$

with the boundary conditions $q(t_f)=q_f$, $q(t_i)=q_i$. However, what we really want to know in quantum mechanics is that particles are first created, interaction each other and finally destroyed by the measurement. Therefore, using "*source*" to create particles, we need vacuum-to-vacuum transition amplitude in the presence of a source [1]. If we note the vacuum state with a source by $|0,t\rangle^J$, we define the vacuum-to-vacuum transition amplitude as

$$Z[J]\equiv\langle 0,\infty|0,-\infty\rangle^J. \tag{4}$$

We assume the source *J (t)* is non-zero only between and $t'$ $(t<t')$. We introduce $T$ and $T'$ for $T<t$ and $t'>T'$ the amplitude becomes

$$\langle Q'T'|QT\rangle^J=\int dq'dq\langle Q'T'|q't'\rangle\langle q't'|qt\rangle^J\langle qt|QT\rangle \tag{5}$$

$$\langle Q'T'|q't'\rangle=\sum_m \phi_m(Q')\,\phi_m^*(q')\exp\left[\frac{i}{\hbar}E_m(t'-T')\right] \tag{6}$$

and
$$\langle qt|QT\rangle=\sum_n \phi_n(q)\,\phi_n^*(Q)\exp\left[-\frac{i}{\hbar}E_n(t-T)\right] \tag{7}$$

Taking the limit $T'\to\infty$ and $T\to-\infty$ in equation (5) we found only the ground state (vacuum) contributes to $\langle Q'T'|QT\rangle^J$

$$Z[J] \equiv \langle 0,\infty | 0,-\infty \rangle^J = \lim_{\substack{T'\to\infty \\ T\to-\infty}} \frac{\langle Q'T' | QT \rangle^J}{\phi_0^*(Q)\phi_0(Q')\exp\left[-\frac{i}{\hbar}E_0(T'-T)\right]}$$

$$= \int dq'dq\, \phi_0^*(q',t')\langle q't' | qt \rangle^J \phi_0(q,t) \tag{8}$$

Where $dq$ is a differential of components in unitary space time given development is then given completely by the *propagator* or *Green's function* [4] thus

$$G(q't',qt) \equiv \langle q't' | qt \rangle^J \tag{9}$$

**3. The Configuration Space for Explaining Characteristic of the Space-Time Domain**

In quantum field theory (QFT) operates is based on concept in special relativity (SR). We characterize space-time by an *event*, which is something that happens in time $t$ and location $(x, y, z)$. The speed of light c can serve in a role as a conversion factor, transforming time into space. Space and time, therefore, from *a configuration space* and we denote coordinates by $(ct, x, y, z)$. We take $ct = x^0$ and $(x, y, z) \to (x^1, x^2, x^3)$. Then an event in space-time is labelled by the coordinates of a vector $x^\mu = (x^0, x^1, x^2, x^3)$.

The quantum mechanics characterization of the complete process, which *begins* and *ends* with the vacuum state has the typical form vacuum-to-vacuum transition amplitude in the presence of the external source $Z[K(x)]$ (the effect of external forces acting on the particle) [6].

$$Z[K(x)] \equiv \langle 0_+ | 0_- \rangle^{K(x)} = \exp\left[\frac{i}{2}\int (dx)(dx')K^*(x)G(x,x')K(x')\right] \tag{10}$$

where, $(dx) = dx^0dx^1dx^2dx^3$ and $G(x,x')$ is the Green's function. The exponential structure is derived from the physical independence of sufficiently remote individual acts of emission propagation and detection, which are embodied in the form of an equation (9). Let $K(x) = K_1(x) + K_2(x)$, where the source $K_2(x)$ is switched on after the source $K_1(x)$ is switched off. $K_1(x)$ will be identified with the emitter and $K_2(x)$ with the detector.

We start from the Green's function of the massive spin zero particles is emitted from an external source $K(x) \equiv \delta(x-x')$ satisfies to the differential equation of Schrödinger's equation.

$$\left[-i\hbar\frac{\partial}{\partial t} - \frac{\hbar^2\nabla^2}{2m}\right]G(x,x') = \delta(x-x') \tag{11}$$

Also, the solution of the differential equation from the equation (11) is given by

$$G(x,x') = \begin{cases} \displaystyle\int \frac{d^3\bar{k}}{(2\pi)^3} e^{i\bar{k}.(\bar{x}-\bar{x}')} \int \frac{dk^0}{2\pi} \frac{e^{-ik^0(x^0-x'^0)}}{-\hbar k^0 + \frac{\hbar^2\bar{k}^2}{2m} - i\varepsilon}, & for\ x^0 > x'^0 \quad (12) \\ 0, & for\ x^0 < x'^0 \quad (13) \end{cases}$$

where $d^4k = (dk^0, d^3\bar{k})$ be differential of components of energy-momentum four vector description in unitary space [7-11]. The problem is simplified as non-relativistic and half-space boundary conditions on the behavior of the massive spin zero particle. Integrate the equation (12) over $k^0, \bar{k}$ to obtain time evolution process $(x^0 > x'^0)$ is then given completely by the Green's function in the space-time domain from the equation (9)

$$\left\langle \vec{x},x^0 \middle| \vec{x}',x'^0 \right\rangle = \left(\frac{m}{2i\pi\hbar T}\right)^{3/2} \exp\left[\frac{im\left|\vec{r}-\vec{r}'\right|^2}{2\hbar T}\right]\left\{\exp\left[\frac{im(z-z')^2}{2\hbar T}\right] - \exp\left[\frac{im(z+z')^2}{2\hbar T}\right]\right\} \tag{14}$$

where, $x=(\vec{x},x^0)=(\vec{r},z,x^0)$, $x'=(\vec{x}',x'^0)=(\vec{r}',z',x'^0)$ and $T=x^0-x'^0$.

## 4. Experimental situation

Let $x''^0$ be any fixed value on the secondary planar surface sources on opaque screen. Suppose that a point source (in Figure 2), situated at $x'$ , emits a particle and the latter was determined to reach and through the secondary planar surface sources in an opaque screen situated at, at the time $T_1$ and finally was determined to reach a detector, situated at $x$, at time $T_2$ and then we get the completeness relation to describe the situation given by

$$\left\langle \vec{x},x^0 \middle| \vec{x}',x'^0 \right\rangle = \left(\frac{m}{2\pi i\hbar(T_2+T_1)}\right)^{3/2} \exp\left[\frac{im\left|\vec{r}-\vec{r}''\right|^2}{2\hbar T_2}\right]\exp\left[\frac{im\left|\vec{r}''-\vec{r}'\right|^2}{2\hbar T_1}\right]$$
$$\times\left\{\exp\left[\frac{im}{2\hbar T_2}(z-z'')^2\right]\exp\left[\frac{im}{2\hbar T_1}(z''-z')^2\right] - \exp\left[\frac{im}{2\hbar T_2}(z+z'')^2\right]\exp\left[\frac{im}{2\hbar T_1}(z''+z')^2\right]\right\} \tag{15}$$

where $x''=(\vec{r}'',z'',x''^0)$, $T_1=x''^0-x^0$ and $T_2=x^0-x''^0$

The transition amplitude that a particle propagate from a point source at space-time coordinate, reach the secondary planar surface sources in an opaque screen at space-time coordinate $x''=\left(\vec{x}'',x''^0\right)$ and end up in an opaque screen at space-time coordinate $x=\left(\vec{x},x^0\right)$The exponential structure is described by vector analysis to be

$$\left\langle \vec{x},x^0 \middle| \vec{x}',x'^0 \right\rangle = \left(\frac{m}{2\pi i\hbar(T_2+T_1)}\right)^{3/2}$$
$$\times\left[\left\{\exp\left[\frac{im}{2\hbar}\left[\frac{\left|\vec{r}-\vec{r}''\right|^2+(z-z'')^2}{T_2}\right]\right]\exp\left[\frac{im}{2\hbar}\left[\frac{\left|\vec{r}''-\vec{r}'\right|^2+(z''-z')^2}{T_1}\right]\right]\right\} - \left\{\exp\left[\frac{im}{2\hbar}\left[\frac{\left|\vec{r}-\vec{r}''\right|^2+(z+z'')^2}{T_2}\right]\right]\exp\left[\frac{im}{2\hbar}\left[\frac{\left|\vec{r}''-\vec{r}'\right|^2+(z''+z')^2}{T_1}\right]\right]\right\}\right] \tag{16}$$

$$\left|\vec{A}\right|^2=\left|\vec{r}-\vec{r}''\right|^2+(z-z'')^2,\ \left|\vec{B}\right|^2=\left|\vec{r}''-\vec{r}'\right|^2+(z''-z')^2$$
$$\left|\vec{C}\right|^2=\left|\vec{r}-\vec{r}''\right|^2+(z+z'')^2,\ \left|\vec{D}\right|^2=\left|\vec{r}''-\vec{r}'\right|^2+(z''+z')^2$$

$$\left\langle \vec{x},x^0 \middle| \vec{x}',x'^0 \right\rangle = \left(\frac{m}{2\pi i\hbar(T_2+T_1)}\right)^{3/2}$$
$$\times\left[\left\{\exp\left[\frac{im}{2\hbar T_2 T_1}\left[T_1\left|\vec{A}\right|^2+T_2\left|\vec{B}\right|^2\right]\right]\right\} - \left\{\exp\left[\frac{im}{2\hbar T_2 T_1}\left[-\left(T_1\left|\vec{C}\right|^2+T_2\left|\vec{D}\right|^2\right)\right]\right]\right\}\right] \tag{17}$$

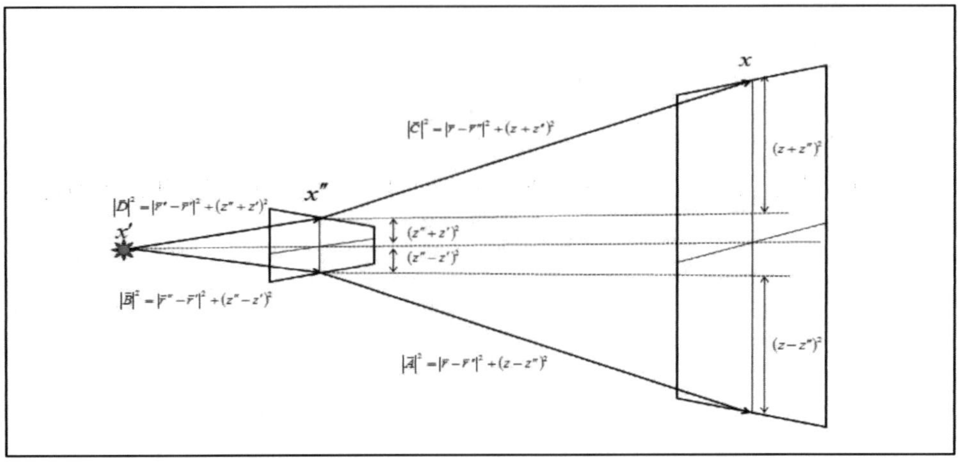


**Figure2**: The following experimental situation of radiation from secondary planar surface sources of a crossed aperture

$$\left\langle \bar{x},x^0 \middle| \bar{x}',x'^0 \right\rangle = \left( \frac{m(2)^{2/3}}{2\pi i\hbar(T_2+T_1)} \right)^{3/2} \times \left[ \frac{\exp\left[ i\alpha(T_1|\bar{A}|^2 - (-T_2|\bar{B}|^2)) \right] - \exp\left[ -i\alpha(T_1\{n|\bar{A}|^2 - (-T_2\{n|\bar{B}|^2\})) \right]}{2} \right] \tag{18}$$

With the functional exponential is defined by $\cos(x) = \dfrac{e^{ix} - e^{-ix}}{2}$

$$\left\langle \bar{x},x^0 \middle| \bar{x}',x'^0 \right\rangle = \left( \frac{m(2)^{2/3}}{2\pi i\hbar(T_2+T_1)} \right)^{3/2} \cos\left[ \alpha(T_1\{n|\bar{A}|^2 - (-T_2\{n|\bar{B}|^2\})) \right] \tag{19}$$

where $\alpha = \dfrac{m}{2\hbar T_1 T_2}$ and $n$ is an integer

$$\left\langle \bar{x},x^0 \middle| \bar{x}',x'^0 \right\rangle = \left( \frac{m(2)^{2/3}}{2\pi i\hbar(T_2+T_1)} \right)^{3/2} \cos\left( \beta_1|\bar{A}|^2 - \beta_2|\bar{B}|^2 \right) \tag{20}$$

where $\beta_1 = n\alpha T_1$ and $\beta_2 = -n\alpha T_2$

We can write of the probability of the general expression for transition amplitude

$$\left| \left\langle \bar{x},x^0 \middle| \bar{x}',x'^0 \right\rangle \right|^2 \cong \left| \cos\left( \beta_1|\bar{A}|^2 - \beta_2|\bar{B}|^2 \right) \right|^2 \tag{21}$$

## 5. Results and Discussion

Development of the Green's function in the space-time domain description of the propagation of massive spin zero particles in equation (15) was based on the configuration space and a time evolution process where particle propagates between emission source and detector. We use the mathematical program for corresponding numerical evaluations between the experimental situation in quantum optics (Figure: 3) and the probability of the general expression in equation (21) for a transition amplitude in quantum field theory situation (Figure: 4).

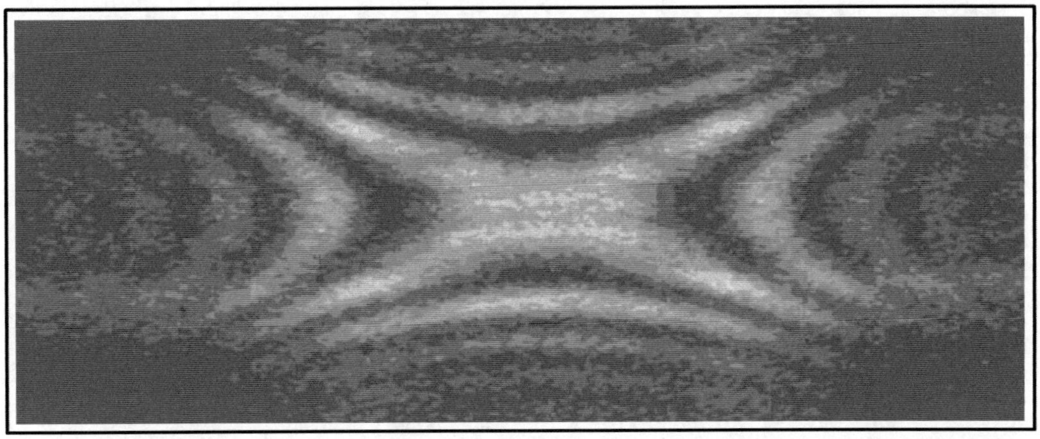

**Figure:3** Radiation from secondary planar surface sources of a crossed aperture. The diffraction pattern of a crossed aperture arrangement illustrating the spatial coherence of a free-electron laser. (Source: DESY in Hamburg; http://mpsd.cfel.de/images/content/e208/e209/index_eng.html).

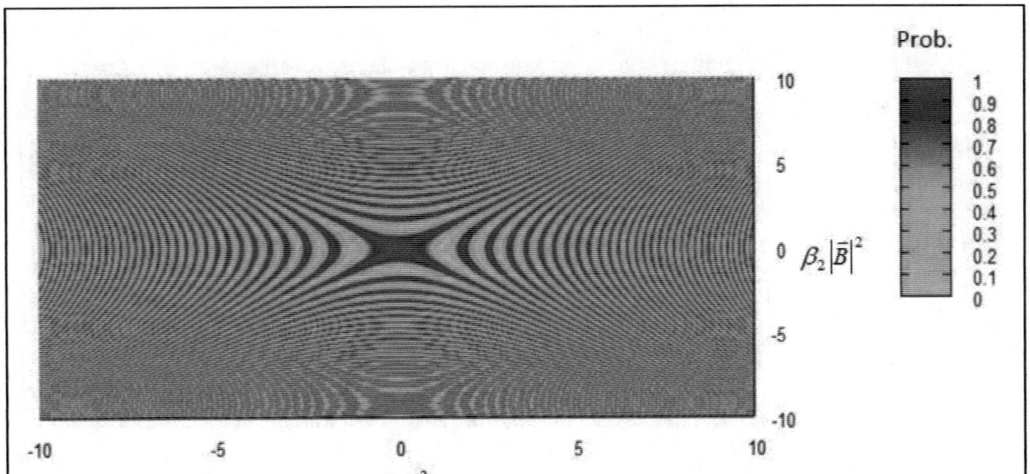


**Figure:4** Radiation from secondary planar surface sources of a crossed aperture. The diffraction pattern of a crossed aperture arrangement illustrating the spatial coherence of the Maxima calculation program (equation (21)).

## 6. Summary

The expression in quantum optics following into the formulation of the radiation in radiant intensity which, analytical solution has not given the time evolution of amplitude. This research was involved with the careful analysis of the propagation of a massive quantum particle with corresponding numerical evaluation in space time as a time evolution process dealing with amplitudes of transitions in the presence of the external source Z [*K (x)*]. By using Feynman's so-called spin-zero electron are technically called Klein-Gordon particles of finite mass between different points in the presence of the emission source and detection source.

We attend and explain the radiation for radiant intensity in space-time domain for the vacuum-to-vacuum transition amplitude. To calculation amplitude transition of the massive spin zero particle propagator stimulated emission by secondary planar surface source in the space-time domain in equation (14). In development of transition amplitude in equation (15) to describe in detail of experimental situations by the diffraction pattern of a crossed aperture arrangement illustrating the spatial coherence of a free-electron laser in quantum optics situation (Figure: 3). Finally, a comparison between our result and experimental in the quantum optics situation by using the mathematical program for corresponding numerical evaluations by the probability of the general expression for a transition amplitude in the equation (21) from quantum field theory situation (Figure:4).

**Acknowledgement**

The author would like to thank Prof. E. B. Manoukian (Ph.D) for his papers activating to research this project and Department of Science and Mathematics (Physics), Faculty of Science and Technology Pathumwan Institute of Technology

## 7. References

[1] Schwinger J. (1970a). *Particles, sources and fields.* Boston: Addison-Wesley.
[2] Feynman R. P. (1985). *QED: The strange theory of light and matter.* Princeton. NJ: Princeton University Press.
[3] Mandel and E Wolf . (1995). *Optical Coherence and Quantum Optics*, Cambridge UK Cambridge University Press.
[4] Manoukian, E. B. (2006). *Quantum theory: A wide spectrum.* New York: Spriger.
[5] Agarwal G. S., Wolf E. (1996). Correlation-induced spectral changes and energy conservation *Physical Review A*, 54 (5).
[6] Schwinger J. (1968). Source and Gravitons, *Physical Review*, 173 (1,264)
[7] E. B. Manoukian (1984a). Particle, sources and statistical distributions. *Hadronic Journal,* 7 (4), 897-910.
[8] E. B. Manoukian (1987e). Reflection-off a reflecting surface in quantum mechanics: Where do the reflections actually occur?. *Nuovo Cimento, 99 (2),* 133-143.
[9] E. B Manoukian. (19870). Reflection-off a reflecting surface in quantum mechanics: Where do the reflections actually occur?. II: The law of reflection. *Nuovo Cimento, 100 (2),* 185-193.
[10] E. B. Manoukian (1989). Theoretical intricacies of the single-slit, the double-slit, and related experiments in quantum mechanics. *Foundations of Physics, 19 (5),* 479-504.
[11] E.B. Manoukian (1989).On Propagation Speeds of Localized Massive Particles in Spacetime in Field Theory, *IL Nuovo Cimento Note Brevi* 101A, N. 6
[12] Maksuwan A. (2014) Quantum Field Theory Viewpoint of Refractive Index. *Advanced Materials Research 979*, 31-34.

# Projectile motion in real-life situation: Kinematics of basketball shooting

**A Changjan**[1] **and W Mueanploy**[1]

[1] Department of Science and Mathematics, Faculty of Science and Technology, Pathumwan Institute of Technology, Bangkok 10330 Thailand.

E-mail: otto_sinkronity@yahoo.com

**Abstract**. Basketball shooting is a basic practice for players. The path of the ball from the players to the hoop is projectile motion. For undergraduate introductory physics course sstudent must be taught about projectile motion. Basketball shooting can be used as a case study for learning projectile motion from real-life situation. In this research, we discuss the relationship between optimal angle, minimum initial velocity and the height of the ball before the player shoots the ball for basketball shooting problem analytically. We found that the value of optimal angle and minimum initial velocity decreases with increasing the height of the ball before the player shoots the ball.

## 1. Introduction

Projectile motion is one topic in an introductory physics course for undergraduate student. The path followed by a projectile is called its trajectory. If we negligible air resistance, projectiles follows a curved trajectory or curved path that is a parabola. One of the classical problems of projectile motion is a basketball shooting problem, What is the optimal arc angle for shooting a ball? Many students would have guessed that the optimal angle is 45 degrees, which not the correct answer. Based on ground-to-ground projectile motion, a projectile launched from the origin and returns to the same horizontal level. The optimal angle that the object can travel from the origin to the maximum horizontal range with a minimum velocity is 45 degrees [1]. But in basketball shooting problem the beginning level is different from final level. Therefore, the optimal angle in basketball shooting problem is different from the optimal angle in ground-to-ground projectile motion.

The basketball shooting problem was studied by many researchers. Tan and Miller [2] studied the kinematics of the two basic styles of shooting in basketball the overhand push shot and the underhand loop shot. It is shown that from a purely kinematics and trajectory point view, the overhand push shot is preferable to the underhand loop shot. The advantages of the underhand loop shot lie in the actual execution of the shot. The optimal arc angles for shooting a basketball were considered by Frank [3], who studied the optimal arc angles that the shooter can send the ball to the center of the hoop with a minimum required amount of speed. He found that the value of this arc angle depends on the shooting height of the player and his distance from the hoop. In another projectile motion situation, Day [4] suggested that the real-life situation of a vehicle vaulting over the side of an elevated roadway and landing on the ground below. Lichtenberg and Will [5] address the problem of the optimum angle at which a putter should release the shot in order to achieve maximum distance.

In this paper discuss the relationship between optimal angle, minimum initial velocity and the height of the ball before the player shoots the ball for basketball shooting problem analytically.

## 2. Model and Calculations

Consider the case when the ball is released from the player as shown in figure 1. In this work air resistance to movement of the ball and the force due to the spinning motion of the ball will be considered negligible. When the ball is released, the trajectory of the ball is a parabola. If the hoop center is $x$ **Error! Bookmark not defined.** m away in horizontal from the player to the hoop is $h'$ m higher than the released height of the ball $(h)$, let us find the minimum initial velocity $(v_0)$ and the associated arc angle $(\theta)$ to send the ball from player to the hoop.

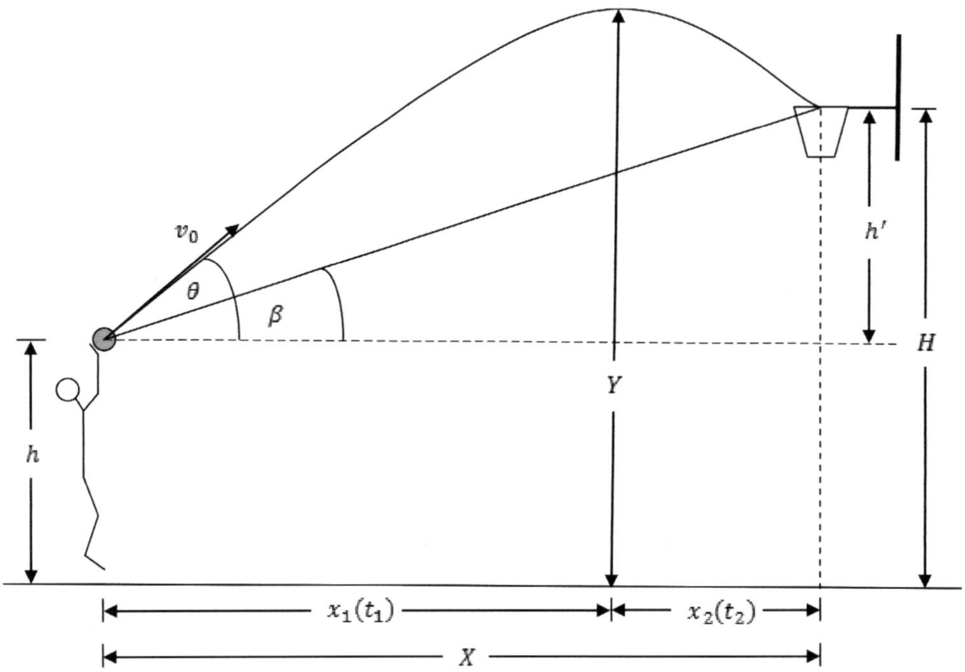


Figure 1 Trajectory of the ball from the player to the hoop

The trajectory of the ball can be defined by the following set of equations.
Horizontal motion:

$$X = (v_0 \cos\theta)\cdot(t_1 + t_2) \tag{1}$$

Vertical motion:

$$v_0 \sin\theta = g t_1 \tag{2}$$

From eq. (2), we got displacement of the ball in vertical motion $(h')$ as

$$h' = (v_0 \sin\theta)(t_1 + t_2) - \frac{1}{2} g (t_1 + t_2)^2$$

$$= (g t_1)(t_1 + t_2) - \frac{1}{2} g (t_1 + t_2)^2 = \frac{1}{2} g (t_1^2 - t_2^2). \tag{3}$$

Here $g$ is acceleration due to gravity at the earth's surface. $t_1$ and $t_2$ are time interval when the ball is released from the player to the maximum height $(Y)$ and the time interval from the maximum height to the hoop respectively.

To find the minimum initial velocity of the ball, the relationship between release angle and initial velocity for a given range must be considered. From eq. (3), we can be rewritten as

$$h' = \frac{g}{2}(t_1 + t_2)[2t_1 - (t_1 + t_2)].$$

Substituting eq. (1) and eq. (2) in eq. (4) we got

$$h' = \frac{g}{2} \cdot \frac{X}{v_0 \cos\theta}\left[2t_1 - \frac{X}{v_0 \cos\theta}\right] = X \tan\theta - \frac{gX^2}{2v_0^2 \cos^2\theta} \quad . \tag{4}$$

We can rearrange eq. (4) as

$$v_0 = \sqrt{\frac{gX^2(\tan^2\theta + 1)}{2(X\tan\theta - h')}}. \tag{5}$$

To solve the problem, we need to differentiate $v_0$ with respect to $\theta$ and set the derivative equal to zero. Then, we got

$$\frac{dv_0}{d\theta} = (X\tan\theta - h')\tan\theta - \frac{X}{2}\sec^2\theta = 0. \tag{6}$$

The trigonometric identities $\sec^2\theta = \frac{1}{\cos^2\theta} = \tan^2\theta + 1$ enable us to write eq. (6) more simply as

$$\frac{dv_0}{d\theta} = X\tan^2\theta - 2h'\tan\theta - X = 0. \tag{7}$$

From quadratic equation $Ax^2 + Bx + C = 0, x = \frac{-B \pm \sqrt{B^2 - 4AC}}{2A}$ where $A, B$ and $C$ are constant. We choose appropriate solution as positive solution, then we get optimal angle as

$$\theta = \tan^{-1}\left(\frac{h' + \sqrt{h'^2 + X^2}}{X}\right). \tag{8}$$

Substituting eq. (8) back into eq. (5), gives an expression for the minimum initial velocity consistent with the ball reaching the hoop. The result is

$$v_0 = \left[g\left(h' + \sqrt{h'^2 + X^2}\right)\right]^{\frac{1}{2}}, \tag{9}$$

where $h' = H - h.$

## 3. Results and Discussions

The relationship between optimal angle $(\theta)$, minimum initial velocity $(v_0)$ and the height of the ball before the player shoots the ball $(h)$ are shown in figure 2. Here, we use $H = 3m$ and $X = 6.75\,m$ that is the hoop distance above the floor and three point field goal distances respectively. We consider the optimal angle and the minimum initial velocity of 1.5-2.1 m in the height of the ball before the player shoots the ball. Figure 2 shows that the value of optimal angle and the value of the minimum initial velocity decrease with increasing the height of the ball before the player shoot the ball. In other words, for basketball shooting problem the optimal angle depends on the height of the ball before the player shoots the ball not fixed at $45^{\circ}$ as in the case of ground-to-ground projectile motion.

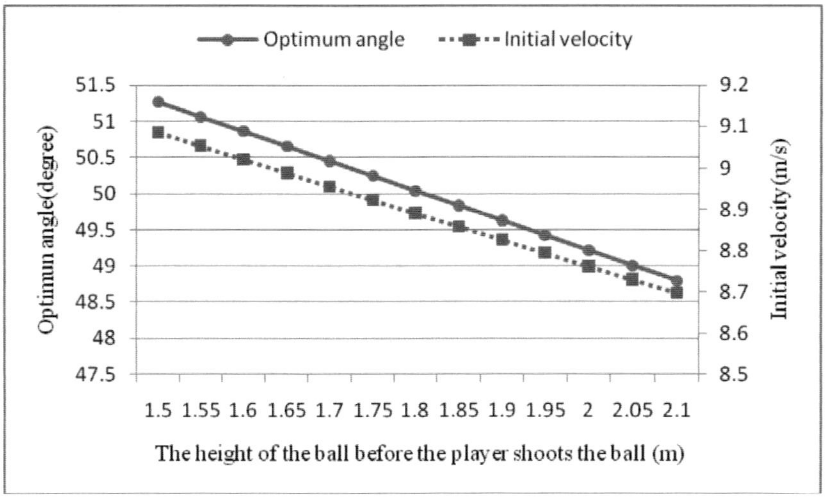


Figure 2 The relationship between optimal angle $(\theta)$, minimum initial velocity $(v_0)$ and the height of the ball before the player shoots the ball $(h)$

## 4. Conclusions

The relationship between the optimal angle, the minimum initial velocity and the height of the ball before the player shoots the ball for basketball shooting problem are calculated analytically. The effect of air resistance to movement of the ball and the force due to the spinning motion of the ball will be considered negligible. We find that the value of optimal angle and the value of minimum initial velocity decreases with increasing the height of the ball before the player shoot the ball. We can use this result as an example to describe the projectile motion that the beginning level is different from final level. An improvement in this research could come from include the effect of air resistance and the force due to the spinning motion of the ball. Which makes the calculation results in a similar to real-life situation increased or applied this result in other projectile motion in real-life situations.

## 5. Acknowledgement

The author would like to thank Pathumwan Institute of Technology for the financial support.

## References

[1] Serway R A and Jewett J W, Jr. 2008 *PHYSICS for Scientists and Engineers with Modern Physics*7th ed. (Thomson Brooks/Cole, CA: Belmout)
[2] Tan A and Miller G 1981 *Am. J. Phys.* **49**(6) 542
[3] Frank K L 2002 *Phys 13 News* **Winter 2002** 8
[4] Day L H 2007 *PHYSICS EDUCATION* **42**(6) 621
[5] Lichtenberg D B and Wills J G 1978 *Am. J. Phys.* **46**(5) 546

# Molecular dynamics simulation of corrosion mitigation of iron in lead-bismuth eutectic using nitrogen as corrosion inhibitor

**Artoto Arkundato[1], Zaki Su'ud[2], Sudarko[3], Mohammad Hasan[4] and Massimo Celino[5]**
[1]Jurusan Fisika, FMIPA, Universitas Jember, Jl. Kalimantan 37 Jember, Indonesia
[2]Jurusan Fisika, FMIPA, ITB, Jl. Ganesha 10 Bandung, Indonesia
[3]Jurusan Kimia, FMIPA, Universitas Jember, Jl. Kalimantan 37 Jember, Indonesia
[4]Jurusan Matematika, FMIPA, Universitas Jember, Jl. Kalimantan 37 Jember, Indonesia
[5]ENEA, CR Casaccia, Via Anguillarese 301, Rome, Italy

E-mail: a.arkundato@gmail.com

**Abstract.** The corrosion of structural materials used in fast nuclear reactor design is a current major problem. It is due to the use of liquid metal as a coolant candidate in the heat transfer system. The liquid metal as lead-bismuth eutectic was found to make high corrosion to structural material as steel. One of the solutions of this problem is to inject some inhibitor into liquid metal. In this current work we simulate the effect of nitrogen injection as inhibitor candidate. The simulation will predict the proper concentration of injected nitrogen and also observe the microscopic structure of the material before and after injection to know the ability of nitrogen as an inhibitor. The simulation follows the molecular dynamics method and for preliminary study we use iron material rather than steel. We also use lennard-jones potential for simplification of the study. It is from our simulation we see nitrogen shows better corrosion mitigation compare with oxygen as in our previous study. The effective inhibition can be achieved by injecting at least 0.056wt.% nitrogen. This amount seems to be able to reduce the corrosion level of iron till about 99.5% for high corrosion at temperature 750 °C.

**Keyword:** lead-bismuth eutectic, corrosion inhibition, nitrogen, molecular dynamics

## 1. Introduction

Corrosion inhibitor is a substance which when added in small concentration to a corroded system will minimize or prevents that corrosion [1]. Corrosion inhibitor can be used to protect metals from corrosion attach and if we can maintain this inhibitor in an exact concentration all the time than we will be able to prevent the corrosion maximally. It is a current issue that the steels used in the fast nuclear reactor installation have been highly corroded if we use liquid lead or lead bismuth eutectic as a coolant material for heat transfer system [2]. The solution of this corrosion problem has two options: (1) finding better steel materials and/or (2) finding a corrosion inhibitor (method) to minimize and

[1] Corresponding author: a.arkundato@gmail.com

prevent corrosion. Especially for corrosion inhibition, there is a direction to use oxygen gas to protect a steel metal from serious corrosion attack of liquid lead or lead bismuth eutectic during high temperature interaction [2]. The degradation of steels by high corrosion effect in liquid metal is for the most part due to the dissolving of the various constituents of the steel metal by liquid metals [3]. Corrosion may be described as mainly the solution of the solid metal in the liquid metal. The control method of corrosion using oxygen is very popular in current days. Impurity such as nitrogen is another choice of inhibitor. By definition the inhibitors are substances which, when added to a chemical system in small amounts (<10 wt.%), interact with the reactants to reduce the reaction rate. Nitrogen dissolved in lead-bismuth eutectic acts as such an inhibitor through the development of a stable film reducing the dissolution rate [4].

The use of computer simulation method based on a proper material model can help us to see the possibility of elements as a promising inhibitor candidate. The molecular dynamics simulation can be used to predict the needed inhibitor concentration to prevent corrosion at maximum level. The simulation result should be able to use as a preliminary study before experiments. In our previous study we were success to show the effect of oxygen for preventing the corrosion of iron (major component of steels) in liquid lead or lead bismuth eutectic [5,6]. We show three conditions: (1) the some atoms of oxygen enter the surface of iron for possibility of protective iron-oxide layer development, (2) some oxygen atoms remain in between of iron surface and liquid metals as a barrier wall for direct interaction, and (3) some oxygen atoms experiences a precipitation in liquid metals. In this current study we want to show the possibility of nitrogen as a corrosion inhibitor in lead bismuth corrosion. We also use the classical molecular dynamics method for this study. We want to predict and compare the ability of corrosion inhibition between oxygen and nitrogen.

From experimental data the degradation of structural materials (steels) due to its direct interaction with liquid metals is very high [2,3,4]. This is not safe for application of fast nuclear reactor design and economically disadvantage. In our previous work we study the corrosion (degradation) mitigation of iron (major component of steels) by using oxygen as inhibitor agent. We also compared with experimental results [5,6]. However the simulation still shows material degradation at the surface of iron (about 30% at temperature 750 °C) [5,6]. In this current work we try to use nitrogen as inhibitor to get more improvements of corrosion mitigation. It was reported that nitrogen can also be used as an inhibitor [3,4]. However the studies of nitrogen as inhibitor of liquid metal corrosion is not much explored yet, computationally and experimentally. Many researches still focused on the oxygen as inhibitor. We hope this preliminary study can give the positive results of using nitrogen as a promising corrosion inhibitor for purpose of nuclear reactor applications.

## 2. Theory

In classical molecular dynamics method we solve the Newton equation of motion for specific phenomenological potential function. We solve this equation for many atoms or molecules of the material system for specific numerical integration algorithm to get the trajectories of the atoms or molecules. Based on the trajectory of all bodies and statistical mechanics then we can predict and calculate all thermodynamics properties of the system as diffusion coefficient and others. Denoting $\vec{f}_{ij}$ is the force exerted by particle $j$ on particle $i$ then the total force acting on particle $i$ should be

$$\vec{F}_i = \sum_j \vec{f}_{ij} \quad (1)$$

When the force is conservative and for specific potential energy $u(\vec{r})$ then it follows

$$\vec{f} = -\nabla u(\vec{r}) \quad (2)$$

Then we have the total force acting on particle $i$ is

$$\vec{F}_i = -\nabla \sum_j u_{ij}(\vec{r}_{ij}) \quad (3)$$

Then dynamics of Newtonian system can be described as

$$m_i \{ d^2 \vec{r}_i / dt^2 \} = \vec{F}_i \quad (4)$$

In this equation $r_i$ is the position of atom i, $m$ is mass of atom i. In molecular dynamics method we solve equation (4) numerically based on specific integration algorithm that we choose. There are many algorithms we can use and one of them is the Beeman algorithm for purpose of accurate calculation. In this work we still use MOLDY molecular dynamics program for our simulation that is an open source and easy-to-use program [7].

Whereas the system of material is usually complex then we approach the interaction of all atoms of simulated material using Lennard-Jones interatomic potential for simplification of the study. In our work we simulate a metal system (iron in liquid metal with nitrogen inhibitor). There is an EAM (embedded atomic method) potential that should be better for metal simulation. However in our work we still use the classical Lennard-Jones potential as a first prediction and preliminary study. The general Lennard-Jones potential is [8]:

$$u(r) = 4\varepsilon\left[(\sigma/r)^{12} - (\sigma/r)^6\right] \quad (5)$$

The (12-6) LJ potential is simple physical model of interaction that approximates the interaction between a pair of neutral atoms. In this equation $\varepsilon$ and $\sigma$ are the potential parameters for pair of atom that can be determined from experimental data. The cross-interactions ($\varepsilon_{AB}$ and $\sigma_{AB}$ ) for pair of different atom (A and B) can be generated by the popular Lorentz-Berthelot mixing rule [9]

$$\sigma_{AB} = (\sigma_{AA} + \sigma_{BB})/2 \quad (6)$$

$$\varepsilon_{AB} = \sqrt{(\varepsilon_{AA} \times \varepsilon_{BB})} \quad (7)$$

Table 1 is summary of the potential parameters that we use in our work.

**Table 1.** Potential parameters for Lennard-Jones Interaction.

| **Pair Interaction** | $\sigma$ **(Å)** | $\varepsilon$ **(eV)** | **References** |
|---|---|---|---|
| Fe-Fe | 0.400 | 2.319 | [8] |
| Pb-Pb | 0.191 | 3.189 | [8] |
| Bi-Bi | 0.059 | 3.050 | fitting data from EAM potential [10] |
| N-N | 0.00853 | 3.656 | [11] |
| Bi-N | 0.02243 | 3.353 | equation (6 and 7) |
| Fe-Pb | 0.2766 | 2.7541 | equation (6 and 7) |
| Fe-Bi | 0.15377 | 2.6847 | equation (6 and 7) |
| Fe-N | 0.05846 | 2.9877 | equation (6 and 7) |
| Pb-Bi | 0.10614 | 3.1194 | equation (6 and 7) |
| Pb-N | 0.04036 | 3.4224 | equation (6 and 7) |

The important physical variable that we want to calculate in this work is diffusion coefficient that can be related to corrosion rate by specific formula. The diffusion coefficient for certain temperature can be calculated from Einstein relation [12]

Journal of Physics: Conference Series **622** (2015) 012009 doi:10.1088/1742-6596/622/1/012009

$$D = \lim_{t\to\infty} \frac{1}{6t}\left\langle \left|\vec{r}(t)-\vec{r}(0)\right|^2 \right\rangle = \lim_{t\to\infty} \frac{MSD}{6t} \qquad (8a)$$

where

$$MSD = \left\langle \left|\vec{r}(t)-\vec{r}(0)\right|^2 \right\rangle \qquad (8a)$$

where $t$ is time, $T$ is temperature, $A$ is activation energy for corrosion to be happen and $\Re$ is a gas constant.

## 3. Simulation procedure and analysis

### *3.1 Details of simulations*

The simulation details of our work can be described as three following steps:

(I) Model of the structure of material.
We model iron in bcc crystal structure, placed in the centre of lead-bismuth eutectic (liquid phase). Nitrogen atoms (position) placed in the liquid metal randomly. All position of atoms of material (Fe, Pb, N, Bi) should be defined as input of molecular dynamic simulation. Other inputs are: mass of atom (as in periodic table of elements), temperature (750 °C), potential parameters ($\sigma$ and $\varepsilon$), number of integration (=160 000 step), number of atoms (Fe, Pb, N, Bi) and mesh time (0.0001 ps). The choice of this temperature is due to there are related experimental results for comparison [5,6]. The material corrosion system was composed from 10745 Fe atoms, 18309 Pb atoms and 22376 Bi atoms and for different percentage of N atoms (from total atoms of Pb, Bi and N): 0.50%, 0.75%, 1.00%, 1.50%, 2.00%, and 2.50%). The lead-bismuth eutectic is composed from 45wt% Pb and 55wt% Bi. In wt% unit then the nitrogen will be injected into lead-bismuth eutectic for different concentration: 0.037wt%, 0.056wt%, 0.074wt%, 0.111wt%, 0.150wt% and 0.187wt%.

(II) Running simulation using MOLDY.
Simulation will be done with MOLDY program. Simulation is in NPT ensemble Anderson mode, zero pressure, Nose-Hoover thermostat and in the modified Beeman integration scheme [7]. At every simulation work (same temperature, same Fe-Pb-Bi concentration and different nitrogen atom) we then will evaluate the effect of different nitrogen concentration for corrosion mitigation.

(III) Calculation of physical properties: After simulation of 160000 integration step then all needed physical properties will be calculated and determined: MSD and D(T). The simulation uses MOLDY code. For our work the diffusion coefficient and positions of atoms can be determined by using MOLDY's facilities such as MSD and Mdavpos [7].

### *3.2 Analysis*

The purpose of physical properties calculation is to find the lowest diffusion coefficient ($D(T)$) of iron diffusion after nitrogen injection with different concentration. High diffusion of iron shows high corrosion of iron and high degradation of iron. Nitrogen is injected to reduce the diffusion of iron in liquid metal. Jmol (http://jmol.sourceforge.net) is used to view the microstructure of material before and after nitrogen injection.

## 4. Results and discussions

*4.1 Diffusion coefficient of iron*

From calculation of iron diffusion coefficient ($D(T$=750°C)) for different concentration of nitrogen (N) then we can compare our current results with our previous work [6] as shown in figure 1. In our previous work we used 44.3wt% Pb and 55.7wt% Bi whereas in our work we used 45wt% Pb and 55wt% Bi for a reason.

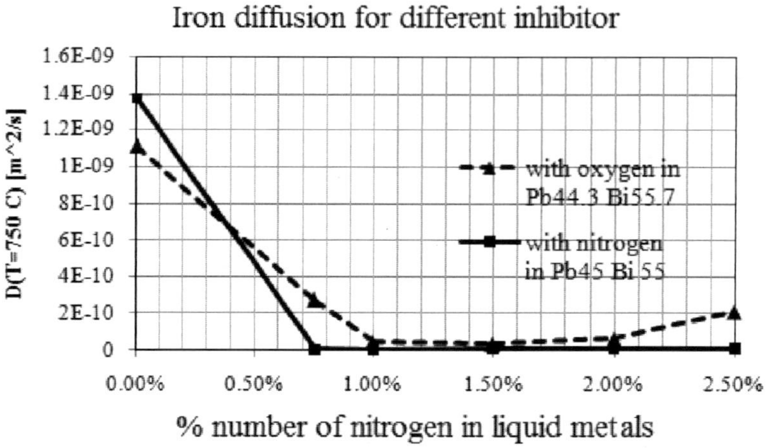


Figure 1. Iron diffusion coefficient in lead-bismuth eutectic.

From figure1 we see that using oxygen [6] and nitrogen as inhibitor then the diffusion coefficient of iron can be reduced at very low level. It means we can reduced the corrosion rate of iron till very low level. It can be seen from that figure that the use of nitrogen seems offering more stability of iron structure compare with oxygen. It seems that is easier to control nitrogen injection compare with oxygen injection because the use of oxygen seems to be controlled more carrefully at certain small concentration for narrow range (between 1.00% and 1.50%). Whereas, from the figure 1, we can see the use of nitrogen showing more flexibly of application where we can control the concentration of nitrogen in more wider ranges starting from 0.75% atoms (or 0.0562wt%) during iron-liquid metal interaction. Morover from that figure 1, we see the injection of 0.75% atoms (= 0.056wt%) of nitrogen can inhibit the corrosion about 99.51% compare without using oxygen (about 96.66%).

*4.2 Structure of iron*

Before simulation at temperature $T$ = 750 °C, the iron is formed in bcc crystal structure (figure 2).

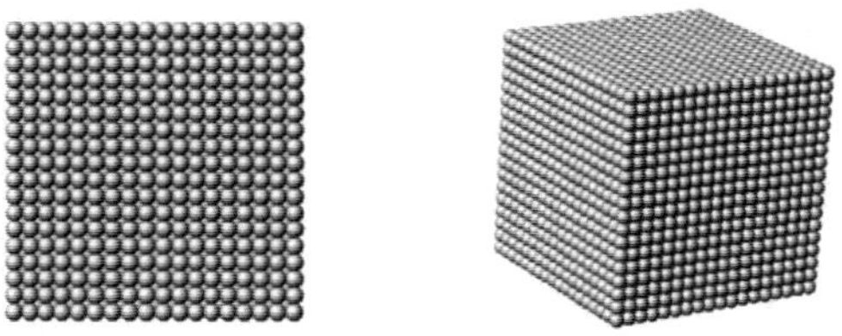

Figure 2. Structure iron before simulation (front and corner view)

After simulation $T$ = 750 °C without nitrogen injection, the structure of iron material is as figure 3.

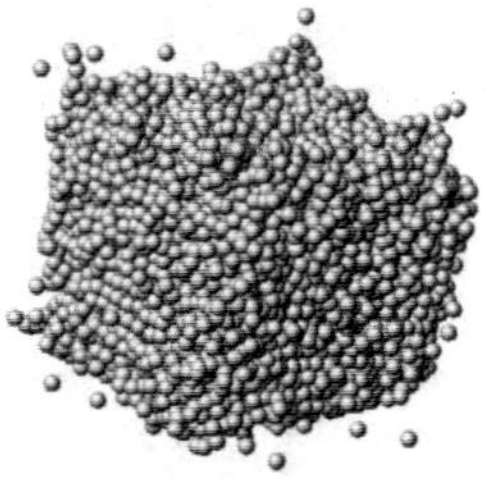

Figure 3. Iron structure without nitrogen injection

After 0.75% (= 0.056wt%) nitrogen injection into lead-bismuth eutectic then the performance of iron is as figure 4 below.

Figure 4. Iron in lead-bismuth eutectic with 0.056wt% nitrogen

Compare with 0.75% oxygen injection [6] then the structure of iron is as figure 5 below.

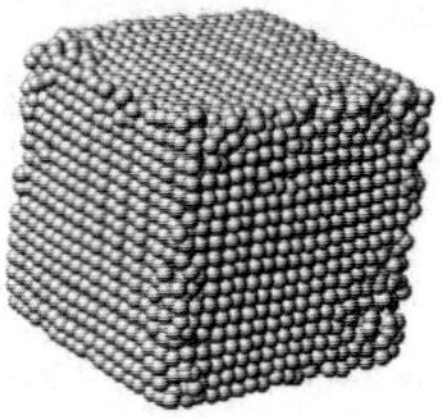

Figure 5. Structure of iron with 0.75% oxygen injection

This current work is preliminary study. The advanced research should be done to know the comprehensive understanding of corrosion mitigation using nitrogen. One of the future works is to run simulation using more accurate potential function for more accurate simulation result. However this current result is still need to be verified with experimental results. Moreover the corrosion simulation should be extended to use steels rather than iron. The application of this mitigation method (using nitrogen as corrosion inhibitor) in nuclear reactor power plant then depend on the next experimental confirmation.

**5. Conclusion and remark**

The corrosion mitigation using nitrogen injection seems offering more flexibility and stability to control, protect and prevent the high corrosion of iron in high temperature lead-bismuth eutectic. We predict the use of at least 0.056wt% of nitrogen for injection it can reduce the corrosion till about 99.51%. These valuable results however need to be checked with experimental results and need to use more accurate inter-atomic potential function such as embedded atomic method potential or others.

## 6. References

[1] Papavinasam S 2000 *Uhlig's corrosion handbook: chapter 59 corrosion inhibitors, second edition* (John Wiley & Sons, inc)

[2] Zhang,J, Li N 2008 Review of the studies on fundamental issues in LBE, *J. Nulc.Mater.* **373** 351-377.

[3] Manly W D, *Fundamentals Of Liquid-Metal Corrosion* (Oak Ridge National Lab). Accessed from web http://www.ornl.gov/info/reports/1956/3445603500959.pdf on Sept 2014

[4] OECD 2007 *Handbook on Lead-bismuth Eutectic Alloy and Lead Properties, Materials Compatibility, Thermal-hydraulics and Technologies* (Nuclear Energy Agency) . Accesssed from web from www.oecd-nea.org/science/reports/2007/pdf/chapter4.pdf on spet 2014

[5] Arkundato A, Zaki Su'ud, Abdullah M, Widayani Sutrisno, Celino M 2013 Inhibition of iron corrosion in high temperature stagnant liquid lead: A molecular dynamics study, *Annals of Nuc. Energy* 62 298–306

[6] Arkundato A, Zaki Su'ud, Abdullah M, Widayani S 2013 Molecular dynamic simulation on iron corrosion-reduction in high temperature molten lead-bismuth eutectic, Turk J. Phys 37: 132-144 doi:10.3906/fiz-1112-12

[7] Refson K, 2000 Moldy: A Portable molecular dynamics simulation program for successive and parallel computers, *Comp. Phys. Commun.*. 126(3) 309-328.

[8] Zhen S, Davies G J 1983 L-J n-m potential energy parameters: calculation of the LJ n-m pot energy parameters for metals, *Phys.Stat.Sol.*(a) 78, 595.

[9] Brodholt J, Wood B 1993 Molecular dynamics simulations of the properties of $CO_2$-$H_2O$ mixtures at high pressures and temperatures, *American Mineralogist,* 78, 558.

[10] D.K. Belashchenko, O.I. Ostrovskii, *Russ. J. Phys. Chem.*, **80,** (2006), 509

[11] Lemmon E W, Jacobsen R T 2004 Viscosity and Thermal Conductivity Equations for Nitrogen, Oxygen, Argon and Air, *Inter. J. of Thermophysics*, 25(1), 21-69

[12] Kupryazhkin A Y, Zhiganov A N, Risovany D V, Nekrassov K A, Golovanov V N 2008 Simulation of diffusion of oxygen and uranium in uranium dioxide nanocrystals, *J. Nucl. Mater.***372** 233

## Acknowledgments

The authors gratefully acknowledge the directorate general of higher education, DP2M, Indonesia for facilitating this research using fundamental research scheme. Thanks also to the ENEA CRESCO supercomputer, Rome, Italy, for supporting a part of calculations.

Journal of Physics: Conference Series **622** (2015) 012010 doi:10.1088/1742-6596/622/1/012010

# Critical Temperature Differences of a Standing Wave Thermoacoustic Prime Mover with Various Helium-Based Binary Mixture Working Gases

**Ikhsan Setiawan**[1], **Makoto Nohtomi**[2], **Masafumi Katsuta**[2]

[1]Physics Department, Faculty of Mathematics and Natural Sciences, Universitas Gadjah Mada, Sekip Utara BLS 21, Yogyakarta 55281, Indonesia

[2]Graduate School of Environment and Energy Engineering, Waseda University, Nishi-tomita 1011, Honjo-city, Saitama pref., Japan

E-mail: ikhsan_s@ugm.ac.id

**Abstract.** Thermoacoustic prime movers are energy conversion devices which convert thermal energy into acoustic work. The devices are environmentally friendly because they do not produce any exhaust gases. In addition, they can utilize clean energy such as solar-thermal energy or waste heat from internal combustion engines as the heat sources. The output mechanical work of thermoacoustic prime movers are usually used to drive a thermoacoustic refrigerator or to generate electricity.

A thermoacoustic prime mover with low critical temperature difference is desired when we intend to utilize low quality of heat sources such as waste heat and sun light. The critical temperature difference can be significantly influenced by the kinds of working gases inside the resonator and stack's channels of the device. Generally, helium gas is preferred as the working gas due to its high sound speed which together with high mean pressure will yield high acoustic power per unit volume of the device. Moreover, adding a small amount of a heavy gas to helium gas may improve the efficiency of thermoacoustic devices.

This paper presents numerical study and estimation of the critical temperature differences of a standing wave thermoacoustic prime mover with various helium-based binary-mixture working gases. It is found that mixing helium (He) gas with other common gases, namely argon (Ar), nitrogen ($N_2$), oxygen ($O_2$), and carbon dioxide ($CO_2$), at appropriate pressures and molar compositions, reduce the critical temperature differences to lower than those of the individual components of the gas mixtures. In addition, the optimum mole fractions of He gas which give the minimum critical temperature differences are shifted to larger values as the pressure increases, and tends to be constant at around 0.7 when the pressure increases more than 2 MPa. However, the minimum critical temperature differences slightly increase as the pressure increases to higher than 1.5 MPa. Furthermore, we found that the lowest critical temperature difference for He-Ar mixture gas is around 66 °C which is achieved in pressure range of 1.5 MPa - 2.0 MPa and mole fractions of helium of 0.55 - 0.65. The He-$N_2$ and He-$O_2$ mixture gases demonstrate almost the same performances, both have the lowest critical temperature difference around 59 °C at pressures of 1.0 MPa - 1.5 MPa and helium's mole fractions of 0.35 - 0.55. For all tested gases, the lowest critical temperature difference of around 51 °C is provided by He-$CO_2$ mixture gas at pressures of 0.5 MPa - 1.0 MPa with helium's mole fractions of 0.15 - 0.40.

## 1. Introduction

Thermoacoustic prime movers (TAPMs) are energy conversion devices which convert thermal energy (temperature gradient) into acoustic work. The devices employ thermoacoustic effects which are mutual thermodynamic interactions between acoustic wave and solid that possesses a temperature gradient. They have simple structure without any moving parts, and therefore they have a good reliability and low construction and maintenance costs. The TAPMs are environmentally friendly because they do not produce any global warming substances like carbon dioxide gas, and they can utilize clean energy such as solar-thermal energy [1][2] or waste heat from internal combustion engines [3] as the heat sources for their operation. The output mechanical work of TAPMs are usually used to drive a thermoacoustic refrigerator [4] or to generate electricity.[5][6]

Thermoacoustic prime movers still have not achieved the efficiencies as high as those of conventional heat engines. The thermal efficiencies of standing wave TAPMs have so far been limited to 0.20 [7][8] because they employ the intrinsically irreversible thermodynamics cycle. On the other hand, travelling wave TAPMs which are based on inherently reversible Stirling cycle can reach thermal efficiency up to 0.3, but the structures are more complex.[8] However, one of the advantages of thermoacoustic devices is that they are able to work with low quality of heat sources. Hence, although they generally have low thermal efficiency, utilization of waste heat or solar-thermal energy to power the TAPMs would be interesting and beneficial.

The heart of a thermoacoustic device is a stack, which is a porous medium in where the thermoacoustic energy conversions take place. The stack is located inside an acoustic resonator filled by a working gas, as shown in Fig. 1 which is an ilustration of a simple standing wave TAPM. By attaching a cold and a hot heat exchangers at both ends of the stack, a temperature gradient along the stack can be built and maintained. A minimum temperature gradient across the stack is required by a TAPM in order to be able to produce acoustic work.

When we intend to utilize the low quality of heat sources, we will need a TAPM which has a low critical temperature gradient. The critical temperature gradient can be significantly influenced by the kinds of working gases inside the resonator and stack's channels. Generally, practitioners have preferred helium as the working gas because it has high sound speed which together with high mean pressure will yield high acoustic power per unit volume of the devices.[9] In addition, by adding a small amount of a heavy gas to a light gas, thereby reducing the Prandtl number of the gas below 2/3 typical of pure gases, can improve the efficiency of thermoacoustic devices. The reduction on Prandtl number lowers the viscous dissipation, and has allowed significant improvement in the efficiency of standing wave thermoacoustic refrigerator.[9]

This paper presents a numerical study and estimation of critical temperature differences of a standing wave TAPM with various helium-based binarymixture working gases. This work provides a method to estimate the the critical temperature difference for binary gas mixture with various compositions in a standing wave TAPM with a stack made of a pile of wire-mesh screens. The calculation results would guide us to appropriately choose the working gas and its composition in applications of harnessing low-quality heat sources to power thermoacoustic prime movers.

## 2. Theory

### *2.1. Onset and critical conditions*

In thermoacoustic prime movers, once the temperature difference across the stack reaches a starting point, the devices will start and maintain an acoustic oscillation by converting heat into acoustic power. This starting point is so-called the onset temperature difference ($\Delta T_{\text{onset}}$). Following Atchley, [10] the onset condition in a standing wave thermoacoustic prime mover can

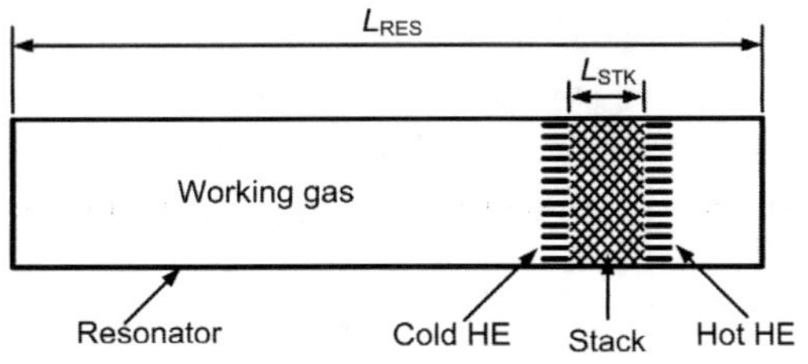


**Figure 1.** Schematic ilustration of a standing wave TAPM. $L_{\text{RES}}$ is resonator length, $L_{\text{STK}}$ is stack length, and HE refers to heat exchanger.

be analyzed by using the quality factor $Q$ which is expressed as

$$Q = -\frac{\omega \dot{E}_{\text{st}}}{\dot{W}} \tag{1}$$

where $\dot{E}_{\text{st}}$ is the stored acoustic energy, $\omega$ is the angular frequency of the oscillation, and $\dot{W}$ is the net power output of the entire device.

As shown in Fig. 1, a thermoacoustic prime mover typically consists of an acoustic resonator, a working gas, a stack, and two heat-exchangers. Hence, $\dot{W}$ can be expressed as [10][11]

$$\dot{W} = \dot{W}_{\text{RES}} + \dot{W}_{\text{HE}} + \dot{W}_{\text{STK}} \tag{2}$$

where $\dot{W}_{\text{RES}}$ and $\dot{W}_{\text{HE}}$ are the dissipations (always negative) in the resonator tube and heat exchangers, respectively, and $\dot{W}_{\text{STK}}$ is the net acoustic power out from the stack which can be either positif or negatif, depending on the temperature difference ($\Delta T$) imposed across the stack and its position in the standing wave.

As $\Delta T$ increases from zero, the thermal dissipations in the stack decrease. When $\Delta T$ increases further until a certain point, the acoustic generation in the stack exactly balances the dissipations in the stack, $\dot{W}_{\text{STK}} = 0$, and the temperature difference is referred as the critical temperature difference ($\Delta T_{\text{crit}}$). At this point, however, $\dot{W}$ is still negative and $Q$ is positif and finite. As $\Delta T$ increases furthermore and reaches the $\Delta T_{\text{onset}}$, the acoustic production overcomes all dissipations, $\dot{W} = 0$ and $Q$ is infinite, the prime mover has zero net loss and comes into the onset condition. From this point, any infinitesimal increases in $\Delta T$ will make $\dot{W}_{\text{STK}}$ positive, the net loss and $Q$ negative, and the prime mover will start the self-oscillation.

From the description mentioned above, it is clear that a low critical temperature difference will lead to a low onset temperature difference. In the next subsection, the derivation of an analytical expression of the critical tempereature difference for standing wave thermoacoustic prime movers based on the linear thermoacoustic theory is presented.

### *2.2. Critical temperature difference*

An expression for the critical temperature difference ($\Delta T_{\text{crit}}$) can be obtained from the linear thermoacoustic theory. The time-averaged acoustic power $d\dot{W}_{\text{STK}}$ produced in the length $dx$ of a stack's channel can be written in complex notation as [9]

$$\frac{d\dot{W}_{\text{STK}}}{dx} = \frac{1}{2}\text{Re}\left[\widetilde{U}_1 \frac{dp_1}{dx} + \widetilde{p}_1 \frac{dU_1}{dx}\right] \tag{3}$$

where $U_1$ is the complex volumetric velocity, $p_1$ is the complex pressure, tilde ($\sim$) sign indicates a complex conjugate, and Re[ ] denotes the real part of a complex number.

For stack's channels which have regular geometries, Eq. (3) can be written as [9]

$$\frac{d\dot{W}_{\rm STK}}{dx} = -\frac{1}{2}r_v|U_1|^2 - \frac{1}{2r_k}|p_1|^2 + \frac{1}{2}{\rm Re}\left[g\widetilde{p}_1 U_1\right] \tag{4}$$

where $r_v$ is the viscous resistance per unit length, $1/r_k$ is the thermal-relaxation conductance per unit length, and $g$ is the complex gain (or attenuation) constant for the volume flow rate, which are respectively defined as

$$r_v = \frac{\omega\rho_{\rm m}}{A}\frac{{\rm Im}[-f_v]}{|1-f_v|^2}, \tag{5}$$

$$\frac{1}{r_k} = \frac{\gamma-1}{\gamma}\frac{\omega A}{p_{\rm m}}{\rm Im}[-f_k], \tag{6}$$

$$g = \frac{f_k - f_v}{(1-f_v)(1-\sigma)}\frac{1}{T_{\rm m}}\frac{dT_{\rm m}}{dx}. \tag{7}$$

In these equations, $\rho_{\rm m}$ is the mean density, $p_{\rm m}$ is the mean pressure, $T_{\rm m}$ is the mean temperature, $\gamma$ is the ratio of specific heats, and $\sigma$ is the Prandtl number of the gas, respectively, $A$ is the cross sectional area of the stack, $f_v$ and $f_k$ are the Rott's thermoviscous functions that enable the description of the three dimensional phenomena in the stack's channel.

For the stack's channel with circular pores geometry, the thermoviscous functions are [9]

$$f_k = \frac{2J_1[(i-1)R/\delta_k]}{J_0[(i-1)R/\delta_k](i-1)R/\delta_k}, \tag{8}$$

$$f_v = \frac{2J_1[(i-1)R/\delta_v]}{J_0[(i-1)R/\delta_v](i-1)R/\delta_v} \tag{9}$$

where $R$ is the pore radius, $\delta_k$ and $\delta_v$ are the thermal and viscous penetration depths, respectively, $J_0$ and $J_1$ are the zeroth- and first-order Bessel functions, respectively. The penetrations dephts are calculated as

$$\delta_k = \sqrt{2k/\omega\rho c_p}, \tag{10}$$

$$\delta_v = \sqrt{2\mu/\omega\rho} \tag{11}$$

where $k$ is the thermal conductivity, $\mu$ is the dynamic viscosity, $c_p$ is the isobaric specific heats, and $\rho$ is the density of the gas.

The critical temperature gradient along the stack $(\nabla_x T_{\rm m})_{\rm crit}$, i.e. $(dT_{\rm m}/dx)_{\rm crit}$, can be obtained by letting $d\dot{W}_{\rm STK}/dx = 0$. By assumming ideal gas and lossless planar acoustic standing wave so that the phase difference between $p_1$ and $U_1$ is $\pi/2$ and hence we have acoustic impedance $Z(x) = p_1/U_1 = (\rho_{\rm m}a/A)\tan(\omega x/a)$, we can get

$$(\nabla_x T_{\rm m})_{\rm crit} = \frac{2\pi}{\lambda}T_{\rm m}\left[\frac{\frac{{\rm Im}[-f_v]}{|1-f_v|^2} + (\gamma-1)\,{\rm Im}[-f_k]\tan^2\left(\frac{\omega x}{a}\right)}{{\rm Im}\left[-\frac{f_k-f_v}{(1-f_v)(1-\sigma)}\right]\tan\left(\frac{\omega x}{a}\right)}\right]. \tag{12}$$

Here, $x$ is the axial location inside the resonator tube measured from the pressure node of the standing wave, $\lambda$ is the wavelength of the sound wave, and $a$ is the sound speed in the gas.

We use the short-stack approximation in where the stack length ($L_{\rm STK}$) is much shorter than the acoustic wavelength ($\lambda$). The stack is short enough so that the pressure and velocity inside the stack do not vary appreciably, and hence the critical temperature gradient can be considered constant along the stack. Then, the critical temperature difference ($\Delta T_{\rm crit}$) across the stack is simply obtained by multiplying $(\nabla_x T_{\rm m})_{\rm crit}$ with the stack length. It can be seen from Eq. (12) that the critical temperature difference is influenced by the geometry of the stack's channels, location of the stack, acoustic wavelength, and properties of the gas which are depending on the gas temperature.

*2.3. Properties of binary-mixture gas*

The spesific heats per unit mass at constant pressure and constant volume for a binary mixture gas can be estimated as

$$c_{p_{\text{mix}}} = \frac{x_1 C_{p_1} + x_2 C_{p_2}}{x_1 M_1 + x_2 M_2} \tag{13}$$

$$c_{v_{\text{mix}}} = \frac{x_1 C_{v_1} + x_2 C_{v_2}}{x_1 M_1 + x_2 M_2} \tag{14}$$

where $x_1$ and $x_2$ are the mole fractions of species 1 and species 2, $C_{p_1}$ and $C_{v_1}$ are the heat capacities for species 1, $C_{p_2}$ and $C_{v_2}$ are the heat capacities for species 2, $M_1$ and $M_2$ are the molecular weights of species 1 and species 2, respectively. The ratio of specific heats for the binary mixture gas ($\gamma_{\text{mix}}$) can thus be computed by the ratio of Eqs. (13) and (14),

$$\gamma_{\text{mix}} = \frac{c_{p_{\text{mix}}}}{c_{v_{\text{mix}}}} \tag{15}$$

The gas density for a binary mixture gas ($\rho_{\text{mix}}$) can be written as a linear combination of the individual component's densities

$$\rho_{\text{mix}} = x_1 \rho_1 + x_2 \rho_2 \tag{16}$$

where $\rho_1$ and $\rho_2$ are the densities of species 1 and species 2, respectively.

The thermal conductivity of a gas mixture ($k_{\text{mix}}$) can be approximated by [12]

$$k_{\text{mix}} = \sum_{i=1}^{n} \frac{x_i k_i}{\sum_{j=1}^{n} x_j \phi_{k,ij}} \tag{17}$$

where

$$\phi_{k,ij} = \frac{1}{2\sqrt{2}} \left(1 + \frac{M_i}{M_j}\right)^{-\frac{1}{2}} \left[1 + \left(\frac{k_i}{k_j}\right)^{\frac{1}{2}} \left(\frac{M_j}{M_i}\right)^{\frac{1}{4}}\right]^2 \tag{18}$$

and $i$ and $j$ are species numbers, $k_i$ is the thermal conductivity of species $i$, $M_i$ is molar mass of species $i$, $n$ is number of species ($n = 2$ for binary mixture). Similarly, the viscosity of a gas mixture ($\mu_{\text{mix}}$) can be approximated by [13]

$$\mu_{\text{mix}} = \sum_{i=1}^{n} \frac{x_i \mu_i}{\sum_{j=1}^{n} x_j \phi_{\mu,ij}} \tag{19}$$

where

$$\phi_{\mu,ij} = \frac{1}{2\sqrt{2}} \left(1 + \frac{M_i}{M_j}\right)^{-\frac{1}{2}} \left[1 + \left(\frac{\mu_i}{\mu_j}\right)^{\frac{1}{2}} \left(\frac{M_j}{M_i}\right)^{\frac{1}{4}}\right]^2 \tag{20}$$

and $\mu_i$ is the viscosity of species $i$. These approximations for thermal conductivity and viscositiy of a gas mixture have a reasonable accuracy.[14] The more accurate calculation methods can be found elsewhere.[15][16]

By using the expressions above, the Prandtl number of the gas mixture ($\sigma_{\text{mix}}$) and the sound speed in the gas mixture ($a_{\text{mix}}$) can respectively be calculated as

$$\sigma_{\text{mix}} = \frac{\mu_{\text{mix}} c_{p_{\text{mix}}}}{k_{\text{mix}}} \tag{21}$$

and

$$a_{\text{mix}} = \sqrt{\frac{\gamma_{\text{mix}} p_m}{\rho_{\text{mix}}}}. \tag{22}$$

## 3. Calculation Procedure

In this section, it is presented a procedure to calculate the critical temperature difference for a standing wave thermoacoustic prime mover with a straight-closed resonator tube which has been shown in Fig 1. The detail description of the calculation method is given below.

### *3.1. Specifying the resonator*

We suppose that the resonator length ($L_{\text{RES}}$) is 1 m. Both ends of the resonator are closed so that a half-wavelength standing acoustic wave can exist with a pressure node at the middle of the resonator tube. The fundamental frequency of the half-wavelength resonator is

$$f = \frac{a_{\text{mix}}}{2L_{\text{RES}}}. \tag{23}$$

### *3.2. Calculating the stack properties*

We consider that the stack is made of a pile of stainless-steel wire-mesh screens with mesh number ($n$) of 30 meshes/inches and wire diameter ($d_{\text{wire}}$) of 0.25 mm. The stack length ($L_{\text{STK}}$) is 4 cm and the location ($x$) of the stack's center is 35 cm from the middle of resonator. The porosity ($\phi$) and hydraulic radius ($r_h$) of the stack can be estimated as [9]

$$\phi = 1 - \frac{\pi n d_{\text{wire}}}{4} \tag{24}$$

$$r_h = d_{\text{wire}} \frac{\phi}{4(1-\phi)} \tag{25}$$

In this work, we use an approximate model of the wire-mesh stack, i.e. it is modeled as an array of circular-cross-section tubes, in where the radius of the tube's cross-section is represented by the effective circular radius [17]

$$r_{\text{eff}} = \frac{\sqrt{D_h d_{\text{wire}}}}{2} \tag{26}$$

where $D_h$ is the hydraulic diameter of the wire-mesh stack. This $r_{\text{eff}}$ is therefore the hydraulic radius of the array of circular-cross-section tubes ($r_{h,\text{tube}}$).

### *3.3. Calculating the properties of the binary-mixture gases*

We investigate four helium-based binary-mixture gases as the working medium of the thermoacoustic prime mover, namely helium-argon (He-Ar), helium-nitrogen (He-$N_2$), helium-oxygen (He-$O_2$), and helium-carbon dioxide (He-$CO_2$). The molar mass ($M$) of the invidual gases are listed in Table 1. We assume that the mean temperature ($T_{\text{m}}$) of the working gas in the thermoacoustic prime mover is 300 K, and the charged pressures (mean pressures, $p_{\text{m}}$) vary from 0.1 MPa until 3.0 MPa. The properties of the individual gases, such as thermal conductivity and dynamic viscosity, at this temperture and in this pressure range are collected from NIST Chemistry Webbook.[18] The properties of the binary mixture gases are estimated by using equations which have been described in Subsection 2.3.

**Table 1.** Molar mass of the individual gases.

| Gas | He | Ar | $N_2$ | $O_2$ | $CO_2$ |
|---|---|---|---|---|---|
| $M$ (g/mol) | 4.00 | 39.95 | 28.01 | 32.00 | 44.01 |

*3.4. Calculating the critical temperature difference*

To calculate the critical temperature difference, first we compute the fundamental frequency of the resonator by using Eq. (23). Then, we calculate the thermal and viscous penetration depths ($\delta_k$ and $\delta_v$) by using Eqs. (10) and (11) with $\omega = 2\pi f$ and adaptations that the thermal conductivity is $k_{\text{mix}}$, the density is $\rho_{\text{mix}}$, the isobaric specific heat is $c_{p_{\text{mix}}}$, and the viscosity is $\mu_{\text{mix}}$. Next, the thermociscous functions ($f_k$ and $f_v$) were calculated by using Eqs. (8) and (9) in where $R$ is $r_{\text{h,tube}} = r_{\text{eff}}$ (see Eq. (26)). After that, by inserting all required quantities which have been found into Eq. (12) we get the critical temperature gradient $(\nabla_x T_{\text{m}})_{\text{crit}}$, and finally by multiplying it with the stack length ($L_{\text{STK}}$), the critical temperature difference $(\Delta T)_{\text{crit}}$ is obtained. These calculations are carried out for various molar compositons and charged pressures of the gas mixtures.

## 4. Results and Discussion

Before we discuss about the calculation results for binary mixture gases, it would be good that we present the results for pure individual gases. Figure 2 shows the calculation results of the $\Delta T_{\text{crit}}$ for each pure individual gases at various charged pressures from 0.1 MPa until 3.0 MPa. It can be seen that for pressures less than 1.0 MPa the $\Delta T_{\text{crit}}$ of pure helium are much higher than those of Ar, $N_2$, $O_2$, and $CO_2$ gases, while at higher pressures, all individual gases have not so different $\Delta T_{\text{crit}}$ which are around 80 °C – 100 °C. As can be seen in Eqs.(8)–(12), the dependence of the critical temperature difference on the gas properties is not as clear. It appears in the ratio of specific heats $\gamma$, Prandtl number $\sigma$, and implicitly in the thermoviscous functions $f_k$ and $f_v$.

The calculation results of $\Delta T_{\text{crit}}$ for binary-mixture gases with various molar compositions at charged pressures from 0.3 MPa until 2.0 MPa are presented in Fig. 3. The left extreme of the abscissa equals 0 is associated with multiple $\Delta T_{\text{crit}}$ of the individual heavier gases (Ar, $N_2$, $O_2$, and $CO_2$), whereas the right extreme equals 1 corresponds to $\Delta T_{\text{crit}}$ of He gas. These all have been summarized in Fig. 2. We can, generally, see in Fig. 3 that the mixing of gases can yield in lower $\Delta T_{\text{crit}}$ than their values of one or both of components of the gas mixtures. In other words, in this pressure range, there is an optimum molar composition of each gas mixture which gives a minimum critical temperature difference $(\Delta T_{\text{crit}})_{\text{min}}$, except in the case of Fig. 3(A). We also

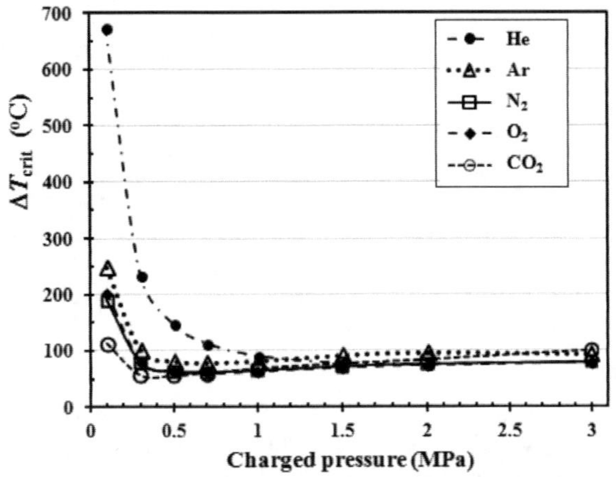


**Figure 2.** The $\Delta T_{\text{crit}}$ for individual pure gases at various pressures.

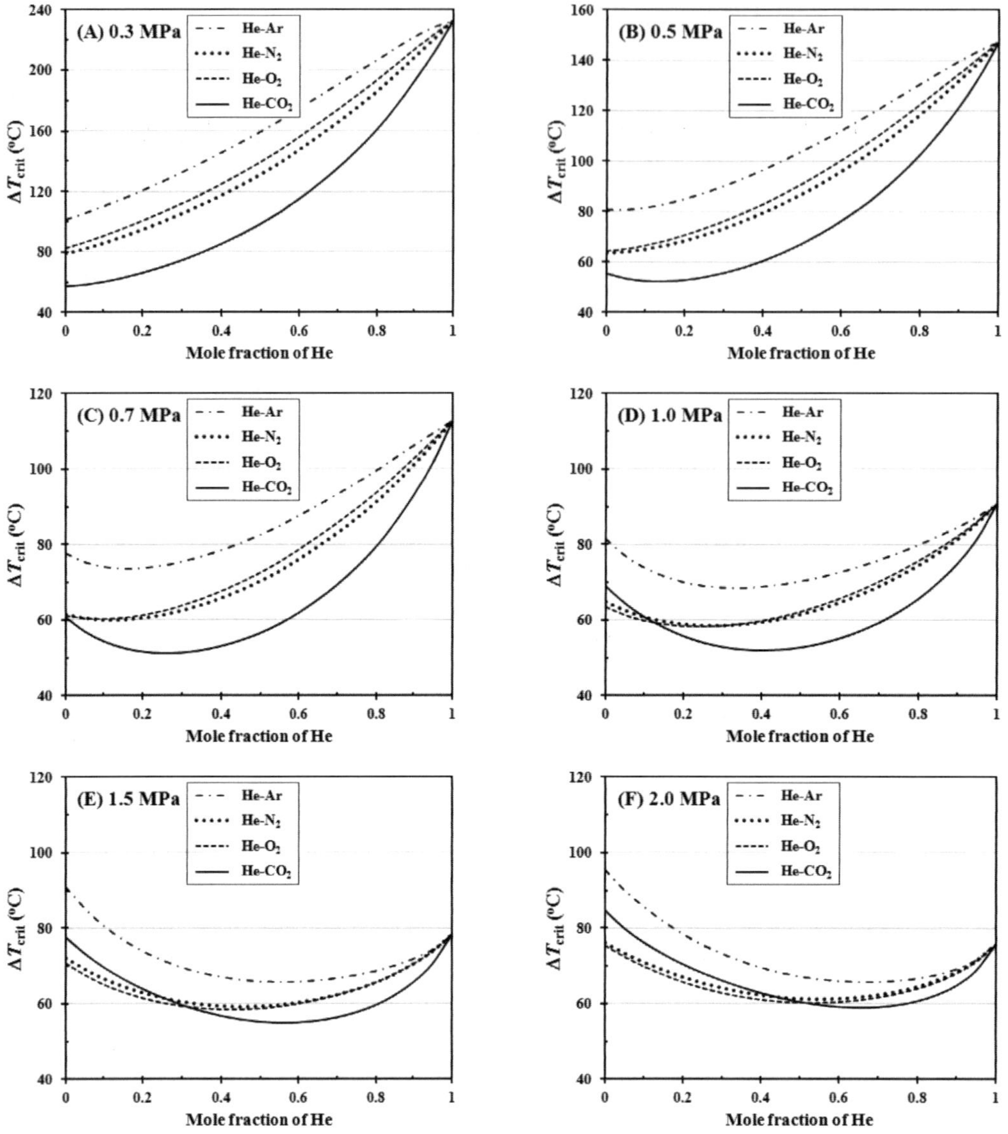


**Figure 3.** The dependence of the $(\Delta T)_{\rm crit}$ of the standing wave TAPM with various binary-mixture gases at charged pressures of (A) 0.3 MPa, (B) 0.5 MPa, (C) 0.7 MPa, (D) 1.0 MPa, (E) 1.5 MPa, and (F) 2.0 MPa.

found that He-Ar mixture gas always has $\Delta T_{\rm crit}$ higher than those of other mixture gases for a given molar composition. On the other hand, He-$CO_2$ gas always gives the lowest $(\Delta T_{\rm crit})_{\rm min}$ at a given charged pressure. In addition, He-$N_2$ and He-$O_2$ gases demonstrate almost the same $\Delta T_{\rm crit}$ againts variation of molar compositions at a given charged pressure.

In Fig. 3(A) the calculation results for gas pressure of 0.3 MPa is presented. The $\Delta T_{\rm crit}$ of

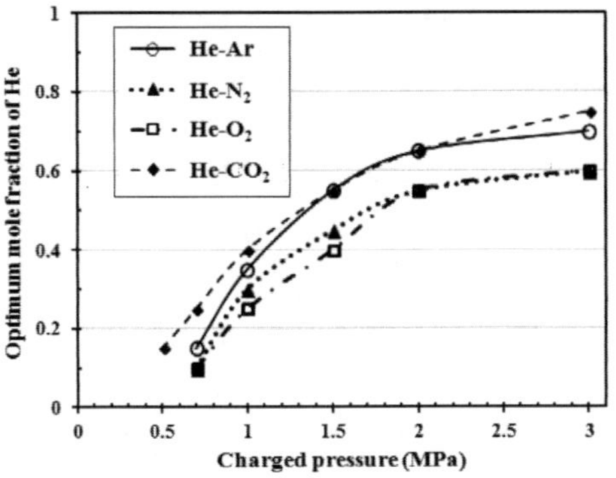


**Figure 4.** The optimum mole fractions of He for various charged pressures.

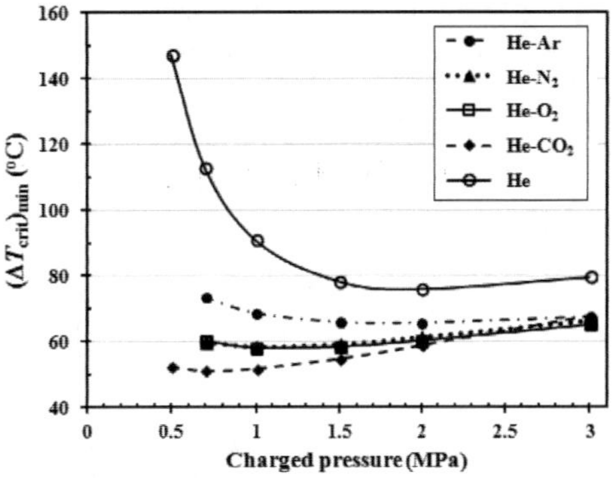


**Figure 5.** The minimum critical temperature differences for various charged pressure.

pure helium is 233 °C, while those of pure Ar, $N_2$, $O_2$, and $CO_2$ gases are 101 °C, 79 °C, 82 °C, and 57 °C, respectively. This figure shows that mixing helium gas with other heavier gases pulls down the $\Delta T_{\text{crit}}$ below that of pure helium gas, but still higher than those of pure heavier gases. At pressure of 0.5 MPa, as shown in Fig. 3(B), only He-$CO_2$ curve has a minimum point, i.e. at mole fraction of He of around 0.15 which gives $(\Delta T_{\text{crit}})_{\text{min}}$ of 52 °C. Moreover, at pressures of 0.7 MPa and higher, each $\Delta T_{\text{crit}}$ curves has a minimum point, at different molar compositions, as can be seen in Figs. 3(C)−(F).

It is also found from Figs. 3(C)−(F) that the optimum mole fractions of helium which give the minimum critical temperature differences are shifted to larger values as the pressure increases, and tend to be constant at around 0.7 when the pressure increases to higher than 2 MPa, as can be seen more clearly in Fig. 4. However, as depicted in Fig. 5, the $(\Delta T_{\text{crit}})_{\text{min}}$ slightly increase when the pressure increases to higher than 1.5 MPa. In addition, Fig. 5 reaffirms that the minimum critical temperature differences of helium-based binary-mixture gases are lower than than those of pure helium gas.

**Table 2.** The lowest $(\Delta T_{\text{crit}})_{\text{min}}$ for different binary-mixture gases and the related optimum ranges of charged pressures, optimum mole fractions of helium, and average percentages of $\Delta T_{\text{crit}}$-decreases from their values of single helium gas at the corresponding charged pressures.

| Gas | Lowest $(\Delta T_{\text{crit}})_{\text{min}}$ (°C) | $p_m$ (MPa) | Mole fraction of He | Average decrease of $\Delta T_{\text{crit}}$ (%) |
|---|---|---|---|---|
| He-Ar | 66 | 1.5 - 2.0 | 0.55 - 0.65 | 15 |
| He-$N_2$ | 59 | 1.0 - 1.5 | 0.35 - 0.55 | 30 |
| He-$O_2$ | 58 | 1.0 - 1.5 | 0.35 - 0.55 | 31 |
| He-$CO_2$ | 51 | 0.5 - 1.0 | 0.15 - 0.40 | 54 |

Table 2 summarizes the lowest$(\Delta T_{\text{crit}})_{\text{min}}$ for various binary-mixture gases together with the related optimum ranges of charged pressures, optimum mole fractions of helium, and average percentages of $\Delta T_{\text{crit}}$-decreases from their values of pure helium gas at the corresponding charged pressures. These results, generally, show that the mixing of helium gas with other common gases yield in significant decreases of the critical temperature differences. The additional advantage of the gas mixing in this case is that we will need smaller amount of helium gas so that reducing the construction cost of the termoacoustic prime mover.

## 5. Future Works and Applications

To follow up these results, we are now constructing a standing wave TAPM with a wire-mesh stack, and experiments will be performed to verify the estimation results. The aim is to find a binary mixture working gas which is experimentally verified giving a low critical temperature difference. The TAPM with a low critical temperature difference can potentially be applied in factory area, for example, to harness the waste heat as the heat source in producing the output work, which in turn can be used to generate electricity or to drive a thermoacoustic refrigerator.

## 6. Conclusions

Mixing helium gas with other common gases, namely argon, nitrogen, oxygen, and carbon dioxide, at appropriate pressures and molar compositions, can reduce the critical temperature differences to lower than those of the individual components of the gas mixtures. The optimum mole fractions of helium which give the minimum critical temperature differences are shifted to larger values as the pressure increases, and tend to be constant at around 0.7 when the pressure increases to higher than 2 MPa. In addition, the minimum critical temperature differences slightly increase when the pressure increases to higher than 1.5 MPa. Moreover, it is found that the lowest critical temperature difference for He-Ar mixture gas is around 66 °C which is achieved in the pressure range of 1.5 MPa - 2.0 MPa and mole fractions of helium of 0.55 - 0.65. The He-$N_2$ and He-$O_2$ mixture gases demonstrate almost the same performances, both have the lowest critical temperature difference around 59 °C at pressures of 1.0 MPa - 1.5 MPa and helium's mole fractions of 0.35 - 0.55. For all tested gases, the lowest critical temperature difference of around 51 °C is provided by He-$CO_2$ mixture gas at pressures of 0.5 MPa - 1.0 MPa with helium's mole fraction of 0.15 - 0.40.

## Acknowledgments

This research was financially supported by Directorate of Higher Education, Dept. of Education and Culture, Republic of Indonesia.

[1] Chen R L and Garret S L 1998 *J. Acoust. Soc. Am.* **103** 2481
[2] Adeff J A and Hoffler T J 2000 *J. Acoust. Soc. Am.* **107** L37
[3] Gardner D L and Howard C Q 2009 it Proc. of Acoust. (Adelaide: Australian Acoust. Soc.)
[4] Yu B, Luo E C, Li S F, Dai W and Wu Z H 2011 *Cryogenics*bf 51 49
[5] Backhaus S, Tward E, and Petach M 2004 *App. Phys. Lett.* **85** 1085
[6] Kitadani Y, Sakamoto S, Sahashi K and Watanabe Y 2010 *Proc. 20th Int. Cong. Acoust.* (Sidney: Australian Acoust. Soc.)
[7] Swift G W 1992 *J. Acoust. Soc. Am* **92** 1551
[8] Backhaus S and Swift G W 2000 *J. Acoust. Soc. Am* **107** 3148
[9] Swift G W 2002 *Thermoacoustics: A Unifying Perspective for Some Engines and Refrigerators*, Acoust. Soc. Am., New York
[10] Atchley A A 1994 *J. Acoust. Soc. Am* **95** 1661
[11] Yu Z, Jaworski A J, and Abduljalil A S 2009 it WASET **25** 722
[12] Mason E A and Saxena S C 1958 *Phys. Fluids* **1** 361
[13] Wilke C R 1950 *J. Chem. Phys.* **18** 517
[14] Ke H, He Y, Liu Y and Cui F 2012 *Int. J. Thermophys.* **33** 1143
[15] Tijani M E H, Zeegers J C H and de Waele A T A M 2002 *J. Acoust. Soc. Am.* **112** 134
[16] Campo A, Papari M M and Abu-Nada E 2011 *Appl. Therm. Eng.* **31** 3142
[17] Ueda Y, Kato T and Kato C 2009, *J. Acoust. Soc. Am.* **125** 780
[18] Linstrom P J and Mallard W G 2013 http://webbook.nist.gov (accessed April)

Journal of Physics: Conference Series **622** (2015) 012011 doi:10.1088/1742-6596/622/1/012011

# Multi-loop Control System Design for Biodiesel Process using Waste Cooking Oil

**Dipesh S Patle[1], Ahmad Z[2,3] and G. P. Rangaiah[4]**
[1]School of Chemical Engineering, Engineering Campus, Nibong Tebal, Penang, 14300, Malaysia.
[2]School of Chemical Engineering, Engineering Campus, Nibong Tebal, Penang, 14300, Malaysia.
[4]Department of Chemical & Biomolecular Engineering, National University of Singapore, 117585, Singapore.

E-mail: chzahmad@usm.my

**Abstract.** Biodiesel is one of the promising liquid fuels for future due to its advantages such as renewability and eco-friendliness. This manuscript describes the development of a multi-loop control system design for a comprehensive biodiesel process using waste cooking oil. Method for controlled variable-manipulated variable (CV-MV) pairings are vital for the stable, effective and economical operation of the process. Liquid recycles, product quality requirements and effective inventory control pose tough challenges to the safe operation of the biodiesel process. A simple and easy to apply effective RGA method [Xiong Q, Cai W J and He M J 2005 A practical loop pairing criterion for multivariable processes *Journal of Process Control* vol. 15 pp 741-747.] is applied to determine CV-MV pairings i.e. control configuration design for the bioprocess. This method uses steady state gain as well as bandwidth information of the process open loop transfer function to determine input-output pairings.

## 1. Introduction

Typically, biodiesel is made by chemically reacting lipids such as vegetable oil with an alcohol to produce fatty acid esters i.e. biodiesel. It can also be produced by pyrolysis, micro-emulsification and transesterification. Among these alternatives, alkali-catalyzed transesterification is commonly used in the industry [1, 2]. Therefore, in this study, a two-step, acid esterification and alkali transesterification, process is considered. Typically, alkali catalyzed transesterification requires the pure oil as it forms soap in the presence of free fatty acids (FFA). Four biodiesel production processes (alkali-catalyzed process using pure vegetable oil, alkali catalyzed process using WCO, acid-catalyzed process using WCO, and acid-catalyzed process using hexane extraction) were proposed by Zhang et al. [3]. As this study deals with biodiesel production from waste cooking oil (WCO), acid esterification of WCO is carried out to convert FFAs into biodiesel.

[3] To whom any correspondence should be addressed.

The complete process with several reactors and distillation columns requires proper input-output (i.e., manipulated variable-controlled variable) pairings for designing an effective control system. In general, the control structure is centralized or decentralized. The latter is more attractive for complex processes with many unit operations; other advantages include its ease in design and implementation, simple tuning and robust behaviour in dealing with faults and uncertainties [4]. Some control loops can be selected based on process knowledge and heuristics. However, for other multivariable processes, as in case of biodiesel process, systematic and reliable methods are needed to determine input-output pairings and to complement the pairings suggested by heuristics. Relative gain array (RGA) is very widely used to determine input-output pairings due to its simplicity. Several other methods have been proposed to determine these pairings, such as dynamic RGA, Effective RGA (ERGA), block relative array, relative disturbance gain array and non-linear RGA [4]. Steady state RGA based loop pairing criterion yields an incorrect interaction assessment in certain cases [5]. Therefore, ERGA is employed in this work to decide the control configuration. The next section describes the process for biodiesel production from WCO.

## 2. Biodiesel production from WCO

The biodiesel process (Figure 1) is taken from our previous work [6], where two alternative processes were optimized using an Excel based multi-objective optimization program. The plant capacity is assumed to be 120 kt per annum, based on potential WCO in Malaysia. Steady-state and dynamic simulations are carried out using Aspen Plus V8.0 and Aspen Plus Dynamics V8.0 respectively. The property model used for these simulations is Dortmund modified UNIFAC [6-8]. To obtain more reliable results, detailed components of palm oil and more realistic kinetics that includes mono- and di-glycerides formation are considered. Esterification and trans-esterification are represented by 10 and 96 reactions, whose details can be seen in [6].

## 3. Effective relative gain array

This method is the extension of the popular RGA into a new method that reflects dynamic interactions of the process under finite bandwidth control. Its main advantages are: it reflects dynamic interaction within the loop without needing the specification of the type of controller and with minimal computation, gives a less conservative controller when the detuning factor design method is employed, and is easy to understand [4]. The loop pairing decision is affected by: steady state gain and response speed. The response speed is proportional to the bandwidth in frequency domain [5]. Therefore, the bandwidth is considered to reflect interactions from finite bandwidth control as well as pairing loops to result in a quick response [5]. ERGA method is summarized as follows (for further details, refer to [5]).

1) Obtain open loop transfer functions (TF)

2) Determine bandwidth from the obtained TF

3) Calculate effective gain matrix $E=G(0)\times\Omega$

where

$$E=\begin{bmatrix} e_{11} & e_{12} & \cdot\cdot & e_{1n} \\ e_{21} & e_{22} & \cdot\cdot & e_{2n} \\ \cdot\cdot & \cdot\cdot & \cdot\cdot & \cdot\cdot \\ e_{n1} & e_{n2} & \cdot\cdot & e_{nn} \end{bmatrix},\ G(0)=\begin{bmatrix} g_{11}(0) & g_{12}(0) & \cdot\cdot & g_{1n}(0) \\ g_{21}(0) & g_{22}(0) & \cdot\cdot & g_{2n}(0) \\ \cdot\cdot & \cdot\cdot & \cdot\cdot & \cdot\cdot \\ g_{n1}(0) & g_{n2}(0) & \cdot\cdot & g_{nn}(0) \end{bmatrix} \text{ and } \Omega=\begin{bmatrix} \omega_{B,11} & \omega_{B,12} & \cdot\cdot & \omega_{B,1n} \\ \omega_{B,21} & \omega_{B,21} & \cdot\cdot & \omega_{B,21} \\ \cdot\cdot & \cdot\cdot & \cdot\cdot & \cdot\cdot \\ \omega_{B,n1} & \omega_{B,n1} & \cdot\cdot & \omega_{B,nn} \end{bmatrix}$$

Thus the bandwidth matrix ($\Omega$) is calculated based on the individual elements of G(s).

4) Find ERGA ($\phi$) ($\phi = E\mathrm{x}E^{-T}$) where

$$\phi=\begin{bmatrix} \phi_{11} & \phi_{12} & \cdot\cdot & \phi_{1n} \\ \phi_{21} & \phi_{22} & \cdot\cdot & \phi_{2n} \\ \cdot\cdot & \cdot\cdot & \cdot\cdot & \cdot\cdot \\ \phi_{n1} & \phi_{n2} & \cdot\cdot & \phi_{nn} \end{bmatrix}$$

Journal of Physics: Conference Series **622** (2015) 012011 doi:10.1088/1742-6596/622/1/012011

Finally, the pairing is selected according to the following rules;

(a) ERGA elements closest to 1 should be chosen.
(b) The positive value of Niederlinski index (NI) $\left(= \frac{det(G(0))}{\prod_{i=1}^{m} g_{ii}(0)}\right)$ is required.
(c) Selected pair (i.e. corresponding ERGA) element should be positive.
(d) The input-output pair having a large ERGA element should not be preferred.

## 4. Application of ERGA to biodiesel process

ERGA is employed to select the input-output pairing in CSTRs and distillation columns in the biodiesel process; see Table 1 for these and other MV-CV pairing. These loops are selected based on the guidelines given in the previous section. Other control loops are selected based on heuristics and process knowledge, as mentioned in Table 1. The resulted control structure is shown in Figure 1. In agreement to the results of ERGA analysis, in all distillation columns, the level in reflux drum and in reboiler are controlled using distillate flow and bottoms flow respectively; this is in accordance with heuristics that suggest the level should be controlled so that the disturbances are directed away from the primary process path [3]. Using reflux rates to control the level in reflux drums is not appropriate as the columns are operating at smaller reflux ratios. This is also in agreement with Richardson's rule that states that an inventory variable should be controlled with the manipulated variable that exercises the largest effect on it within that unit [3]. Similarly, the liquid level in reactors and phase separators are controlled using liquid outlets.

To demonstrate the effectiveness of these pairings, the performance of the control loops for ±10% variation in WCO is investigated. Controllers have been tuned using autotuner in Aspen Plus Dynamics. Figure 2 suggests that the control system with the CV-MV pairings in Table 1 works satisfactorily. Note that a complete PWC structure has not been developed in this study as the main focus is on the selection of input-output pairings only. A complete PWC for the process and its performance can be found in our recent work [8].

## 5. Conclusions

A simple, effective and easy to apply ERGA is applied to determine CV-MV pairings in the biodiesel production process from WCO. Using steady state gain and bandwidth information of the process open loop transfer functions, control loop pairings are identified that reflects the dynamic loop interactions under finite bandwidth control. The developed process model with the derived control structure can be used for further studies, such as evaluation of PWC structure and operator training simulator development.

## Acknowledgments

"RU grant (1001/PJKIMIA/814155) and IPS graduate fund" for the financial support. Thanks to Prashant Wankhede of Reliance India Limited, Chandrakant Bende of Mylan Pharmaceutical limited and Chetan Sayankar, Consultant of Inventive Engineering & Technology Pvt. Ltd. & SRS Biodiesel and Engineering Pvt. Ltd. for giving us important insights into the biodiesel process.

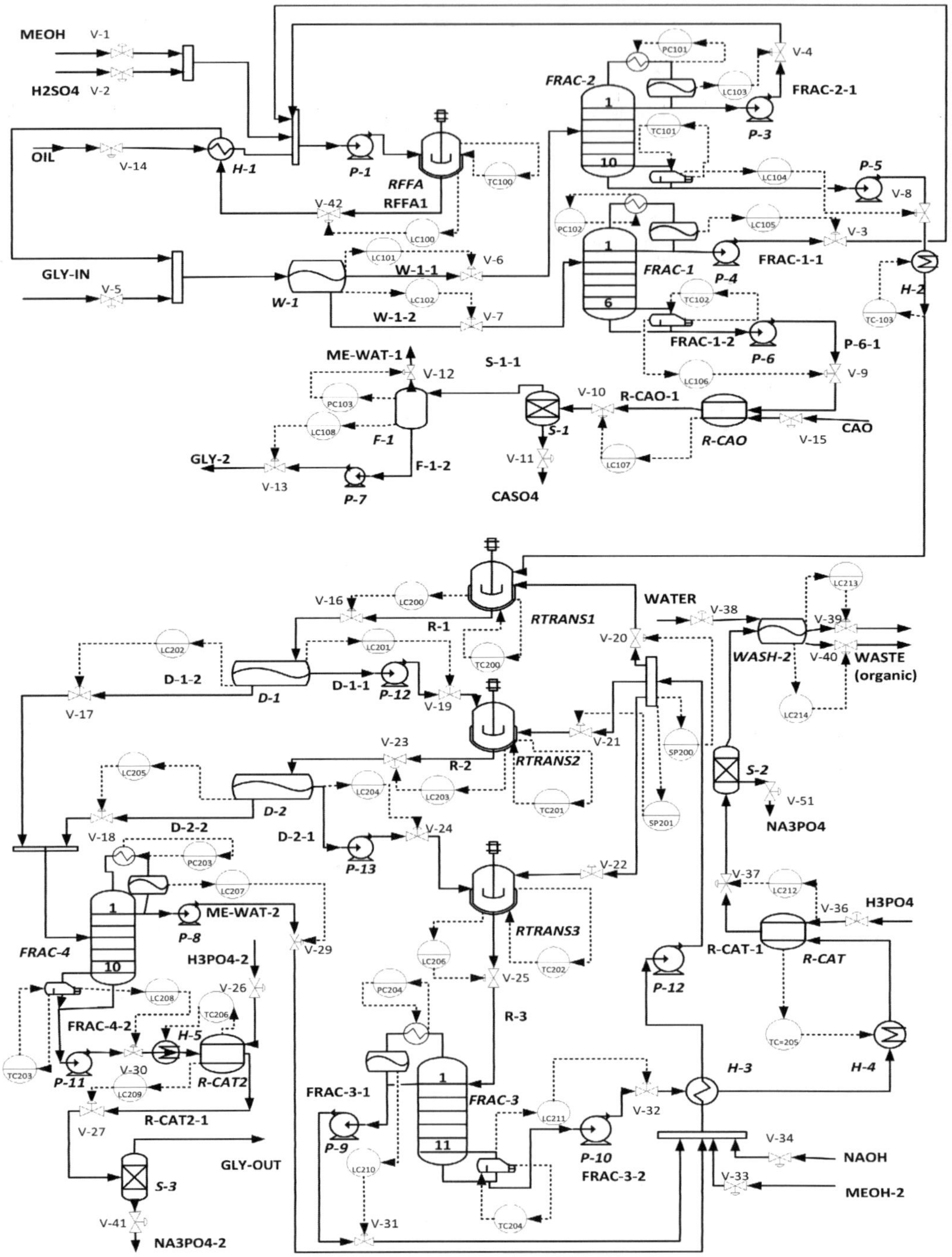


**Figure 1.** Control scheme designed for the biodiesel plant using waste cooking oil.

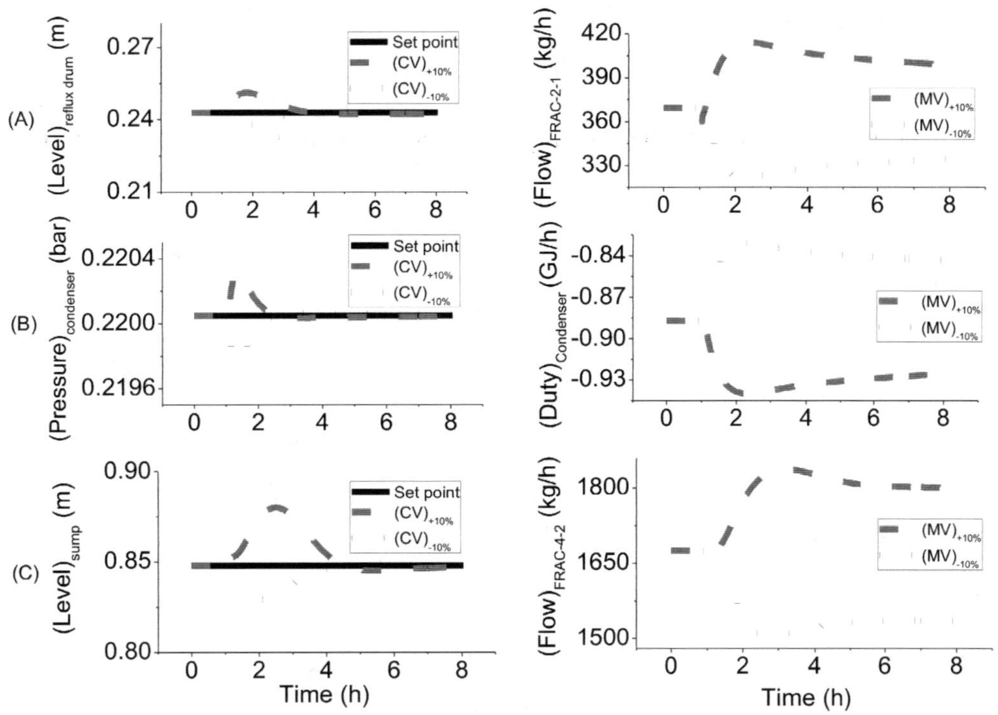


**Figure 2.** Dynamic behaviour of controllers for ±10% change in WCO: (A) Reflux drum level control in FRAC-2, (B) Pressure control in FRAC-2, (C) Sump level control in FRAC-4. The corresponding manipulated variable is shown in the plots on the right side.

**Table 1.** Control structure (i.e., MV-CV pairings) and controller parameters for the biodiesel process

| Controllers | Controlled variable (CV) | Manipulated variable (MV) | Controller parameters [Kc (%/%), $\tau_i$ (min)] |
|---|---|---|---|
| PC101 | Condenser pressure in FRAC-2 | Condenser duty in Frac-2 | 2; 10 |
| PC102 | Condenser pressure in FRAC-1 | Condenser duty in Frac-1 | 2; 10 |
| PC103 | Pressure in F-1 | Vapor flowrate [V-12] | 2; 10 |
| TC100 | Temperature in RFFA | Utility flow | 6.56; 13.2 |
| TC101 | Bottom stage temperature in Frac-2 | Reboiler duty in Frac-2 | 21.71; 13.56 |
| TC102 | Bottom stage temperature in Frac-1 | Reboiler duty in Frac-1 | 33.3; 11.3 |
| TC103 | Temperature of heater 'H-2' outlet | Heat duty in 'H-2' | 22.9; 12.6 |
| LC100 | Level in RFFA | Liquid outlet flow [V-42] | 10; 60 |
| LC101 | Light phase level in W-1 | Light phase outlet flow [V-6] | 5 |
| LC102 | Heavy phase level in W-1 | Heavy phase flow [V-7] | 5 |
| LC103 | Reflux drum level in FRAC-2 | Distillate flow [V-4] | 2 |

| | | | |
|---|---|---|---|
| LC104 | Column base level in FRAC-2 | Bottoms flow [V-8] | 2 |
| LC105 | Reflux drum level in FRAC-1 | Distillate flow [V-3] | 2 |
| LC106 | Column base level in FRAC-1 | Bottoms flow [V-9] | 2 |
| LC107 | Level in R-CAO | Liquid outlet flow [V-10] | 10; 60 |
| LC108 | Level in F-1 | Liquid outlet flow [V-13] | 5; 60 |
| PC203 | Condenser pressure in FRAC-4 | Condenser duty in Frac-4 | 2; 10 |
| PC204 | Condenser pressure in FRAC-3 | Condenser duty in Frac-3 | 2; 10 |
| TC200 | Temperature in RTRANS1 | Utility flow | 13.87; 19.8 |
| TC201 | Temperature in RTRANS2 | Utility flow | 12.11; 13.2 |
| TC202 | Temperature in RTRANS3 | Utility flow | 15.54; 26.4 |
| TC203 | Bottom stage temperature in Frac-4 | Reboiler duty in Frac-4 | 21.22; 9.62 |
| TC204 | Bottom stage temperature in | Reboiler duty in Frac-3 | 23.23; 13.83 |
| TC205 | Temperature in R-CAT | Heat duty in 'H-4' | 9.63; 17.23 |
| TC206 | Temperature in R-CAT2 | Heat duty in 'H-5' | 13.63; 7.92 |
| LC200 | Level in RTRANS1 | Liquid outlet flow [V-16] | 10; 60 |
| LC201 | Light phase level in D-1 | Light phase outlet flow [V-19] | 5; 60 |
| LC202 | Heavy phase level in D-1 | Heavy phase flow [V-17] | 5; 60 |
| LC203 | Level in RTRANS2 | Liquid outlet flow [V-23] | 10; 60 |
| LC204 | Light phase level in D-2 | Light phase outlet flow [V-24] | 5 |
| LC205 | Heavy phase level in D-2 | Heavy phase flow [V-18] | 5 |
| LC206 | Level in RTRANS3 | Liquid outlet flow [V-25] | 10; 60 |
| LC207 | Reflux drum level in FRAC-4 | Distillate flow [V-29] | 2 |
| LC208 | Column base level in FRAC-4 | Bottoms flow [V-30] | 2 |
| LC209 | Level in R-CAT2 | Liquid outlet flow [V-27] | 10; 60 |
| LC210 | Reflux drum level in FRAC-3 | Distillate flow [V-31] | 2 |
| LC211 | Column base level in FRAC-3 | Bottoms flow [V-32] | 2 |
| LC212 | Level in R-CAT | Liquid outlet flow [V-37] | 10; 60 |
| LC213 | Light phase level in WASH-2 | Light phase outlet flow [V-39] | 5 |
| LC214 | Heavy phase level in WASH-2 | Heavy phase flow [V-40] | 5 |

## References


[1] Biodiesel solutions. Malaysian Palm Oil Board, Washington, DC, http://www.palmoilworld.org/biodiesel.html (accessed March 2014).

[2] Lurgi, acquired by Air Liquide Global E&C Solutions in 2007, Paris, France, http://www.engineering-solutions.airliquide.com/en/welcome.html (accessed March 2014).

[3] Zhang Y, Dubé MA, McLean DD and Kates M 2003 Biodiesel production from waste cooking oil: 1. Process design and technological assessment Bioresource Technol **89(1)** 1–16.

[4] Moaveni B and Kariwala V 2012 Input-Output pairing selection for design of decentralized controller in *Plantwide control - recent developments and applications* Edited by G P Rangaiah and V Kariwala, Wiley.

[5] Xiong Q, Cai W J and He MJ 2005 A practical loop pairing criterion for multivariable processes *J of Process Control* **15** 741-747.

[6] Patle DS, Sharma S, Ahmad Z and Rangaiah GP 2014 Multi-objective optimization of two alkali catalyzed processes for biodiesel from waste cooking oil *Energy Convers and Manage* **85** 361-372.

[7] Aspen Plus: Aspen Plus Biodiesel Model (Examples); Aspen Technology: Cambridge, MA, 2012.

[8] Patle DS, Ahmad Z and Rangaiah GP 2014 Plantwide control of biodiesel production from waste cooking oil using integrated framework of simulation and heuristics *Ind Eng Chem Res:* **53** 14408-14418.

Journal of Physics: Conference Series **622** (2015) 012012 doi:10.1088/1742-6596/622/1/012012

# Modeling Of The Pores Form Influence On The Hydraulic Resistance Of Membranes And Their Permeability

**Zhanat Umarova[1], Saule Botayeva[2], Zlikha Makhanova[3], Anar Makhatova[4]**

[1] an associate professor of Information systems Department, M.Auezov South Kazakhstan State University, Tauke khan avenue, 5, Shymkent, Kazakhstan
[2] an associate professor of Information systems Department, M.Auezov South Kazakhstan State University, Tauke khan avenue, 5, Shymkent, Kazakhstan
[3] an associate professor of Information systems Department, M.Auezov South Kazakhstan State University, Tauke khan avenue, 5, Shymkent, Kazakhstan
[4] senior lecturer of Computer Science Department, M.Auezov South Kazakhstan State University, Baytursinov Street, Shymkent, Kazakhstan

Zhanat-u@mail.ru

**Abstract**. Until the present time, modeling of the pores form influence on the hydraulic resistance of membranes and their permeability has not been analyzed. The present study suggests the determination of the optimum form of pores and the thorough consideration of the issue on the productivity of polymer membranes with the profile pores.

## 1. Introduction

The widespread notion about the structure of porous adsorption layers are based on the possibility of describing pores as formation, to which definite geometric sizes and form can be ascribed [1]. In correspondence with such notions, the classification of porous systems is usually based on likening the real systems to the artificial ones, constructed on the models with the organized structure by several rules [2].

The model with various types of spheres packs and other bodies of the same size is the most wide-spread. On the basis of this model, the model of capillaries of various form and length is constructed as well as the model of pores as conventional canals between the contacting spheres [3].

## 2. The main part

The work [4] introduces the experiments, where glass capillaries have been chosen as the model objects, correlations between geometric characteristics of which (diameter, length of the cylindrical section, length of the conic section and so on) are close to such ones for pores in the track-etched membranes. The configuration with sharp contractions ('zero' thickness of the wall) has been received by the following way. The metallic diaphragm with round hole 1,0 mm was attached to the end of the cylindrical capillary. The diaphragm thickness comprised 200 mcm, that is it was essentially less than the hole diameter.

The volume speed of the flow of the semi synthetic motor oil at the pressure differential is P = 353 Pa. The corresponding calculation values show that the fault of measuring time and amount of flowing liquid introduced an error to the end result less than 1%. The coefficient of the dynamic viscosity of the oil was taken equal to 0,293 Pa·s (measured at $18{,}5^{o}$C). Under the given conditions, the liquid flow corresponded to Reynolds numbers of the order $10^{-3} - 10^{-2}$.

The wholly satisfactory fit of the measured and calculated values is being observed. The differences are partially stipulated by the errors of defining viscosity and pressure differential. Therefore, relations of flow speeds for various capillaries are more demonstrative. These relations, calculated separately from measured and predicted values, conform much better [7].

This is an advantage in the productivity, which has been achieved by the transition from the cylindrical form to the cone-cylindrical one. Let us note that in spite of the sufficiently gross simplications, the experiment with capillaries doesn't show essential divergence with the calculation data. These results allow to conclude that the above mentioned method is sufficiently adequate and can be employed for the optimization of pores form in the membranes.

Using the derivative relations, we will calculate specific productivity of the membranes, having various configurations of pores. Within the same configurations, we will vary the correlation between *R* and *r*, as well as between *L* and *l*. Our task will be the finding of such combination of membrane parameters, at which the volume speed of the flow for the given fixed *r* will be maximum.

We calculate the specific productivity of the membrane, related to the ordinary porosity. In this case, the calculations results can be presented in two-dimensional system of coordinates, exactly in the form of functions $Q_m/\varepsilon\ (R)$ at varied values of the parameters *l* and *r* and fixed *L*.

## 3. The experimental part

Let's consider the speed of the liquid flow through the membranes with pores of various configurations.

Let the pore be a twisting canal with the sharp contractions at the ends, which can be considered as the holes with zero thickness of the wall. The pressure fall in the internal site territory of the membrane pore by the width L can be found from Hagen-Poiseuille formula:

$$\Delta P_1 = \frac{8q\mu LU}{R^2} \tag{1}$$

where $\Delta P_1$ is pressure differential, μ – a coefficient of dynamic viscosity of the medium, $U$ - speed of the flow inside the pore, $R$ – pore radius, $q$ - a coefficient of pores tortuousness.

Let's define the tortuousness coefficient by the formula

$$q = \frac{l}{L} \tag{2}$$

$l$ - average length of the pores.

The pressure differential on the hole in the partition of zero thickness can be expressed in the following way:

$$\Delta P_2 = \frac{3\alpha\mu\pi R^2 U}{r^3} \tag{3}$$

Here $r$ – a hole radius at the output of the pore on the membrane surface, $\alpha$ - a coefficient, taking into account the asymmetric arrangement of the exit hole of the pore relating to its axis.

In the first approximation, we will accept that the pressure differential in all pores will be equal to the total of the pressure differentials in the cylindrical section and at the end contractions:

$$\Delta P = \Delta P_1 + \Delta P_2^i + \Delta P_2^e \tag{4}$$

Where $\Delta P_2^i$ - differential at the internal edge; $\Delta P_2^e$ - differential at the external edge.

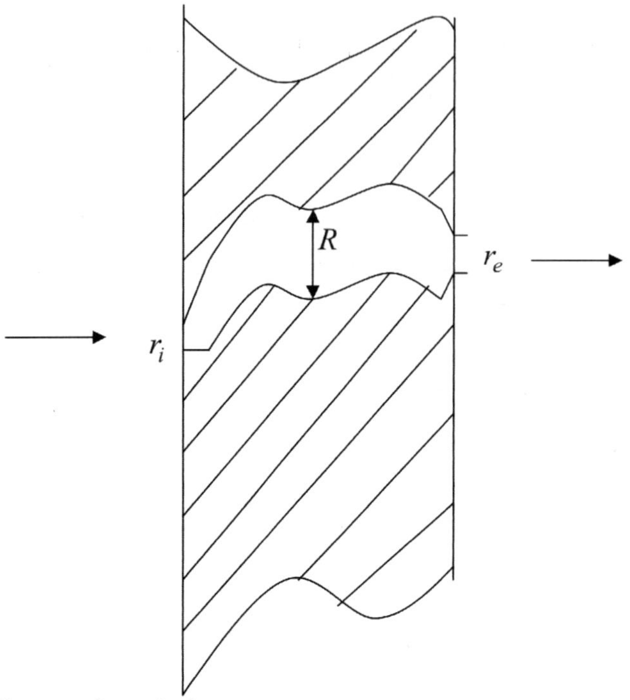


**Figure 1. Flow diagram through the pore.**

Let $N$ – a number at the area unit of the membrane surface.

Then the coefficient of the free cross-section of the membrane or surface porosity is defined by the formula:

$$p = \pi r^2 N \tag{5}$$

The connection between volume porosity $\varepsilon$ of the membrane with pores of the given configuration and surface porosity is calculated in the following way:

$$\varepsilon = pq \tag{6}$$

From here, we get the expressions for the differential at the membrane edge:

$$\Delta P_2 = \frac{3\alpha\mu\pi\varepsilon\ U}{pr} \tag{7}$$

$$\Delta P_2 = \frac{3\alpha\mu\pi q U}{r} \tag{8}$$

Then the interconnection between the flow, filtered through the liquid membrane $Q$ and its geometric characteristics will look like:

$$\Delta P = \frac{\mu q^2 Q}{\varepsilon F}\left[\frac{8L}{R^2} + 3\pi\left(\frac{\alpha_i}{r_i} + \frac{\alpha_e}{r_e}\right)\right] \tag{9}$$

where $F$ - membrane area.

2. Let the pore have conic contraction at the ends that is close to the real form of the pores, which can be received in the practice [6].

We assume that the cylindrical section by the length $L$ at the ends transits to the conic contraction by the length $l_k$, ended by the holes of the radius $r$.

The pressure fall $\Delta P_k$ in the conic site territory of the pore at the narrow angles of conicity $\gamma$ can be found from the following correlation [5]:

$$\Delta P_k = \frac{8q\mu QR^2 l_k}{3r^4 \varepsilon F}\left(z + z^2 + z^3\right) \tag{10}$$

where

$$z = 1 - \frac{\gamma l_r}{R} \tag{11}$$

Summing the pressure falls in the internal and conic site territories; we will receive an expression for the resulting pressure differential through the asymmetric membrane:

$$\Delta P = \frac{8q\mu Q}{\varepsilon F}\left[\frac{Lq}{R^2} + \frac{R^2}{3}\left(\frac{l_{ki}}{r_i^4}\left(z_i + z_i^2 + z_i^3\right) + \frac{l_{ke}}{r_e^4}\left(z_e + z_e^2 + z_e^3\right)\right)\right] \tag{12}$$

If the pore has an asymmetric configuration, there is the contraction only on one side of the membrane and we will receive:

$$\Delta P = \frac{8q\mu Q}{\varepsilon F}\left[\frac{Lq}{R^2} + \frac{R^2}{3}\left(\frac{l_{ki}}{r_i^4}\left(z_i + z_i^2 + z_i^3\right)\right) + \frac{3\pi q \alpha_e}{r_e}\right] \tag{13}$$

## 4. Results

Analysis of the results leads to the following conclusions:

1. Even at zero thickness of the selective layer, the membrane productivity growth is constrained by the factor 4-5.

2. With the growth of the internal radius of pores related to the ordinary porosity, the specific productivity of membranes passes through maximum. The state of the maximum depends on the thickness of the selective layer.

3. At the thickness of the selective layer 3 mcm, the advantage in the productivity is insignificant. At the thickness of the selective layer 1 mcm, the growth of the productivity is essential, whereby the optimum correlation between internal radius of pores and pores radius in the selective layer comprises 3- 3,5.

## 5. Conclusion

In our calculations, we assume that the volume porosity should be constrained by the size 0,15. The choice of this boundary is somewhat conventional; it is mainly imposed by the requirement to the mechanic strength of the membrane. As it is known from the experiment, track-etched membranes based on the films are sufficiently durable at the porosity not more than 0,15-0,25. More thorough procedure of optimization should take into consideration that the strength depends not only on the volume porosity but also on film thickness as well as on the pores density [8]. At equal values of the volume porosity, the membranes with higher density of pores have slightly worst mechanic characteristics than the membranes with less density of pores [9]. Therefore the final selection of optimum parameters of membranes can be realized only experimentally.

The carried out theoretical consideration allows calculating the specific productivity of membranes with thin selective layers. In spite of the approximate character of model calculations, they are essential for the optimization of the membranes structure and the further processing of the technology of their production. The following step in our research will be the production of membranes with pores of various profiles and comparison of their properties with the calculated ones.

**References**

[1] M. Nakagaki 1991 *Physical Chemistry membranes* (Moscow) p.256.
[2] N.I. Nikolaev 1980 *Diffusion in membranes* (Moscow) pp.134-167
[3] A.V. Smirnov, A.S. Bartov 2005 *Probabilistic approach in assessing the selectivity of the membrane to separate ions* Membranes (№ 2 (26)) pp. 23-30
[4] P.Yu.Apel and S.N.Dmitriyev 2004 *Optimization of track membranes formes* Critical technologies, Membranes, №3 (23), pp. 32-37
[5] Zh.Umarova, Sh.Eskendirov 2012 *Mechanism of Gases Transfer through Polymer Membranes and Membrane Module Calculation Model to Separate Gases* Procedia Technology Volume 1 pp. 356–361
[6] Zh.Umarova 2011 *Algorithm for Solving the Model of Multistage Hollow Fiber Membrane Systems and the Identification of Model Parameters* Informatics Engineering and Information Science, Volume 251 pp 466-476
[7] Zh.Umarova, S.Botayeva, A.Yegenova, A.Usenova 2014 *Mathematical Model and Calculation Algorithm of Micro- and Meso Levels of Separation Process of Gaseous Mixtures in Molecular Sieves* International Conference on Mathematics, Engineering & Industrial Applications (ICoMEIA 2014) proceedings book (Malaysia) p.74
[8] A.V. Smirnov, A. Bartov 2005 *Probabilistic approach in assessing the selectivity of the membrane to separate ions* Membranes № 2 (26) pp. 23-30
[9] E.R.Volkova, E.A.Denysiuk 2003 *New method for estimating the diffusion coefficient of the penetrant in polymeric materials* Aerospace and high technology (Perm) p.45

ScieTech 2015 IOP Publishing
Journal of Physics: Conference Series **622** (2015) 012013 doi:10.1088/1742-6596/622/1/012013

# Automatic Structures – Recent Results and Open Questions

**Frank Stephan**

Department of Mathematics and Department of Computer Science, National University of Singapore, 10 Lower Kent Ridge Road, Block S17, Singapore 119076, Republic of Singapore

E-mail: fstephan@comp.nus.edu.sg

**Abstract.** Regular languages are languages recognised by finite automata; automatic structures are a generalisation of regular languages where one also uses automatic relations (which are relations recognised by synchronous finite automata) and automatic functions (which are functions whose graph is an automatic relation). Functions and relations first-order definable from other automatic functions and relations are again automatic. Automatic functions coincide with the functions computed by position-faithful one-tape Turing machines in linear time. This survey addresses recent results and open questions on topics related to automatic structures: How difficult is the isomorphism problem for various types of automatic structures? Which groups are automatic? When are automatic groups Abelian or orderable? How can one overcome some of the limitations to represent rings and fields by weakening the automaticity requirements of a structure?

## 1. Introduction

Automata theory [13] deals to a large extent with finite automata. These devices process an input symbol by symbol and have a finite memory which they update in reach round of processing an input symbol; the memory also indicates whether the word – as seen so far – satisfies the condition to be checked. For example, the well-known algorithm to check whether a number is a multiple of 3 consists of adding up each digit one after the other and in between only to store the remainder by 3; thus the memory consists only of taking one of the choices 0, 1 and 2. If after the processing, the result is 0 then 3 divides the number just processed. So one can write the algorithm as a table or diagramme of its states and the memory transitions on each input symbol.

| state | type | succ at 0,3,6,9 | succ at 1,4,7 | succ at 2,5,8 |
|---|---|---|---|---|
| 0 | start, accepting | 0 | 1 | 2 |
| 1 | rejecting | 1 | 2 | 0 |
| 2 | rejecting | 2 | 0 | 1 |

For example, when processing input 22876, the automaton is initially in the state 0, after reading the first 2 in state 2, after reading the next 2 in state 1, after reading the 8 in state 0, after reading the 7 in state 1 and after reading the 6 in state 1; thus the number is not a multiple of 3 and the input is "rejected". Such mechanisms allow to check various properties of integers $x$ (given as sequences of their decimal digits): does $x$ have remainder $a$ when divided by $b$ (for constants $a, b$); does $x$ have an even number of decimal digits; does $x$ have exactly five decimal digits with value 3; is $x$ the sum of two powers of 10? One can also make easy syntactic checks and also process other information than numbers.

This concept of a finite automaton turned out to be very fruitful in the theory of computation and is also used in many applications: mainly due to the fact that it is though primitive still quite expressive and that automata can be compared and manupulated easily by computer programs without running into decidability-problems.

Important mathematical concepts are often independently discovered by various researchers at various places in the world. For example the concept of a 0 is due to Indian mathematicians as well as to mathematicians of the ancient Maya population in Mezoamerica. Similarly, automatic structures which are applying automata theory to algebraic structures were formalised and developed independently by three groups of researchers: The first work is due to Hodgson [11, 12]; the second team of researchers where Khoussainov and Nerode [16] and the third team of discoverers are Blumensath and Grädel [4, 5]. These teams all spotted the natural idea to carry over the ideas of automata theory from sets to relations and functions; all teams also had in common the idea of going a more restrictive path compared to the alternative path of finite transducers and rational relations and functions [3]; one of the reasons for their choice is to maintain the decidability of the first-order theory of the so defined structures what is not the case for finite transducers and their structures. Khoussainov and Minnes [15] as well as Rubin [22] provide surveys about automatic structures.

There are several equivalent though similar approaches to formalise automatic relations: One way is to consider a finite automaton reading in parallel several input-words, each word by one symbol in each step, where exhauseted input-words provide a special symbol # from outside the alphabet; the automaton accepts iff it is in an accepting state after having processed all the tuple of inputs not entirely consisting of #. The second approach makes this a bit more implicit: One first defines the convolution of several strings and then says that a relation $R$ of several, say three, inputs is automatic iff the set $\{\text{conv}(x, y, z) : (x, y, z) \in R\}$ is regular. Here the convolution of strings combines the symbols of $x, y, z$ at the same position into a new character from the convoluted alphabet where, again, exhausted strings are coded using a special symbol like #. For example, $\text{conv}(0010, 0123456, \varepsilon) = \langle 0, 0, \# \rangle, \langle 0, 1, \# \rangle, \langle 1, 2, \# \rangle, \langle 0, 3, \# \rangle, \langle \#, 4, \# \rangle, \langle \#, 5, \# \rangle, \langle \#, 6, \# \rangle$.

For automatic functions one can either define that their graph is an automatic function or one can say that there is a non-deterministic automaton which reads from each input one symbol in each round and produces for each output one symbol in each round and which goes into an accepting state after all inputs and outputs have become a #; such a function must be unique in the sense that no tuple of inputs has two different tuples of output; due to the pumping lemma it therefore holds that there is a constant $c$ such that no output can be longer than $n + c$ if the longest member of the input tuple has length $n$. Here non-determinism is essential, as certain automatic functions cannot be computed by a deterministic finite automaton. An example of such a function is the function which exchanges the first and last symbol: $a_1a_2a_3 \ldots a_{n-1}a_n \mapsto a_na_2a_3 \ldots a_{n-1}a_1$: when processing $a_1$, the automaton guesses and memorises $a_n$ and outputs it; it then reads and copies $a_2, \ldots, a_{n-1}$ until — when receiving the input $a_n$ — it non-deterministically outputs $a_1$ and compares checks that the swap of $a_1$ and $a_n$ was executed correctly. If so the dfa goes into an accepting else into a rejecting state; if symbols different from # follow after this swap then the automaton goes into a rejecting state forever and invalidates the complete computation. Brough, Khoussainov and Nelson [7] studied the more restrictive model of automatic functions computed by a dfa in this way and the corresponding structures.

In general, a structure $(A, f_1, \ldots, f_n, r_1, \ldots, r_m)$ of $n$ functions and $m$ relations is *automatic* iff $A$ is a regular set and $f_1, \ldots, f_n$ are automatic functions and $r_1, \ldots, r_m$ are automatic relations. Here the domains of these functions and relations are either $A$ or, for some $k$, the $k$-fold Cartesian product $\{\text{conv}(x_1, x_2, \dots, x_k) : x_1, x_2, \dots, x_k \in A\}$. Furthermore, to simplify notation, one also calls structures which are isomorphic to automatic structures just "automatic".

Normally one assumes that the inputs in automatic relations of structures have a joint begin and end depending on the string-length; one can also code structures such that they have an aligned middle (say the position whether the dollars are separted from the cent) and then the numbers start somewhere (after perhaps some #) and end somewhere (with trailing #). This second presentation is in some cases useful and can be mapped, by coding, back to the first one. One can show that, for example, the structure of positive finite fractional decimal numbers (some digits before and some digits after the decimal dot, but not an infinite sequence of digits) with addition and $\leq$ form an automatic structure. For example, given three numbers $x, y, z$, an automaton to check whether $x + y = z$ would process from the back to the front and the states would be "correct and carry to next digit (c)", "correct and no carry to next digit (n)" and "incorrect (i)":

```
Correct Addition      Incorrect Addition    Missing Leading Digit
 # 2 3 5 8 . 2 2 5     3 3 3 3 . 3 3 #       1 2 3 . 4 5 6
 # 9 1 1 2 . # # #     # # 2 2 . 2 2 2       9 8 7 . 6 5 4
 1 1 4 7 0 . 2 2 5     # 1 5 5 . 5 5 2       1 1 1 . 1 1 #
n c n n c n n n n n   i i n n n n n n n     c c c c c c c n
```

In these three examples, $x$ stands in the top, $y$ in the second and $z$ in the last row. The states of the automaton are from the back to the front, they have processed the digits behind them but not those before them. The filling symbol # is identified with 0. As long as all dots are aligned, they are just ignored by the automaton and do not cause a state-change. The state "n" is the starting and accepting state, the state "c" can go to an accepting state if the next column is correctly aligned. Once a state "i" is reached, the automaton will reject the computation as incorrect, independently of what will come next. One can also easily make an automaton which goes from the front to the back and compares two numbers and which ends in an accepting state whenever the first number is larger than the second.

This concept can easily be generalised to deal with other digit systems (like binary) or to deal with positive and negative numbers (including 0). Similarly the integers with addition and $\leq$ form an automatic structure. One can also combine both structures by having an extra predicate which tells whether a number is an integer, that is, has no nonzero digits after the decimal dot. These type of structures are, roughly, the most complicated subgroups of the rationals which are automatic; in particular Tsankov [24] showed that the structures of the rationals with addition itself is not automatic. Other non-automatic structures are the ring of integers $(\mathbb{Z}, +, \cdot)$, any infinite field and the free group generated by two or more generators. Finite structures as well as their eventually constant sequences (which component-wise operations on these sequences) are automatic. There are infinite automatic rings: Taking a finite ring $(R, +, \cdot)$, one now considers $(R_\omega, +, \cdot)$ to be the set of all eventually constant sequences $a_0a_1a_2\dots$ with component-wise operations: $(a + b)_k = a_k + b_k$, $(a \cdot b)_k = a_k \cdot b_k$. Such a sequence $a_0a_1a_2\dots$ can be represented by the shortest prefix $a_0a_1\dots a_n$ such that $a_m = a_n$ for all $m \geq n$ in order to get a finite representation. In the following, a few topics on automatic structures will be highlighted more: The characterisation when a function is automatic, orderings, groups, ordered groups, semiautomatic structures.

## 2. Characterising Automatic Functions

Turing [25] developped a machine model which – now named after him – became a basic model for computations in complexity theory, as it permitted to formalise complexity classes in an easy and clear way. A one-tape Turing machine is a Turing machine which has finitely many states (like a finite automaton) and which has a read-write-head on a tape of symbols where, in each step, depending on the symbol and the state, it replaces the current symbol by a new one, it updates to a new state and it moves the tape by $-1$, 0 or $+1$ positions. A position-faithful Turing machine has on the tape a start-position with a special symbol which it never crosses. The input is written initially starting after this start-position and the computation replaces it by the output starting also after this start-position. After the end of the input / output are on the tape only blanc symbols which are not used for any other purpose. The run-time of the Turing machine is the number of steps it uses to convert the input into the output; "linear time" means that there is a constant $c$ such that the machine runs on input of length $n$ for at most $(n+1)\cdot c$ steps and has then replaced the input by the output and reached a halting state after which it does not carry out any further activity. Note that a Turing machine can use an alphabet much larger than just the alphabet used for input and output; these additional symbols might be used to store some data on the tape. This is essential for the theorem below. Furthermore, a Turing machine is deterministic iff it in each situation can do only one activity and there is no choice between several alternatives how to proceed. A non-deterministic Turing machine, however, can choose between several alternatives and those can lead to various runs; as different runs might result in different output, the machine might do at the end some checks. In the case that the checks go through, the machine goes into a special "accepting halting state" and the computation is valid; in the case that the checks do not go through, the machine goes into a special "rejecting halting state" and the computation is invalid. For the same input, all valid computations have to give the same output; furthermore, in the case of a time-bound, all computations, whether valid or invalid, have to keep this time-bound, that is, must terminate within $(n+1)\cdot c$ steps for a linear-time Turing machine and input of length $n$. The next theorem gives a Turing machine characterisation of automatic functions – here with one input mapped to one output; if several inputs or outputs are involved, one has to reduce them to one by using convolutions.

**Theorem 2.1** (Case, Jain, Seah and Stephan [8]). *For a function $f$ from strings to strings, the following conditions are equivalent:*

- *$f$ is automatic;*
- *$f$ is computed in deterministic linear time by a position-faithful one-tape Turing machine;*
- *$f$ is computed in non-deterministic linear time by a position-faithful one-tape Turing machine.*

This result works for deterministic and non-deterministic linear time. It would be interesting to know whether it also works for position-faithful machines using alternating linear time; here the alternating machine first would write non-deterministically the output onto the tape which consists of at each time of a convolution of three words $(x, y, z)$ where $x$ is the input, $y$ is the guessed output and $z$ is some work-content; in the case that the machine accepts the pair $(x, y)$ according to the convention of alternating Turing machines then $y$ is a valid output. Furthermore the machine is required to keep the time-limit on all computation pathes involved, even on those for wrongly guessed output.

Note that in Theorem 2.1, the notion of position-faithfulness is needed. To see this consider the function which erases all leading 0 of binary strings, that is, which maps $x1y$ to $1y$ and $x$ to $\varepsilon$ for $x \in 0^*$ and $y \in \{0,1\}^*$. This function can be computed by a one-tape Turing machine which is not position-faithful; however, the function is not automatic as the corresponding position-faithful Turing machine would have to move a large amount of information by a non-constant amount of cells. Similarly, two-tape Turing machines can in linear-time do things which automatic functions cannot do, for example to realise the above movement of cells, thus also those are more powerful than automatic functions.

Journal of Physics: Conference Series **622** (2015) 012013 doi:10.1088/1742-6596/622/1/012013

## 3. Decidability Questions

One important property of automatic structures is that one can use first-order logic to define new relations and functions from given automatic relations and functions; this construction is effective and hence one can even use the well-known automata-theoretic tools in order to decide whether a subset of the domain of an automatic structure defined such a way is empty, finite or coincides with the domain. This permits to decide the first-order theory of a given automatic structure [16]. An application is to get an alternative proof of the decidability of Presburger arithmetics (first-order formulas on the integers involving addition and order) and on the other hand, one also knows that structures with undecidable first-order theory are not automatic. Another important question is the question whether one can effectively decide whether two automatic structures are isomorphic. This is in general impossible, but for some special cases, it is possible: One can decide whether two automatic Boolean algebras are isomorphic and one can decide whether two automatic ordinals (viewed as regular sets of lower ordinals with the corresponding automatic order-relation) are isomorphic [22]. This stands in contrast with the following results.

**Theorem 3.1** (Kuske, Liu and Lohrey [17]). *It is impossible to decide whether two given automatic structures are isomorphic for the following types of automatic structures:*

- *It is* $\Pi^0_1$*-complete to decide whether two automatic equivalence relations are isomorphic;*
- *It is* $\Sigma^1_1$*-complete to decide whether two automatic linear orders are isomorphic.*

These and similar results show that only few families of structures have a decidable isomorphism problem. Barany asked whether the following family of structures has a decidable isomorphism problem, where succ is the successor function on the natural numbers and $A, B \subseteq \mathbb{N}$ (in the corresponding representation).

**Question 3.2** (Barany [6]). *Given two structures* $(\mathbb{N}, A, succ)$ *and* $(\mathbb{N}, B, succ)$*, can one decide whether the structures are isomorphic, that is, whether* $A = B$*?*

A further natural question is on how difficult it is to know whether a given structure (described by computer programs computing the corresponding operations) has an automatic copy which is isomorphic to it.

**Question 3.3** (Kuske). *Given a recursive structure by Turing machines or computer programs deciding the domain and relations and graphs of functions involved, is it a* $\Sigma^1_1$*-complete problem to decide whether the so-described structure is isomorphic to any automatic structure?*

As the corresponding isomorphism problem is $\Sigma^1_1$-complete, this problem might also be a candidate for this complexity level.

## 4. Groups

Groups have been studied under various constraints of automaticity. One of the first frameworks to study groups within automatic structures is the approach by Epstein et al. [10] who taylormade their definition in order to capture a large class of finitely generated group while maintaining a natural coding of the representatives of the group elements. Their approach is orthogonal to the automaticity notion of Khoussainov and Nerode [16], though automaticity by the latter on finitely generated groups implies automaticity by the first, however the concept of automatic groups by Khoussainov and Nerode works also for groups which are not finitely generated. In the present paper, "automatic" always means "automatic as defined by Khoussainov and Nerode"; the reader should be aware that also Epstein et al. [10] call their groups just "automatic".

Recall that a group is a set with a binary operation such that this operation is associative, that there is a neutral element and that every member of the group has an inverse; groups are

Journal of Physics: Conference Series **622** (2015) 012013 doi:10.1088/1742-6596/622/1/012013

quite common in mathematics with the integers $(\mathbb{Z}, +)$, rationals $(\mathbb{Q}, +)$ and reals $(\mathbb{R}, +)$ being examples for infinite groups. Such groups where $x + y = y + x$ for all $x, y$ are called Abelian or commutative. There are also a multitude of finite groups. When these groups are small, they can be given by a table as in these examples (addition modulo 4, multiplication modulo 5 of non-zero numbers, double-bit exclusive or):

| + | 0 | 1 | 2 | 3 |
|---|---|---|---|---|
| 0 | 0 | 1 | 2 | 3 |
| 1 | 1 | 2 | 3 | 0 |
| 2 | 2 | 3 | 0 | 1 |
| 3 | 3 | 0 | 1 | 2 |

| · | 1 | 2 | 3 | 4 |
|---|---|---|---|---|
| 1 | 1 | 2 | 3 | 4 |
| 2 | 2 | 4 | 1 | 3 |
| 3 | 3 | 1 | 4 | 2 |
| 4 | 4 | 3 | 2 | 1 |

| ⊕ | 00 | 01 | 10 | 11 |
|---|---|---|---|---|
| 00 | 00 | 01 | 10 | 11 |
| 01 | 01 | 00 | 11 | 10 |
| 10 | 10 | 11 | 00 | 01 |
| 11 | 11 | 10 | 01 | 00 |

Groups do not need to be Abelian. For example, the group of permutations of members of a set $X$ with at least 3 elements has the group operation $\circ$ with $(f \circ g)(x) = g(f(x))$ for all $x \in X$. This group is not Abelian; in the case that $X$ is finite, also the group of permutations of $X$ is finite. All finite groups are automatic.

A group is called Abelian-by-finite iff it has a subgroup of finite index which is Abelian. Finitely generated automatic groups are closely related to those which are Abelian-by-finite and to a certain degree, one can also say something about groups which itself are not finitely generated.

**Theorem 4.1** (Oliver and Thomas [21]). *A finitely generated group $(G, \circ)$ is automatic iff it is Abelian-by-finite.*

**Theorem 4.2** (Nies and Thomas [20]). *If $(G, \circ)$ is an automatic group then each finitely generated subgroup is Abelian-by-finite.*

These results can, for example, be used to show that Thompson's group $F$ is not automatic (in the sense of Khoussainov and Nerode); a famous open problem [1] is whether $F$ is automatic in the sense of Epstein et al. [10].

**Example 4.3.** Thompson's group $F$ is given by generators $x_0, x_1, \ldots$ satisfying $x_n x_m = x_m x_{n+1}$ whenever $m < n$. Hence $x_1(x_0)^n = (x_0)^n x_{n+1}$ and $x_{n+1} = (x_0)^{-n} x_1 (x_0)^n$, thus Thompson's group $F$ is finitely generated. Thompson's group $F$ can indeed be described by only finitely many equations:

$$
\begin{aligned}
(x_0 x_1^{-1}) \circ ((x_0)^{-1} x_1 x_0) &= ((x_0)^{-1} x_1 x_0) \circ (x_0 x_1^{-1}); \\
(x_0 x_1^{-1}) \circ ((x_0)^{-2} x_1 (x_0)^2) &= ((x_0)^{-2} x_1 (x_0)^2) \circ (x_0 x_1^{-1}).
\end{aligned}
$$

So it is a finitely presented group. Now Thompson's group $F$ is not Abelian-by-finite: Assume by way of contratdiction that $F$ has an Abelian subgroup $K$ of finite index $k$. Then for every $y \in F$, two powers $y^i, y^j$ with $1 \le i < j \le k+1$ go into the same set of the form $zK$ where $z \in F$ – index $k$ means that there are only $k$ of these sets. Then $y^i = zv$ and $y^j = zw$ for $v, w \in K$ and $y^{-i+j} = v^{-1} \circ z^{-1} \circ z \circ w = v^{-1} \circ w \in K$. So $y^{k!} \in K$ for every $y \in F$. In particular, $(x_1)^{k!}$ and $(x_0)^{k!}$ are in $K$ and

$$
(x_{k!+1})^{k!} = ((x_0)^{-k!} x_1 (x_0)^{k!})^{k!} = (x_0)^{-k!} \circ (x_1)^{k!} \circ (x_0)^{k!} = (x_1)^{k!}
$$

what contradicts known facts about this group: The $k!$-th powers of different $x_i$ and $x_j$ are different. Hence Thompson's group $F$ is not automatic.

Two important questions related to automatic groups are still open.

**Question 4.4** (Nies [19]). Is there an infinite and simple automatic group?

**Question 4.5** (Nies [19]). Is every torsion-free automatic group Abelian-by-finite?

Here a simple group is a group $(G, \circ)$ which does not have a proper subgroup $H$ of at least two elements such that $x \circ H = H \circ x$ for all $x \in G$. Infinite Abelian groups are never simple.

A group $(G, \circ)$ with neutral element $e$ is torsion-free when $x \circ x$, $x \circ x \circ x$, ... are all different from $e$ for every $x \in G - \{e\}$. Note that in Nies' question the restriction to "torsion-free" is needed; for example if one takes a finite group which is not Abelian (like the permutation group of a set with 3 elements), then one can just consider all functions $f$ from $\mathbb{N}$ into this finite group which are eventually constant with a component-wise operation for each member; this group is also automatic, but not Abelian-by-finite.

The above questions mainly asked on how to describe the automatic groups; an important and well-known question is also on when two automatic groups are isomorphic. In particular, the question whether one can decide the isomorphy of Abelian automatic groups.

**Question 4.6.** *Is there an algorithm which decides whether two given automatic Abelian groups are isomorphic?*

Given a group $(G, \circ)$, one can fix a subset $F \subseteq G$ of generators of $G$ and consider the edge set $E = \{(x, y) \in G : \exists z \in F\, [x \circ z = y]\}$. Such a graph is connected. Furthermore, the graph is *transitive*, that is, for every $x, y \in G$ there is a graph-isomorphism which maps $(G, E)$ to itself and maps $x$ to $y$. Berdinsky and Khoussainov [2] investigated connected and transitive automatic graphs and showed that some of them are not derived from a group in the above mentioned way.

## 5. Groups and Order

An important topic is to combine groups with orders. There are two notions: the ordered automatic group which satisfy both implications $x < y \Rightarrow x \circ z < y \circ z$ and $x < y \Rightarrow z \circ x < z \circ y$ for all $x, y, z$ and the right-ordered / left-ordered groups which satisfy only the respective condition.

**Theorem 5.1** (Jain, Khoussainov, Stephan, Teng and Zou [14]). *Every automatic ordered group is Abelian, even if only the group operation and not the ordering is automatic. However, there is a non-Abelian left-ordered group.*

This non-Abelian automatic left-ordered group is the Klein bottle group with two generators $a, b$ where $a \circ b = b^{-1} \circ a$ and $a^i b^j < a^{i'} b^{j'}$ iff $i < i'$ or ($i = i'$ and $j < j'$). It should be noted that there are many ways to introduce an order on $(\mathbb{Z}^2, +)$ and one would be to choose real numbers $a, b$ and to look at the ordering which $a \cdot \mathbb{Z} + b \cdot \mathbb{Z}$ inherits from the real numbers. The next result shows that there are non-trivial choices where the resulting ordered group is automatic.

**Theorem 5.2** (Jain, Khoussainov, Stephan, Teng and Zou [14]). *For every square-root $\sqrt{n}$ of a non-square natural number $n$, the ordered group $(\mathbb{Z} + \sqrt{n} \cdot \mathbb{Z}, +, <)$ with the ordering $<$ inherited from the real numbers is an automatic ordered group where both, $+$ and $<$, are automatic.*

The following arguments give an idea on how to prove the automaticity of $(\mathbb{Z} + \sqrt{3} \cdot \mathbb{Z}, +, <)$; this case is easier than many others. Choose a real number $u = 2 + \sqrt{3}$, this number is a member of the given group and satisfies $u^2 = 4u - 1$. Now one represents every member of the group by a polynomial of the form

$$a_n u^n + a_{n-1} u^{n-1} + \ldots + a_1 u + a_0 + a_{-1} u^{-1} + \ldots + a_{-m} u^{-m}$$

where $a_k \in \{-3, -2, -1, 0, 1, 2, 3\}$ for all $k$. These polynomials are always members of the group: By $u^{k+1} = 4u^k - u^{k-1}$, it follows from $u^{k-1}, u^k$ being in the group that also $u^{k+1}$ is in the group and from $u^k, u^{k+1}$ being in the group that also $u^{k-1}$ is in the group. As $u^0 = 1$ and $u$ are in the group, one can then conclude using induction that also all larger and all smaller powers of $u$ are also in the group. Furthermore, due to replacing each $4u^k$ by $u^{k+1} + u^{k-1}$ and $-4u^k$ by $-u^{k+1} - u^{k-1}$ until no coefficient has an absolute value above 3, one can reduce every polynomial into this form without changing the value it represents; the process terminates as every replacement reduces the sum of the absolute values of the coefficients.

Note, that one can check with a finite automaton whether such a polynomial is positive, negative or equal to 0: Starting in the front, one has two variables $v, w$ initialised with 0. When processing $a_k$, $vu+w$ represents $a_n u^{n-(k+1)} + \ldots + a_{k+1}$ and $v, w$ are updated to values $v', w'$ by $v' = 4v+w$ and $w' = -v+a_k$ so that $v'u+w' = (vu+w)u+a_k = a_n u^{n-k} + \ldots a_{k+1}u + a_k$. These updates are aborted when it is secure that the number is properly positive or properly negative; this is the case when $v'u+w' > 3 \cdot (1+u^{-1}+u^{-2}+\ldots)$ or $v'u+w' < -3 \cdot (1+u^{-1}+u^{-2}+\ldots)$ (while $vu+w$ was in the interval). Indeed, one can chose a constant $c$ such that whenever one of the new values $v', w'$ is outside the interval $\{-c, -c+1, \ldots, c-1, c\}$ for the first time then $vu+w$ passes the threshold to indicate that the number is positive or negative. Note that when $w' > c$ then $w' \leq c+3$ and $v < 3-c$ and $v' < 4v+w \leq 9-3c$ so that $vu+w < 6-2c$ due to $u > 1$. Hence one can argue that whenever one of $v', w'$ is out of the range while $v, w$ were in the range then $v'$ is out of the range itself and $w'$ is at most 3 away from the range, thus $v'u+w'$ is, for $c$ large enough, so large that the sign of $v'$ determines the outcome of the test. In the case that the threshold is not passed through the whole processing of the polynomial, then there are only finitely many choices left and a table tells the automaton which of these choices correspond to a positive number, negative number or to 0.

This key idea can also be adjusted to check whether a sum of two or three polynomials is 0 and so one can compare polynomials or check whether one polynomial is the sum of two others. Furthermore, using the length-lexicographic ordering permits to identify among the various representatives for any number in the group a least representative which can be taken into the domain of the structure.

More precisely, if one has an automatic structure $(A, +, <)$ with $\equiv$ being an automatic equivalence relation representing equality, then one can make the set $B = \{x \in A : \forall y \in A\ [y \equiv x \Rightarrow x \leq_{length-lexicographic} y]\}$ and $(B, +, <)$ is then an automatic structure with exactly one representative for each class of equivalent elements in $A$. This completes the proof-sketch that the ordered group $(\mathbb{Z} + \sqrt{3}\mathbb{Z}, +, <)$ is automatic.

For the natural numbers, one can define the order from the group operation: $x < y \Leftrightarrow \exists z\,[x+1+z = y]$. Therefore whenever $+$ is automatic, so is $<$. For the integers, this is no longer true, leaving the following open question.

**Question 5.3** (Jain, Khoussainov, Stephan, Teng and Zou [14]). *Is there a automatic copy of $(\mathbb{Z}, +)$ such that the ordering $<$ in this copy is not automatic?*

Although this main question is open, there is a related result obtainable by storing fractions between 0 and 1 with denominators being a power of two and a power of three, respectively, in different components of convoluted numbers where the first one uses binary and the second one ternary digits. While an automatic addition can survive this split representation, no automaton for the order can deal with it. This idea permits to prove the following result.

**Theorem 5.4** (Jain, Khoussainov, Stephan, Teng and Zou [14]). *There is an automatic copy of $(\{x \cdot 2^y \cdot 3^z : x, y, z \in \mathbb{Z}\}, +)$ in which the addition is automatic but not the order.*

## 6. Semiautomatic Structures

Automatic structures are limited in the way that they can neither be used to represent infinite fields nor some important groups like $(\mathbb{Q}, +, <)$. Therefore, efforts have been undertaken to overcome this problem by making the structures more general. The key idea is to relax the automaticity requirement on some of the functions and relations involved. Here one says that a function $f$ is semiautomatic iff every induced function obtained by fixing all but one inputs is automatic; similarly for relations. One might also consider to allow multiple representatives for elements of the structure and only require that each equivalence class in the representation is regular, but no longer that the equivalence relation induced by the equality of structure elements is automatic. As one might want to relax this only for some and not all notions, at semiautomatic structures one separates by a semicolon the automatic relations and functions from those which are only semiautomatic. An example is the ring $(\mathbb{Z}, +, <, =; \cdot)$ where the multiplication $\cdot$ is only semiautomatic while the other operations and relations are automatic; as the equality is only semiautomatic in some semiautomatic structures, it is here indicated that here the equality is automatic as well. This structure can be made using any of the usual representations of the additive groups of integer with order, as the multiplication with a constant $c$ can then be realised by $c$ times adding the input to 0. Tsankov [24] provides a result which can be formulated as follows in the language of semiautomatic structures; subrings contain 0, 1 and are closed under addition and multiplication.

**Theorem 6.1** (Tsankov [24]). *$(A, +, <, =; \cdot)$ is a semiautomatic subring of the rationals iff $A = \{x \cdot p^y : x, y \in \mathbb{Z}\}$ for some $p \in \{1, 2, 3, \ldots\}$; the case $p = 1$ captures the ring of integers.*

If one abstains from requiring + to be automatic but only requires it to be semiautomatic, then one gets quite a multitude of examples of semiautomatic groups.

**Theorem 6.2** (Jain, Khoussainov, Stephan, Teng and Zou [14]). *The groups $(\mathbb{Q}, =; \cdot)$ and $(\mathbb{Q}, <, =; +)$ are semiautomatic.*

Furthermore, when allowing that all relations and functions are only semiautomatic, one can get ordered semiautomatic fields.

**Theorem 6.3** (Jain, Khoussainov, Stephan, Teng and Zou [14]). *The field $(\mathbb{Q}+\sqrt{n}\mathbb{Q}; +, \cdot, <, =)$ is semiautomatic for every non-square integer $n$.*

However, not every countable structure is semiautomatic. For example, if one looks at vector spaces of dimension 16 over the vector space of the rationals and postulates that all linear functions from the vectorspace to itself are automatic, then this structure is automatic, however, it becomes non-automatic, when the following additional operations are required to be automatic: inverting one component (mapping $x$ to $1/x$ for the case that $x \neq 0$ and mapping 0 to 0); testing whether the coordinates are actually from $\mathbb{Z}^{16}$. Note that if one looks just at the field of rationals itself, then $(\mathbb{Q}; +, \cdot, /, <, =)$ and $(\mathbb{Q}, \mathbb{Z}; +, \cdot, <, =)$ are both semiautomatic. On one hand, the ways to obtain the two semiautomatic structures use different representations, they cannot be combined. On the other hand, the result on the non-semiautomaticity of $\mathbb{Q}^{16}$ with the above mentioned operations was proven by coding Matiyasevich' theorem [18] that it is undecidable to check whether a polynomial in several integer variables takes the value 0. This coding cannot be done when using only one in place of 16 dimensions and therefore the question whether the corresponding structure is semiautomatic is left open, see item (d) below. The other items also stem from limitations of the known constructions where there is also no known proof that these limitations cannot be overcome.

**Question 6.4** (Jain, Khoussainov, Stephan, Teng and Zou [14]).

**(a)** *Are the structures $(\mathbb{Q}, <, =; +, \cdot)$ or $(\mathbb{Q}, =; +, \cdot)$ semiautomatic?*

**(b)** *Is the polynomial ring $(\mathbb{Q}[x]; +, \cdot, =)$ semiautomatic?*

**(c)** *Is there a transcendental field extension of the rationals which is semiautomatic?*

**(d)** *Is $(\mathbb{Q}, \mathbb{Z}; +, -, \cdot, /, <, =)$ semiautomatic?*

## 7. Conclusion

The present survey gives an overview over various recent results from the field of automatic structures and also points to open questions which might be interesting for further study.

These questions include decidability questions of isomorphism problems or, more general, their complexity in the arithmetical or analytical hierarchy; the question to which extent automatic groups are Abelian-by-finite and how automaticity of groups relates to the order of these – a line of research initiated by Nies, Oliver and Thomas in three papers [19, 20, 21] – and on the general limits of automata-theory when dealing with structures.

**Acknowledgments**
The author would like to thank Bakhadyr Khoussainov, Dietrich Kuske, André Nies, Sasha Rubin, Pavel Semukhin and Yue Yang for correspondence; he thanks the NUS and MOE for support through grants R146-000-181-112 and R146-000-184-112.

## References

[1] Belk J M 2004 *Thompson's Group F* PhD Thesis, Cornell University
[2] Berdinsyky D and Khoussainov K 2014 On automatic transitive graphs *Developments in language theory – Eightteens International Conference*, DLT 2014 (Ekaterinburg) Proceedings. *Springer LNCS* **8633** 1–12
[3] Jean Berstel J 1979 *Transductions and Context-free Languages* Teubner-Verlag
[4] Blumensath A 1999 *Automatic structures* Diploma thesis, RWTH Aachen
[5] Blumensath A and Grädel E 2000 Automatic structures. *Fifteenth Annual IEEE Symposium on Logic in Computer Science*, LICS 2000 (Santa Barbara) p. 51-62 (Los Alamitos: IEEE Computer Society Press)
[6] Barany V 2007 *Automatic Presentations of Infinite Structures.* PhD Thesis, Rheinisch-Westfälische Technische Hochschule Aachen
[7] Brough M, Khoussainov B and Nelson P 2008 Sequential automatic algebras *Fourth Conference on Computability in Europe*, CiE 2008 (Athens) Proceedings *Springer LNCS* **5028** 84–93
[8] Case J, Jain S, Seah S and Stephan F 2013 Automatic functions, linear time and learning *Logical Methods in Computer Science* **9** 3
[9] Delhommé C 2004 Automaticité des ordinaux et des graphes homogènes. *Comptes Rendus Mathematique* **339** 5–10
[10] Epstein D, Cannon J, Holt D, Levy S, Peterson M and Thurston W 1992 *Word Processing in Groups.* (Boston: Jones and Bartlett Publishers)
[11] Hodgson B R 1976 *Théories décidables par automate fini.* Ph.D. thesis, University of Montréal
[12] Hodgson B R 1983 Décidabilité par automate fini *Annales des sciences mathématiques du Québec* **7** 39–57
[13] Hopcroft J E, Motwanti R and Ullman J D 2007 *Introduction to Automata Theory, Languages, and Computation* Third Edition (Reading, MA: Addison-Wesley Publishing)
[14] Jain S, Khoussainov B, Stephan F, Teng D and Zou S 2014 Semiautomatic structures. Computer Science – Theory and Applications – Ninth International *Computer Science Symposium in Russia*, CSR 2014 (Moscow) Proceedings. *Springer LNCS* **8476** 204–217
[15] Khoussainov B and Minnes M 2010 Three lectures on automatic structures *Proceedings of Logic Colloquium 2007, Lecture Notes in Logic* **35** 132–176
[16] Khoussainov B and Nerode A 1995 Automatic presentations of structures *Logical and Computational Complexity* LCC 1994 Proceedings *Springer LNCS* **960** 367–392
[17] Kuske D, Liu J and Lohrey M 2013 The isomorphism problem on classes of automatic structures with transitive relations *Transactions of the American Mathematical Society* **365** 5103–5151
[18] Matiyasevich Yu V 1970 Diofantovost' perechislimykh mnozhestv *Doklady Akademii Nauk SSSR* **191** 279–282 (Russian) Enumerable sets are Diophantine *Soviet Mathematics Doklady* **11** 354–358 (English)
[19] Nies A 2007 Describing groups *Bulletin of Symbolic Logic* **13** 305–339
[20] Nies A and Thomas R M 2008 FA-presentable groups and rings *Journal of Algebra* **320** 569–585
[21] Oliver G and Thomas R M 2005 Automatic presentations for finitely generated groups *Twentysecond Annual Symposium on Theoretical Aspects of Computer Science* (STACS 2005) (Stuttgart) Proceedings *Springer LNCS* **3404** 693–704
[22] Rubin S 2008 Automata presenting structures: a survey of the finite string case *The Bulletin of Symbolic Logic* **14** 169–209
[23] Tan W Y 2010 *Reducibilities between regular languages* Master Thesis, Department of Mathematics, National University of Singapore
[24] Tsankov T 2011 The additive group of the rationals does not have an automatic presentation *The Journal of Symbolic Logic* **76** 1341–1351
[25] Turing Alan 1936 On Computable Numbers, with an Application to the Entscheidungsproblem *Proceedings of the London Mathematical Society 2* **42** 230–265

Journal of Physics: Conference Series **622** (2015) 012014 doi:10.1088/1742-6596/622/1/012014

# Survival Analysis of Patients with End Stage Renal Disease

**J D Urrutia, W S Gayo, L A Bautista and E B Baccay**
Faculty Members, Department of Mathematics and Statistics, College of Science, Polytechnic University of the Philippines, Sta. Mesa, Manila, Philippines

E-mail: jackieurrutia20@gmail.com, math_urrutia@yahoo.com.ph, jdurrutia@pup.edu.ph, willygayo@gmail.com, wsgayo@pup.edu.ph, labautista1971a@yahoo.com, labautista@pup.edu.ph, math_baccay@yahoo.com, ebbaccay@pup.edu.ph

**Abstract.** This paper provides a survival analysis of End Stage Renal Disease (ESRD) under Kaplan-Meier Estimates and Weibull Distribution. The data were obtained from the records of V. L. Makabali Memorial Hospital with respect to time t (patient's age), covariates such as developed secondary disease (Pulmonary Congestion and Cardiovascular Disease), gender, and the event of interest: the death of ESRD patients. Survival and hazard rates were estimated using NCSS for Weibull Distribution and SPSS for Kaplan-Meier Estimates. These lead to the same conclusion that hazard rate increases and survival rate decreases of ESRD patient diagnosed with Pulmonary Congestion, Cardiovascular Disease and both diseases with respect to time. It also shows that female patients have a greater risk of death compared to males. The probability risk was given the equation $R = 1 - e^{-H(t)}$ where $e^{-H(t)}$ is the survival function, H(t) the cumulative hazard function which was created using Cox-Regression.

## 1. Introduction

In 2012, the National Kidney and Transplant Institute (NKTI) cited kidney failure as the 9th leading cause of death among Filipinos. Most people know that a major function of the kidneys is to remove waste products and excess fluid from the body. These waste products and excess fluid are removed through the urine. The production of urine involves highly complex steps of excretion and re-absorption. This process is necessary to maintain a stable balance of body chemicals.

A diagnosis of End Stage Renal Disease (ESRD) means that you are in the final stage of kidney disease and your kidneys are not functioning well enough to meet the needs of daily life. It also used to be an automatic death sentence; however, advancements in treatments now allow patients to live much longer than ever before. Without dialysis or a kidney transplant, death will occur. The outcome of the treatment depends on the patient.

Having kidney disease increases your chances of also having cardiovascular disease, heart attacks, and strokes. Also, the body can hold in too much fluid which could lead to swelling in your arm and legs, high blood pressure, or fluid in your lungs called pulmonary edema. To analyse the random cause of the morbidity among Filipino people, the researchers generated a mathematical survival model that will predict the absolute risk of death with End Stage Renal Disease patients. This study aimed to determine what variables affect the probability of survival of Chronic Kidney Disease patients diagnosed with ESRD and to assess the effectiveness of the developed survival model to provide reasonably accurate failure analysis and failure forecasts of the conditional status of patient.

## 2. Statement of the Problem

This study aimed to undertake a survival analysis of ESRD patients. Specifically, it attempted to find answers to the following questions:

1) What is the risk profile of ESRD patients diagnosed with Pulmonary Congestion and Cardiovascular Disease that leads to death?
2) What are the survival functions of patients with ESRD, ESRD with Cardiovascular Disease, Pulmonary Congestion, and Both Cardiovascular Disease and Pulmonary Congestion according to age and gender group using Kaplan–Meier Product Limit Estimates and Weibull Distribution?
3) What are the hazard functions of patients with ESRD, ESRD with Cardiovascular Disease, Pulmonary Congestion, and Both Cardiovascular Disease and Pulmonary Congestion according to age and gender group using Kaplan–Meier Product Limit Estimates and Weibull Distribution?
4) Is there a significant difference between Kaplan-Meier Limit Estimates and Weibull Distribution according to age and gender group when generating survival and hazard rates?
5) What are the probability risks of death of patients with ESRD, ESRD with Cardiovascular Disease, Pulmonary Congestion, and Both Cardiovascular Disease and Pulmonary Congestion using Cox-Regression and Weibull Distribution?

## 3. Methodology

This section describes the data and methods used in this study to come up with the development of a survival model and estimate the probability of surviving all causes of death for a specified time interval calculated from the cohort of ESRD cases.

### *3.1. Data Gathering Procedure*

In this study, all of the data needed by the researchers were obtained from the records of V. L. Makabali Memorial Hospital. The data collected are from the total number of Chronic Kidney Disease patients diagnosed with ESRD from 2008 to 2013. These consist of the age, gender, and final diagnosis of ESRD patients that led them to death.

To compute the risk of death on CKD patients diagnosed with ESRD, the status of each patient are needed to be observed. The status of the patients is 1 if the person failed, meaning the person died and 0 if the person was censored or alive over the given time. Thereafter, the data must be organized to fit in statistical packages like Number Cruncher Statistical System (NCSS), Statistical Analysis and Graphics and Statistical Package for the Social Sciences (SPSS).

### *3.2. Kaplan-Meier Method*

The Kaplan–Meier (KM) method is a nonparametric survival analysis that is used to estimate the two survival probabilities curves dealing with differing survival times especially when not all of the subjects continue in the study. The assumptions of KM method are:

1. The subjects are a random sample from the population of interest, so that they are independent of each other.
2. Censored individuals have the same prospect of survival as those who continue to be followed. This cannot be tested for and can lead to a bias that artificially reduces $S$.
3. The Kaplan-Meier survival estimate is a step function that changes at every distinct survival time, but does not change at censoring times (unless a survival time happens to be tied to a censoring time).
4. The event studied (e.g. death) happens at the specified time. Late recording of the event studied will cause artificial inflation of $S$.

### *3.3. Log-Rank Test*

The log–rank test is a large-sample chi-square test that uses as its test criterion a statistic that provides an overall comparison of the KM curves being compared. The main objective is to compare two survival curves by treatment group. Here we consider two groups. The hypotheses of interest are stated as follows:

$$H_0:\ S_1(t) = S_2(t)$$
$$H_1:\ S_1(t) \neq S_2(t)$$

where the null hypothesis $H_0$ states that the survival curves are identical in the two populations and the alternative hypothesis $H_1$ states otherwise.

*3.4. Weibull Distribution*

The Weibull distribution is worldwide used to model life data. The distribution can handle increasing, decreasing or constant failure-rates and can be created for data with and without suspensions (non-failures). The Weibull distribution is flexible and fits to a wide range of data, including Normal distributed data. Only log–normal data does not fit in the Weibull distribution and needs separate analyses. For creating the plot you need to record the time to failure that can be expressed in mileage, cycles, minutes, strength, stiffness or similar continuous parameters. The Weibull cumulative distribution function (CDF), has the explicit equation:

$$F(t) = 1 - e^{-(t/\eta)^{\beta}}$$

where: $F(t) \coloneqq$ probability of failure at time $t$;
$t \coloneqq$ time, cycle, miles or any appropriate parameters;
$\eta \coloneqq$ characteristic life or scale parameter; and
$\beta \coloneqq$ slope or shape of the parameter.

*3.4.1. Anderson–Darling Test.* The Anderson-Darling test is used to test if a sample of data came from a population with a specific distribution. The Anderson-Darling test is an alternative to the chi-square and Kolmogorov-Smirnov goodness of fit tests. The hypotheses of interest are stated as follows:

$H_0$: The data follows Weibull distribution.
$H_1$: The data do not follows Weibull distribution.

*3.5. Cox Regression*

It is nonparametric in the sense that it involves an unspecified arbitrary baseline hazard function. It is parametric because it assumes a parametric form for the covariates. The baseline hazard function is scaled by a function of the model's (time–independent) covariates to give a general hazard function. It is usually written in terms of the hazard model formula,

$$H(t,X) = h_0(t)\, e^{\sum_{i=1}^{n} \beta_i X_i}$$

An important feature of this formula, which concerns the proportional hazards assumption, is that the baseline hazard is a function of t, but does not involve the $X$'s. In contrast, the exponential expression shown here, involves the $X$'s, but does not involve t. the X's here is called time–independent $X$'s.

*3.5.1. Graphical Method.* There are two types of graphical techniques to check the proportional hazard assumptions. The most common technique is by comparing the estimated $\ln(-\ln(S(t)))$. If the two survival curves do not intersect and are parallel, it clearly provides evidence against the assumptions. Another graphical approach is by comparing observed with predicted survival curves. If the two survival curves are close, then the proportional hazard assumption is plausible.

*3.5.2. Test of Goodness Fit.* This approach provides a Chi–Square statistics where the computed values for each variable rely on p–values. This p–value is used for evaluating the PH assumption for each variable. If the p–values are large then PH is satisfied, whereas small p–values provide violation of PH assumption.

*3.5.3. Probability Risks.* The risk of Coronary Heart Disease is given by

$$R(X, Death) = 1 - e^{-H(t,X)}.$$

where $e^{-H(t,X)}$ is the survival function in terms of its cumulative hazard rate function $H(t, X)$. Since there were more covariates involved in the study, the researchers used the Cox Regression to estimate the hazard rate of the event on ESRD patients. The formula is given by

$$H(t, X) = h_0(t)\, e^{\sum_{i=1}^{n} \beta_i X_i}$$

where $H_0(t)$ is the cumulative baseline hazard function. This term depends on time but not the covariates, $X_i$ is possibly time–independent covariables, and $\beta$ is the regression coefficients.

## 4. Results and Discussion

The study examined the data collected for trends and differences between independent variables such as ESRD, Pulmonary Congestion, Cardiovascular Disease, age and gender. This section will also interpret the data for further statistical analysis regarding the effects of chosen predictors on death.

### *4.1. Risk Profile of V.L. Makabali Memorial Hospital Patients*

Table below presents the risk profile of 289 V.L. Makabali Memorial Hospital patients with End Stage Renal Disease (ESRD). Of these 97 have complications such as Pulmonary Congestion and Cardiovascular Disease that leads to death.

**Table 1.** Risk Profile of V.L. Makabali Memorial Hospital Patients

| **Category of Patients according to Diagnosis** | **Number of Individuals** | **Male** | **Female** | **Greater than 52** | **Less than or equal to 52** |
|---|---|---|---|---|---|
| ESRD | 148 | 78 (53.79%) | 70 (48.61%) | 97 (50.78%) | 51 (52.04%) |
| ESRD with Pulmonary Congestion | 45 | 21 (14.48%) | 24 (16.67%) | 28 (14.66%) | 17 (17.34%) |
| ESRD With Cardiovascular Disease | 54 | 22 (15.17%) | 32 (22.22%) | 38 (19.90%) | 16 (16.32%) |
| ESRD with both Pulmonary congestion and cardiovascular disease | 42 | 24 (16.55%) | 18 (12.5 %) | 28 (14.66%) | 14 (14.28%) |
| **TOTAL** | **289** | **145** | **144** | **191** | **98** |
| ESRD that leads to death | 6 | 2 (4.76%) | 4 (7.27%) | 3 (4.68%) | 3 (9.09%) |
| ESRD with Pulmonary Congestion that leads to death | 26 | 11 (26.19%) | 15 (27.27%) | 15 (23.44%) | 11 (33.33%) |
| ESRD With Cardiovascular Disease that leads to death | 37 | 15 (27.27%) | 22 (40%) | 28 (43.75%) | 9 (27.27%) |
| ESRD with both Pulmonary congestion and cardiovascular disease that leads to death | 28 | 14 (33.33%) | 14 (33.33%) | 18 (28.13%) | 10 (30.30%) |
| **TOTAL** | **97** | **42** | **55** | **64** | **33** |

Table 1 also shows separate gender data for males and females and for age greater than 52 and less than or equal 52 (expressed in counts and %). Thus, for 148 ESRD patients without complications, 78 (or 53.79%) are males while 70 (or 48.61%) are females. In addition, 97 (or 50.78%) are with ages greater than 52 while 51 (or 52.04%) with ages less than or equal 52.

It also shows that there are more female ESRD patients with cardiovascular disease (22.22%) than patients with pulmonary congestion (14.48%); and there are 16.55% of male ESRD patients with both complications. While for the female ESRD patients, 48.61% have no complications. There are more female ESRD patients affected by cardiovascular disease (22.22%) than those with pulmonary congestion (16.67%); and there are only 12.5% of female ESRD patients with both complications.

In patients of ages greater than 52, there are 50.78% with End Stage Renal Disease only, 14.66% with Pulmonary disease, 19.90% with Cardiovascular disease, and 14.66% with both complications while patients with ages less than or equal to 52, 52.04% have end stage renal disease only. 17.34% of the patients have pulmonary congestion and 16.32% have cardiovascular disease, while 14.28% of the patients have both complications.

From the group of male ESRD patients, 4.76% had died with only ESRD while 26.19% died having ESRD with pulmonary congestion. There are 27.27% of male ESRD patients who died having cardiovascular disease while there are 33.33% of male patients died having both complications. From the female group, 7.27% died while having ESRD. There are more female ESRD patients that died with cardiovascular than patients with pulmonary congestion with a percentage of 27.27% and 40%. From the female group, 33.33% of them died while having both complications; it is the same with the male patients.

On the other hand, for patients with ages greater than 52 years, 4.68% died with only End Stage Renal Disease, 23.44% died with pulmonary congestion, 43.75% died with cardiovascular congestion and 28.13% died with both complications. For patients with ages less than 52 years, 9.09% died with end stage renal disease only while 33.33% died with pulmonary congestion. Moreover 27.27% of patients died with cardiovascular disease and also 30.30% died with both complications.

*4.2. Survival functions of patients with ESRD only, ESRD with Cardiovascular Disease only, ESRD with Pulmonary Congestion only, and ESRD with both Cardiovascular Disease and Pulmonary Congestion using Kaplan-Meier Product Estimates and Weibull Distribution.*

*4.2.1. Age Group.*

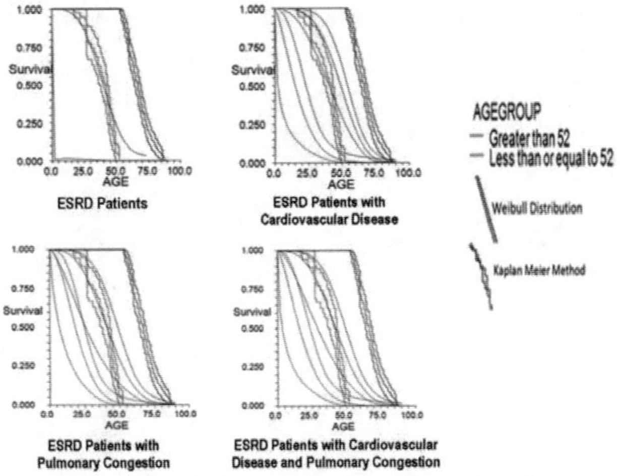


**Figure 1.** Survival Functions According to Age Group

Figure 1 shows the survival functions according to age of the patients with ESRD, ESRD with Cardiovascular disease, ESRD with Pulmonary Congestion, and ESRD patients with both complications. Kaplan-Meier Estimates is represented by a step function while the Weibull distribution is represented by an exponential function. The vertical axis stands for the probability of survival of patients with respect to age and its complications. The figures show that as the age of the patient increases the probability of its survival decreases.

*4.2.2. Gender Group.*

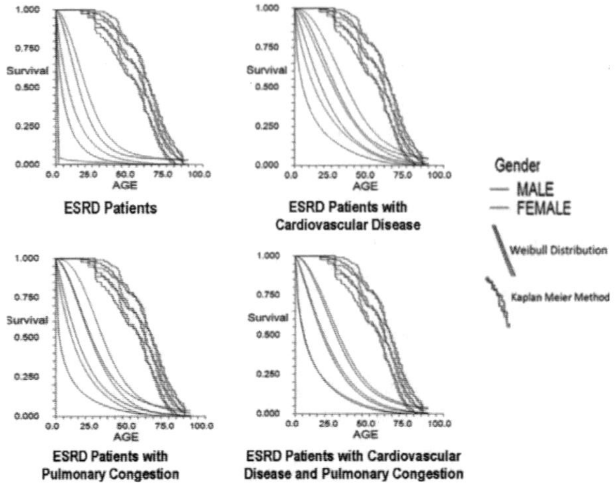


**Figure 2.** Survival Functions According to Gender Group

Figure 2 shows the survival functions according to gender of the patients with ESRD, ESRD with Cardiovascular disease, ESRD with Pulmonary Congestion, and ESRD patients with both complications. Kaplan-Meier Estimates is represented by a step function while the Weibull distribution is represented by an exponential function. The vertical axis stands for the probability of survival of patients with respect to age and its complications. The probability of survival of male and female patient decreases with respect to age.

*4.3. The hazard functions of patients with ESRD, Pulmonary Congestion, and Cardiovascular Disease and with both Pulmonary Congestion and Cardiovascular Diseases using Kaplan-Meier Limit Estimates and Weibull Distribution*

*4.3.1. Age Group.*

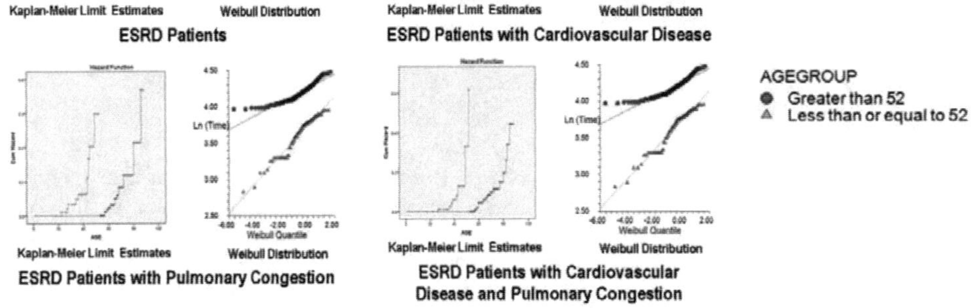


**Figure 3.** Hazard Functions of Patient According to Age Group

Figure 3 shows the hazard functions according to age group of the patients with ESRD, ESRD with Cardiovascular disease, ESRD with Pulmonary Congestion, and ESRD patients with both complications using Kaplan-Meier Limit and Weibull Distribution. Note that under the age group, it shows that patients whose ages are either greater or less than 52 follow linearity.

*4.3.2. Gender Group.*

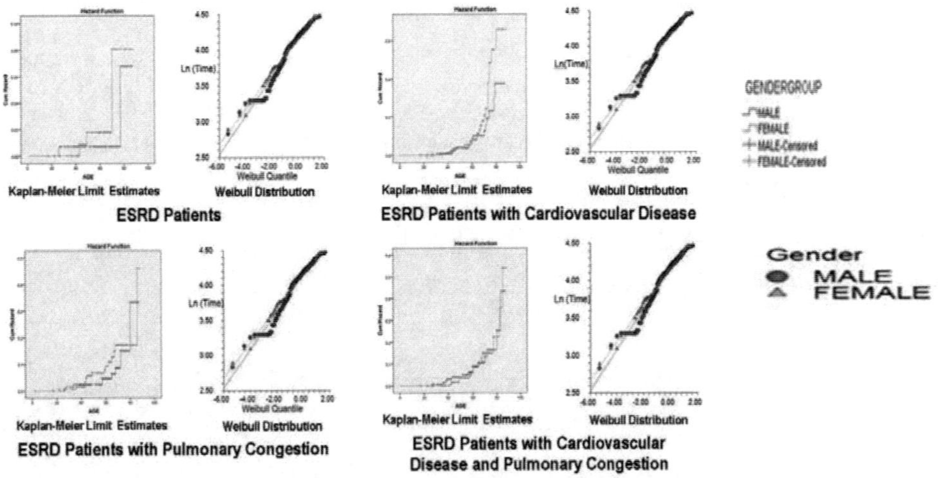


**Figure 4.** Hazard Functions of Patient According to Gender Group

Figure 4 shows the hazard functions according to gender group of the patients with ESRD, ESRD with Cardiovascular disease, ESRD with Pulmonary Congestion, and ESRD patients with both complications using Kaplan-Meier Limit and Weibull Distribution. Note that under the gender group, it shows that patients whose ages are either male or female follow linearity.

*4.4. Is there a significant difference between Kaplan-Meier Limit Estimates and Weibull Distribution when generating survival and hazard rates?*

*4.4.1. Age Group.*

**Table 2.** Significant Difference of the Survival and Hazard Rates under Age Group between Kaplan-Meier Limit Estimates and Weibull Distribution

| Paired Sample Test | | | | | |
|---|---|---|---|---|---|
| | | | Computed t-value | p-value | Interpretation |
| Survival Rates | Less than or equal to 52 years old | ESRD | 2.364 | 0.142 | Not Significant |
| | | ESRD with Cardiovascular Disease | 9.040 | 0.000 | Significant |
| | | ESRD with Pulmonary Congestion | 4.323 | 0.003 | Significant |
| | | ESRD with both complication | 7.762 | 0.000 | Significant |
| | Greater than 52 years old | ESRD | 5.168 | 0.122 | Not Significant |
| | | ESRD with Cardiovascular Disease | 8.306 | 0.000 | Significant |
| | | ESRD with Pulmonary Congestion | 5.098 | 0.001 | Significant |
| | | ESRD with both complication | 6.796 | 0.000 | Significant |

| | | | | | |
|---|---|---|---|---|---|
| Hazard Rates | Less than or equal to 52 years old | ESRD | -1.928 | 0.194 | Not Significant |
| | | ESRD with Cardiovascular Disease | -5.738 | 0.001 | Significant |
| | | ESRD with Pulmonary Congestion | -3.246 | 0.012 | Significant |
| | | ESRD with both complication | -3.892 | 0.006 | Significant |
| | Greater than 52 years old | ESRD | -2.377 | 0.254 | Not Significant |
| | | ESRD with Cardiovascular Disease | -4.672 | 0.000 | Significant |
| | | ESRD with Pulmonary Congestion | -2.090 | 0.066 | Not Significant |
| | | ESRD with both complication | -3.420 | 0.004 | Significant |

Table 2 shows that there is no significant difference between the survival rates generated by Kaplan-Meier Estimates and Weibull Distribution under age groups of patients with ESRD for both survival and hazard rates and also the hazard rates for ESRD with Pulmonary Congestion whose age is greater than 52 years old. And the other combinations are all significant.

*4.4.2. Gender Group*

**Table 3.** Significant Difference of the Survival and Hazard Rates under Gender Group between Kaplan-Meier Limit Estimates and Weibull Distribution

| **Paired Sample Test** | | | | | |
|---|---|---|---|---|---|
| | | | Computed t-value | p-value | Interpretation |
| Survival Rates | Female | ESRD | 2.619 | 0.120 | Not Significant |
| | | ESRD with Cardiovascular Disease | 5.051 | 0.000 | Significant |
| | | ESRD with Pulmonary Congestion | .979 | 0.347 | Not Significant |
| | | ESRD with both complication | 4.146 | 0.002 | Significant |
| | Male | ESRD | 1.074 | 0.477 | Not Significant |
| | | ESRD with Cardiovascular Disease | 7.172 | 0.000 | Significant |
| | | ESRD with Pulmonary Congestion | 2.588 | 0.036 | Significant |
| | | ESRD with both complication | 3.099 | 0.009 | Significant |
| Hazard Rates | Female | ESRD | -1.730 | 0.226 | Not Significant |
| | | ESRD with Cardiovascular Disease | -3.004 | 0.008 | Significant |
| | | ESRD with Pulmonary Congestion | -1.278 | 0.225 | Not Significant |
| | | ESRD with both complication | -2.209 | 0.047 | Significant |
| | Male | ESRD | -1.029 | 0.491 | Not Significant |
| | | ESRD with Cardiovascular Disease | -6.179 | 0.000 | Significant |
| | | ESRD with Pulmonary Congestion | -2.939 | 0.022 | Significant |
| | | ESRD with both complication | -1.930 | 0.080 | Significant |

Table 3 shows that there is no significant difference between the survival rates generated by Kaplan-Meier Estimates and Weibull Distribution under gender groups of patients with ESRD for both survival and hazard rates and also the survival and hazard rates for ESRD with Pulmonary Congestion female patients. And the other combinations are all significant.

*4.5. What are the probability risks of death of patients with ESRD, Pulmonary Congestion, Cardiovascular Disease and with Both Cardiovascular Disease and Pulmonary Congestion using Cox-Regression and Weibull Distribution?*

*4.5.1. Probability risk of Death on ESRD patients under Cox Regression.* The probability risk of Death is identified by the model $R(x, Death) = 1 - e^{-H(t,X)}$ where $e^{-H(t,X)}$ is the survival probability in terms of a cumulative hazard rate function $H(t,X)$ which is derived from Cox Proportional Hazard Model and can be verified in the appendices.

Substituting the Cox proportional hazard rate function to the model, the following probability risk functions:

1. ESRD with Pulmonary Congestion

$$\boldsymbol{R(x,Death)} = \mathbf{1} - \boldsymbol{e}^{-(\mathbf{0.000715}\boldsymbol{e}^{\mathbf{0.065}t-\mathbf{2.639}x_1-\mathbf{0.206}x_2})}$$

where $t$ = age of individual

$$x_1 = \begin{cases} 1, & \text{if affected by ESRD} \\ 0, & \text{if affected by ESRD with Pulmonary Congestion} \end{cases}$$

$$x_2 = \begin{cases} 1, & \text{if male} \\ 0, & \text{if female} \end{cases}$$

2. ESRD with Cardiovascular Disease

$$\boldsymbol{R(x,Death)} = \mathbf{1} - \boldsymbol{e}^{-(\mathbf{0.000258}\boldsymbol{e}^{\mathbf{0.086}t-\mathbf{2.546}x_3-\mathbf{0.262}x_4})}$$

where $t$ = age of individual

$$x_3 = \begin{cases} 1, & \text{if affected by ESRD} \\ 0, & \text{if affected by ESRD with Cardiovascular Disease} \end{cases}$$

$$x_4 = \begin{cases} 1, & \text{if male} \\ 0, & \text{if female} \end{cases}$$

3. ESRD with both Cardiovascular Disease and Pulmonary Congestion

$$\boldsymbol{R(x,Death)} = \mathbf{1} - \boldsymbol{e}^{-(\mathbf{0.00045}\boldsymbol{e}^{\mathbf{0.075}t-\mathbf{2.654}x_5-\mathbf{0.204}x_6})}$$

where $t$ = age of individual

$$x_5 = \begin{cases} 1, & \text{if affected by ESRD} \\ 0, & \text{if affected by ESRD with Cardiovascular Disease and Pulmonary Congestion} \end{cases}$$

$$x_6 = \begin{cases} 1, & \text{if male} \\ 0, & \text{if female} \end{cases}$$

*4.5.2. Probability Risk of Death on ESRD Patients under Weibull Distribution.* The probability risk of Death in Weibull Distribution is identified by the model $R(x,Death) = 1 - e^{-H(t,X)}$ where $e^{-H(t,X)}$ is the survival probability in terms of a cumulative hazard rate function $H(t,X)$.

Substituting the Weibull Distribution hazard rate function to the model, the researchers' got the general probability risk functions of patients with Cardiovascular Disease, Pulmonary Congestion and with both Disease according to the gender of the patients. For female the model is

$$\boldsymbol{R(t,Death)} = \mathbf{1} - \boldsymbol{e}^{-\left(\frac{t}{\mathbf{63.98523}}\right)^{\mathbf{4.009278}}}$$

And for male the model is

$$\boldsymbol{R(t,Death)} = \mathbf{1} - \boldsymbol{e}^{-\left(\frac{t}{\mathbf{63.71656}}\right)^{\mathbf{3.615808}}}$$

where $t$ = age of individual.

## 5. Conclusion

According to the results, the researchers conclude that having End Stage Renal Disease (ESRD) with complications increases the probability of death. In addition, female patients have greater risk of death compared to males and patients with ages greater than 52 years have a higher risk of death compared to patients with ages less than 52. The estimated survival and hazard rates of ESRD patients under age and gender group by Kaplan-Meier Method and Weibull Distribution has no significant difference with p-values greater than 0.05 and also there is no significant difference between the hazard rates of male ESRD patients with both Cardiovascular Disease and Pulmonary Congestion with p-value equals to 0.08. The developed probability risk of death in ESRD patients generated by Cox regression and Weibull Distribution at certain age will help medical experts to improve the quality of medications and technologies that will lessen the risk of death in ESRD and to patients with or without ESRD as a base of healthy lifestyle living.

## 6. Recommendation

End-stage renal disease is often preventable with the proper control of blood pressure, blood lipids, and blood glucose, in combination with appropriate medications. For patients who have progressed to the need for dialysis, morbidity and mortality can be reduced and quality of life enhanced through adherence to an appropriate dietary and medical regimen, along with regular physical activity. Future researchers may include other factors such as Body Mass Index, blood counts or tests, blood pressure and urine test as independent variables for the risk of death. The researchers suggest the use of more samples from different hospitals that would make the model a general representation for estimating the risk of death.

## 7. References

[1] Mandayam S, Shahinian VB. Are chronic dialysis patients at increased risk for cancer? J Nephrol. 2008;21:166-174.
[2] http://www.philippineembassy-usa.org/news/3246/300/PH-Hemodialysis-Center-Recognized-among-World-s-Best-PPP-Projects/d,phildet/
[3] Danguilan, Romina. The burden of kidney disease in the philippines. ABS- CBN News; National Transplant and Kidney Institute
[4] https://www.bridgeoflifemm.org/our-missions/philippines/
[5] http://medobserver.com/specialtyarticle.php?ArticleID=224
[6] http://medobserver.com/specialtyarticle.php?ArticleID=126
[7] http://www.philippineembassy-usa.org/news/3246/300/PH-Hemodialysis-Center-Recognized-among-World-s-Best-PPP-Projects/d,phildet/
[8] http://www.ncbi.nlm.nih.gov/pmc/articles/PMC3055978/
[9] http://www.buzzle.com/articles/kidney-dialysis-life-expectancy.html
[10] http://www.bridgeoflifemm.org/our-missions/philippines/
[11] http://r08.pia.gov.ph/index.php?article=1141338790357
[12] http://prezi.com/gp04ypz-fxpi/case-study-chronic-kidney-disease-secondary-to-diabetes-mel/
[13] http://www.nkti.gov.ph/kidney_health.do
[14] http://business.inquirer.net/104331/on-kidney-failure-and-transplantation
[15] http://www.livingkidneydonorsearch.com/living-donation/dialysis-a-poor-option/

# Monitoring the Water Quality in the Recycling Process

**A Antonyová[1], P Antony[2] and B Soewito[3]**
[1]University of Prešov in Prešov, Faculty of Management, Department of Mathematical Methods and Managerial Informatics, Konštantínova 16, Prešov 08001, Slovak Republic
[2]Apmikro, Pavlovičovo námestie 19, Prešov 08001, Slovak Republic
[3]Graduate Program, Bina Nusantara University, Jakarta 11530, Indonesia

E-mail: antonyova@gmail.com

**Abstract.** Specific water contamination requires the recycling process prior to its discharge into the public sewerage network. Electro-flotation technology was used for cleaning of waste water contaminated with the disperse colorants. Dispersion colorants were used to decorate the boxes, made of corrugated board, in the company for the production of packaging. The objective of this paper is to present a method of optimization to determine the length of the time interval for electro-flotation process. Interval should be set so as to achieve the degree of cleaning the water that is the maximum possible in the process of electro-flotation. The measurement of the light passing through the measuring the translucent tube determines the actual degree of the water purity. The measurement is carried out by means of a photodiode in different wavelengths. The measured values in the measuring tube are compared with the nominal value, which corresponds to pure distilled water. Optimization the time interval to clean the water using electro-flotation was determined for yellow color. The optimum interval for the water contaminated with the yellow color was set to 1800s.

## 1. Introduction

Nowadays maintaining the ecosystem requires recycling the waste water with the specific contamination prior to its discharge to the municipal sewage treatment plants. For instance genotoxicity testing of domestic waste prior to its discharge to municipal sewage treatment plants was designated as essential in [3].

Government with its legislative-economic rules also drives the firms toward industrial ecology. To fulfill all the requirements, the company for the packaging production designed the special equipment for treatment the waste water that was contaminated with the dispersion colorants. The dispersion colorants are used for decoration the paper boxes. When cleaning the machines, a huge amount of contaminated water is created. This has created the need to use new, environmentally acceptable and affordable technologies for waste water treatment. The technology of the waste water purifier is based on the flotation process. The sophisticated electronics controls the system from the mixing the waste water with an oxidizing agent to its discharge to the flotation tank. The flotation itself takes place in the flotation tank wherein during the electrolysis reaction; the formed bubbles are associated with particles of the colorant. Electrodialysis is applied as the technology for treatment the water contaminated with the heavy metals [4]. The article [7] deals with the clarification the water in the rivers of Dnieper basin. Ultrafiltration is used instead of traditional coagulation followed by a mechanical filtration.

During the flotation process, bubbles with the colorant travel toward the surface of the wastewater, where they form foam which is swept at regular time intervals. Flotation as a technology was proved as suitable within the experimental operation, as well as in practice. However the whole process needs monitoring the degree of the water clarification during the flotation process.

There are a few ways how to monitor the water quality or the whole process of the water treatment. Michael Allen with his coauthors in [1] introduces the system of real-time monitoring the amount and quality of drinking water in the distribution infrastructure. Contamination warning based on decision-support system with on-line control was build in collaboration of the The United States Environmental Protection Agency (USEPA) and the towns of state Ohio. Mesi and Kopliku collected water from the surfase of the Shkodra Lake in Albania and evaluate the level of contamination that was cost with uncontrolled releases of industrial and urban waste. They used analysis of variance and Student Newman-Keuls (SNK) tests in [5]. Greek editors Kostas Voudouris and Dimitra Voutsa collected in [8] the set of articles devoted to the monitoring the water quality in several countries, especially surface water and groundwater that are the main sources of fresh water for drinking purposes. The water testing is based on the conventional hydrochemical techniques and statistical analysis, such as trend and prediction methods as well as decision support systems with the risk analysis. Authors of different countries illustrate the specific situation with the water sources and assessment of water pollution in their countries. Underwater robot was constructed in [10] to allow more relevant access to the water quality data.

Andrew Moxey collaborated with the experts across OECD member countries on categories related to ecosystem quality and its impact to the human health. Special attention is paid in [6] to agriculture, which is considered to be both environment polluter with the chemicals as well as a victim of pollution. The concentration of polycyclic aromatic hydrocarbons (PAHs) in agricultural topsoil that was contaminated with the wastewater in the particular regions of China is described in [9]. Then correlation was constructed between the irrigation groundwater and topsoil in the levels of contamination. Bhandari and Abrol in [2] discuss their experiences with testing harmful effect of industrial waste water that was discharged as untreated. They used pH electrode (PE03), conductivity electrode (k=1), purity sensor and thermistor to check the water properties and subsequently their impact on human as well as aquatic life.

To check the degree of the waste water treatment during the flotation process we decided to use the special equipment based on the photodiode properties.

## 2. Methodology to monitor the degree of the waste water treatment

Monitoring the degree of the waste water treatment is done using the photodiode that is expressed in Figure 1. Figure 1 shows a pipe in which the receiving photodiode is situated on the right. In the middle of the pipe, syringe with a sample of treated wastewater is of enlightened through. The syringe is passed through the opening in the pipe. Placing the pipe is seen in the Figure 3, where the bottom right is a view of the PIN photodiode.

The measurement of the loss of light in different wavelengths is carried out passing the light through the measuring chamber in the shape of the transparent tube. The tube is filled with contaminated water, which is subject to measurement.

The excitation LED powered via a resistor (trimmer would not be sufficiently stable) and resistors are only available in certain values. Therefore the values are not equal, but are sufficiently stable. For example, the voltage of 320mV is determined on the photodiode with the red light from the LED of wavelength 630nm.

The measured values in the measuring tube are compared with the nominal value, which corresponds to pure distilled water as follows:

For wavelength of 950nm (the infrared light), the pure destilled water represents 330mV, what is the voltage on the photodiode, which is loaded with resistor.

Similarly:
for wavelength of 630nm (the red light), the pure destilled water represents 320mV,
for wavelength of 595nm (orange light), the pure destilled water represents 331mV,
for wavelength of 580nm (yellow light), the pure destilled water represents 324mV,
for wavelength of 530nm (green light), the pure destilled water represents 338mV,
for wavelength of 470nm (blue light), the pure destilled water represents 385mV,
for wavelength of 400nm (ultraviolet light), the pure destilled water represents 309mV.

In the evaluation of the actual measurement by the human eye, it should be recognized that the change in light transmission of the contaminated water is a logarithmic dependence, that is, the change in value of light transmission of 25% (3 dB decibel) is also detectable with the human eye. That is, if the purity of the solution is higher than 75% of the value measured in pure distilled water, then the human eye does not detect it as dirty, it seems therefore clear. It is therefore necessary to establish the degree of purity of water through precision measuring electronic devices.

The water is stored after purification, by means of flotation, so there is a spontaneous sedimentation that is caused by physical-chemical process of flotation. Thus, a further 48 hours there is completion of the cleaning process, wherein the transparency of treated water is comparable with pure distilled water.

Calibration according to the pure distilled water when measuring the voltage on the photodiode, which is equipped with a working resistor, as well as actual measurements drafts of the contaminated water during its clarification were performed using an accurate digital voltmeter with A / D converter with 24 bit resolution.

Direct measurement using optics directly in the flotation vessel is not possible due to large amounts of microbubbles of oxygen and hydrogen, which arise in the process of electro-flotation. The effect of the bubbles had to be eliminated also after pumping a new sample into the external chamber (Figure 3) to measure in such a way that the sample that has been pumped, has been left to stand for 90s. And also the time interval of the measuring cycle had to be adapted accordingly.

### 3. Equipment description

Figure 1 shows a representation of the measuring light tube, wherein one end comprises an exact insert that is made of textolite as a cover material. Seven pieces of LED, whose wavelengths are indicated in the text, are stored in the cover.

**Figure 1.** Orange LED is active. Transmitted light illuminates into the photodiode.

LEDs are positioned so that the axis of the light emitted by them directed in each case exactly into the middle of the opposite insert, which is expressed on the right in the Figure 1. The receiving photodiode PIN type SFH203, which is located in the center of the cover, measures the intensity of the light going from the diode through the transparent measuring chamber shown in Figure 3.

Figure 2 shows also the connection of the device for measuring the water clarity to the flotation vessel. In the center of Figure 3 is seen a modified tubing connected to a syringe. Pumping of the dirty water into the syringe provides engine driving a miniature pump (Figure 2).

Figure 3 shows a view on the attachment of the light measuring tube with photodiode that is shown in Figure 1, to the workbench. In front of the tube is shown receiving photodiode, which is from the back of the shade with the black rubber, to avoid any possible interfering influence of the light from the ambient lighting. Translucent polypropylene tube is inserted vertically into the working tube loophole. The tube is hermetically sealed firmly at the top with plastic insert and connected together using silicone tubing to the flotation container. Purified water is pumped from the flotation vessel in the selected time intervals.

In the middle of Figure 2, the measuring syringe is shown. The syringe is always filled with contaminated water that is measured. The syringe is supported by the small prism so that it keeps the same position. Linking provides the silicone tubing. Below the center of Figure 3 the outlets of photodiode are shown. The photodiode is fixed and it is covered with the rubber ring. Leads are connected to the measurement device to determine the voltage in mV.

The outputs of the photodiode are connected in parallel with load resistor. Thus, it is connected to an electronic measuring system which measures the voltage of the photodiode in mV. Resistor creates a defined load for a photodiode in order to measure the linear light intensity for each LED that means the measurements are possible in the different wavelengths.

**Figure 2.** Connecting the device to measure water purity to the flotation vessel.

Flotation process takes place in a flotation vessel continuously but small pump is used for periodic pumping with measuring tact 150s. The actual pumping to fill the transparent measuring chamber is set to 3s. So the initial solution that was measured prior to 150s is replaced during 3s by a new, cleaner solution from the current flotation vessel. The whole method of measurement is thus constructed so that the original solution prior 150s does not contaminate the new solution for the actual measurement. So the whole process of measurement samples for individual light wavelengths must be done, including reserve of 90s to stabilize the sample, rather than during the 150s, before the drawing of new samples. So it is left for measurement itself up to 60s.

**Figure 3.** Detail of the measurement system.

## 4. Experimental results and analysis

The results of measurements of voltage, in mV, at the photodiode for the individual wavelengths in the water that is polluted with the dispersion colorant of yellow tone 2201 are shown in Figure 4.

The difference in the degree of water pollution between each measurement was determined as follows:

$$Difference / s = \frac{|LT_{n+1} - LT_n|}{150} \tag{1}$$

where $LT_i$ means the value of the Light Transmittance [%] that was obtained during the $i$-th measurement.

Light Transmittance [%] was set as comparison of the voltage [mV] on photodiode during the measurement of the clarified waste water with the nominal value [mV], which corresponds to clear distilled water.

Statistical method of least squares was used to fit the curve of the Light Transmittance for the individual wavelengths. Least square method is applied using the program product EXCEL with the formula for the index of determination:

$$R^2 = \frac{\sum_{i=1}^{n}(\hat{y}_i - \overline{y})^2}{\sum_{i=1}^{n}(y_i - \overline{y})^2} \quad (2)$$

where

$y_i$ ... the values of Light Transmittance [%] for the individual wavelengths,

$\hat{y}_i$ ... the functional values for the polynomial of the sixth degree to fit the individual curves,

$\overline{y}$ ... the average of the values Light Transmittance [%] for the individual wavelengths,

$n$ ... number of measurement as amount of the values obtained during the laboratory experiment for the individual wavelengths.

The individual functions to fit the curve:

For the wavelength of 950nm (Figure 5):

$$y = -0.0002x^6 + 0.0101x^5 - 0.2448x^4 + 3.0349x^3 - 20.446x^2 + 72.606x - 25.675 \quad (3)$$

with the value for the index of determination:

$R^2 = 0.9999$

For the wavelength of 630nm:

$$y = -0.00002x^6 + 0.0016x^5 - 0.0426x^4 + 0.4176x^3 - 1.3431x^2 + 6.7199x - 4.2717 \quad (4)$$

with the value for the index of determination:

$R^2 = 0.9994$

For the wavelength of 595nm (Figure 6):

$$y = -0.00000007x^6 + 0.0009x^5 - 0.0327x^4 + 0.3719x^3 - 1.1952x^2 + 5.3398x - 3.2731 \quad (5)$$

with the value for the index of determination:

$R^2 = 0.999$

For the wavelength of 580nm (Figure 7):

$$y = 0.0002x^6 - 0.0095x^5 + 0.1556x^4 - 1.2025x^3 + 5.0441x^2 - 7.9634x + 4.3516 \quad (6)$$

with the value for the index of determination:

$R^2 = 0.9992$

For the wavelength of 530nm (Figure 8):

$$y = 0.0001x^6 - 0.0062x^5 + 0.1066x^4 - 0.8375x^3 + 3.4545x^2 - 5.8004x + 3.3307 \quad (7)$$

with the value for the index of determination:

$R^2 = 0.999$

For the wavelength of 470nm:

$$y = 0.00005x^6 - 0.0027x^5 + 0.0475x^4 - 0.385x^3 + 1.4921x^2 - 2.5315x + 1.4138 \quad (8)$$

with the value for the index of determination:

$R^2 = 0.9993$

For the wavelength of 400nm:

$$y = 0.00003x^6 - 0.0017x^5 + 0.0308x^4 - 0.2428x^3 + 0.9053x^2 - 1.4636x + 0.7858 \quad (9)$$

with the value for the index of determination:

$R^2 = 0.999$

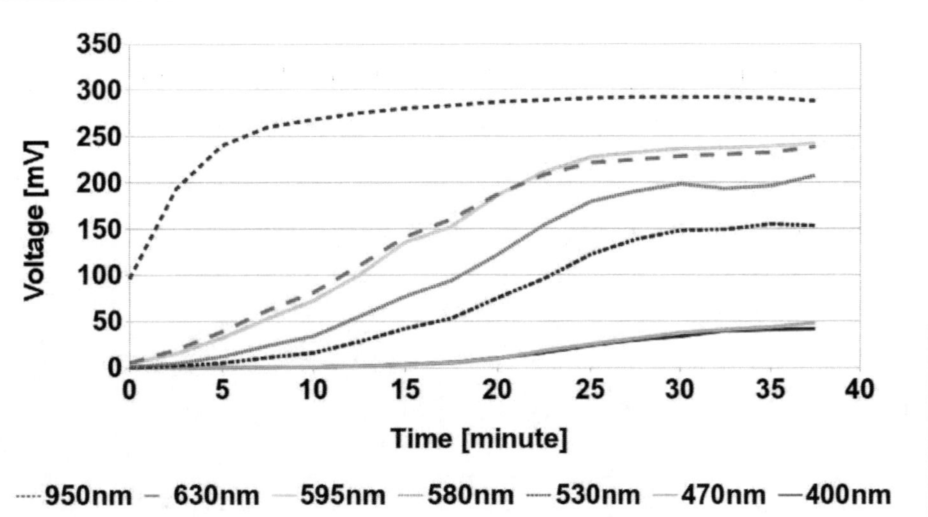


**Figure 4.** Measurements of voltage in mV at the photodiode for the individual wavelengths.

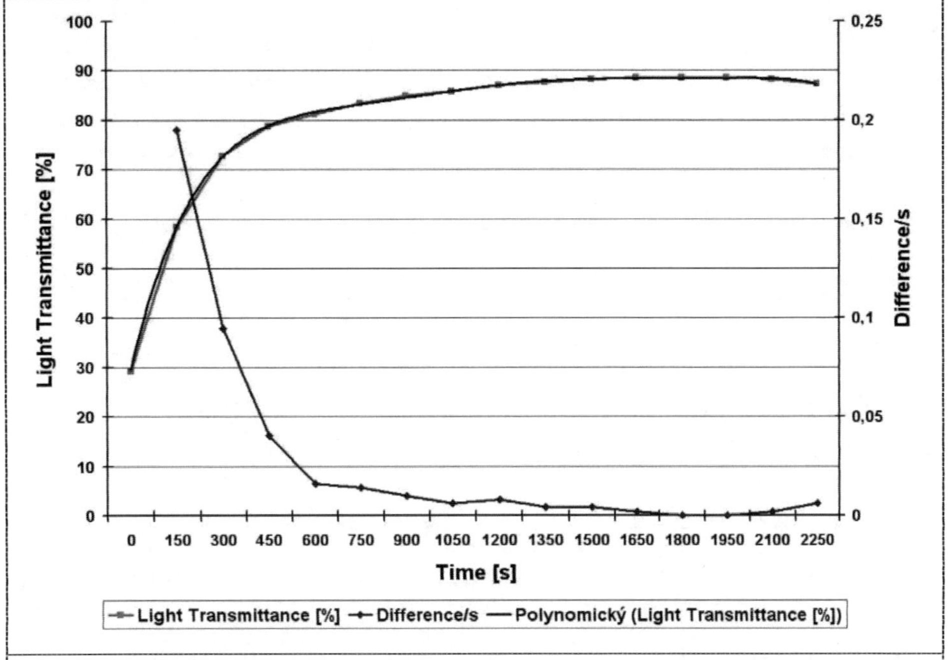


**Figure 5.** Light Transmittance and Difference/s for the wavelength 950nm.

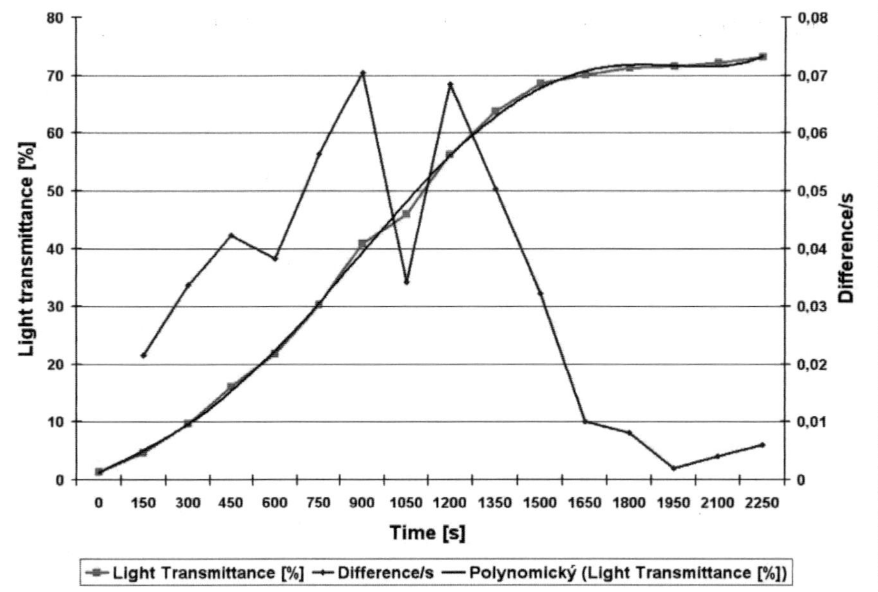


**Figure 6.** Light Transmittance and Difference/s for the wavelength 595nm.

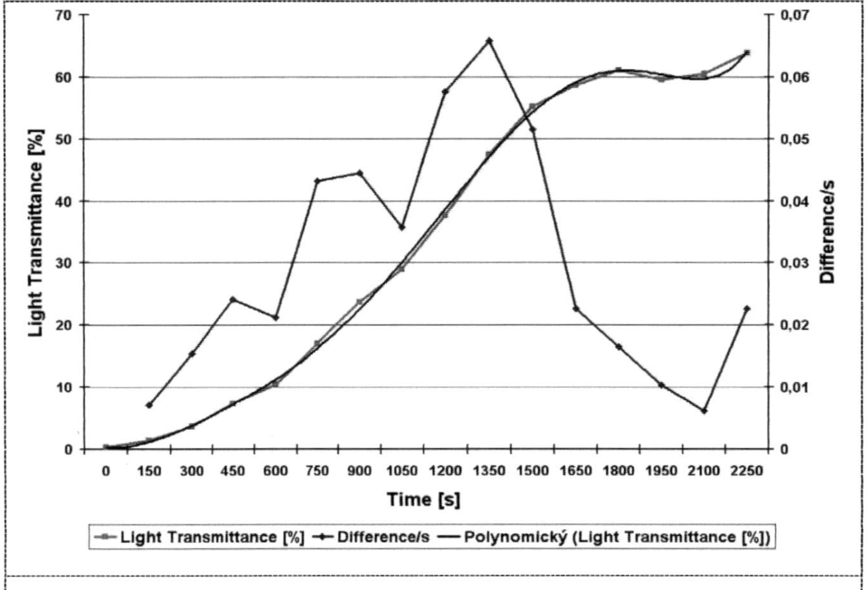


**Figure 7.** Light Transmittance and Difference/s for the wavelength 580nm.

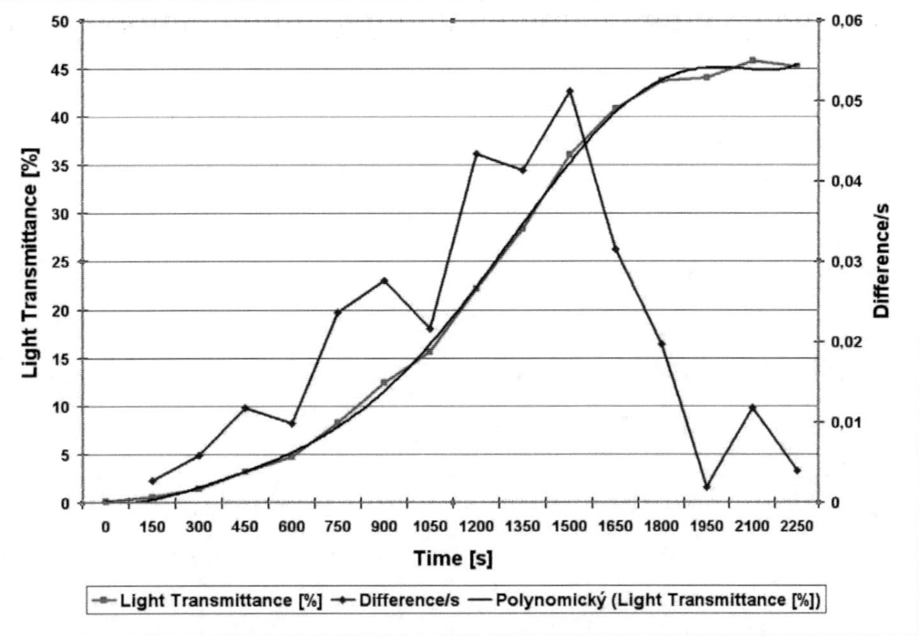


**Figure 8.** Light Transmittance and Difference/s for the wavelength 530nm.

## 5. Conclusions

The paper presents the technology that has been developed to determine the degree of clarity for contaminated water during the active electro-flotation process so that just the acting electrolytic flotation reaction does not affect the actual measured sample of treated contaminated water nor the measurement process itself. The measurement process for the individual sample must be done during 120s between the insertion of new sample. Determination of Light Transmittance [%] and Difference/s is used to determine the optimal time interval for the maximum possible cleaning of wastewater contaminated with the dispersion colorants in the process of electro-flotation. The water is subsequently clarified by spontaneous sedimentation within the next 48 hours.

Experiments confirmed that the required period of electro flotation depends on the color tone of the dispersion colorant, which caused the water pollution. After this period, which is required for flotation, purity of the water already so clear, that further it cannot be improved more using the same technology. The further improvement the water quality occurs only after a certain time period when spontaneous sedimentation of the water that has already been treated in the flotation process. In practice there is a mixture of different tones colors in the waste water and, therefore, time for the electro-flotation was set to 45 minutes (2700 seconds).

Analysis of the data obtained was performed by application the statistical method of least squares. The optimal time interval for the maximum possible cleaning of waste water contaminated with a yellow dispersion colorant of tone 2201 was set 1800s.

In our following research we will try to determine the optimal time intervals for the maximum possible cleaning up the wastewater contaminated with the dispersion colorants using the electro-flotation technology for variety of colors.

## 6. Acknowledgments

The contribution is based on the scientific research that was organized as a part of the international scientific grant 4363-6-14/16 "Modeling of environmental management processes" of the research institute in Dubna, Federal Republic of Russia and Faculty of Management at the University of Prešov in Prešov in Slovak Republic.

## 7. References

[1] Allen M, Preis A, Iqbal M, Srirangarajan S, Lim H B, Girod L and Whittle A 2011 Real-time in-network distribution system monitoring to improve operational efficiency (*Journal American Water Works Association* vol. 103(7)) pp 63-75

[2] Bhandari V and Abrol P 2013 Field Monitoring of Treated Industrial Waste Water (*International Journal of Electrical and Computer Engineering* vol. 3(5)) pp 629-634

[3] Kaur K, Mathur N and Bhatnagar P 2012 Comparative Study of Usage of Microbial Strains for Monitoring Waste Water Treatment Plants (*Universal Journal of Environmental Research and Technology* vol. 2(2)) pp 26-37

[4] Klishchenko P E, Chebotareva R D and Goncharuk V V 2012 Extraction of zinc from aqueous solutions by electrically polarized cation exchange fibers (*Journal of Water Chemistry and Technology* vol. 34, Issue 6) pp 253-257

[5] Mesi A and Kopliku D 2013 Toxicity Bio-Monitoring of Shkodra Lake Surface Water Using a Higher Plant Assay (*Academic Journal of Interdisciplinary Studies* vol. 2(8)) pp 133-140

[6] Moxey A 2012 Agriculture and Water Quality: Monetary Costs and Benefits across OECD Countries OECD (*Publishing Edinburgh*) pp 1-68

[7] Svetleishaya E M, Mitchenko T E and Astrelin I M 2014 Removal of natural organic matter by ultrafiltration (*Journal of Water Chemistry and Technology* vol. 36, Issue 1) pp 25-30

[8] Voudouris K and Voutsa D (Ed.) 2012 Water Quality Monitoring and Assessment (*InTech* Rijeka) pp 1-614

[9] Wang N, Li H, Long J, Cai Ch, Dai J, Zhang J and Wang R 2012 Contamination, source, and input route of polycyclic aromatic hydrocarbons in historic wastewater-irrigated agricultural soils (*Journal of Environmental Monitoring* vol 14) pp 3076-3085

[10] Xu L, Gu H, Li Ch, Shi A and Shen J 2013 System Design of Water Quality Monitoring Robot with Automatic Navigation and Self-test Capability (International Journal of Control and Automation vol. 6(5)) pp 67-82

Journal of Physics: Conference Series **622** (2015) 012016 doi:10.1088/1742-6596/622/1/012016

# Improving The Performance of Customer Loyalty of Online Ticketing in Indonesia's Showbiz Industry

**M Dachyar**[1] **and E S Athory**
Industrial Engineering Department, Universitas Indonesia

E-mail: mdachyar@yahoo.com

**Abstract.** Currently the entertainment industry is adopting online ticketing for supporting business from main products to profitability improvement. E-loyalty needs to be examined because are deals with less loyal customer characteristic. The sample are 249 customers whom have purchased for tickets online at least once. Data was gathered by questionnaires and analyzed by Path Analysis and Importance – Performance Analysis. The findings of the research indicate that satisfaction has the strongest relationship to e-loyalty. This study provides four improvement alternatives as a form of new business development strategy for showbiz industry in Indonesia.

## 1. Introduction

The use of Internet are rapidly changed the way companies doing business. E-commerce sales and penetration of online shoppers has increased globally in 2013. The number of digital buyers in Indonesia are expected to reach 7.4 million in 2015 [1]. Demographic forecast in 2020, are trends to be increasing to over two times of middle class and affluent consumer (Boston Consulting Group, 2012). This can affects to consumption behavior and lifestyles. Showbiz activities; going to music concerts, festivals, and sporting events are lifestyle which are increasingly prevalent in recent years. Showbiz industry are adopting online ticketing for improving its profitability. However, there are services that have not yet facilitated compared to facilities abroad. This indicates that there are many spaces for service improvement. Online customers have the character that less loyal to a product. E-loyalty are identified can improve service quality. The service improvement is a strategy increasing customer loyalty which impacted to company profitability.

E-loyalty is defined as the favorable attitude of customers towards e-business which leads to duplication of purchase [2]. Loyalty and satisfaction are the main goals for online companies to increase profitability, gain, and maintain competitiveness advantage [3]. Not the same as traditional business, online customers are not specifically interact with individuals. Customer can interact with companies through the users interface that allows them to initiate the desired transaction. This study add to equip some previous studies [4] [5] [6] [7] on e-loyalty to on-line purchase of goods in other countries, so the e-ticketing application being developed in Indonesia are considered to be known.

---

[1] Industrial Engineering Department, Faculty of Engineering, Universitas Indonesia, DEPOK-16424.

## 2. Research Methodology

Based on previous research about e-loyalty which conducted by Semeijn in 2005 [4], Safa on 2013 [5], Yee in 2010 [6] and Yang in 2004 [7], the relationship between factors can be made to get an assessment of 5 experts by questionnaires. The results of the expert assessment with a scale of 1 to 5, then calculated using Geomean method, to produce an initial baseline models as can be seen in Figure1.

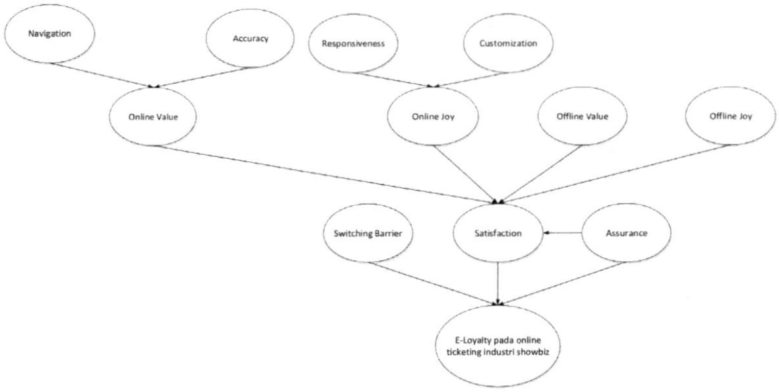


Figure 1. Initial Model

Based on the initial model above, some hypothesis needs to be tested in accordance to the initial model created in figure 1. Some relationships that need to be tested.

H1 : Is Navigation factor has a positive effect on Online Value
H2 : Is Accuracy factor has a positive effect on Online Value
H3 : Is Responsiveness factor has a positive effect on Online Joy
H4 : Is Customization factor has a positive effect on Online Joy
H5 : Is Online Value factor has a positive effect on terhadap Satisfaction
H6 : Is Online Joy factor has a positive effect on Satisfaction
H7 : Is Offline Value factor has a positive effect on Satisfaction
H8 : Is Offline Joy factor has a positive effect on Satisfaction
H9 : Is Assurance factor has a positive effect on Satisfaction
H10 : Is Assurance factor has a positive effect on E-loyalty
H11 : Is Satisfaction factor has a positive effect on E-loyalty
H12 : Is Switching Barrier factor has a positive effect on terhadap E-loyalty

To determine the relationship between factors in figure 1 we have used statistical methods Path Analysis. Questionnaire are then distributed to people who ever make purchase in showbiz online ticket at least once. Service industry companies provide ticketing services include music concert, sporting events, performing arts, and festivals tickets. The website is the primary distribution network for products owned by the service provider company. The 249 repondents as the valid data was collected and are processed.

To identify the drivers of e-loyalty, path analysis are used for assessing and modifying theoretical models. Path analysis is a theory that is associated with the development of multiple regression that are usually used to examine the relationship of correlation matrix with causal model. The purpose of path modelling is to analyse the causal relationships of sets of variables in minimizing differences between the covariance sample and theoretical models [8]. For a comprehensive review of customer satisfaction, Importance-Performance Analysis method, which was first proposed by Martilla and James in 1977, this is a popular technique to gain the customer satisfaction and prioritize service quality improvement [9]. This method are also used to eliminate the observed factors to attain priority improvement.

### 3. Data Analysis

Path analysis are used to find the significance of variables, correlational between variables and evaluate the hypothesized model. The data that has proven for its validity are then proceed to find the significance of each corellation. A corellation is well significant tested by calculating the P-value indicates the value of significance. Based on the model after modification significance value of P-values below (Table 1). Therefore, all hypotheses can be accepted

Table 1. Estimate & Significance Value

| | | | Estimate | S.E. | C.R. | P |
|---|---|---|---|---|---|---|
| Online_Value | <--- | Navigation | .243 | .046 | 5.245 | *** |
| Online_Value | <--- | Accuracy | .673 | .117 | 5.744 | *** |
| Online_Joy | <--- | Customization | .281 | .068 | 4.112 | *** |
| Online_Joy | <--- | Responsiveness | .349 | .063 | 5.519 | *** |
| *Satisfaction* | <--- | Offline_Joy | .577 | .085 | 6.794 | *** |
| *Satisfaction* | <--- | Online_Joy | .897 | .116 | 7.717 | *** |
| *Satisfaction* | <--- | Offline_Value | .184 | .052 | 3.561 | *** |
| *Satisfaction* | <--- | Online_Value | .209 | .098 | 2.131 | .033 |
| E_loyalty | <--- | Switching_Barrier | .162 | .034 | 4.835 | *** |
| E_loyalty | <--- | *Satisfaction* | .477 | .081 | 5.875 | *** |
| E_loyalty | <--- | Assurance | .139 | .064 | 2.159 | .031 |

In addition to the relationships that has been mentioned in hypotheses, it can be seen that the modified model (Figure 2) provides new relationships (Table 2).

The initial hypotheses that assumed e-loyalty are directly influenced by satisfaction, switching barriers, and assurance. Satisfaction does have the strongest relationship to e-loyalty. This means the most loyal customer are satisfied customer. Switching barrier is ranked 4th in influencing customer loyalty, this suggests that switching from online ticketing service still requires efforts. Buying tickets online is easier and more comfortable to customers, but the price and online transaction service are still barriers for customers. Assurance or trust factor is ranked seventh. It indicates that online customers are more concerned with the product than they earn about privacy and security.

In addition to the initial hypotheses, it turns out that there are additional relationships which have an even greater correlation. Those relationships showed effects that were not predicted earlier, such as online and offline joy effects. Both of these factors are respectively ranked second and third as the driver of e-loyalty. Pleasure is becoming the focus of customer than the value given in the online and offline phases. This is also due to the type of products offered by the service providers. The product offered is a ticket show that is associated with joy or pleasure.

Responsiveness, or the company's response to customers has the fifth largest value. Basically, customers need empathy as part of service experience. Therefore, this is such an important thing although interaction in online environment does not take place physically. Offline value is associated with the company's service quality exclude the website service. Offline order fulfillment system (after payment) also be customer's attention, but not strongly related. This is again because customer is more concerned with the joy they receive when they get the product. In addition, the service offered by provider of showbiz online ticketing in Indonesia is relatively similar. Thus, customers do not have

Journal of Physics: Conference Series **622** (2015) 012016 doi:10.1088/1742-6596/622/1/012016

benchmark to other kind of service. This causes customers' expectations in customization service are not too high.

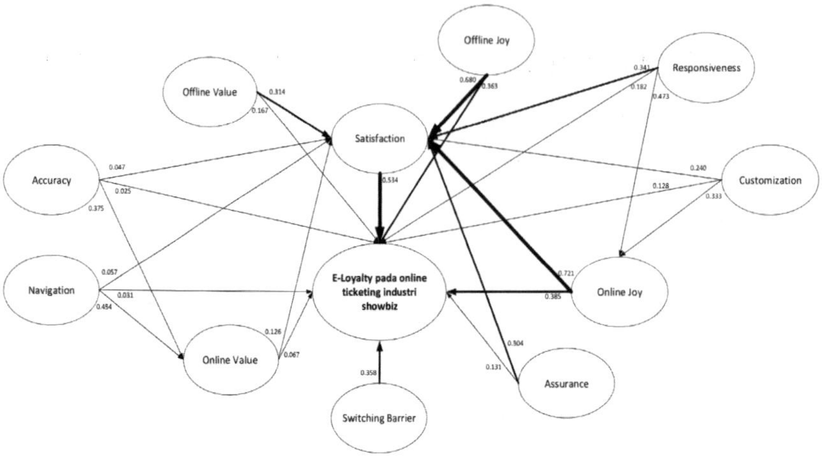


Figure 2. Modified Model

Table 2. Relationship Among Factors (All)

| | Online Joy | Online Value | Satisfaction | E-loyalty |
|---|---|---|---|---|
| Assurance | 0 | 0 | 0.304 | 0.131 |
| Switching Barrier | 0 | 0 | 0 | 0.358 |
| Offline Value | 0 | 0 | 0.314 | 0.167* |
| Offline Joy | 0 | 0 | 0.680 | 0.363* |
| Customization | 0.333 | 0 | 0.240* | 0.128* |
| Responsiveness | 0.473 | 0 | 0.341* | 0.182* |
| Accuracy | 0 | 0.375 | 0.047* | 0.025* |
| Navigation | 0 | 0.454 | 0.057* | 0.031* |
| Online Joy | 0 | 0 | 0.721 | 0.385* |
| Online Value | 0 | 0 | 0.126 | 0.067* |
| Satisfaction | 0 | 0 | 0 | 0.534 |

*emerging new correlation

Navigation problems such as website interface, structure, and content has direct influence to satisfaction even to e-loyalty, nevertheless the relationship is relatively weak. Issues regarding to the accuracy of information both company and products also has a direct relationship to satisfaction and loyalty with a relatively weak relationship.

A mathematical formulation is derived from this model to calculate the value of customer loyalty. The mathematical model for loyalty and satisfaction are:

E-loyalty = 0.534 Satisfaction + 0.385 Online Joy + 0.363 Offline Joy + 0.358 Swithing Barrier + 0.182 Responsiveness + 0.167 Offline Value + 0.131 Assurance + 0.128 Customization + 0.067 Online Value + 0.031 Navigation + 0.025 Accuracy

Satisfaction = 0.721 Online Joy + 0.68 Offline Joy + 0.341 Responsiveness + 0.314 Offline Value + 0.304 Assurance + 0.24 Customization + 0.126 Online Value + 0.057 Navigation + 0.047 Accuracy

Based on the results of path analysis method, satisfaction is a factor which contributes to e-loyalty. Furthermore, Importance-Performance Analysis methods are used to review customer satisfaction. Importance - Performance Analysis method examined mapping of the performance (x) and expectations (y) value which establish four quadrants matrix. Each quadrant describes priority in taking policy either improving performance or maintaining performance. Data distribution of performance and importance from the customer point of view is shown in Figure 3

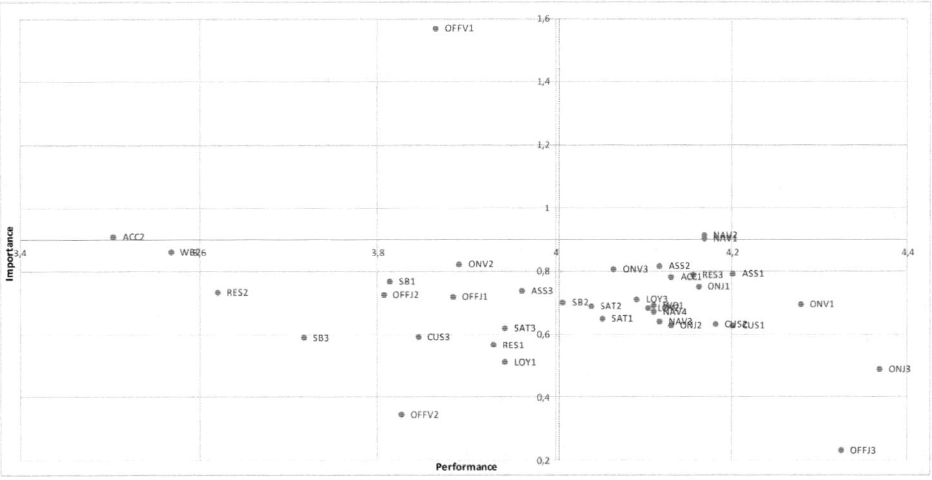


Figure 3. Importance-Performance Matrix

Intrepretations of the matrix are as follows:

1. Quadrant I (Concentrate Here) ; Quadrant I shows the factors or attributes that have high value in building customer loyalty but do not perform properly. Attributes included in this quadrant are :

- The service provider / company responds to my request quickly (RESP 3)
- I feel comfortable with the place reserved when I redeem ticket voucher (OFFJ 1)
- I feel comfortable with the queue when I redeem ticket voucher (OFFJ2)
- Through website, I am able to interact with the service provider / company (RES 1)
- Feedback provided by the service provider /company is in accordance with my expectations (RES 2)
- If there is an offline service that offers ticket I was looking for at a lower price, I would still buy it online (SB 1)
- If there is an offline service that provides a good service, I would still buy it online (SB 2)
- Voucher redemption system applied by the company has met my expectation (OFFV 1)

Attributes that fall into this quadrant represent key area that is becoming top priority for company to be improved.

2. Quadrant II (Keep Up the Good Work) ; Quadrant II shows the factors or attributes that have high value and has been properly implemented. Attributes included in this quadrant are:

- Website page can be accessed quickly (NAV 1)
- I can make a search on website easily (NAV 3)
- Structure and content of the website enables me towards the things I want (NAV 4)

- I feel safe with my personal information that I give to company (ASS 2)
- I believe in the security and privacy that are promised by company (ASS 3)
- The time I spend for online transaction is useful (ONV 2)
- Online ticketing website improve my shopping effectiveness (ONV3)
- I feel comfortable doing online purchases (ONJ 2)

All attributes that fall into this quadrant are the strength and pillar of the company so that it needs to be maintained.

3. Quadrant III (Low Priority) ; Quadrant III shows the factors that have low value and did not perform too well. Attributes included in this quadrant are:

- I always get the ticket I want on website (CUS 3)
- There are things I do not like when buying tickets offline than online (SB 3)
- System for ticket refund or event cancellation has met my expectations (OFFV 2)

Items in this quadrant have low priority for improvement and pose no threat to the company or service provider.

4. Quadrant IV (Possibly Overkill) ; Quadrant IV is quadrant that indicates the attributes which has low value but has good performance.

- Online services on the website provide more value (ONV 1)
- Information about service provider can be found on the website (ACC 1)
- Information about the event is completely available on website (ACC 2)
- Payment option that are provided has met my expectation (CUS 1)
- Voucher download system has already ease me (CUS 2)
- Service provider / online ticketing company is known for its trust and clarity of services to customers (ASS 1)
- I feel happy when I get the ticket voucher (ONJ 3)
- I feel happy when I get the original ticket (OFFJ 3)
- Website display has user-friendly interface (NAV 2)

Items in this quadrant have good performance, but do not really have high value regarding to loyalty. So it is better for he company or service provider to allocate resources to another higher priority items.

Average value of the overall performance is 3.9996, which are fairly good. However, there are several factors considered as important things by customer but have below-average performance. Thus, the items are becoming the focus for improvement. Based on Importance-Performance Analysis, factors that have high value but low performance are the items with greatest priority for improvement. After classifying factors into 4 quadrants, quadrant 1 is obtained to become the focus of improvement. Literature study and experts discussion are conducted to determine the appropriate improvement alternative. Solutions for each problem in quadrant 1 are shown in table 3

Table 3. Improvement Category Based on Its Function

| No. | Improvement | Function |
|---|---|---|
| 1. | Offline service facility (distribution channel, RFID membercard) | Eliminates discomfort (place, queues) that occurs in offline order fulfillment process. |
| 2. | Interactivity feature (live chat, quality control) | Provides communication media between the customer and the provider. |
| 3. | Value price (membership reward, promosi buy&get) | Gives customers more value than the price they pay. |
| 4. | Additional services (sell your ticket atau ticket | Service innovation that is not owned by competitor which also can be a barrier. |

| No. | Improvement | Function |
|---|---|---|
| | assurance) | |

Further step is to combine results from both Importance-Performance Analysis method and Path Analysis method. The combination of these two methods are expected to be more comprehensive to determine a new business development strategy, also accommodates customer's perspective.
The calculation of mathematical formulation was conducted to determine the effects of each alternative also combination of several alternatives. The calculation are expected gives company insight to allocate resources in the right area. The loyalty value of each condition is obtained from the calculation based on the performance value. Loyalty values based on mathematical formulation for each condition are shown in table 4.

The comparison above describes condition if one or more combination improvements are implemented. Offline service facility is the most optimal choice for improvement. Two combination improvement are most optimal implemented by provisioning offline service facility and interactivity features. Three combination are the most optimal implemented by provisioning offline service facility, interactivity features, and value price. Alternatives, the greatest impact on increasing loyalty is implemented all improvement options (offline service facility, interactivity features, value price, and additional service). Implementation of improvement, requires in-depth review and feasibility study for appropriate order.

Table 4. Loyalty Value at Various Condition

| Offline service | Interactivity feature | Value price | Additional service | Loyalty Value | Percentage of loyalty increasing |
|---|---|---|---|---|---|
| - | - | - | - | 25.176 | 0% |
| √ | - | - | - | 26.239 | 4.22% |
| - | √ | - | - | 25.876 | 2.78% |
| - | - | √ | - | 25.606 | 1.71% |
| - | - | - | √ | 25.409 | 0.92% |
| √ | √ | - | - | 26.940 | 7.01% |
| √ | - | √ | - | 26.669 | 5.93% |
| √ | - | - | √ | 26.472 | 5.15% |
| - | √ | √ | - | 26.306 | 4.49% |
| - | √ | - | √ | 26.109 | 3.71% |
| - | - | √ | √ | 25.838 | 2.63% |
| √ | √ | √ | - | 27.369 | 8.71% |
| √ | √ | - | √ | 27.172 | 7.93% |
| √ | - | √ | √ | 26.902 | 6.86% |
| - | √ | √ | √ | 26.539 | 5.41% |
| √ | √ | √ | √ | 27.602 | 9.64% |

Improvements were made to improve service quality, which can improve customer satisfaction, raise customer loyalty and increase market share [10]. Improving services to increase customer loyalty needs to be done because of an increase 5% loyalty can result 30-85% profitability [11]. The results of calculation can provide insight for companies to mix satisfaction initiatives and loyalty programs.

**4. Conclusion**

This research has proven that e-loyalty are respectively can influenced by satisfaction, online joy, offline joy, switching barrier, responsiveness, offline value, assurance, customization, online value, navigation, and accuracy. The data that are processed support all initial hypotheses in path analysis method. Further analysis gives 12 additional relationships among factors.

The results of path analysis showed that satisfaction has strongest relationship in affecting e-loyalty. Further review about satisfaction is analyzed by Importance-Performance Analysis. The problems are shown from the customer's perspective that adjusted by weighted factors. Estimated increase of loyalty value is derived for consideration for new business strategy planning. Based on loyalty value calculation, strategy that provide the greatest impact is the combination of 4 improvement alternatives (offline service, interactivity features, value price, additional service).

This study shows that the pleasure customer from online phase is the most important factor influencing satisfaction and e-loyalty. Followed by pleasure of offline phase are not much different value than online joy factor. This supports previous theory about positive relationship between satisfaction and loyalty; online phase has greater effect than offline [12]. As well as the theory offline order fulfillment are important as the quality of online service.

Customer loyalty is a good indicator to measure performance of company. Thus, e-loyalty becomes important to company. Implementation of business development strategies are more effective by combining the customer's perspective with driver's value. The results of this study are expected to provide insight for online ticketing service providers of showbiz industry, identify the drivers of e-loyalty, understand customer needs and improve service performance.

**Reference**

[1] eMarketer 2013 http://www.emarketer.com/newsroom/index.php/emarketer-review-key-2013-trends-coverage-areas-platform-growth
[2] Anderson, R. E. 2003 E-satisfaction and e-loyalty: A contingency framework. Psychology & marketing, 123-138.
[3] Liong, L. S. 2011 Relationship between service quality, satisfaction, and loyalty of Google users. International Journal of Electronic Commerce Studies, 35–56.
[4] Semeijn, J. V. 2005 E-services and offline fulfilment: how e-loyalty is created. Managing Service Quality, 182-194.
[5] Safa, N. S. 2013 A customer loyalty formation model in electronic commerce. Economic Modelling, 559-564.
[6] Yee, B. Y. 2010 Factors affecting customer loyalty of using Internet banking in Malaysia. Journal of Electronic Banking Systems, 21
[7] Yang, Z. 2004 Customer perceived value, satisfaction, and loyalty: the role of switching costs. Psychology & Marketing, 799-822.
[8] Stage, F. K. 2004 Path analysis: An introduction and analysis of a decade of research. The Journal of Educational Research, 5-13.
[9] Matzler, K. F. 2004 Employee satisfaction: does Kano's model apply? Total Quality Management and Business Excellence, 1179-1198
[10] Verona, G. 2002 A dynamic model of customer loyalty to sustain competitive advantage on the web. European Management Journal, 299-309.
[11] Reichheld, F. F. 1990 Zero defections: quality comes to services. Harvard Business Review, pp. 105-111.
[12] Shankar, V. S. 2003 Customer satisfaction and loyalty in online and offline environments. International journal of research in marketing, 153-175.

# Analysis of Urban Comprehensive Carrying Capacity of the prefecture-level city in Gansu province

**JING Li[12], MA Ling[3], Ma Shaohu[3], CHEN Xingpeng[1]**
1.College of Earth and Evironmental Sciences, Lanzhou University, Lanzhou 73000,
2.School of Urban Economics and Tourism culture, Lanzhou City University, Lanzhou 730070, Gansu China;
3.School of Education, Lanzhou City University, Lanzhou 730070, Gansu China

E-mail: jinglilzu@163.com

**Abstract.** Based on the theory of urban carrying capacity, this paper put forward twenty-two indexed applied to assess the problems of the prefecture-level city in Gansu province. The results show that resource carrying capacity is the most important factor that influences city development; the way of improving the urban comprehensive carrying capacity is the improvement of pressure-bearing index; the supply index does no tmatch demand index; the urban comprehensive carrying capacity presents stair-step decline; the leadership of two biggest cities are weaker relatively. To solve these problems, some measures are put forward such as changing the speed and mode l of development, improving the pattern of the human behaviors, etc.

## 1. Introduction

China's urbanization has developed rapidly in the 21st century with an average annual rate of 3.1% growth and the urbanization rate reached 53.73% in 2013. The increasing scale of cities and the rapidly accumulating population have caused changes in the process of social and economic development. These changes not only represent an improving civilization, but also have brought great pressure to the urban resource, environment, transportation, and further affecting the further improvement of the quality of urbanization.

The word "capacity" first appeared in ecology, then gradually permeated into the environmental science, demography, economics. Now, it has become a concept which is closely related with resource endowment, technology, social choice and values with a relative limit connotation of ethical character [1].The concept of urban carrying capacity first appeared in 2002. K.O defined the concept of urban carrying capacity as  as the material development of human activities, population growth and land utilization, which can realize the sustainable development of living environment system without causing degradation or irreversible damage [2].The concept was clearly put forward by the ministry of construction in January,2005,in "*the notification about strengthening the revision and approval of overall urban planning*." .The study about urban comprehensive carrying capacity is relatively late. It mainly started after 2000, but it has produced certain achievements. Some scholars of have carried on quantitative analysis on the urban land resources, water resources, environment resources, mineral resources and other single factor of bearing capacity, and put forward the corresponding countermeasure [3] [4] [5] [6] [7].While, some scholars comprehensively evaluate the urban (land) bearing capacity from the perspective of multi-factors, construct evaluation index system, carry out empirical analysis and quantitative evaluation, and put forward the measures to improve the bearing capacity. [8] [9] [10].At the same time, some scholars expand the research scope to the comprehensive carrying capacity of urban agglomeration and have probed into the methods and ideas.The study mainly concentrates on the study of single urban bearing capacity factor and involves less comprehensive research that regards the city as a system..(3) the quantitative method has not formed a general calculation model, thus the evaluation results are diversified.(4) the study area is given priority to the southeast coastal developed areas, and the city carrying capacity related problems are rarely involved. To sum up, this paper argues that the urban comprehensive carrying capacity refers to the coordination and balance between population and human activities under certain resource constraints as well as the organic combination of human activity and natural resources, environment, economic and social conditions. At the same time, it should be realized that natural resource abundance and the city's economic development level do not have complete positive correlation and any social and economic development of a city is built on the basis of the synergy of many factors.[14].

As the pressure of the cities in the southeast coast continues to increase, western cities will become one important position of accepting transferring agricultural population, therefore the research of their comprehensive carrying capacity is especially important. This paper takes 12 district cities in the Gansu province as the research object, then establishes the index system of the urban comprehensive carrying capacity to conduct objective measurement and evaluation on the comprehensive carrying capacity of those 12 cities, with the purpose of analyzing the methods of improving the urban comprehensive carrying capacity and promoting the coordinated development of the urban population, resources and environment.

## 2. Establish the evaluation index system of urban comprehensive carrying capacity

The data mainly comes from *the 2012 development yearbook of Gansu province*. The research object of this study is the 12 prefecture-level cities of Gansu province(Lin Xiazhou, Gan Nanzhou will not be evaluated due to the lack of data), including Lanzhou, Jiayuguan, Jinchang, Baiyin, Tianshui, Wuwei, Zhangye, Pingliang, Jiuquan, Qingyang, Dingxi, Longnan. Starting from the connotation of urban comprehensive carrying capacity and according to the regional characteristics, and the principles of being scientific, systematic, feasible, operational, the comprehensive carrying capacity index system of those 12prefecture level cities is established (figure 1), including target layer, base layer, criterion layer and index layer. Among them, the criterion layer includes two parts, the pressure index and pressure-bearing index.The pressure index refers to the demand pressures the city bears. Because the resources are limited and irrefragable, the greater the demand of the city for a certain resource, the more pressure the resource bears and the more limitation the urban development may face. Pressure bearing index refers to supply ability of the city. The more resources the city provides, the more development space the city has and the greater comprehensive bearing capacity the city has [15].

## 3. Modeling Method

### *3.1. The normalization of evaluation index*

Testing matrix method was used to calculate coefficients. According to the criteria of every testing matrix, two neighboring elements can be compared. The importance and degrees of importance will be judged by reciprocal scale methods from 1 to 9. The pairwise comparison matrix can be determined by the comparison of the average score for each element. The data collected by evaluation index has different dimensions and units, thus needing to be standardized. Suppose that, X is the index for the urban comprehensive carrying capacity,corresponding to the sample matrix of city j and i evaluation index, namely the equation is $X = x_{ij}$ .This paper divides the evaluation index into two major categories, namely the cost index and the efficiency index. The cost index refers to the indicators, the attribute value of which is negatively related to the urban comprehensive carrying capacity and the attribute value should be as small as possible. The efficiency index refers to the indicators, the attribute value of which is positively related to the urban comprehensive carrying capacity and the attribute value should be as big as possible [16].In order to eliminate the influence of the dimensions and dimensional units, this paper uses the fuzzy subjection function to nondimensionalize these two types of indicators..

**Table.1** Evaluation index systems of urban comprehensive carrying capacity of the refecture-level city in Gansu Province

| Target layer | Base layer | Criterion layer | Index layer |
|---|---|---|---|
| Urban comprehensive carrying capacity a | Resources carrying capacity b1 0.3075 | Pressure index 0.1944 | Per capita daily water consumption c1 0.1001 |
| | | | Comprehensive energy consumption c2 0.0943 |
| | | Pressure bearing index 0.1131 | Per capital construction land area c3 0.0415 |
| | | | Total water resources c4 0.0728 |
| | Environmental carrying capacity b2 0.1847 | Pressure index 0.034 | The proportion of the days when the air quality reaches or above the second level in one year c5 0.0216 |
| | | | Sewage emissions c6 0.0124 |
| | | Pressure bearing index 0.1507 | Green coverage ratio of construction land area c7 0.0809 |
| | | | Processing capacity of municipal sewage plant per day c8 0.0698 |
| | Economic capacity b3 0.2142 | Pressure index 0.1297 | Per capita gdp pc9 0.0705 |
| | | | Proportion of primary industry c10 0.0592 |
| | | Pressure bearing index 0.0845 | Proportion of tertiary industry c11 0.0413 |
| | | | Growth rate of gross value of production c12 0.0432 |
| | Infrastructure capacity b3 0.1643 | Pressure index 0.0759 | Per capita road area c13 0.0501 |
| | | | Transportation amount of household refuse c14 0.0258 |
| | | Pressure bearing index 0.0884 | The proportion of classified highway c15 0.0438 |
| | | | The number of buses per ten thousand persons c16 0.0446 |
| | Public service capacity b4 0.1293 | Pressure index 0.0578 | Number of bed in health agency per ten thousand persons c17 0.0251 |
| | | | The proportion of the person who have the lowest living ensure in the cities and towns c18 0.0148 |
| | | | Primary and secondary school enrollment c19 0.0179 |
| | | Pressure bearing index 0.0715 | The proportion of the person participating in new rural cooperative medical system c20 0.0098 |
| | | | The proportion of the medical treatment and public health |

| | | | |
|---|---|---|---|
| | | | expenditure in each area c21 0.0286 |
| | | | The proportion of the education expenditure in each area c22 0.0331 |

The cost indicators use the glower semi-trapezoid fuzzy subjection function to nondimensionalize

$$x_{ij}^{'} = \begin{cases} 0 & x_{ij} = Mx_{ij} \\ \dfrac{Mx_{ij} - x_{ij}}{Mx_{ij} - mx_{ij}} & mx_{ij} < x_{ij} < Mx_{ij} \\ 1 & x_{ij} = mx_{ij} \end{cases} \tag{1}$$

The efficiency indicators use the glower semi-trapezoid fuzzy subjection function to nondimensionalize

$$x_{ij}^{'} = \begin{cases} 0 & x_{ij} = mx_{ij} \\ \dfrac{x_{ij} - mx_{ij}}{Mx_{ij} - mx_{ij}} & mx_{ij} < x_{ij} < Mx_{ij} \\ 1 & x_{ij} = Mx_{ij} \end{cases} \tag{2}$$

In the equation: $x_{ij}^{'}$ is the result of normalization and its value is between zero and one; $x_{ij}$ is the original value of the index; $mx_{ij}$ is the minimum value of the original values of different areas at the same period; $Mx_{ij}$ is the maximum value of the original values of different areas at the same period

*3.2. The determination of index weight*

The entropy value method is a kind of objective weighting method. It can better avoid the deviation of the subjective assignment and determine the index weight according to the information provided by the original values of the indicators.And the indicators actually involved are expressed by the standardized value.The ultimate evaluation value and the level of the comprehensive carrying capacity is the product of both parts.The weight of the evaluation index system determined by the entropy method can fully reflect the description required by the level of urban comprehensive carrying capacity in the stress levels, pressure levels and coordination and the statistics also meet the requirements of entropy value method.Therefore, it is scientific to use the entropy method to evaluate comprehensive carrying capacity [17].Calculation steps are as follows:

1.Calculate the weight of index layer

First step:suppose that the index of the base layer(or criterion layer) is $i$, the index of the target layer is $j$, then the index weight $p_{ij}$ of the Jth sample under the ith index is $p_{ij} = \dfrac{x_{ij}^{'}}{\sum_{i=1}^{n} x_{ij}^{'}}$, $n = 12$

The second step:calculate the entropy $e_j$ of the jth index: $e_j = -\dfrac{\sum_{j=1}^{n} p_{ij} \ln p_{ij}}{\ln p_{ij}}$

The third step:calculate the utility value of the jth indicator: $g_j = 1 - e_j$

The fourth step:calculate the weight of the jth index: $w_j = \frac{g_j}{\sum_{j=1}^{n} g_j}$

2.Calculate the weight of the criterion layer and the base layer

The evaluation system of the multilayer structure can, according to the additivity of the entropy, take advantage of the utility value of indexical information of lower to determine the weight value $w_j$ corresponding to the upper structure value in proportion.In the previous steps, the utility value of each index has been calculated. Calculate sums of the utility value of each index of the lower structure, then get the utility value of all kinds of index,with the detonation $G_j,(j=1,2,\cdots)$. And subsequently obtain the sums of the utility value of all index: $G=\sum G_i$ ;the weight of the corresponding index: $w_i = \frac{G_i}{G}$

*3.3. The construction of index model*

For further research of the bearing capacity level used to evaluate the city, this paper adopts the urban comprehensive carrying capacity evaluation model. The comprehensive carrying capacity index systematically reflects the overall level of urban development ability under certain resources and environment constraints.At the same time we can also find out the short slab factors influencing the comprehensive carrying capacity, thus put ting forward the corresponding countermeasures.

The evaluation value of subsystem of the base layer adopts the following evaluation model: $Y_i = \sum_{i=1}^{m} w_j x_{ij}$ (3)

In equation: $Y_i$ is the integrated assessment value of the ith subsystem in the base layer; $m$ is the number of the evaluation index contained in the $i$ th subsystem; $w_j$ is the weight of the jth indicator in the ith subsystems; $x_{ij}$ is the evaluation value of the jth indicator in the ith subsystem.

The urban comprehensive carrying capacity value A of the base layer adopts the following evaluation model: $A = \sum_{i=1}^{n} w_i Y_i$ . Obviously,the bigger A is, the better A is. Therefore, as the weight is known, we can easily determine every scheme or sort them according to the value of urban comprehensive carrying capacity.

## 4. The analysis of the comprehensive carrying capacity of the cities in Gansu province

*4.1. The resources bearing capacity has the greatest impact on the comprehensive carrying capacity of the prefecture-level cities in Gansu province.*

It can be seen from table 1 that among the weights of all subsystems which are determined by the entropy method, the bearing capacity of resources has the greatest influence,with a weight of 0.3075.The following are respectively economical capacity, environmental capacity, infrastructure capacity, and public service capacity.The daily water consumption per capital is the single index, far more than the second index, which illustrates that the problems of water resources is the main factor influencing the comprehensive carrying capacity of the prefecture level cities in the Gansu province.

*4.2. The analysis and calculation of indexes*

Step 1 The analysis of pressure bearing index

It can be seen from table 2 that, the pressure bearing index of the prefecture-level cities in Gansu province are not identical.In general, the comprehensive capacity pressure pressuring bearing indexes of three cities are more than 1, which respectively are Jiayuguan, Jinchang and Jiuquan.The pressure bearing indexes of other nine cities including the provincial capital Lanzhou are less than 1, which shows that for most of the prefecture-level cities in Gansu province , the main method of improving the urban comprehensive carrying capacity in the future is to improve the bearing capacity.Seen from the perspective of resources pressure bearing index,the jiayuguan city is the highest and longnan city is the lowest;seen from the perspective of environmental pressure bearing index, the highest is the Dingxi city, the lowest is the Jinchang city; From the perspective of economic pressure bearing index, the highest is Dingxi city, the lowest is the Jinchang city; from the perspective of infrastructure pressure bearing index, the pressure index of each city is similar;from the perspective of public service pressure index, it can be seen that most of the cities have low pressure bearing index. Except fo the Jiayuguan city which is above 1, the dat of 10 cities are below 0.5.

**Table.2** The pressure-bearing index of urban comprehensive carrying capacity of the refecture-level city in Gansu Province

| Area | resources pressure bearing index | Environmental pressure bearing index | economic pressure bearing index | infrastructure pressure bearing index | public service pressure bearing index | urban comprehensive carrying capacity pressure bearing index |
|---|---|---|---|---|---|---|
| Lanzhou | 0.48812 | 0.49395 | 0.21186 | 1.16654 | 0.28120 | 0.92730 |
| Jiayuguan | 1.20267 | 0.11739 | 0.07546 | 1.15813 | 1.65863 | 1.84320 |
| Jinchang | 0.89133 | 0.08263 | 0.07828 | 1.15716 | 0.96980 | 1.53113 |
| Baiyin | 0.57339 | 0.22621 | 0.15509 | 1.16210 | 0.30970 | 0.96405 |
| Tianshui | 0.26019 | 0.51621 | 0.51679 | 1.15967 | 0.20215 | 0.68492 |
| Wuwei | 0.32454 | 1.42605 | 0.29218 | 1.15917 | 0.22667 | 0.75078 |
| Zhangye | 0.62585 | 2.44142 | 0.18114 | 1.15621 | 0.29181 | 0.95697 |
| Pingliang | 0.46513 | 0.85420 | 0.36989 | 1.16057 | 0.22434 | 0.79130 |
| Jiuquan | 0.73667 | 0.33471 | 0.13035 | 1.15341 | 0.38058 | 1.11419 |
| Qingyang | 0.37449 | 0.79551 | 0.20158 | 1.16175 | 0.25355 | 0.80511 |
| Dingxi | 0.35204 | 3.32301 | 0.79569 | 1.15869 | 0.19820 | 0.70291 |
| Longnan | 0.23569 | 2.09464 | 0.66957 | 1.15590 | 0.19533 | 0.66407 |

**Table.3** The analysis of factors carrying capacity and urban comprehensive carrying capacity of the refecture-level city in Gansu Province

| Area | resources carrying capacity | Environmental carrying capacity | economic carrying capacity | infrastructure carrying capacity | public service carrying capacity | urban comprehensive carrying capacity | Ranking |
|---|---|---|---|---|---|---|---|
| Lanzhou | 0.08681 | 0.05465 | 0.16252 | 0.05517 | 0.10979 | 0.46894 | 6 |
| Jiayuguan | 0.11377 | 0.05226 | 0.34437 | 0.04686 | 0.03281 | 0.59008 | 3 |
| Jinchang | 0.10456 | 0.05394 | 0.34704 | 0.04877 | 0.04328 | 0.59760 | 2 |
| Baiyin | 0.09976 | 0.06085 | 0.21475 | 0.05606 | 0.10316 | 0.53458 | 4 |
| Tianshui | 0.06117 | 0.04509 | 0.06271 | 0.04153 | 0.10754 | 0.31804 | 10 |
| Wuwei | 0.07059 | 0.04996 | 0.10496 | 0.04616 | 0.10870 | 0.38037 | 8 |
| Zhangye | 0.10195 | 0.05995 | 0.18564 | 0.05531 | 0.10681 | 0.50966 | 5 |
| Pingliang | 0.07709 | 0.04981 | 0.08760 | 0.04604 | 0.10934 | 0.36988 | 9 |

| | | | | | | | |
|---|---|---|---|---|---|---|---|
| Jiuquan | 0.11921 | 0.06403 | 0.26595 | 0.05948 | 0.09427 | 0.60294 | 1 |
| Qingyang | 0.08024 | 0.05481 | 0.15273 | 0.04988 | 0.10724 | 0.44490 | 7 |
| Dingxi | 0.06500 | 0.04445 | 0.04723 | 0.04075 | 0.10720 | 0.30463 | 11 |
| Longnan | 0.05827 | 0.04384 | 0.05133 | 0.04011 | 0.10683 | 0.30037 | 12 |

Step 2 The analysis of the comprehensive carrying capacity index
It can be seen from the table 3 that the comprehensive carrying capacity of the the prefecture-level cities in Gansu province are echeloned, reflecting in that the west is high and the east is low. The standard deviation of the urban comprehensive carrying capacity of the five cities,Jiuquan, Jiayuguan, Jinchang, Baiyin, Zhangye is above the mean value of 0.5, thus belonging to the higher bearing capacity. And the comprehensive carrying capacity of other cities is below the mean value of 0.5. The cities with medium bearing capacity are Lanzhou, Qingyang ($0.42 < A < 0.48$). The urban comprehensive carrying capacity of the five cities, Tianshui, Wuwei, Pingliang, Dingxi, Longnan , are below 0.4 with more than two standard deviations, which shows that these cities belong to low bearing capacity.

Seen from the figure 1, the contribution of each factor system for comprehensive -carrying in each city of Gansu province is not the same.The improvement of the urban comprehensive carrying capacity is a systematic and coordinated process.The weakening of the function of any elements will have great impact on the coordination degree of the whole city system and the decline of the comprehensive carrying capacity.

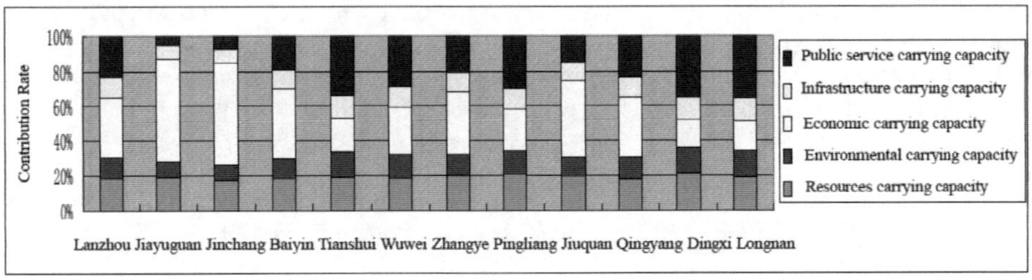


**Figuer.1** The rate of contribution of factors carrying capacity to urban comprehensive carrying capacity of the refecture-level city in Gansu Province

Step 3 The Lanzhou city and Tianshui city carrying capacity among the urban agglomeration
The Lanzhou city and Tianshui city are the first and second largest city of Gansu province. They also respectively are the center of the Lanzhou-Xining urban agglomeration the Guanzhong - Tianshui economic zone, playing a leading role in the region.But seen from table 3, the comprehensive ability of Lanzhou city only ranks fourth in the province,and the resource carrying capacity is its short board factor. What is different other big cities is that, the development of Lanzhou city is more limited by the shortage of land resources. While the development problem of Tianshui city is different from Lanzhou's. Tianshui's economic bearing capacity has a large gap with other cities'with a low contribution rate. And its problems of resource bearing capacity are mainly manifested in the shortage of water resources.

## 5. Conclusion

The composition of urban comprehensive carrying capacity is not the simple sum of the carrying capacity of various factors but the organic combination of them. At the same time, it also should be recognized that the short board factors of the urban comprehensive carrying capacity reflect in congenital aspect and acquired aspect. And the starting points of breaking through these two constraint factors are different. Congenital short factors is difficult to make a breakthrough, such as land resources, water resources; Acquired short factors have a relative big room to improve, including public services and infrastructure construction which can make a breakthrough through government planning, institutional innovation.

### *5.1. Make full use of resources and improve the urban comprehensive carrying capacity*

Take the demand for resources and the utilization of resources as breakthrough, reasonably guide demands, and improve the efficiency. Transform the concept of resource utilization of enterprises and residents in daily production and life, implement the system that encourages and rewards saving and gid their reasonable resource demand orientation. Encourage to develop circular economy to realize the efficient utilization of resources, strive to build a conservation-minded society, improve the urban comprehensive carrying capacity.

### *5.2. Improve the bearing capacity of system factor according to local conditions and improve urban comprehensive carrying capacity*

Seen from the above analysis, the factors affecting the urban comprehensive carrying capacity of the prefecture-level cities in Gansu province are not the same, thus it is necessary to improve the bearing capacity of system factor according to local conditions so as to eventually realize the purpose of improving the urban comprehensive carrying capacity. Take the Lanzhou city as an example, improving the urban comprehensive carrying capacity needs to take promoting resources bearing capacity as the breakthrough and carry out order construction of new city zone through further revitalization and deep mining of the existing land resource.To sum up, when a certain element of the city comprehensive carrying capacity system becomes short slab, not only its breakthrough needs to be considered, the further optimization of the system structure also should be considered.

### *5.3. Steadily push forward the urbanization process, improve the quality of urbanization, guide the reasonable distribution of population, form reasonable structure of the urban comprehensive carrying space carrier*

The prefecture-level cities of Gansu province have become the important front for acceptance of the transferring rural population. But it can be seen from the analysis that the urban resource utilization efficiency is low with the problems of waste of land expansion, environmental problems, lack of infrastructure and public service ability existing, which shows that the urban development pattern has certain problems. It is suggested that the future population distribution not be too concentrated in the big cities so as to alleviate the pressure of the big city. Under policy guidance,the population transfer can be concentrated in small and medium-sized cities, thus enhancing the connotation of the city.

**Acknowledgement**

Urbanization of Institutional transformation and Institutional Innovation in Gansu Province of Minorities Areas (1111-06); Mechanism of new-type promoted Urbanization in Gansu Province of Minorities Areas (2011-GSCFY - KJ08); Ecological Civilization intensive Perspectives for intensive development strategies in middle-sized and small cities of Northwest Minorities Regions (2013-GSCFY- RW30)

**Reference**

[1] Sun Chunhua. Research on Comprehensive Evaluation of Carrying Capacity about Prefecture-level City in Inner Mongolia. Journal of Finance and Economics Theory. 2013(5):81-87

[2] Oh K, Jeong Y, Lee D, et a.l An Intergrated Framework for the Assessment of Urban Carrying Capacity . Korea Plan Assoc,2002,37(5):7-26.

[3] United States Environmental Protection Agency. Four Townships Environmental Carrying Capacity Study. 2002,

[4] Yu Weidong Min Qingwen. Some Progresses and Perspectives in Study on the Carrying Capability of Water Resources. ARID ZONE RESEARCH,2003,20(1):60-66.

[5] CHENG Li-li L(U) Cheng-wen XU Guo-lin. Study on Population Supporting Capacity of Land Resources in Anhui Province. RESOURCE DEVELOPMENT & MARKET ,2006,22(4): 318-320.

[6] Zhu Yizhong Xia Jun. Assessment of water resources carrying-capacity with multi-criteria scenario analysis method: a case study in Zhangye region. GEOGRAPHICAL RESEARCH,2005, 24(5): 732-740.

[7] Song Xiaomeng, Kong Fanzhe, Zhan Chesheng. Assessment of water resources carrying capacity in Tianjin City of China.Water Resources Management, 2011, 25: 857-873.

[8] Cang Hui. Carrying capacity, population equilibrium, and environment's maximal load. Ecological Modelling, 2006, 192:317-320.

[9] LV Guangming HE Qiang. A Study on Urban Comprehensive Carrying Capacity under Sustainable Development Theory. URBAN STUDIES,2009,16(4):1-3.

[10] LONG Zhihe, REN Tongxian, LI Min and HU Guiping. The research of city carrying capacity of Cuangzhou. SCIENCE AND TECHNOLOGY MANAGEMENT RESEARCH,2010,30(5):204-207.

[11] L IU Xiaoli and FANG Chuanglin. Progress and Prospect of Study on Carrying Capacity of Resource and Environment of City Clusters. PROGRESS IN GEOGRAPHY ,2008,27(5):35-42.

[12] OU Chao-min and LIU Ren-yang.Evaluation of Urban Synthetic Carrying Capacity of the Chang-Zhu-Tan Urban Agglomerations. JOURNAL OF NATURAL SCIENCE OF HUNAN NORMAL UNIVERSITY, 2009,32(3):108-112.

[13] CHEN Juan LI Jing-bao QING Xiong-zhi. Evaluation on Urban Comprehensive Carrying Capacity of "3+5" Urban Agglomeration of Hunan Province. TERRITORY & NATURAL RESOURCES STUDY 2010,(3):69-74.

[14] Buckley R. An ecological perspective on carrying capacity[J].Annals of Tourism Research,1999,(03):705-708.

[15] GAO Hong-li TU Jian-jun and YANG Le. Evaluation of the Comprehensive Carrying Capacity of Cities. JOURNAL OF SOUTHWEST UNIVERSITY(NATURAL SCIENCE),2010.

[16] Gu Chaolin, Zhen Feng. Synthesized Evaluation of Capacity of Sustainable Development of Cities in Jiangsu Province. JOURNAL OF NANJING UNIVERSITY(NATURAL SCIENCES). 2001 7(3)

[17] YANG Li-hua HE Xue-feng. On the Sustainable Development Index of Trade Based on Entropy Method-Exemplified by the Empirical Research of Zhejiang Province. JOURNAL OF NINGBO UNIVERSITY(LIBERAL ARTS EDITION),2007(1):22-24.

[18] SUN Li, L(U) Bin and HU Jun. Urban Carrying Capacity Evaluation of Zhongyuan city Agglomeration. AREAL RESEARCH AND DEVELOPMENT.2008(6):16-20

# Statistical analysis of the strength and lifetime under tension of crystalline polymeric solids

**C Y Li[1,2] and K H Nitta[1]**

[1]Institute of Science and Engineering, Kanazawa University, Kakuma Campus, Kanazawa, 920-1192, JP

[1]E-mail: nitta@se.kanazawa-u.ac.jp

**Abstract.** The ductile fracture behavior under uniaxial tension of melt-crystallized isotactic polypropylene specimens at room temperature was investigated from a statistical point of view. Each tensile test was performed more than one hundred times and statistical data for the breaking point were obtained under each tensile condition. The probability distribution curves of the fracture time and strength approximately followed Gaussian statistics at lower tensile speeds, but changed to a Weibull function at higher-speed tests. Additionally, with increasing tensile speed the mean and standard deviation of the fracture time decreased linearly. The toughness, which is the total area under the stress-strain curves, was found to be independent of the tensile conditions, indicating that fracture toughness is a criterion for fracture under tension.

## 1. Introduction

Characterization of tensile fracture behavior in commodity semicrystalline plastics is important for their extension to engineering and industrial applications. Semicrystalline polymers show high ductility with yielding, necking and strain hardening under typical tensile conditions. In accordance with the amount of plastic deformation involved in various tensile conditions, fracture characteristics can be broadly categorized as brittle or ductile. The two categories can be distinguished by whether or not yielding and necking is involved. However, the fracture behavior of semicrystalline polymers cannot easily be classified as either ductile or brittle. Nitta et al [1] found that high density PE materials can be broken in a brittle fashion at higher strain rates and lower temperatures but show a clear yield peak. In this case, they are called semi-brittle.

Over the past decades, statistical analyses of fracture processes in plastics, rubbers, metals, ceramics, fibers and composites have received attention because the statistical data play a critical role in designing materials with high reliability and long life time for engineering applications [2]. The characteristics and corresponding structural models for brittle or quasi-brittle fractures have been comprehensively and extensively studied; in particular by Bazant [3] and Pheonix [4]. Bazant et al. [3,5] suggested that the specimen size of brittle and/or quasi-brittle materials strongly affects the

[1] To whom any correspondence should be addressed.
[2] Present address: Toray Battery Separator Film Co., Ltd. Nasu-Shiobara, Tochigi, 329-2763, JP

strength distribution function. Their work focused on probabilistic formulations for practical assessments of uncertainties in strength predictions using the classical weakest link theory [6]. The behavior of brittle fracture was analyzed using Weibull distribution functions [7]. However, few statistical studies have been performed on the ductile fracture behavior of commodity plastics in typical tensile tests [8].

In this work, isotactic polypropylene (iPP) was chosen as a model plastic material. The ability to predict the fracture points of such iPP materials under uniaxial extension is of critical importance to determine reliability and efficiency for practical use. Therefore, it is useful to find the characteristics of fracture behavior of iPP that relate directly to the service performance of the material. Mechanical experiments such as the uniaxial elongation test are useful to determine optimal conditions for drawing of the plastics. In this work, we examined the probability functions for tensile breakage data such as fracture time, strength, and toughness of iPP films under various tensile conditions.

## 2. Experimental

Commercial grade iPP with $M_w=4.0\times10^5$ and $M_w/M_n=5$ was used in this work. The iPP films were prepared by compression molding at 230°C for 5 min, and quenching in boiling water (100°C). The film thickness was controlled to be about 200 μm. The crystallinity of the films was 59% (estimated from density) and their melting temperature was 162°C (estimated from differential scanning calorimetry). Double-edge notched specimens were cut from films for the tensile measurements. We employed a double-edge notched specimen with gauge length $L_0 = 4$ mm, 4-mm width and 200-μm thickness. The uniaxial elongation tests for these specimens were repeated more than 100 times at room temperature under each test condition. The tensile speed was varied from 15 mm/min to 70 mm/min. Additionally, tensile tests were performed at higher tensile speeds of 140 and 170 mm/min. The time to fracture was read out from the tensile load-time data. The tensile strength was determined by dividing the tensile load at the breaking point by the initial cross-sectional area.

The data from the tensile tests were analyzed statistically by taking the class size to be 10, corresponding to the square of the sample size. Distribution analysis of the fracture data was carried out using OriginPro 8 commercial software (OriginLab Co. Ltd, USA).

## 3. Results and discussion

### *3.1 Probability distribution functions of fracture time and strength*

All stress-elongation time curves were measured more than 100 times with fixed crosshead speed. The ten stress-time curves measured at 15-mm min$^{-1}$ crosshead speed are shown in figure 1. The iPP specimens showed a distinct yield point as a maximum in the stress-strain curves beyond the elastic limit. When using notched specimens, the neck constitutes the entire portion of the gauge length immediately after the neck is initiated [1]. Thus, the necked region in the stress-strain curves can be neglected. Therefore, the yield region after the initial elastic region is followed by a virtually complete strain-hardening deformation, and as a result the stress linearly increases as elongation proceeds after the yield process. As seen in this figure, the fracture time has a wide scatter, and thus the probability distribution data for the fracture time as well as the fracture stress (strength) were fit to a typical Gaussian distribution (figure 2).

The shape of the curve of the tensile stress versus elongation time (figure 1) is roughly composed of a yield peak, hardening region, and a negligible necking region. Therefore, we found a linear relationship between stress and elongation time in the hardening region near the breaking point:

$$\sigma = G_p t + \omega_p , \tag{1}$$

where $G_p$ is the slope of the stress-time curve in the hardening region and $\omega_p$ is the intercept on the stress axis. The fluctuation of $G_p$ is very small compared with the time to fracture, $t_B$, and $\omega_p$, with $G_p =$

$0.128 \pm 0.01$ MPa $s^{-1}$.

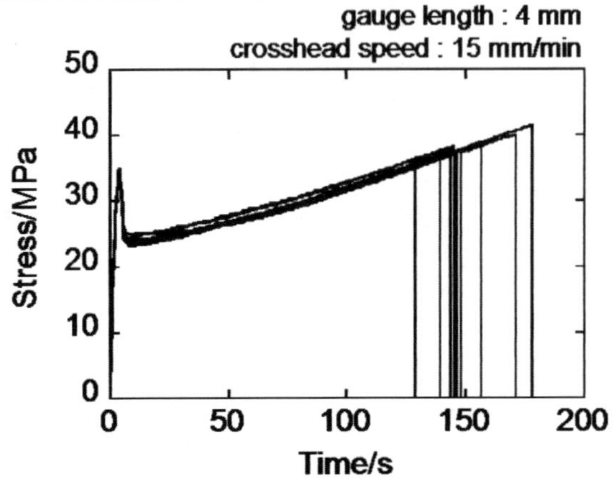


**Figure 1.** Stress-strain curves for iPP specimens with 4-mm gauge length, measured at 15-mm/min crosshead speed at room temperature .

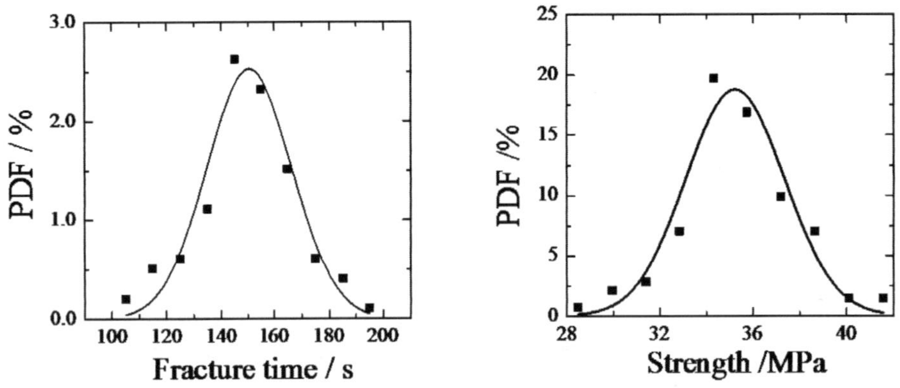


**Figure 2.** Probability density functions of fracture time and fracture strength for 15-mm/min crosshead speed. The solid line is the fit to a Gaussian distribution function.

Consequently, the tensile strength is linearly transferred over the time to fracture, $t_B$, offset by the $\omega_p$ values. When we consider the fluctuation of $\omega_p$ as a noise fluctuation, this process can be treated as a well-known Kalman filter problem [9]. Because the noise in $\omega_p$ is a result of fluctuation of the load cell and specimen thickness, $\omega_p$ is independent of $t_B$. Furthermore, the fact that the probability density function of the fracture strength can be fit by a Gaussian distribution implies that $\omega_p$ is Gaussian. If the relation between the stress and elongation time closely followed a concave or convex curve, the distribution would have negative or positive skewness, respectively, leading to a derivative Weibull fit [10].

*3.2 Tensile speed dependence*

With increasing tensile speed, fracture time decreased and the dispersion in the PDF curves narrowed. Furthermore, the mean fracture time $\bar{t}_B$ and its standard deviation $\Delta_t$ are proportional to the inverse of the tensile speed $v$, given by $v\bar{t}_B = \lambda_0$ where $\lambda_0$= 37 mm, and $\Delta_t = \delta_0/v$ where $\delta_0$=3.2 mm (see figure 3). Consequently, the PDF curves of the fracture time can be written in terms of the reciprocal quantity $\tau = 1/v$ as

$$p_t(t_B;\tau)=\frac{1}{\sqrt{2\pi(\delta_0\tau)^2}}\exp\left[-\frac{(t_B-\lambda_0\tau)^2}{2(\delta_0\tau)^2}\right]. \tag{2}$$

Here $\tau$ is the intrinsic time. In the tensile tests, the elongation velocity $v$ is above zero so that the intrinsic time $\tau$ has a time horizon. Hence, as $v \to \infty$, the limit of the intrinsic time is given by $\tau = 0$. Here we can assume the initial condition $t_B = 0$ when $\tau = 0$, according to the straight line to the origin. The PDF satisfies the partial differential equation

$$\frac{\partial}{\partial\tau}p_t(t_B;\tau)=\left\{-\lambda_0\frac{\partial}{\partial t_F}+\delta_0^2\tau\frac{\partial^2}{\partial {t_F}^2}\right\}p_t(t_B;\tau). \tag{3}$$

This is a *Kolmogorov forward equation* [11], in which the fracture behavior in the ductile experiment can be regarded as a diffusion process in which fracture events correspond to particle diffusion, intrinsic time $\tau$(=1/$v$) corresponds to the diffusion time, and the fracture time $t_B$ corresponds to the diffusion distance.

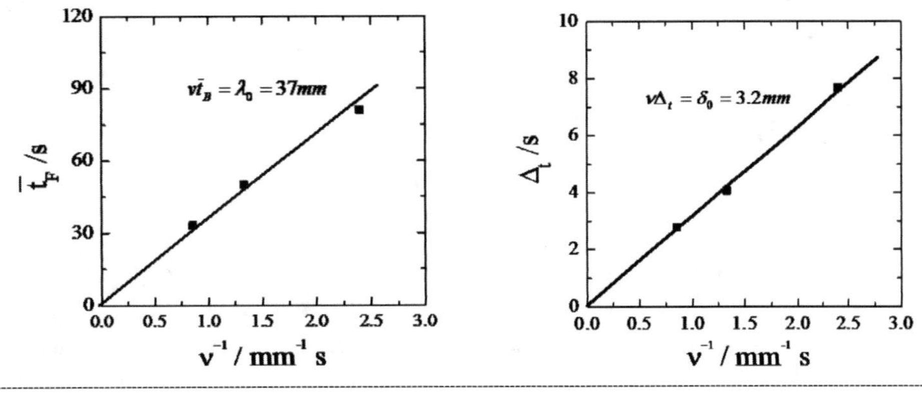


**Figure 3.** Mean fracture time for iPP specimens (left) and standard deviation (right) plotted against the reciprocal value of tensile speed.

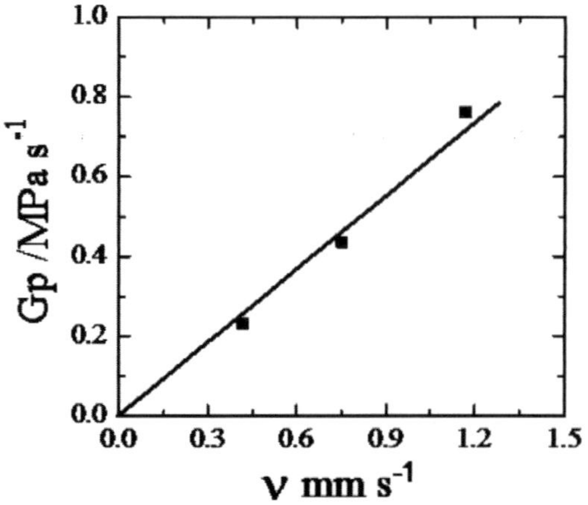


**Figure 4.** Strain-hardening slope Gp of iPP specimens plotted against tensile speed *v*.

Figure 4 shows the dependence of the strain-hardening slope $G_p$ on tensile speed. We found a linear proportionality between $G_p$ and $v$ with the empirical relation $G_p = \gamma_0 v$, where $\gamma_0$ = 670 MPa/m. As a result, equation (1) can be rewritten as

$$\sigma = \gamma_0 vt + \omega_p = \gamma_0 L_0 \frac{vt}{L_0} + \omega_p = \gamma_0 L_0 \varepsilon + \omega_p, \tag{4}$$

where $L_0$ is the initial gauge length. Because the product of $v$ and $t$ is the increment of the gauge length under tension, the strain $\varepsilon$ can be given by $vt/L_0$, and thus the stress value is proportional to the strain $\varepsilon$, and Haward's stress-hardening modulus [12] corresponds to $\gamma_0 L_0 / 3$. The stochastic behavior of stress is obtained from the stochastic fracture time plus the Gaussian noise $\omega_p$ via the linear relation of equation (1), revealing the mean and variance of strength:

$$\bar{\sigma}_B = \mathrm{E}\left[G_p t_B + \omega_p\right] = \gamma_0 \varepsilon_0 L_0 + \bar{\omega}_p, \tag{5}$$

$$\Delta_\sigma^2 = \mathrm{E}\left[(\sigma_F - \bar{\sigma}_F)^2\right] = \gamma_0^2 \delta_{\varepsilon_0}^2 + \Delta_\omega^2. \tag{6}$$

Considering that the noise fluctuation $\omega_p$ can be caused by experimental errors such as fluctuations of the load cell and specimen thickness, $\bar{\omega}_p$ and $\Delta_\omega^2$ are likely to be independent of $L_0$. Hence, the mean of the strength is predicted to be proportional to the gauge length and its variance is independent of the gauge length.

The concept of toughness is defined in several ways, one of which is the tensile toughness defined as

$$U_F = \int_0^{\varepsilon_F} \sigma(\varepsilon) d\varepsilon. \tag{7}$$

The tensile toughness, which is the energy per volume required to break a specimen, is a measure of the ability of a material or fabricated article to withstand load application in the form of tensile resistance. Interestingly, not only the mean value but also the standard deviation of the fracture toughness were almost independent of the tensile speed as shown in figure 5. This independence implies that fracture toughness is an intrinsic characteristic and a criterion for fracture under the present tensile conditions.

As shown in figure 3, the mean fracture time and its standard deviation were proportional to the reciprocal tensile speed. The fracture events at various speeds were determined by the fracture toughness, and the toughness values were constant, independent of the tensile speed. So as speed is increased, the fracture time required to reach the criteria of toughness will be shorter. Therefore, the mean fracture time is proportional to the reciprocal of the tensile speed. The stretched sample consists of a number of oriented cavities (see figures 9 and 10). As the tensile speed is increased, the time period between fracture events shortens, therefore the fracture time standard deviation becomes narrower with higher tensile speed.

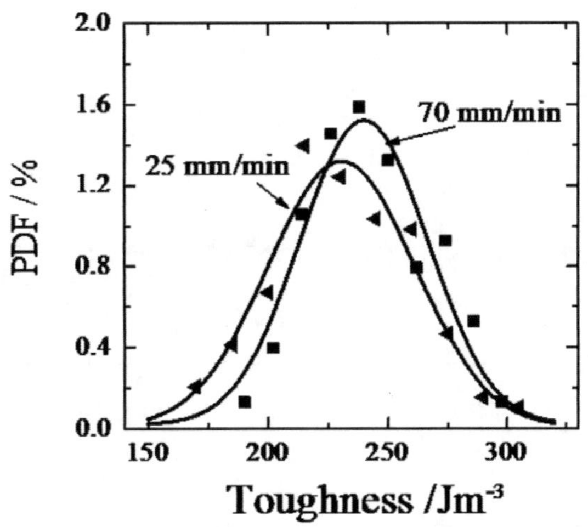


**Figure 5.** Dependence of toughness distribution of iPP specimens on tensile speed, at tensile speeds of 25 and 70 mm/min.

*3.3 Higher tensile speed*

Figure 6 compares the PDF curves of fracture time measured at higher tensile speeds of 140 and 170 mm/min. The fracture time data at a crosshead speed of 140 mm/min were symmetric and fit to a typical Gaussian distribution with a skewness value around zero, similar to the fracture time data measured at tensile speeds ranging from 15 to 70 mm/min. The symmetric fracture time scatter for the 170-mm/min data leans to the left with a skewness value of 1.39, and the solid curve is the nest normalized Weibull function given by

$$p_t(t_B) = \frac{m}{\eta}\left(\frac{t_B}{\eta}\right)^{m-1} \exp\left\{-\left(\frac{t_B}{\eta}\right)^{m}\right\}, \quad (8)$$

where $m$ is the shape parameter and $\eta$ is the scale parameter of the distribution. The transition of the PDF curve from a Gaussian to a Weibull function occurred at a tensile speed between 140 and 170 mm/min. This transition seems to be associated with the transition from ductile to semi-brittle [13]. Also, the mean values of the fracture time at higher tensile speeds fall on the same line $\bar{t}_B = \lambda_0 / v$ plotted by the inverse of the tensile speed $v^{-1}$, with the empirical relationship obtained at the tensile speed range from 15 to 70 mm/min.

The standard deviation of fracture toughness data measured at speeds of 170 and 140 mm/min are 43 and 51 $Jm^{-3}$, respectively; larger than all the values of toughness measured at lower speeds (about 25 $Jm^{-3}$). However, there is no clear difference in the mean values of the fracture toughness for the specimens stretched at lower and higher tensile speeds.

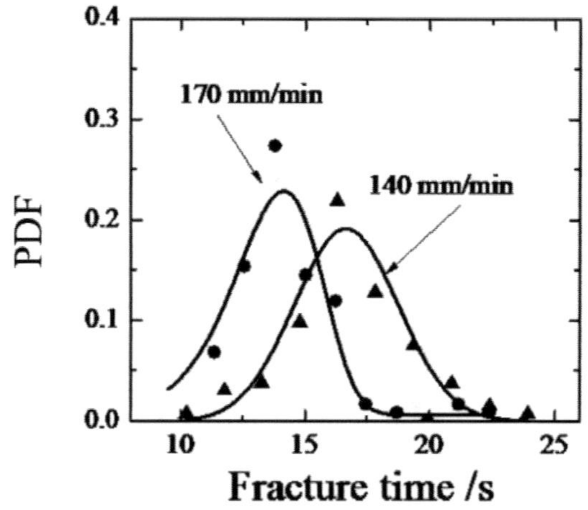


**Figure 6.** PDF curves of the time distribution of iPP specimens, showing dependence on tensile speed. PDF curve at 140-mm/min tensile speed was fit to a Gaussian distribution and PDF at 170-mm/min tensile speed was fit by a Weibull distribution with m = 8.605 and $\eta$ = 14.33.

*3.4 Cavity and void observations*

The intensity of small angle X-ray scattering (SAXS) in deformed semicrystalline polymers such as iPP is known to be a superposition of contributions from scattering on periodic structures (crystalline-amorphous) and much more intense scattering from voids [14]. For specimens stretched beyond the yield point, the scattering profile ascribed to the periodic structure is concealed behind the scattering signals from voids [15].

Figure 7 shows SAXS 2D patterns for the drawn specimens stretched beyond the yield point in the stress-time curve at speeds of 45 and 170 mm/min. All the drawn directions are horizontal. According to these data, for all specimens the signals ascribed to a large fraction of voids suddenly appeared at smaller angles, consisting of two intense round lobes centered on the median in a peanut shell shape. This pattern corresponds to the signature from larger heterogeneities elongated with the long axis oriented perpendicular to the applied load [16]. Moreover, the intensity of scattering tends to increase at lower tensile speeds, indicating that the void size and volume show a positive dependence on the tensile speed. Figure 8 shows SAXS patterns for fracture specimens taken from the zone in the vicinity of the fracture cross-section drawn up to 170% of nominal strain. A rhombus or elliptical scattering pattern is seen in the vertical direction for the fracture specimens drawn at 45 or 170 mm/min tensile speed, respectively. According to figures 7 and 8, we find that as elongation proceeds, the orientation of the voids changed from perpendicular to parallel to the stretching direction. Further deformation led the cavities to align along the deformation direction (maximum scattering is in the direction perpendicular to deformation). These patterns show that the cavities have been reoriented at a local strain of around 170%. [17].

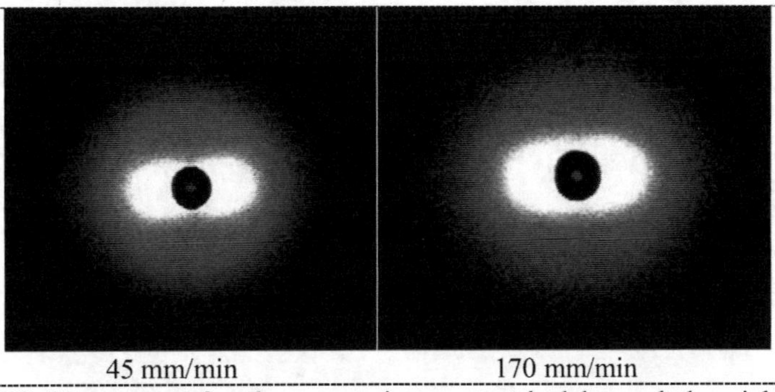


**Figure 7.** SAXS 2D patterns for drawn specimens stretched beyond the yield point in the stress-time curve, at speeds of 45 and 170 mm/min. The drawn direction is horizontal.

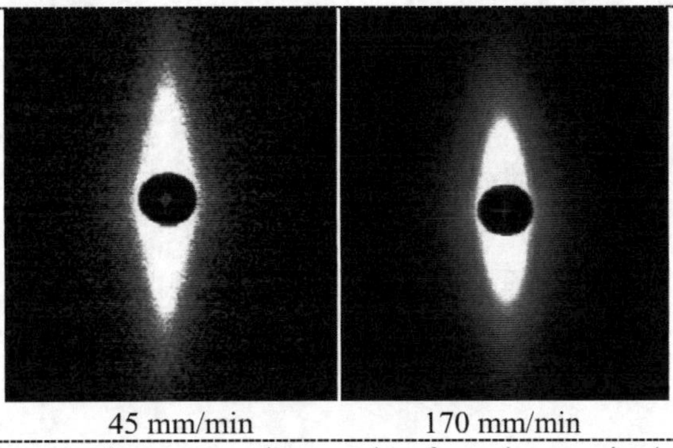


**Figure 8.** SAXS patterns for fracture specimens taken from the zone in the vicinity of the cross section. The drawn direction is horizontal.

To characterize the difference in features between the ductile and semi-brittle fractures, we performed scanning electron microscopic (SEM) imaging of the surface zone of specimens failed at speeds of 170 and 45 mm/min. To distinguish the different influences of intense and low tensile speeds, figure 9 shows SEM images of the surface zone of specimens failed at speeds of 170 and 45 mm/min. Similar void clusters separated by damaged and undamaged walls are detected perpendicular to the tensile direction on the surfaces for both kinds of specimens. There are more numerous and broader voids over the surface for the specimen stretched at 170 mm/min, compared with the one stretched at 45 mm/min. The same results can be visualized by polarized optical microscopy (figure 10). This means that, for specimens stretched at various tensile speeds, the fracture behavior is dominated by the propagation of voids and cavities. For similar void types, the opening and propagation speeds of the voids and cavities is proportional to the tensile speed, and thus the mean fracture time is inversely proportional to the tensile speed. When the specimen is stretched at extremely high tensile speed, voids and cavities coalescence as large voids appear and then propagate instantly, reducing the specimen lifetime. Thus, the probability of fracture increases, and the fracture time probability density distribution leans to the left.

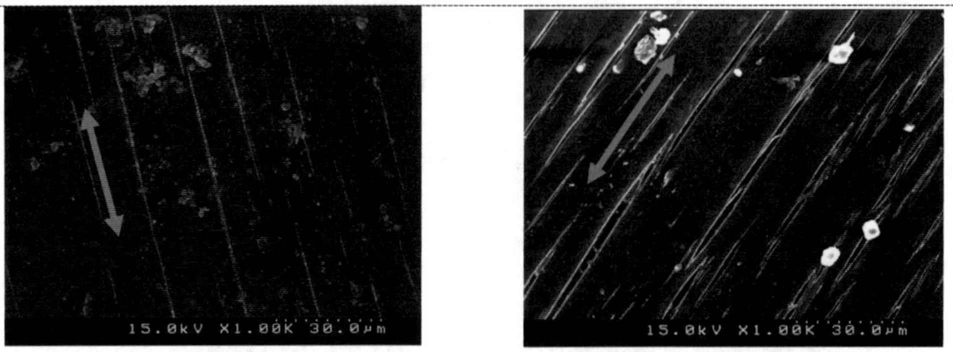


45 mm/min 170 mm/min

**Figure 9.** SEM images of the surface zone of specimens failed at speeds of 45 and 170 mm/min. The red arrows indicate the stretching direction.

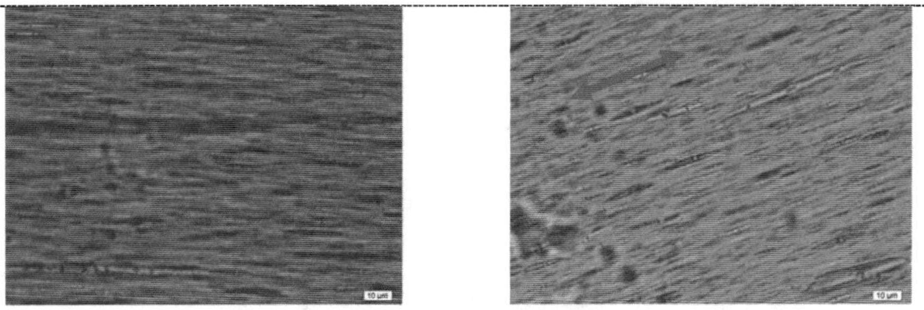

45 mm/min 170 mm/min

**Figure 10.** Polarized optical microscopic images of the surface zone of specimens failed at speeds of 45 and 170 mm/min. The red arrows indicate the stretching direction.

## 5. Conclusions

In this work, the statistical nature of tensile fracture for iPP materials under different tensile speeds was sd. Data collected from more than one hundred tensile tests were analyzed using Gaussian and Weibull distribution curves to quantify the nature of the data scattering. We have drawn the following conclusions:

1: The probability distribution curves of fracture time, strength, strain and toughness can be fit by Gaussian distribution functions.

2: The fracture time is inversely proportional to the tensile speed. The fracture time monotonously increases with increasing gauge length of the specimen.

3: The average fracture strength is more sensitive to gauge length than to tensile speed. Longer gauge length leads to larger average strength.

4: Toughness is an intrinsic characteristic, independent of tensile speed and gauge length, and is a possible criterion for fracture.

5: At crosshead speeds above 170 mm/min, the probability distribution function of the fracture time changed from a Gaussian to a Weibull distribution.

These fracture time distributions make it possible to predict precise safety factors based on the probability of fracture events. Consequently, we can define the service limitations of the material. This will provide important information for materials design, to improve the reliability and lifetime of the materials under various conditions.

**References**

[1] Nitta KH and Ishiboro T 2002 *J. Polym.Sci. Polym. Phy.***40** 2018
[2] Gorus A 2012 *Analysis of reliability and quality control: fracture mechanics 1*(New York: Wiley)
[3] Bazant ZP 2004 *Proc. National Acad. Sci. USA* **101** 13400
[4] Phoenix SL 1978 *Int. J. Fracture* **14** 327
[5] Bazant ZP and Pang SD 2004 *Fracture Mechanics Concrete Structures* 189
[6] Freudenthal AM and Gumbel EJ 1959 J. Amer. Statist. Assol. **49** 575
[7] Weibull W 1951 *J. Appl. Mech.* 293
[8] Ben PY and Xu J 2011 *Polym. Eng. Sci.* **51** 574
[9] Kalman RE 1960 *Trans ASME J Basic Eng.* 35
[10] Rousu DN 1973 *Technometrics* **15** 927
[11] Kolmogoroff A 1931 *Math Annal* **104** 415
[12] Haward RN 1993 *Macromolecules* **26** 5860-5869.
[13] Zhang XC, Butler MF, Cameron RE 2000 *Polym* **41** 3797
[14] Allan P, Bevis M 1977 *Philos Mag* **35** 405
[15] Butler MF, Donald AM , Ryan AJ 1998 *Polymer* **39** 39
[16] Zhang XL, Botler MF and Cameron RE 2000 *Polym.* **41** 3797
[17] Pawlak A 2007 *Polym.* **48** 1397

# New Attacks on RSA with Modulus $N = p^2q$ Using Continued Fractions

**M A Asbullah**[1] **and M R K Ariffin**[2]

[1,2]Al-Kindi Cryptography Research Laboratory, Institute for Mathematical Research, Universiti Putra Malaysia, Serdang, 43400, Malaysia
[2]Department of Mathematics, Faculty of Sciences, Universiti Putra Malaysia, Serdang, 43400, Malaysia

E-mail: [1]`ma_asyraf@upm.edu.my`, [2]`rezal@upm.edu.my`

**Abstract.** In this paper, we propose two new attacks on RSA with modulus $N = p^2q$ using continued fractions. Our first attack is based on the RSA key equation $ed - \phi(N)k = 1$ where $\phi(N) = p(p-1)(q-1)$. Assuming that $q < p < 2q$, $2p^{5/3}|p^{1/3} - q^{1/3}| < \frac{1}{3}N^{\beta}$ and $d < N^{\frac{1-\beta}{2}}$, we show that $\frac{k}{d}$ can be recovered among the convergents of the continued fraction expansion of $\frac{e}{N-(2N^{2/3}-N^{1/3})}$. Our second attack is based on the equation $eX - \left(N - \left(ap^2 + bq^2\right)\right)Y = Z$ where $a, b$ are positive integers satisfying $\gcd(a, b) = 1$, $|ap^2 - bq^2| < N^{1/2}$ and $ap^2 + bq^2 = N^{2/3+\alpha}$ with $0 < \alpha < \frac{1}{3}$. Given the conditions $|Z| < \frac{1}{3}N^{1/3+\alpha}Y$ and $1 \le Y \le X < \frac{1}{2}N^{\frac{1}{6}-\frac{\alpha}{2}}$, we show that one can factor $N = p^2q$ in polynomial time.

## 1. Introduction

Prior to 1970's, the encryption and decryption were done only in symmetrical ways. It is only until 1978, the RSA cryptosystem [13] went public and it is regarded now by the cryptographic community as the first realization of the public key cryptosystem. The security of RSA is based on the intractability to solve for the following four key equations; $N = pq$ for large primes $p$ and $q$ of the same size, the Euler totient $\phi(N) = (p-1)(q-1)$; the key equation $ed - \phi(N)k = 1$ such that $e$ is known and finally the modular $e^{th}$-root problem $c \equiv m^e \pmod{N}$.

The RSA cryptosystem is likely to have faster decryption if the secret exponent $d$ is relatively small. Nevertheless, in 1994 [10] observe that if $d$ is small, say $d < N^{1/4}$ so by exploiting the key equation $ed - \phi(N)k = 1$, he establishes that using a mathematical tool, namely the continued fraction, then such secret exponent can efficiently be recovered. This is one of major concern to implement RSA appropriately, because the knowledge of secret exponent $d$ lead to factoring $N$ in polynomial time. Since then, many researchers have pursued the same direction (i.e. using convergent by continued fraction method, for instance ([5], [6], [14]) purposely to improve the bound $\delta > \frac{1}{4}$ for $d < N^{\delta}$. Besides the continued fraction approach, the usefulness of Coppersmith's theorem [7] widely applied for cryptanalyzing RSA key equations, such as ([1],[3],[8]). A survey on the mathematical cryptanalysis of the RSA cryptosystem can be found in [4].

As mentioned in [1] such that moduli of the form $N = p^2q$ are frequently used in cryptography and therefore they represent one of the most important cases. This assertion is confirmed by the fact that many RSA-like cryptosystem are designed using such modulus. For example, ([9],[11],[12], [15]).

**Our contribution.** Hence, in this paper, we contribute two new attacks on RSA-type modulus of $N = p^2q$ using continued fractions. Our first attack is motivated from some previous attacks on RSA using a good approximation of $\phi(N) = p(p-1)(q-1)$. We observe that for $N = p^2q$ with $q < p < 2q$, if we use the term $N - (2N^{2/3} - N^{1/3})$ as an approximation of $\phi(N)$ satisfying the key equation $ed - \phi(N)k = 1$ then $\frac{k}{d}$ is one of a convergent of the continued fraction $\frac{e}{N-(2N^{2/3}-N^{1/3})}$ satisfying $2p^{5/3}|p^{1/3} - q^{1/3}| < \frac{1}{3}N^{\beta}$ and $d < N^{\frac{1-\beta}{2}}$.

The second attack is extending the result of [3] for $N = p^2q$ and $e$ satisfying the variant equation $eX - (N - (ap^2 + bq^2))Y = Z$ where $a, b$ are positive integers with $\gcd(a, b) = 1$ such that $|ap^2 - bq^2| < N^{1/2}$, $ap^2 + bq^2 = N^{2/3+\alpha}$ with $0 < \alpha < \frac{1}{3}$. We show tha if $|Z| < \frac{1}{3}N^{1/3+\alpha}Y$ and $1 \leq Y \leq X < \frac{1}{2}N^{\frac{1}{6}-\frac{\alpha}{2}}$, then $N$ can be factored in polynomial time using the second attack.

The rest of this paper is organized as follows. In Section 2 we give an introductory to continued fraction and some previous result regarding to the RSA cryptanalytic method using continued fraction. Section 3 and Section 4 give details around our two approaches, respectively. Finally, Section 5 concludes this paper.

## 2. Preliminaries

We start with the definition and an important theorem regarding to a continued fraction. We then provide some previous results on cryptanalyzing RSA utilizing such theorem. In this work, the symbol $\phi(N)$ is a notation for Euler totient function for its respective modulus $N$.

### 2.1. Continued Fraction

**Definition (Continued Fraction).** The continued fraction of a real number $R$ is an expression of the form

$$R = a_0 + \cfrac{1}{a_1 + \cfrac{1}{a_2 + \cfrac{1}{a_3 + \dots}}}$$

where $a_0 \in \mathbb{Z}$ and $a_i \in \mathbb{N} - \{0\}$ for $i \geq 1$. The numbers $a_0, a_1, a_2, \ldots$ are called the partial quotients. We use the notation $R = [a_0, a_1, a_2, \ldots]$. For $i \geq 1$ the rational $r_i/s_i = [a_0, a_1, a_2, \ldots]$ are called the convergents of the continued fraction expansion of $R$. If $R = \frac{a}{b}$ is a rational number with $\gcd(a, b) = 1$, then the continued fraction expansion is finite.

There are various results and applications of continued fraction. A key role in all our arguments is played by the following result.

**Theorem 1 (Legendre).** Let $a, b, x, y$ be integer such that $\gcd(a, b) = \gcd(x, y) = 1$. Suppose $|\frac{a}{b} - \frac{x}{y}| < \frac{1}{2y^2}$. Then $\frac{x}{y}$ is a convergent of the continued fraction expansion of $\frac{a}{b}$.

### 2.2. Previous Attacks On RSA Using Good Approximation of $\phi(N)$

In this section we provide some previous attacks on RSA using the convergents of good approximations of $\phi(N)$.

*2.2.1. Wiener's Attack* Firstly, let us consider the public exponent $e$ and the private exponent $d$ of an RSA instance related by the key equation $ed - k\phi(N) = 1$ rewritten as

$$\left|\frac{e}{\phi(N)} - \frac{k}{d}\right| = \frac{1}{\phi(N)d}$$

Suppose that $\left|\frac{e}{\phi(N)} - \frac{k}{d}\right| < \frac{1}{2d^2}$. Then by Theorem 1, $\frac{k}{d}$ is a convergent of the continued fraction expansion of $\frac{e}{\phi(N)}$. However, the value of $\phi(N)$ is unknown. Suppose we replace $\phi(N)$ with any known value $\varphi$ such that it is an approximation of $\phi(N)$ and satisfies $\left|\frac{e}{\varphi} - \frac{k}{d}\right| < \frac{1}{2d^2}$. This idea was firstly introduced by Wiener [10]. Since the fact that $\varphi$ is approximate to $N$, then he uses $\varphi = N$ which is publicly known. With the additional condition $d < \frac{1}{3}N^{1/4}$, he shows that $\left|\frac{e}{N} - \frac{k}{d}\right| < \frac{1}{2d^2}$, which by Theorem 1 implies that $\frac{k}{d}$ is a convergent of the continued fraction expansion of $\frac{e}{N}$. This leads to the following result.

**Theorem 2 [10].** Let $N = pq$ be an RSA modulus with $q < p < 2q$. Let $e < \phi(N)$ and $d$ be a public and private exponent, respectively. If $d < \frac{1}{3}N^{1/4}$, then $\left|\frac{e}{\phi(N)} - \frac{k}{d}\right| < \frac{1}{2d^2}$.

#### *2.2.2. de Weger's Generalization*

Later, de Weger [5] observes that if the difference between two RSA primes (i.e. $|p-q|$) is small, then $N-2\sqrt{N}+1$ is a better approximation to $\phi(N)$ instead of $N$. Replacing $\varphi = N-2\sqrt{N}+1$, combining with $\varphi > \frac{3}{4}N, N > 8d$ then $\frac{k}{d}$ is a convergent of the continued fraction expansion of $\frac{e}{N-2\sqrt{N}+1}$ which is satisfying the Theorem 1, as stated in the following result.

**Theorem 3 [5].** Let $N = pq$ be an RSA modulus with $q < p < 2q$ with $N > 8d$ and $|p-q| < N^{\beta}$. Let $e < \phi(N)$ with $\phi(N) > \frac{3}{4}N$ and $d < N^{\delta}$ be a public and private exponent, respectively. If $\delta < \frac{3}{4} - \beta$, then $\left|\frac{e}{N-2\sqrt{N}+1} - \frac{k}{d}\right| < \frac{1}{2d^2}$.

#### *2.2.3. Maitra and Sarkar's Attack*

In [5], de Weger considered the situation that $p$ and $q$ are too close. This implies that $p - q$ is small. Later on, Maitra and Sarkar [14] consider the case that $p$ and $2q$ are too close. This means that $|2q-p|$ is small. Assuming this, they showed that $N - \frac{3}{\sqrt{2}}\sqrt{N}+1$ is a better approximation to $\phi(N)$ instead of $N$.

Replacing $\varphi = N - \frac{3}{\sqrt{2}}\sqrt{N}+1$ with $N > 8d$, [14] shows that $\frac{k}{d}$ is a convergent of the continued fraction expansion of $\frac{e}{N-\frac{3}{\sqrt{2}}\sqrt{N}+1}$ satisfying Theorem 1, as follows.

**Theorem 4 [14].** Let $N = pq$ be an RSA modulus with $q < p < 2q$ with $N > 8d$ and $|2q-p| < N^{\gamma}$. Let $e < \phi(N)$ and $d < N^{\delta}$ be a public and private exponent, respectively. If $\delta < \frac{3}{4} - \gamma$, then $\left|\frac{e}{N-\frac{3}{\sqrt{2}}\sqrt{N}+1} - \frac{k}{d}\right| < \frac{1}{2d^2}$.

#### *2.2.4. Chen et.al's Attack*

In 2009, Chen et al. [6] generalize the result of [5] and [14] using the difference between two multiples of primes $|aq - bp|$ with $a > b$ and assuming that the ratio of the RSA primes $\frac{p}{q}$ is close to a simple fraction $\frac{b}{a}$ such that $(b(a^2+1)q - a(b^2+1)p)(aq-bp)) > 0$. Let $|aq-bp| < N^{\gamma}$. Replacing $\varphi = N - \frac{a+b}{\sqrt{ab}}\sqrt{N} + 1$, [6] shows that $\frac{k}{d}$ can be obtained from the continued fraction expansion of $\frac{e}{N-\frac{a+b}{\sqrt{ab}}\sqrt{N}+1}$ which is satisfying the Theorem 1. The work of [6] obtain the same result as [5] if $a = b = 1$, and get the same result as [14] when $\frac{b}{a} = \frac{1}{2}$. We present the result of [6] as follows.

**Theorem 5** [6]. Let $N = pq$ be an RSA modulus with $q < p < 2q$ with $N > 8d$ and $|aq - bp| < N^{\gamma}$. Let $e < \phi(N)$ and $d < N^{\delta}$ be a public and private exponent, respectively. If $\delta < \frac{3}{4} - \gamma$, then $\left| \frac{e}{N - \frac{a+b}{\sqrt{ab}}\sqrt{N} + 1} - \frac{k}{d} \right| < \frac{1}{2d^2}$.

## 3. Our First Attack

In this section we prove our first attacks. We begin with a simple lemma fixing the sizes of the prime factor of the RSA modulus.

**Lemma 1**. Let $N = p^2 q$ with $q < p < 2q$. Then

$$2^{-1/3} N^{1/3} < q < N^{1/3} < p < 2^{1/3} N^{1/3}.$$

*Proof.* Suppose $q < p < 2q$. Multiplying by $p^2$, we get $N < p^3 < 2N$. Hence $N^{1/3} < p < 2^{1/3} N^{1/3}$. Multiplying $q < p < 2q$ by $q^2$, we get $q^3 < N < 2q^3$ implies $2^{-1/3} N^{1/3} < q < N^{1/3}$. This terminates the proof. ■

Let $N = p^2 q$. Then

$$\phi(N) = p(p-1)(q-1) = p^2 q - p^2 - pq + p = N - (p^2 + pq - p).$$

The following result gives a interval for $N - \phi(N) = p^2 + pq - p$ in terms of $N$. It shows that if $p \approx q$, then $N - \left(2N^{2/3} - N^{1/3}\right)$ is a good approximation to $\phi(N)$, while if $p \approx 2q$, then $N - \left(\left(2^{2/3} + 2^{-1/3}\right) N^{2/3} - 2^{1/3} N^{1/3}\right)$ is a good approximation to $\phi(N)$.

**Lemma 2**. Let $N = p^2 q$ with $q < p < 2q$. Then

$$2N^{2/3} - N^{1/3} < N - \phi(N) < \left(2^{2/3} + 2^{-1/3}\right) N^{2/3} - 2^{1/3} N^{1/3}.$$

*Proof.* Suppose $N = p^2 q$ with $q < p < 2q$. We have $N - \phi(N) = p^2 + pq - p = p^2 + \frac{N}{p} - p$. Define the function $f$ such that $f(p) = p^2 + \frac{N}{p} - p$. Then, the derivative is such that

$$f'(p) = 2p - \frac{N}{p^2} - 1 = 2p - q - 1 > 0.$$

Then the function $f$ is strictly increasing for the interval of $p$, that is $(N^{1/3}, 2N^{1/3})$ by Lemma 1. Hence $f\left(N^{1/3}\right) < f(p) < f\left(2N^{1/3}\right)$, which leads to

$$2N^{2/3} - N^{1/3} < p^2 + pq - p < \left(2^{2/3} + 2^{-1/3}\right) N^{2/3} - 2^{1/3} N^{1/3}.$$

Since $p^2 + pq - p = N - \phi(N)$, this terminates the proof. ■

**Lemma 3**. Let $N = p^2 q$ and $\phi(N) = N - (p^2 + pq - p)$ with $q < p < 2q$. Then

$$\left| N - \left(2N^{2/3} - N^{1/3}\right) - \phi(N) \right| < 2p^{5/3} |p^{1/3} - q^{1/3}|.$$

*Proof.* Let $N = p^2q$. Using $\phi(N) = p(p-1)(q-1) = p^2q - p^2 - pq + p = N - (p^2 + pq - p)$, we get

$$\begin{aligned}
\left|N - \left(2N^{2/3} - N^{1/3}\right) - \phi(N)\right| &= \left|p^2 + pq - p - \left(2N^{2/3} - N^{1/3}\right)\right| \\
&= \left|p^2 + pq - p - \left(2\left(p^2q\right)^{2/3} - \left(p^2q\right)^{1/3}\right)\right| \\
&= |p^{1/3} - q^{1/3}| \times p^{2/3}\left(p + p^{2/3}q^{1/3} - p^{1/3}q^{2/3} - 1\right) \\
&< |p^{1/3} - q^{1/3}| \times p^{2/3}\left(p + p^{2/3}q^{1/3}\right) \\
&< |p^{1/3} - q^{1/3}| \times p^{2/3} \times 2p \\
&= 2p^{5/3}|p^{1/3} - q^{1/3}|.
\end{aligned}$$

This terminates the proof.■

**Theorem 6.** Let $N = p^2q$ with $q < p < 2q$. Let $1 < e < \phi(N) < N - \left(2N^{2/3} - N^{1/3}\right)$ satisfying $ed - k\phi(N) = 1$ for some unknown integers $\phi(N), d, k$. Assume $\phi(N) > \frac{2}{3}N$ and $N > 6d$. Let $2p^{5/3}|p^{1/3} - q^{1/3}| < \frac{1}{3}N^\beta$ and $d < N^\delta$. If $\delta < \frac{1-\beta}{2}$ then $\left|\frac{e}{N-(2N^{2/3}-N^{1/3})} - \frac{k}{d}\right| = \frac{1}{2d^2}$.

*Proof.* We transform the equation $ed - k\phi(N) = 1$ to

$$ed - k(N - (p^2 + pq - p)) = 1$$

$$ed - k(N - (N - \phi(N))) = 1$$

$$ed - k(N - (2N^{2/3} - N^{1/3}) + (2N^{2/3} - N^{1/3}) - (N - \phi(N)) = 1$$

$$ed - k(N - (2N^{2/3} - N^{1/3})) = 1 - k(N - \phi(N) - (2N^{2/3} - N^{1/3}))$$

Dividing by $d(N - (2N^{2/3} - N^{1/3}))$, on the right hand side we get

$$\begin{aligned}
\left|\frac{e}{N - (2N^{2/3} - N^{1/3})} - \frac{k}{d}\right| &= \left|\frac{e}{N - (2N^{2/3} - N^{1/3})} - \frac{e}{\phi(N)} + \frac{e}{\phi(N)} - \frac{k}{d}\right| \\
&\le \left|\frac{e}{N - (2N^{2/3} - N^{1/3})} - \frac{e}{\phi(N)}\right| + \left|\frac{e}{\phi(N)} - \frac{k}{d}\right| \\
&\le e\left|\frac{\phi(N) - (N - (2N^{2/3} - N^{1/3}))}{\phi(N)(N - (2N^{2/3} - N^{1/3}))}\right| + \left|\frac{ed - k\phi(N)}{\phi(N)d}\right| \\
&\le e\left|\frac{\phi(N) - (N - (2N^{2/3} - N^{1/3}))}{\phi(N)(N - (2N^{2/3} - N^{1/3}))}\right| + \left|\frac{ed - k\phi(N)}{\phi(N)d}\right| \\
&\le e\left|\frac{N - (2N^{2/3} - N^{1/3}) - \phi(N)}{\phi(N)(N - (2N^{2/3} - N^{1/3}))}\right| + \left|\frac{ed - k\phi(N)}{\phi(N)d}\right|
\end{aligned}$$

Since $e < N - (2N^{2/3} - N^{1/3})$ and $ed - k\phi(N) = 1$ , then we have $\left|\frac{e}{N-(2N^{2/3}-N^{1/3})} - \frac{k}{d}\right| < \left|\frac{N-(2N^{2/3}-N^{1/3})-\phi(N)}{\phi(N)} + \frac{1}{\phi(N)d}\right|$. Using $2p^{5/3}|p^{1/3} - q^{1/3}| < \frac{1}{3}N^\beta, d < N^\delta$ and $\phi(N) > 4d$, we get

$$\left|\frac{N-(2N^{2/3}-N^{1/3})-\phi(N)}{\phi(N)}+\frac{1}{\phi(N)d}\right|<\frac{1}{2}N^{\beta-1}+\frac{1}{4d^2}$$
$$<\frac{1}{2}N^{\beta-1}+\frac{1}{2}N^{-2\delta}$$

In order to satisfy the Theorem 1, it is suffice when we take $\beta-1<-2\delta$, then $\delta<\frac{1-\beta}{2}$.■

**Corollary 1.** Consider Theorem 6. If the unknown integer $d$ be discovered from the continued fraction expansion of $\frac{e}{N-(2N^{2/3}-N^{1/3})}$, then $N$ can be factored in polynomial time.

*Proof.* Suppose we obtained the secret exponent $d$ according to Theorem 6. Observe that from the relation of $\frac{ed-1}{k}=\phi(N)=p(p-1)(q-1)$, therefore computing $\gcd(\frac{ed-1}{k},N)$ should give the prime $p$.■

Hence, we design the following algorithm to further recovering the prime factorization of the modulus $N=p^2q$.

Algorithm 1.

INPUT: The public key modulus $(N,e)$ satisfying $N=p^2q$ and Theorem 6.
OUTPUT: The prime factors $p,q$.

1. Compute the continued fraction $\frac{e}{N-(2N^{2/3}-N^{1/3})}$.
2. For each convergent $\frac{k_i}{d_i}$ of $\frac{e}{N-(2N^{2/3}-N^{1/3})}$, compute $\phi(N)_i=\frac{ed_i-1}{k_i}$.
3. Compute $g=\gcd(\phi(N)_i,N)$.
4. If $1<g<N$, then stop.

## 4. Our Second Attack

In this section we describe our second attack which is based on the work of Nitaj [3]. We extend the method to the modulus $N=p^2q$ instead of $N=pq$. We firstly provide Nitaj's result as follows.

**Theorem 7.** Let $N=pq$ with $q<p<2q$. Let $\frac{a}{b}$ be an unknown convergent of the continued fraction expansion of $\frac{q}{p}$ with $a\geq 1$ and $|ap-bq|<\frac{1}{2}N^{1/2-\alpha}$. Let $e$ be a public exponent satisfying the equation $eX-(N-(ap+bq))Y=Z$ with $\gcd(X,Y)=1$. Set $ap+bq=N^{1/2+\alpha}$ with $0<\alpha<\frac{1}{2}$. If $|Z|<N^{1/2+\alpha}X$ and $1\leq Y\leq X<\frac{1}{2}N^{1/4-\alpha/2}$, then $N$ can be factored in polynomial time.

*Proof.* Refer [[3], Theorem 3].

We are now ready to give our second attack. Let the integer closest x denoted as [x].

**Lemma 4.** Let $N=p^2q$ with $q<p<2q$. Let $a,b$ be integers with $\gcd(a,b)=1$. Let $ap^2+bq^2=N^{2/3+\alpha}$ with $\alpha<\frac{1}{3}$. If $|ap^2+bq^2-S|<\frac{1}{3}N^{1/3-\alpha}$ then $abq=\left[\frac{S^2}{4N}\right]$.

*Proof.* Set $S = ap^2 + bq^2 + x$. Observe that

$$\begin{aligned}(ap^2 - bq^2)^2 &= (ap^2)^2 - 2(ap^2bq^2) + (bq^2)\\ &= (ap^2)^2 + 2(ap^2bq^2) - 2(ap^2bq^2) - 2(ap^2bq^2) + (bq^2)\\ &= (ap^2 + bq^2)^2 - 4(ap^2bq^2)\\ &= (ap^2 + bq^2)^2 - 4abqN.\end{aligned}$$

Hence we get

$$(ap^2 - bq^2)^2 = (ap^2 + bq^2)^2 - 4abqN. \tag{1}$$

Now, consider

$$\begin{aligned}S^2 - 4abqN &= (ap^2 + bq^2 + x)^2 - 4abqN\\ &= (ap^2 + bq^2)^2 + 2x(ap^2 + bq^2) + x^2 - 4abqN.\end{aligned}$$

Using (1), we can rewrite the above equation as

$$S^2 - 4abqN = (ap^2 - bq^2)^2 + 2x(ap^2 + bq^2) + x^2. \tag{2}$$

Suppose $|ap^2 - bq^2| < N^{1/2}$. Hence, using $|x| < \frac{1}{3}N^{1/3-\alpha}$, then (2) becomes

$$\begin{aligned}|S^2 - 4abqN| &< (N^{1/2})^2 + 2|x|N^{2/3+\alpha} + x^2\\ &< N + 2(\frac{1}{3}N^{1/3-\alpha})N^{2/3+\alpha} + (\frac{1}{3}N^{1/3-\alpha})^2\\ &= N + \frac{2}{3}N + \frac{1}{9}N^{2/3-2\alpha}\\ &= N(1 + \frac{2}{3} + \frac{1}{9}N^{-1/3-2\alpha})\\ &< 2N\end{aligned}$$

Thus, we have $|S^2 - 4abqN| < 2N$. If we divide $|S^2 - 4abqN|$ by $4N$, then $\frac{S^2}{4N} - abq < \frac{2N}{4N} = \frac{1}{2}$. It follows that $abq = \left[\frac{S^2}{4N}\right]$. ∎

**Lemma 5.** Let $N = p^2q$ with $q < p < 2q$. Let $a, b$ be integers with $\gcd(a, b) = 1$ such that $ap^2 + bq^2 = N^{2/3+\alpha}$ with $\alpha < \frac{1}{3}$. Let $e$ be a public exponent satisfying the equation $eX - (N - (ap^2 + bq^2))Y = Z$ with $\gcd(X, Y) = 1$. If $|Z| < \frac{1}{3}N^{1/3+\alpha}Y$ and $1 \le Y \le X < \frac{1}{2}N^{\frac{1}{6}-\frac{\alpha}{2}}$, then $\frac{Y}{X}$ is a convergent of continued fraction $\frac{e}{N}$.

*Proof.* Set $ap^2 + bq^2 = N^{2/3+\alpha}$ with $\alpha < \frac{1}{3}$. Rearrange the equation $eX - (N - (ap^2 + bq^2))Y = Z$ as

$$eX - NY = Z - (ap^2 + bq^2)Y. \tag{3}$$

Now, suppose $|Z| < \frac{1}{3}N^{1/3+\alpha}Y$ and $1 \le Y \le X < \frac{1}{2}N^{\frac{1}{6}-\frac{\alpha}{2}}$. If we divide (3) by $NX$ then we get

$$
\begin{aligned}
\left|\frac{e}{N}-\frac{Y}{X}\right| &= \left|\frac{Z}{NX}-\frac{(ap^2+bq^2)Y}{NX}\right| \\
&\leq \left|\frac{Z}{NX}\right| + \frac{(ap^2+bq^2)Y}{NX} \\
&\leq \left|\frac{Z}{NX}\right| + \frac{(ap^2+bq^2)X}{NX} \\
&< \frac{\frac{1}{3}N^{1/3+\alpha}X}{NX} + \frac{N^{2/3+\alpha}X}{NX} \\
&< N^{-2/3+\alpha} + N^{-1/3+\alpha} \\
&< 2N^{-1/3+\alpha}
\end{aligned}
$$

Observe that, for $X < \frac{1}{2}N^{\frac{1}{6}-\frac{\alpha}{2}}$ then $\left|\frac{e}{N}-\frac{Y}{X}\right| < \frac{1}{2X^2}$. Hence by Theorem 1, $\frac{Y}{X}$ is a convergent of continued fraction $\frac{e}{N}$.■

**Theorem 8.** Let $N = p^2q$ with $q < p < 2q$. Let $a, b$ be integers with $\gcd(a,b) = 1$ such that such that $|ap^2 - bq^2| < N^{1/2}$. Let $e$ be a public exponent satisfying the equation $eX - (N - (ap^2 + bq^2))Y = Z$ with $\gcd(X, Y) = 1$. Set $ap^2 + bq^2 = N^{2/3+\alpha}$ with $\alpha < \frac{1}{3}$. If $|Z| < \frac{1}{3}N^{1/3+\alpha}Y$ and $1 \leq Y \leq X < \frac{1}{2}N^{\frac{1}{6}-\frac{\alpha}{2}}$, then $N$ can be factored in polynomial time.

*Proof.* Suppose $|Z| < \frac{1}{3}N^{1/3+\alpha}Y$ and $1 \leq Y \leq X < \frac{1}{2}N^{\frac{1}{6}-\frac{\alpha}{2}}$ with $\gcd(X,Y) = 1$. Hence by Lemma 5, $\frac{Y}{X}$ is a convergent of continued fraction $\frac{e}{N}$. Rearrange (3) as $(ap^2+bq^2)Y - NY + eX = Z$. Dividing by $Y$, we get

$$(ap^2 + bq^2) - N + \frac{eX}{Y} = \frac{Z}{Y}. \tag{4}$$

Set $S = \frac{N-eX}{Y}$. Then (4) can be written as $(ap^2 + bq^2) - S = \frac{Z}{Y}$. Hence, we have $|(ap^2 + bq^2) - S| = \frac{|Z|}{Y} < |Z| < \frac{1}{3}N^{1/3+\alpha}$. By Lemma 4, if $|ap^2 - bq^2| < N^{1/2}$ and $|ap^2 + bq^2 - S| < \frac{1}{3}N^{1/3-\alpha}$ then $\left[\frac{S^2}{4N}\right] = abq$. It follows that we obtained $q = \gcd\left(\left[\frac{S^2}{4N}\right], N\right)$.■

According to Theorem 8, we design the following algorithm to obtain the prime factor of $N = p^2q$ as follows.

Algorithm 2.

INPUT: The public key modulus $(N, e)$ satisfying $N = p^2q$ and Theorem 6.
OUTPUT: The prime factors $p, q$.

1. Compute the continued fraction $\frac{e}{N}$.
2. For each convergent $\frac{Y}{X}$ of $\frac{e}{N}$, compute $S = \frac{N-eX}{Y}$.
3. Compute $\left[\frac{S^2}{4N}\right]$.
4. Compute $h = \gcd\left(\left[\frac{S^2}{4N}\right], N\right)$.
5. If $1 < h < N$, then stop.

## 5. Conclusion

This paper shows two new attacks on RSA-type modulus of $N = p^2q$ using continued fractions. The first attack we note that $1 < e < \phi(N) < N - (2N^{2/3} - N^{1/3})$ is a good approximation of $\phi(N)$ leading to get $d$ satisfying $2p^{5/3}|p^{1/3} - q^{1/3}| < \frac{1}{3}N^{\beta}$ and $d < N^{\delta}$. With this result, we come up with an algorithm for factoring $N = p^2q$, as we described in Algorithm 1.

Concerning to the second attack with $e$ satisfying the equation $eX - (N - (ap^2 + bq^2))Y = Z$ such that $|ap^2 - bq^2| < N^{1/2}$, then we show that this result also can find the primes factor of $N$ in polynomial time fulfilling the condition $ap^2 + bq^2 = N^{2/3+\alpha}$ with $0 < \alpha < \frac{1}{3}$, $|Z| < \frac{1}{3}N^{1/3+\alpha}Y$ and $1 \leq Y \leq X < \frac{1}{2}N^{\frac{1}{6}-\frac{\alpha}{2}}$.

**Acknowledgement**
The authors would like to thank the Ministry of Education, Malaysia for research funding. The authors also would like to thank Prof. Dr. Abderrahmane Nitaj from the Laboratoire de Mathématiques, Université de Caen, France, for valuable comments and discussion.

## References

[1] May A 2004 Secret exponent Attacks on RSA-type schemes with moduli $N = p^r q$ In PKC 2004 *LNCS* **2947** Springer-Verlag pp. 218-230.
[2] Menezes A, van Oorschot C, and Vanstone 1997 *A Handbook of Applied Cryptography* CRC Press Washington.
[3] Nitaj A 2009 Cryptanalysis of RSA using the Ratio of the Primes In Preneel B. (ed.): Progress in Cryptology-AFRICACRYPT 2009 *LNCS* **5580** Springer-Verlag pp. 98-115.
[4] Nitaj A 2013 *Diophantine and Lattice Cryptanalysis of the RSA Cryptosystem* Springer-Verlag pp. 139-168.
[5] de Weger B 2002 Cryptanalysis of RSA with Small Prime Difference *Applicable Algebra in Engineering Communication and Computing* **13**(1) pp. 1728.
[6] Chen C Y, Hsueh C C and Lin Y F 2009 A Generalization of de Wegers Method In the *5th IAS/IEEE International Conference on Information Assurance and Security (IAS 2009)* pp. 344-347.
[7] Coppersmith D 1997 Small Solutions to Polynomial Equations and Low Exponent RSA Vulnerabilities *J. of Cryptology* **10**(4) Springer-Verlag pp. 233-260.
[8] Blömer J and May A 2004 A Generalized Wiener Attack on RSA *Practice and Theory in Public Key Cryptography PKC 2004 LNCS* **2947** Springer-Verlag pp. 113.
[9] Asbullah M A and Ariffin M R K 2014 Rabin-$p$ Cryptosystem: Practical and Efficient Method for Rabin based Encryption Scheme *arXiv preprint (2014)*. Available at http://arxiv.org/abs/1411.4398.
[10] Wiener M J 1990 Cryptanalysis of Short RSA Secret Exponents *IEEE Trans. on Information Theory* **IT-36** pp. 553-558.
[11] Nishioka M, Satoh H and Sakurai K 2001 Design and Analysis of Fast Provably Secure Public-Key Cryptosystems based on a Modular Squaring *Information Security and CryptologyICISC 2001 LNCS* **2288** Springer-Berlin Heidelberg pp 81-102.
[12] Ariffin M R K, Asbullah M A, Abu N A and Mahad Z 2013 A New Efficient Asymmetric Cryptosystem Based on the Integer Factorization Problem of $N = p^2q$ *Malaysian Journal of Mathematical Sciences* **7**(S) pp. 19-37.
[13] Rivest R L, Shamir A and Adleman L M 1978 A Method for Obtaining Digital Signatures and Public-Key Cryptosystems *Comm. of the ACM* **21**(2) pp. 120-126.
[14] Maitra S and Sarkar S 2008 Revisiting Wieners Attack - New Weak Keys in RSA *ISC 2008 LNCS* **5222** Springer-Verlag pp. 228-243.
[15] Takagi T 1998 Fast RSA-type Cryptosystem Modulo $p^k q$. *Advances in Cryptology Crypto98 LNCS* **1462** Springer-Verlag pp. 318 326.

Journal of Physics: Conference Series **622** (2015) 012020 doi:10.1088/1742-6596/622/1/012020

# Monitoring Actuarial Present Values of Term Life Insurance By a Statistical Process Control Chart

**M. Hafidz Omar**[1]
Assoc Professor of Statistics and Actuarial Science, *Dept of Mathematics & Statistics*, King Fahd University of Petroleum & Minerals, Dhahran 31261, Saudi Arabia

E-mail: omarmh@kfupm.edu.sa

**Abstract.** Tracking performance of life insurance or similar insurance policy using standard statistical process control chart is complex because of many factors. In this work, we present the difficulty in doing so. However, with some modifications of the SPC charting framework, the difficulty can be manageable to the actuaries. So, we propose monitoring a simpler but natural actuarial quantity that is typically found in recursion formulas of reserves, profit testing, as well as present values. We shared some simulation results for the monitoring process. Additionally, some advantages of doing so is discussed.

## 1. Introduction

Tracking the performance of insurance policy for a group of policyholders is part and parcel of any insurance contract. Typically such work is done laboriously without much help from charting techniques such as a statistical process control (SPC) chart. In this paper, we explore the possibility of using SPC charts to simplify the meticulous work of monitoring performance of insurance policy.

One major challenge for applying SPC charts to insurance applications is the issue of what attribute or measure is of interest to track. Another is how to index this attribute or measure. Yet another is whether the attribute or measure is constant or variable and whether or not it is independent across time. All these are complex questions that complicate a straightforward use of an SPC charting technique.

To understand the challenges, we now review the literature and put these challenges in its context.

## 2. Literature

Methods of Statistical Process Control (SPC) have been used quite extensively in industrial settings to produce goods to certain quality specifications and to maintain product quality.

The first SPC chart was introduced by [1] which was used as in-stream control tool to guide the production process. Later, [2] popularized the SPC techniques within his Total Quality Management (TQM) idea. Japanese industries extensively espoused Deming's TQM idea where they found that technicians with little or no training in statistics can track product quality fairly well because all they

[1] Correspondence should be addressed to this author.

had to do was to graphically check if the means randomly fall in a certain band of tolerance to signify acceptability of the product produced by the process [3]-[4].

SPC charts have been used not only in the industrial settings but also in the education setting [5] to track and monitor the consistency of ratings across time. It has also been applied in accounting by [6] to monitor financial reporting of public companies. The SPC technique was discussed by [7] for its applicability in finance to monitor stock market returns. To our knowledge, SPC techniques have yet to find its application in the field of insurance and actuarial science.

### *2.1. Rules When Tracking with SPC Chart*

SPC charts involve tracking some statistical quantities from random samples across time. To provide a clear procedure to identify processes that are out of control, some rules were suggested by [8] as follows:

1) Any point is either beyond the upper or lower 3 sigma control limit.
2) At least eight consecutive points are on only one side of the control chart.
3) Two or three consecutive points are inside the control limit but outside the upper or lower 2 sigma warning limit.
4) Four or five consecutive points are outside the 1-sigma limits.
5) A non-random or unusual pattern in the data (such as a cyclic pattern).
6) One or more points near a warning or control limit.

Originally only the first rule was introduced but over the years more rules were introduced to increase the sensitivity of the SPC charts. Many researchers such as [9] have discussed the sensitivity of these rules.

### *2.2. Statistical Process Control in Actuarial Science?*

Typically SPC charts involve tracking some statistical quantities from random samples across time. The inherent variation shown by the chart is known to be coming from random sampling of these statistical quantities and any departure there from is thought of as coming from some process shift.

Actuarial quantities to be monitored, however, pose some difficulties in applying the SPC charting technique. First, the quantities to be track such as actuarial present values of insurance portfolios are not necessarily random samples across time and most often are not. Second, for each time point, a sample loss observation such as death or accident may or may not be observed. Third, for longer insurance contracts, the actuarial present values can be perceived as a weighted average of the random loss event across several time points. Fourth, tracking a non-suitable quantity will result in futility rather than utility of the SPC technique. Thus, some ingenuity is needed to find a suitable quantity to track.

## 3. Procedure

### *3.1. Selection of Appropriate Actuarial Quantity*

Two main quantities that are often at the heart of insurance policies are (1) interest rate risk and (2) probability of the loss event. One may track the first quantity fairly well with SPC charting techniques assuming that variable interest rates are fairly stationary around some risk-free target values. In this approach, the observed quantities are thought of as random samples across time. As for the second quantity, the probability of a loss event is typically not independent of tracking time. As such, if such quantity were to be tracked, it cannot be summed or averaged across time nor can the same tracking limits be used.

To avoid the difficulties discussed above and to be relevant to insurance applications, we propose tracking the actuarial present value of a 1-year insurance policy across time. Embedded in this singular quantity is both the interest rate risk and loss probability mentioned earlier. That is, over successive time periods $h$, we will monitor the observed quantity

$$\hat{A}_{\stackrel{1}{x+h}:\overline{1}|} = (1 + i_h)^{-1}\hat{q}_{x+h} \tag{1}$$

against the target

$$A_{\stackrel{1}{x+h}:\overline{1}|} = vq_{x+h} \tag{2}$$

where $v = (1+i)^{-1}$ . As can be seen in the above equations, not only does the observed 1-year present values change with time, the target in equation (2) can also change.

To track the performance of 1-year insurance policies, we need to set up 2-sigma warning and 3-sigma control limits (see [8]). The following variance of the 1-year insurance policy is instrumental in setting this up.

$$Var(v^{K(x)+1}) = {}^2A_{\stackrel{1}{x+h}:\overline{1}|} - \left(A_{\stackrel{1}{x+h}:\overline{1}|}\right)^2 = v^2 p_{x+h} q_{x+h} \tag{3}$$

The 2-sigma warning and 3-sigma control limits are then defined as follows:

$$A_{\stackrel{1}{x+h}:\overline{1}|} \pm k\sqrt{Var(v^{K(x)+1})/n_h} = vq_{x+h} \pm kv\sqrt{p_{x+h}q_{x+h}/({}_hp_x \times n)}$$

$$= v\left(q_{x+h} \pm k\sqrt{p_{x+h}q_{x+h}/\left(n\prod_{j=0}^{h-1} p_{x+j}\right)}\right) \tag{4}$$

where $k$ = 2 or 3 for the two types of limits respectively and $n_h$ represents the size of the group expected to survive at time $h$.

### *3.2. Data Sources*

For this paper, we use the 1989-91 US life Table for female lives as available in [10]. In particular and for illustration, we concentrate on a group of adult females of initial working age of 25 and consistently monitor the 1-year term life insurance for them. A short excerpt of this single-decrement US life Table is given below:

Table 1. 1998-91 U.S. Life Table (for female lives)

| $x$ | $l_x$ | $d_x$ | $p_x$ | $q_x$ |
|---|---|---|---|---|
| 25 | 98325 | 58 | 0.99941 | 0.00059 |
| 26 | 98267 | 59 | 0.99940 | 0.00060 |
| 27 | 98208 | 61 | 0.99938 | 0.00062 |

For simplicity, we consider the case where benefits are payable only at the end of the year of death. The target and 2- and 3-sigma limits come from the table as described in equations (2) and (4) earlier. The simulated random data of size $n_h$ 25+$h$ year old adult females come from the following population:

1) The random samples of loss events of size $n_h = 1000\prod_{j=0}^{h-1} p_{25+j}$ (at ages 25+$h$, $h$= 0, 1, 2, …, 25) comes from Bernoulli $b(1,q_{x+h})$. Note that $n_0 = n = 1000$.
2) The random sample of 25 interest spot-rate risk $i_h$ comes from a Normal distribution with mean $i$ = 0.06, and standard deviation $\sigma$ = 0.015. That is, $i_h$ comes from a $N(0.06,0.015)$ distribution.

We will refer to this data set as scenario 1.

The simulated observations of loss events above are then used to define the new probability of loss (such a death) events $q_{x+h}$. The interest spot-rate risk $i_h$ provide the discounting function specific to period $h$. The loss probability and the discounting function together define the sample observations in period $h$ as described in (1) above.

To provide comparison when the portfolio is beyond expectation, we provide also the following scenarios when the simulated observation data come from the following population:

1) Scenario 2 (process shift with disturbances in life Table probabilities): In addition to the random samples from scenario 1, the random samples of loss events of size $n_h = 1000\prod_{j=0}^{h-1} p_{25+j}$ (at ages 25+$h$, $h$= 16, 17,...,25) comes from Bernoulli $b(1,q_{x+h}+W)$ where $W$ comes from $N(0.007,0.003)$.
2) Scenario 3 (process shift with larger disturbances in the spot-rate interest risk): The random sample of 25 interest spot-rate risk $i_h$ as in scenario 1 before except when $h$= 16 to 25, the spot-rates come from a Normal distribution with mean $i = 0.09$, and standard deviation $\sigma = 0.03$.
3) Scenario 4 (combined disturbances in loss probability and spot-rates): a combination of scenarios 1 and 2 above.

We now track the simulated observation across the $h$ time periods.

## 4. Results and Discussions

Here we share the results of the simulation study to illustrate the potential use of the SPC charts in Actuarial applications.

Figure 1 below provides a simulated result for a 1-year term life insurance portfolio for 1000 participants with end-of-year benefit of 1 unit. As can be seen in figure 1, the observations are all within the warning and control limits. This signifies that there are no unexpected violations in the statistical process for the insurance portfolio.

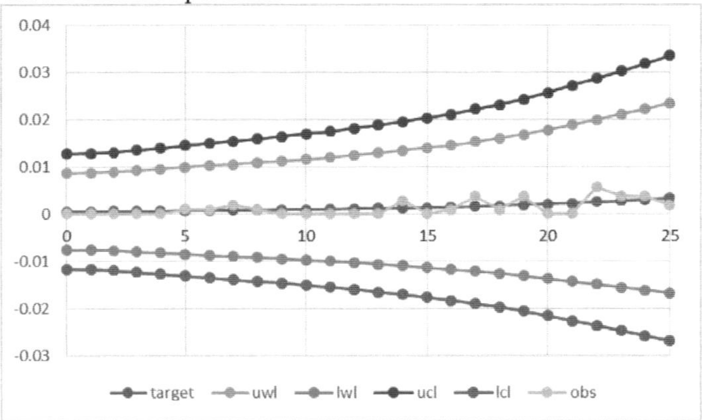


**Figure 1.** Statistical Process Control Chart for a 1-year term Life Insurance

Figure 2 below provides the result when there is a simulated upward shift in the loss probability.

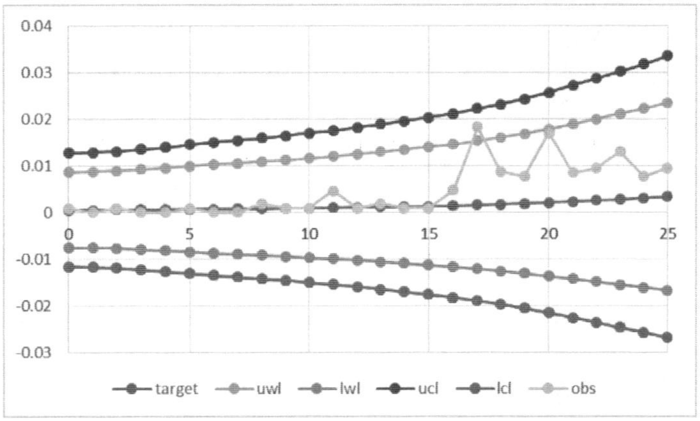


**Figure 2**. SPC Chart for a 1-year term Life Insurance with some disturbance in loss probabilities

As shown by figure 2, we observe a violation of the upper warning limit at time $h$=17. We also have an observation nearing the upper warning limit at $h$=20. It is also apparent that the process has shifted from the target values since all observations after time $h$=16 were oscillating around a higher mean than the target values. Figure 2 shows four of the six characteristics of an "out of control" process described by [8] earlier. That is, although no points are outside the 3 sigma limits and less than 2 consecutive points are outside the 2 sigma warning limits, the other four rules described by [8] earlier were violated.

Figure 3 below provides the results when there is a simulated shift in the interest spot-rates. As shown by figure 3, although there are some disturbances in spot-rates, there does not seem to be any noticeable disturbance in the process as described by [8].

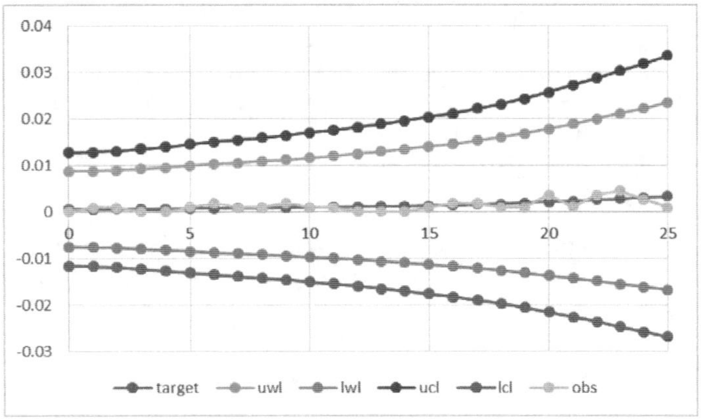


**Figure 3.** SPC Chart for a 1-year term Life Insurance with some disturbance in spot-rates

Some explanation of the results can be offered here. Since we are monitoring 1-year term life actuarial present values, the discounting function does not show much effect over a short period of one year. If the contracts we monitor are more than 1-year, the compound effects of the spot-rates on the observed actuarial present values may have been more noticeable.

Figure 4 below provides the results when there is a simulated shift in the interest spot-rates as well as a shift in the loss probabilities.

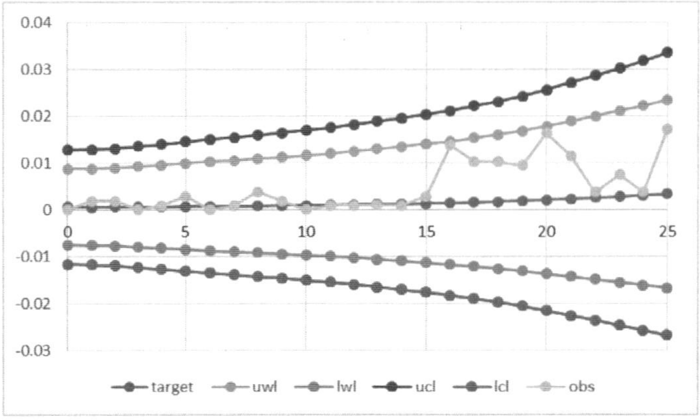


**Figure 4.** SPC Chart for a 1-year term Life Insurance with some disturbance in spot-rates and some disturbances in loss probabilities

As shown by figure 4, we have observations nearing the upper warning limit at $h$=16 and $h$=20. It is also apparent that the process has shifted since all observations after time $h$=16 were oscillating around a higher mean than the target values. Figure 4 shows four of the six characteristics of an "out of control" process described by [8]. That is, although no points are outside the 3 sigma control limits and no points are outside the 2 sigma warning limits, the other four rules described by [8] earlier were violated.

## 5. Conclusion

The use of SPC charts for actuarial applications have been explored in this study. It is found that 1-year term life insurance contracts are suitable to be monitored across time. It is beneficial to monitor such short-term contracts for several reasons. One, profit or loss on insurance portfolios are measured at least annually and could be measured as frequent as monthly or quarterly. Two, although longer-term contracts are not monitored here, the recursive relationships in actuarial present values can be used on the one-year contracts to derive the effects on the longer-term contracts.

Our simulation results suggest that the effect of changes in loss probabilities affect the observed actuarial present values more than the same effects in spot-rates. This could be partly due to our focus on short-term 1-year contracts where spot-rates for one-year may not provide much of a difference in short term actuarial present values.

## 6. Study Limitations

Our results indicate that the effect of changes in the process in terms of loss probabilities may affect the observed actuarial present values more than the same effects on spot-rates. If we were able to monitor longer-term contracts rather than 1-year terms, our results may have been different. In particular, if we were able to monitor more than 1-year contract terms, we would have been able to see the compounding effects of investments based on the interest spot-rates and the long-term compounding effects of the mortality rates. However, this is not only difficult as described earlier but also since profit determinations for insurance companies must be done at least once a year, tracking over longer periods practically does not help companies determine their level of annual profits.

In this paper, we have provided some simulated results to demonstrate the use of SPC in actuarial science where such methods have not been used before. More research is however needed to know more about the promise of such method.

Moreover, more extensive simulation work than the one done here is needed if the focus is to tease out which of the two components (loss probability or spot-rates) of the observed actuarial present values is more susceptible to disturbances.

## 7. References

[1] Shewhart, W A 1929 Quality Control Charts, *Bell System Technical Journal*, 5(11), 593-603,.
[2] Deming, W E 1951 *Elementary Principles of the Statistical Control of Quality – Series of Lectures*. Tokyo: Nippon Kagaku Gijutsu Remnei.
[3] Deming, W E 1967 What happened in Japan? *Industrial Quality Control*, (August), 91-92.
[4] Deming, W E 1981 *Japanese Methods for Productivity and Quality*, Washington DC: George Washington University Press.
[5] Omar, M H 2010 Statistical Process Control Charts for Measuring and Monitoring Temporal Consistency of Ratings, *J. of Ed. Meas*, 7 (1), 18 -35.
[6] Dull R B and Tegarden D P 2004 Using control charts to monitor financial reporting of public company, *Int. J. Accounting Information System*, 109-127.
[7] Severin T and Schmid W 1998 Statistical Process Control and its application in Finance" in Risk Management, Econometrics and Neural Networks: Selected Articles of the 6th *Econometric Workshop in Karlsruhe*, Germany., G. Bol, G. Nakhaeizadeh, and K. Vollmer, Ed.Heidelberg: Germany: Physica-Verlag HD, pp. 83-104.
[8] Montgomery D C 1991 *Introduction to Statistical Quality Control*, 2nd ed. Singapore: John Wiley & Sons.
[9] Nelson LS 1988 Control Charts. In Kotz S, Johnson N L and Read C B (Eds.), *The Encyclopedia of Statistical Sciences*, 2nd ed. New York: Wiley, 2, 176-183.
[10] Cunningham R J, Herzog T N, and London R L 2012 *Models for Quantifying Risk*, 5th ed. Winsted, CT: ACTEX publications.

**Acknowledgments**
The author appreciates KFUPM for providing research facilities for this study.

# Production of cellulose phosphate from oil palm empty fruit bunch: Effect of chemical ratio

**R Rohaizu and W D Wanrosli**

School of Industrial Technology, Universiti Sains Malaysia, 11800 Minden, Pulau Pinang, Malaysia

E-mail: roh_aizu@yahoo.com

**Abstract.**
Cellulose phosphate was synthesized from oil palm biomass residue that has the potential to represent a considerable added value product for the oil palm biomass utilization. Cellulose phosphate (CP) is prepared via a phosphorylation process using the $H_3PO_4/P_2O_5/Et_3PO_4$/hexanol sequence using oil palm empty fruit bunch microcrystalline cellulose (OPEFB-MCC) as the starting material. Various factors affect its synthesis; one of them which is the subject of this investigation is the orthophosphoric acid ($H_3PO_4$) to triethylphosphate ($Et_3PO_4$) ratio which have the capability to increase the phosphorus content of CP. It is believed that during this reaction, the esterification of the free hydroxyl groups of the cellulose occurred. The $H_3PO_4/Et_3PO_4$ ratios applied were 0.16, 1.00, and 1.84. The effect of the $H_3PO_4/Et_3PO_4$ ratio on phosphorus content, yield, water swelling and molecular structure of CP are discussed.

**Keywords:** cellulose phosphate; oil palm empty fruit bunch microcrystalline cellulose; $H_3PO_4/P_2O_5/Et_3PO_4$/hexanol sequence; $H_3PO_4/Et_3PO_4$ ratio; phosphorus content

## 1. Introduction

In 2011, Malaysia had 4.908 million hectares of oil palm plantation, which covers approximately 73 % of the Malaysia's agricultural land [1]. Despite the obvious benefits, for 2005 alone, an estimated 56 million tons of oil palm biomass are generated in Malaysia and annual production is expected to increase in line with the growing worldwide demand for palm oils [2]. At present, there is a massive quantity of oil palm biomass residue still perceived as waste by-products. However, despite this issue, oil palm biomass are now being utilized as an alternative towards economically useful. Later effort, oil palm empty fruit bunch (EFB) is now viewed as feasible alternatives to sources of fibers, which can be converted into paper making pulp that would have good printing properties and a good formation within papermaking, yet to be extensively commercialized.

Preferably, a solution that utilizes these plentiful by-products to replace wood sources should be realized. This will be a positive impact towards a more sustainable development. Oil palm biomass residue is a reliable resource because of its availability, continuity, non-hazardous biodegradable material [3], and do not compete its food production in terms of land destination [4]. Furthermore, oil palm fibers are versatile, stable, and can be processed into various dimension grades to fit specific applications.

Realizing the potential of this oil palm biomass, recently, there have been numerous ongoing efforts to increase profits from oil palm biomass fiber by producing fuel and biobased chemicals, with an emphasis on its major component, cellulose. Cellulose such an important material is among the potential output of use for industrial exploitation and can represent a considerable value added product for the oil palm biomass utilization.

Currently products made from lignocellulosic biomass feedstock represent only minor fraction of the chemical industries product, however, the interest in the bio-based products has increased because of the rapidly rising barrel cost, an increasing concern about the environment issue and the depletion of the fossil resources in the near future [4].

Besides oil palm biomass, cellulose can be produced from others lignocellulosic biomass and some potential feedstocks including wood waste such as timber residues from saw mills, agriculture waste, such as straw and corn stover, grasses such as miscanthus and algae [4]. For industrial, most of the cellulose fiber derived from wood (softwood and hardwood) and relatively small amount in volume of non-woody fibre, which mainly from bagasse, bamboo, jute, ramie, hemp, flax and cotton. Nevertheless, non-woody fibers are important particularly in the developing world to reduce the spent on importing costly wood pulp and in advantages since non-woody fibers produce an annual crop with higher yield than wood [5].

There are great numbers of potential uses of cellulose within different industries. Until now, cellulose is the major constituent in the production of material in some of the largest industries in particular in pulp, paper, and textile industries, and expected will continue to be so in the future. However, the limited usage of pure cellulose has led to develop a number of cellulose surface modification strategies to enrich several key properties and characteristic of cellulose fibers that do not exist in the parent cellulose [6]. Chemical modification via etherification and esterification of polysaccharides are interesting routes represent the most versatile transformations for preparing new types of polymers based on polysaccharides, with advances in performance and application [7]. The unique characteristic of cellulose obtained from cellulose modification (cellulose derivatives) therefore accord it more advantages for the application purpose.

For example, the incorporation of cellulose into phosphate group alters the significance of cellulose application by endowing the material advantageous phosphate characteristics. Specific effect may thus be obtained by the phosphate group are the reduction in the flammability of the material and an increase in ion exchange capability [8, 9]. Flame resistance provides an advantage to cellulose phosphate especially for utilization in the manufacturing of textile and paper. This property has an adequate capability to execute thermal insulation of materials resulting in increased material protection upon exposure to sufficient heat. Cellulose phosphate is also identified as a cation-exchange material [9, 10]. The ion-exchange capability of phosphate advances the application of cellulose phosphate as an absorbent material. With this property, cellulose phosphate constituent a very important part of the world market as an absorbent material.

Significant improvements in surface properties and functionalities will be possible, allowing the development of many more new products that could replace a wide range of material because of their biocompability, biodegradation and non-toxicity [11]. Furthermore, the usage of cellulose and its derivatives in a diverse array of other applications, such as biosensors, bioengineering, pharmaceutics, smart and responsive material with polysaccharides and thus continue to grow on a worldwide basis [6]. Moreover, cellulose phosphate from oil palm empty fruit bunch can be classified as a value added biomaterial [12-14], that can provide benefits and extend well beyond discipline, which is the key to exploiting the full potential of oil palm biomass residue.

In the previous study, it has been shown that cellulose phosphate (CP) can be successfully synthesized from oil palm empty fruit bunch (OPEFB) [12, 13]. However, for the intent of application purposes, a higher degree of OPEFB-CP substitution should be achieved. The objective of this study is to investigate the effect of orthophosphoric acid ($H_3PO_4$) to triethylphosphate ($Et_3PO_4$) ratio in cellulose phosphate synthesis from OPEFB. For this purpose, OPEFB first has to be converted to pure cellulose pulp by a pulping process followed by an environmentally bleaching process using a totally chlorine free (TCF) bleaching sequences to eliminate the remaining lignin that gives a brownish color to the pulp. Microcrystalline cellulose (MCC) is prepared via a hydrolysis process of the TCF pulp. Various characterizations of the resulting CP were carried out.

## 2. Materials and Methods

### 2.1. Preparation of microcrystalline cellulose

OPEFB cellulose was prepared using the environmentally benign process as described by Wan Rosli *et al.* (2003) [15] and Leh *et al.* (2008) [16]. The procedure involved water prehydrolysis of the OPEFB, which was then followed by soda pulping. The unbleached pulp was afterwards bleached using a TCF sequence of oxygen (O), ozone (Z) and peroxide (P) to a Kappa no. of 1.4.The resultant (OZP) bleached pulp was later hydrolyzed to microcrystalline cellulose (MCC) with 2.5 M HCl while maintaining a solid to liquid ratio of 1:20 and refluxing at 105 ± 2 °C for 15 min. After hydrolysis, the material was thoroughly washed with distilled water before air-drying. Powdered OPEFB-MCC was then obtained by grinding the sample in a ball mill, and was subsequently kept in a desiccator over phosphorous pentoxide before further use [12].

### 2.2. Synthesis of cellulose phosphate

4.0 g of MCC powder was swollen consecutively in distilled water, ethanol and hexanol for 24 hours to activate the cellulose surface. Phosphorylation reaction was carried out in the four-neck round bottom flask equipped with a nitrogen inlet, a condenser, a thermometer, and a mechanical stirrer. MCC was then dispersed in 29 mL 1-hexanol. Then a solution of 50 g phosphorus pentaoxide ($P_2O_5$) in various ratios of $H_3PO_4$ to $Et_3PO_4$ was added portion wise to the suspension. The reaction was allowed to proceed under constant stirring and a nitrogen atmosphere at 50 °C for 72 h. After chemical treatment, modified MCC were rinsed with hexanol, ethanol, and then extraction against distilled water to remove excessive reagents. The resulting yield was then dried at room temperature.

### 2.3. Characterization method

#### *2.3.1. Determination of phosphate content*

Quantification of phosphorous requires its conversion into dissolved orthophosphate. This is then followed by colorimetric determination of the dissolved orthophosphate. A sulfuric acid-nitric acid mixture was used as oxidants for the Kjeldahl digestion procedure to oxidize organic matter and thereby release phosphorus as orthophosphate. Phosphorus concentration was determined spectrometrically at 880 nm by the ascorbic acid colorimetric method.

#### *2.3.2. Water sorption*

For water sorption measurement, 10 mg of samples was allowed to hydrate in excess distilled water at room temperature in centrifuge polyethylene tubes. After centrifugation at 4000 rpm for 10 min following removal of the excess water, the samples were then periodically weighed several times. Water sorption as expressed by % swelling was determined via the difference of weight before and after swelling.

#### *2.3.3. Differential Scanning Calorimetry (DSC)*

DSC thermograms were obtained using a Perkin Elmer DSC6 with approximately 10 mg of material in sealed aluminium pans that were heated from ambient temperature to 300 °C at a heating rate 10 °C /min under nitrogen atmosphere.

*2.3.4. Fourier Transform Infrared Spectroscopy (FTIR)*
FTIR was performed using a Nicolet Avatar 360 under transmission mode from 400 to 4000 $cm^{-1}$ with a 4 $cm^{-1}$ resolution, using the potassium bromide (KBr) method.

## 3. Results and discussion

Fig. 1 shows the FTIR spectra between MCC and its phosphorylated form. After phosphorylation, a few peaks that were earlier present in the MCC spectrum disappeared and concurrently the appearance of several new bands which are attributed to the phosphate group. New absorption peaks which characterized CP were indicated by the vibration bands of PO4 at 1027 $cm^{-1}$, P=O at 1377 $cm^{-1}$, and P-H at 2360 $cm^{-1}$ (Fig. 1 b, c, d) [17]. Although the reaction was carried out under different $H_3PO_4/Et_3PO_4$ ratio it seemed that each phosphorylated sample showed similar changes in absorption peaks. The resulting spectrums suggest that cellulose phosphate (CP) was successfully synthesized from oil palm empty fruit bunch via MCC and the molecular structures of the derived CP products are probably comparable.

The effect of the $H_3PO_4/Et_3PO_4$ ratio on the phosphorus content, yield and water swelling of the cellulose phosphate are shown in Table 1. It can be seen that by varying the $H_3PO_4/Et_3PO_4$ ratio significantly influenced the phosphorus content and percentage yield of CP. As the $H_3PO_4$ concentration increases, the phosphorus content of the samples also increases, but it was the opposite for the percentage yield. These results suggest that by increasing the concentration of the $H_3PO_4$, it will enhance the substitution reactivity of the hydroxyl groups to the phosphate group in anhydroglucose units of the cellulose molecules, thus increasing the phosphate content.

On the contrary, it seems that as the phosphorylation process progresses, degradation of MCC also took place resulting in less CP (yield) was obtained. This indicates that there was probably a competitive reaction between the phosphorylation and hydrolysis of the resulted product from the reaction [9]. Our previous study [12, 13] showed that the phosphorylation process is affected by the duration and temperature of the reaction, while in this study it is shown that the phosphorylation process is also sensitive to the $H_3PO_4/Et_3PO_4$ ratio.

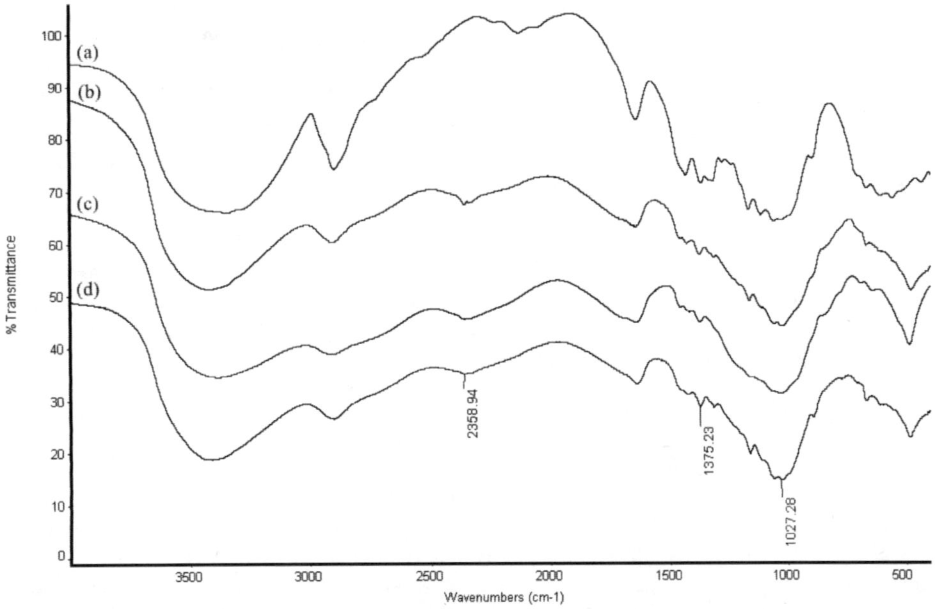


Fig. 1: FTIR spectra of (a) MCC and CP with $H_3PO_4/Et_3PO_4$ ratio (b) 0.16 (c) 1.00 (d) 1.84

Table 1: Effect of $H_3PO_4/Et_3PO_4$ ratio on of phosphorus content, yield and water swelling of CP

| $H_3PO_4/Et_3PO_4$ | P (%) | Yield (%) | Swelling (%) |
|---|---|---|---|
| 0.16 | 10.34 | 72.17 | 172 |
| 1 | 12.80 | 62.44 | 468 |
| 1.84 | 20.66 | 45.56 | 1142 |

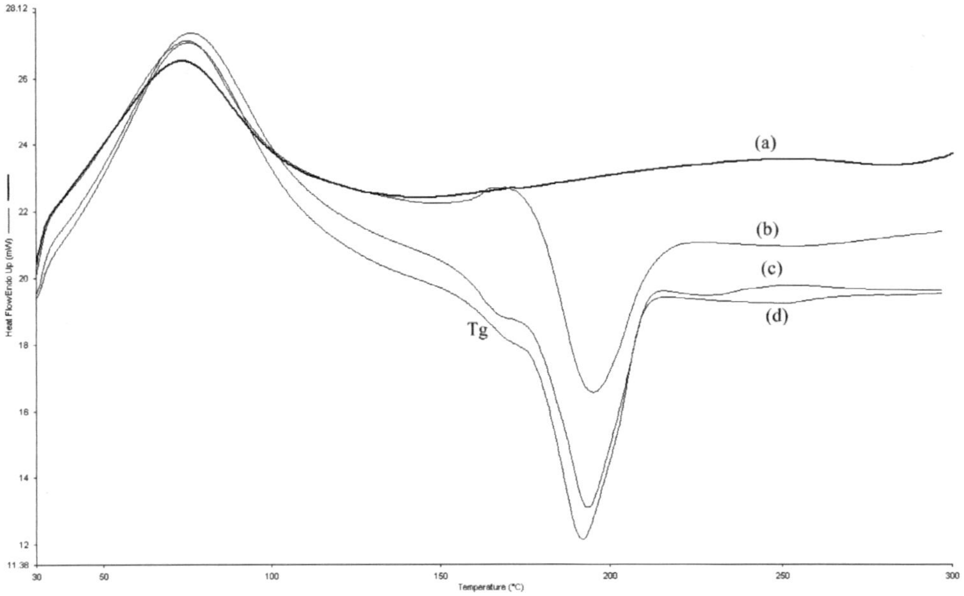


Fig. 2: DSC thermograms of (a) MCC and CP with $H_3PO_4/Et_3PO_4$ ratio (b) 0.16 (c) 1.00 (d) 1.84

Macroscopically, the water swelling ability of CP was significantly higher compared to MCC (3 %). This observation indicates that CP swells considerably in water with the higher the phosphorus content, the higher is the swelling ability. This was possible due to the incorporation of the phosphate group within the MCC which formed an ion exchange material.

The solubility of cellulose was relatively low in a many types of solvent; this is due to the hydrogen bonding between cellulose molecules and the crystalline region of molecular structure [18]. However, after the phosphorylation, the solubility of the resulted product increases. This is because of the substitution of the hydroxyl group with the phosphate group in the cellulose molecule. The presence of a larger group of phosphate as compared to the hydroxyl group in the cellulose molecule distracts the arrangement of crystalline pattern in the molecular structure. In this process, the phosphate group inside the cellulose granules would be able to create a repulsive force that might enhance inter- and intramolecular space and allow more water molecules to be absorbed. As a consequence, the phosphorylation process changed the molecular structure of the amorphous region and increased the swelling ability of modified cellulose in water.

To confirm this explanation, the crystallinity and thermal stability of MCC and CPs were examined using differential scanning calorimetry (DSC). According to the results (Fig. 2), what is observed could have been the phenomena related to the morphological changes of the MCC and CPs molecular structure. DSC curves of MCC and CP illustrate that phosphorylation strongly influence cellulose thermal degradation behavior. The first endothermic peak was observed at ca. 85 °C due to evaporation of adsorbed water in the samples. Results indicated that CP has a higher water holding capacity as observed from the area of evaporation peak of CP which is slightly bigger compared to MCC itself. This results support the water swelling ability of the CP as discussed earlier. The additional exothermic peak of CP observed at ca. 180 °C appears to be mainly due to the thermal decomposition point of phosphate group [19, 20].

Results also indicate that part of the crystalline region of MCC molecule was transformed to the amorphous region, which can be observed as a Tg. When the phosphate group was introduced to the MCC as a main chain, Tg can be observed in the CPs thermograms around 162°C which is not present in the MCC thermogram (Fig. 2). The Tg value for each sample showed no large difference among them even though the $H_3PO_4/Et_3PO_4$ ratio and the phosphate content of each sample were different. From this result, we can conclude that the Tg value for each sample was not affected by the $H_3PO_4/Et_3PO_4$ ratio and the value of DS.

## 4. Conclusion

The FTIR analysis for CPs confirm the existence of a phosphate group. The results obtained illustrates that the synthesis of cellulose phosphate from OPEFB-MCC had been significantly influenced by the $H_3PO_4/Et_3PO_4$ ratio. Experiment data demonstrated that phosphorus content, yield and water swelling ability of CP change variably with increasing of $H_3PO_4/Et_3PO_4$ ratio. As the phosphate group was incorporated into the MCC chain, the formation of an amorphous region and the water swelling ability of the modified MCC also were observed. The OPEFB-CP synthesized from oil palm fruit bunch fibers has the potential to be utilized to generate more advanced materials in various biomaterial applications.

## 5. Acknowledgement

The authors express their gratitude to Universiti Sains Malaysia for the research university grant scheme 1001/PTEKIND/814122, 1001/PTEKIND/845027 and Minister of Higher Education Malaysia for sponsoring postgraduate studies of Rohaizu, R.

## References

[1] Wendy P Q N *et al* 2012 *Journal of Cleaner Production* **34** 57-65
[2] Bazmi A A, Zahedi G and Hashim H 2011 *Renewable & Sustainable Energy Reviews* **15** (1) 574-83
[3] Abdullah N and Sulaiman F 2013 *Biomass Now - Sustainable Growth and Use*, ed Matovic M D (InTech) *The Oil Palm Wastes in Malaysia* pp 75 -100
[4] Verardi A *et al* 2012 *Hydrolysis of Lignocellulosic Biomass: Current Status of Processes and Technologies and Future Perspectives* ed Lima M A P (InTech) *Bioethanol* pp 95 - 122
[5] Jones J P E and Wegner T H 2009 *Structure-Property Relationships in Biomineralized and Biomimetic Composites* **1187** 57-66
[6] Habibi Y and Vignon M R 2008 *Cellulose* **15** (1) 177-85
[7] Praskalo J *et al* 2009 Carbohydrate Polymers **77** (4) 791-98
[8] Li W *et al* 2002 *Analytica Chimica Acta* **464** (2) 331-9
[9] Granja P L *et al* 2001 *Journal of Materials Science* **36** (9) 2163-72.
[10] Granja P L *et al* 2011 *Journal of Applied Polymer Science* **82** (13) 3341-53.
[11] Sun B *et al* 2005 *Cellulose* **12** (1) 59-66
[12] Wanrosli W D, Rohaizu R and Ghazali A 2011 *Carbohydrate Polymers* **84** (1) 262-7
[13] Wanrosli W D, et al 2013 *Industrial Crops and Products* **50** 611-7
[14] Wan Rosli W D, Kassim M H M and Seeni A 2011 *Bioresources* **6** (2) 1719-40
[15] Wan Rosli W D et al 2003 *Holzforschung* **57** (1) 106-13
[16] Leh C P et al 2008 *Industrial Crops and Products* **28** (3) 260-7
[17] Kaputskii V E, Komar V P and Skornyakov I V 1988 *Journal of Applied Spectroscopy* **48** (2) 176-9
[18] Credou J and Berthelot T 2014 *Journal of Materials Chemistry B* **2** (30) 4767-88

[19] Scheirs J, Camino G, and Tumiatti W 2001 *European Polymer Journal* **37**(5) 933-42
[20] Beyler C L and Hirschler M M 2002 *SFPE Handbook of Fire Protection Engineering* (3$^{rd}$ ed) ed DiNenno P J *et al* (Quincy, Massachusetts 02269: National Fire Protection Association) Thermal Decomposition of Polymers.pp 110-31

Journal of Physics: Conference Series **622** (2015) 012022 doi:10.1088/1742-6596/622/1/012022

# Modeling Philippine Stock Exchange Composite Index Using Time Series Analysis

**W S Gayo, J D Urrutia, J M F Temple, J R D Sandoval and J E A Sanglay**
Faculty Members, Department of Mathematics and Statistics, College of Science, Polytechnic University of the Philippines, Sta. Mesa, Manila, Philippines

E-mail: willygayo@gmail.com, wsgayo@pup.edu.ph, jackieurrutia20@gmail.com, math_urrutia@yahoo.com.ph, jdurrutia@pup.edu.ph

**Abstract.** This study was conducted to develop a time series model of the Philippine Stock Exchange Composite Index and its volatility using the finite mixture of ARIMA model with conditional variance equations such as ARCH, GARCH, EGARCH, TARCH and PARCH models. Also, the study aimed to find out the reason behind the behavior of PSEi, that is, which of the economic variables – Consumer Price Index, crude oil price, foreign exchange rate, gold price, interest rate, money supply, price-earnings ratio, Producers' Price Index and terms of trade – can be used in projecting future values of PSEi and this was examined using Granger Causality Test. The findings showed that the best time series model for Philippine Stock Exchange Composite index is ARIMA(1,1,5) – ARCH(1). Also, Consumer Price Index, crude oil price and foreign exchange rate are factors concluded to Granger cause Philippine Stock Exchange Composite Index.

## 1. Introduction

Stock market is a set of connections that gives a platform for about each economic transaction in the business world at a dynamic rate entitle the stock value that is based on the market equilibrium. It is an extremely nonlinear vibrant system whose performance is manipulated by a number of factors namely inflation rates, interest rates, economic atmosphere, political issues, and many more. In particular, stock markets are characterized by uncertainty. Some sections of the associated stock market are measured using the stock market indices. There are indices that were used in specific collection of stocks that satisfy the set criteria while there are so-called broad-based indices that measure the general movement of the market. Philippine Stock Exchange Composite Index (PSEi) is the sole broad-based index used by Philippine Stock Exchange. By estimating the changes in the stock prices of the chosen listed companies, the PSEi provides a "snapshot" of the market's overall condition.
Forecasting is tricky and may lead you to uncertainty for it needs too much risk in deriving the desire outcome. Inconsistency of some factors such as interest rates and foreign exchange rates, is one of the reasons why forecasting is a sort of dilemma.

## 2. Objectives

This study has two (2) main goals, to construct a time series model for PSEi and to determine which of the nine (9) economic variables can be used in predicting PSEi. To do this, the study sought to answer the following questions.

1) What is the movement of time series of PSEi?

2) What are the time series models formulated using the mixture of ARIMA model with ARCH models and some of its extensions?
3) Among the constructed models, what is the best model for Philippine Stock Exchange Composite index?
4) Among the factors, what is/are Granger cause(s) the PSEi?

## 3. Methodology

This section discusses the techniques used in the research method, the data gathering procedure, and the analytical method. The purpose of the study is to formulate a mathematical model PSEi using a time series process and to find out some economic variable influencing the PSEi.

### *3.1. Data Gathering Procedure*

The researchers collected data about the PSEi from January 2, 2002 to December 27, 2013. The daily closing prices of PSEi are only considered. These data came from PSE

Data of economic variables, pertaining to the factors, were gathered in different sectors in the Philippines, precisely National Statistics Office (also available at www.census.gov.ph) – for Consumers Price Index (CPI), Export and Import or Term of trade (TOT), and Producers Price Index (PPI); and Bangko Sentral ng Pilipinas (also available at www.bsp.gov.ph) – for foreign exchange rate (FE), interest rate (IR), money supply (MS) and price-earnings ratio (PE). The respective departments are the main offices just located in the cities of Manila. While, crude oil price (COP) and Gold price (GP) in the international market are available at www.indexmundi.com.

### *3.2. Time Series*

The time series $\{y_1, y_2, \dots, y_t\}$ data consists of stretch of (roughly) equidistant chronologically ordered observation which are being collected and monitored over a regular period of time. The time series, $\{y_t\}$, is said to be strictly stationary if the joint distribution of the random vector $\{y_t, y_{t+1}, \cdots, y_{t+k}\}$ is equal to the one of the $\{y_s, y_{s+1}, \cdots, y_{s+k}\}$ for all combinations of $t, s$ and $k$. The time series, $\{y_t\}$, is said to be weakly stationary or simply stationary if (1) $E(y_t) = \mu$, (2) $Var(y_t) = \sigma^2$, and (3) $Cov(y_t, y_{t+h}) = \gamma_h$ for all lags $h$.

*3.2.1. White Noise Process.* A stochastic process $\{\varepsilon_t\}$ is called a white noise process if it is a sequence of uncorrelated random variables from a fixed distribution with (1) $E[\varepsilon_t] = 0$ for all t, (2) $Var[\varepsilon_t] = \sigma^2$ for all t, and (3) $Cov(\varepsilon_t, \varepsilon_{t+k}) = \gamma_k = 0$ for all $k \neq 0$.

*3.2.2. Unit Root Test.* It is used study to determine if the time series is stationary or not. The researchers used Augmented Dickey-Fuller test statistics. The Augmented Dickey-Fuller test is an improved Dickey-Fuller test, derived from the regression model $y_t = \alpha\, y_{t-1} + \varepsilon_t$. Where the null hypothesis is unit root exists and hence the series is non-stationary versus the alternative hypothesis, unit root does not exist and hence stationary.

*3.2.3. Autoregressive (AR) Process.* Let the time series $\{\mathbf{y}_t\}$ be stationary, the autoregressive process of order *p*, AR(*p*), is given by

$$\mathbf{y}_t = \phi_1\, \mathbf{y}_{t-1} + \phi_2\, \mathbf{y}_{t-2} + \cdots + \phi_p\, \mathbf{y}_{t-p} + \varepsilon_t.$$

It is very common model for time series. It is a stochastic process used in statistical calculation in which future values are estimated based on a weighted sum of past values.

*3.2.4. Moving Average (MA) Process.* Let the time series $\{\mathbf{y}_t\}$ be stationary, the moving average process of order *q*, MA(*q*), is given by

$$\mathbf{y}_t = \theta_1\, \varepsilon_{t-1} + \theta_2\, \varepsilon_{t-2} + \cdots + \theta_p\, \varepsilon_{t-q} + \varepsilon_t.$$

It is widely used as indicator in technical analysis that helps smooth out price action by filtering out the "noise" from random price fluctuations. MA is a trend-following or lagging indicator because it is based on past prices.

*3.2.5. Autoregressive Moving Average (ARMA) Process.* Let the time series $\{\mathbf{y}_t\}$ be stationary, the autoregressive moving average process of order $p$ and $q$, ARMA($p,q$), is given by

$$\mathbf{y}_t = \phi_1 \mathbf{y}_{t-1} + \phi_2 \mathbf{y}_{t-2} + \cdots + \phi_p \mathbf{y}_{t-p} + \theta_1 \varepsilon_{t-1} + \theta_2 \varepsilon_{t-2} + \cdots + \theta_q \varepsilon_{t-q} + \varepsilon_t.$$

*3.2.6. Autoregressive Conditional Heteroskedasticity (ARCH) Test.* ARCH test is a test statistics that is computed from an auxiliary regression

$$e_t^2 = \beta_0 + \beta_1 e_{t-1}^2 + \beta_2 e_{t-2}^2 + \cdots + \beta_q e_{t-q}^2 + \varepsilon_t.$$

This examines the null hypothesis that there is no ARCH effect, that is, the variance of the residual term is constant through time against the alternative hypothesis that the variance is not constant.

*3.2.7. Autoregressive Conditional Heteroskedasticity (ARCH) Process.* The Autoregressive Conditional Heteroskedasticity, ARCH(1) models are discrete time representation of the conditional variance,

$$\hat{\sigma}_t^2 = c + \alpha_1 \varepsilon_{t-1}^2.$$

ARCH process was introduced by Robert F. Engle. This process allows the conditional variance to change over time as a function of past errors leaving the unconditional variance constant.

*3.2.8. Generalized ARCH (GARCH) Process.* The Generalized ARCH, denoted by GARCH(1,1) is an equation of conditional variance of the form,

$$\hat{\sigma}_t^2 = c + \alpha_1 \varepsilon_{t-1}^2 + \beta_1 \hat{\sigma}_{t-1}^2 .$$

GARCH process is an extension of the ARCH process which allows for a longer memory and a much more flexible structure. According to this model, conditional volatility is dependent on the past error terms and on its previous own lags.

*3.2.9. Exponential GARCH (EGARCH) Process.* The Exponential GARCH, denoted by EGARCH(1,1) is of the form,

$$\log(\hat{\sigma}_t^2) = c + \alpha_1 \left| \frac{\varepsilon_{t-1}^2}{\sqrt{\hat{\sigma}_{t-1}^2}} \right| + \beta_1 \log(\hat{\sigma}_{t-1}^2) + \gamma_1 \left[ \frac{\varepsilon_{t-1}}{\sqrt{\hat{\sigma}_{t-1}^2}} - \sqrt{\frac{2}{\pi}} \right].$$

EGARCH process is one of the extensions of ARCH model that accounts for an asymmetric response to a "shock". EGARCH eliminates the non-negativity constraints of GARCH models by formulating the conditional variance equation in logarithm.

*3.2.10. Threshold ARCH (TARCH) Process.* The Threshold ARCH, denoted by TARCH(1,1), is given by

$$\hat{\sigma}_t^2 = c + \beta_1 \hat{\sigma}_t^2 + \alpha_1 \varepsilon_{t-1}^2 + \gamma_1 \varepsilon_{t-1}^2 I_{t-1}^- \text{ where } I_{t-1}^- = \begin{cases} 1 & \text{if } \varepsilon_{t-1} < 0 \\ 0 & \text{otherwise} \end{cases}$$

TARCH process is also one of the extensions of ARCH models that capture the important asymmetric impact of news on volatility. Also, TARCH considers the past square residuals and past volatility.

*3.2.11. Power ARCH (PARCH) Process.* The Power ARCH, denoted by PARCH(1,1,1), is the equation

$$\hat{\sigma}_t^\delta = c + \beta_1 \hat{\sigma}_{t-1}^\delta + \alpha_1 (|\varepsilon_{t-1}| - \gamma_1 \varepsilon_{t-1})^\delta, \text{ where } \delta > 0 .$$

PARCH process is an extension of ARCH model that uses the conditional standard deviation as a measure of volatility instead of the variance. Also, PARCH model includes the past residuals, previous conditional standard deviation and leverage effect.

*3.3. Engle – Granger Methodology for Testing Co-integration*
The researchers decided to use the proposed method of Engle – Granger in testing co-integration between PSEi and each variable. The following is the two steps process of Engle – Granger test of co-integration:

i. Estimate a regression using ordinary least square on *I(1)* data; and
ii. Apply staionarity test to the residual of the estimated regression.

*3.3.1. Granger Causality Test.* The Granger Causality test investigates whether a time series can help in predicting another time series. In this study, researchers will examine which of the economic variables can be used in forecasting Philippine Stock Exchange Composite Index.

After satisfying stationarity and co-integration, lag selection is identified. The number of lags which has least Akaike Information Criterion (AIC) will be chosen. To apply Granger Causality test, it is necessary to assume a particular autoregressive process of order *p*,

$$\mathbf{y}_t = \alpha_0 + \alpha_1 \mathbf{y}_{t-1} + \alpha_2 \mathbf{y}_{t-2} + \cdots + \alpha_p \mathbf{y}_{t-p} + \beta_1 \mathbf{x}_{t-1} + \cdots + \beta_2 \mathbf{x}_{t-p} + \varepsilon_t \, .$$

where $\mathbf{y}_t$ and $\mathbf{x}_t$ are stationary time series. An F-test is conduct to the null hypothesis,

$$\mathrm{H}_0: \beta_1 = \beta_2 = \cdots = \beta_p = 0$$

that $\mathbf{x}_t$ does not Granger cause $\mathbf{y}_t$ against the alternative hypothesis that $\mathbf{x}_t$ does Granger cause $\mathbf{y}_t$. To be able to reject null hypothesis, F-test statistics probability must not exceed 0.05.

## 4. Results and Discussion

This section explains the nature and summary statistics of PSEi and the nine (9) economic variables. In addition, this section presents the test for the assumption, model specification and verification, forecasting of PSEi, and the test for the causality between the PSEi with nine economic variables.

*4.1. PSEi*
The daily closing price of Philippines Stock Exchange Composite index (PSEi) is used in this study. The study period covers from January 2, 2002 to December 27, 2013 which contains 2944 trading days.

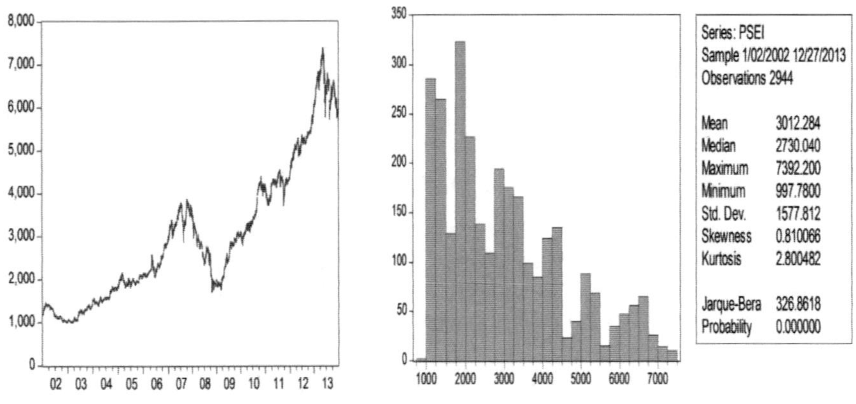


**Figure 1.** PSEi Daily Closing Price

Figure1 shows the time series, histogram and summary statistics of PSEi from 2002 to 2013. For the past years, PSEi reached Php 7,392.20, its highest closing price last May 15, 2013 and attained its lowest closing price on December 19, 2002 with Php 997.78. Figure 1 also reveals that PSEi is leptokurtic, kurtosis greater than 0.263, and is positively skewed. As per Jarque-Bera statistics, PSEi is non-normal, since the probability, 0.00, rejects the null hypothesis, the series is normally distributed.

*4.2. Time Series Modelling*

One of the objectives of this study is to construct a model Philippine Stock Exchange Composite index using Autoregressive Integrated Moving Average (ARIMA) and Autoregressive Conditional Heteroskedasticity (ARCH) models.

*4.2.1. Unit Root Test.* In order to test for the existence of unit root, the researchers used Augmented Dickey-Fuller (ADF) test

**Table 1.** Unit Root Test for PSEi

| Include in Test Equation | ADF Test Statistics | Mackinnon Critical Values | | |
|---|---|---|---|---|
| | | 1% level | 5% level | 10% level |
| Intercept | -0.056326 | -3.432385 | -2.862325 | -2.567232 |
| Trend and Intercept | -2.145997 | -3.961161 | -3.411334 | -3.127512 |
| None | 1.772261 | -2.565749 | -1.940932 | -1.616627 |

Table 1 reveals the daily closing price of PSEi is non-stationary at level, since the absolute values of the computed ADF test statistics are less than the absolute value of Mackinnon critical values. In order to achieve stationarity, we need to transform the data to its first difference logarithm.

**Table 2.** Unit Root Test for DLOG(PSEi)

| Include in Test Equation | ADF Test Statistics | Mackinnon Critical Values | | |
|---|---|---|---|---|
| | | 1% level | 5% level | 10% level |
| Intercept | -47.45259 | -3.432382 | -2.862324 | -2.567231 |
| Trend and Intercept | -47.44452 | -3.961161 | -3.411334 | -3.127512 |
| None | -47.39008 | -2.565748 | -1.940932 | -1.616627 |

Table 2 summarizes the unit root test of PSEi at first difference logarithm. Since the absolute values of the computed ADF test statistics are greater than the absolute value of Mackinnon critical values, first difference logarithm of PSEi is already stationary.

*4.2.2. Lag Specification.* The appropriate autoregressive and moving average terms are identified using the Partial Autocorrelation Function (PACF) and Autocorrelation Function (ACF) of the time series, which indicates the significant lags for the AR and MA terms.

Significant spikes of ACF and PACF are the candidates for the terms of MA and AR, respectively. Lags 1, 3, 4, 5, 11 and 12 are the candidates for moving average term. While lags 1, 3, 4, 5, and 11 are the candidates to be the autoregressive term. To check if there are other significant lags, we need to check the correlogram of residuals. According to the correlogram of residuals, it shows that there are no other significant spikes. And since there are no other significant terms, we can proceed in identifying the ARIMA model. Only the ARIMA model with significant terms are considered and the model with highest $R^2$ and log-likelihood and least Akaike Information Criterion (AIC) and Schwarz Bayesian Criterion (SBC) will be chosen.

**Table 3.** Summary of Candidates for ARIMA Models

| Model | $R^2$ | Log-likelihood | AIC | SBC |
|---|---|---|---|---|
| ARIMA(3,1,0) | 0.002408 | 8523.566 | -5.796984 | -5.792911 |
| ARIMA(4,1,0) | 0.001655 | 8519.123 | -5.795932 | -5.791859 |
| ARIMA(5,1,0) | 0.003335 | 8518.229 | -5.797296 | -5.793222 |

| | | | | |
|---|---|---|---|---|
| ARIMA(11,1,0) | 0.001579 | 8496.309 | -5.794208 | -5.792738 |
| IMA(1,1) | 0.017461 | 8555.649 | -5.812877 | **-5.808808** |
| IMA(1,3) | 0.002401 | 8533.265 | -5.797666 | -5.793597 |
| IMA(1,4) | 0.001591 | 8532.071 | -5.796854 | -5.792785 |
| IMA(1,5) | 0.003290 | 8534.576 | -5.798557 | -5.794488 |
| IMA(1,11) | 0.001723 | 8532.265 | -5.796986 | -5.792917 |
| ARIMA(1,1,3) | 0.019635 | 8555.534 | -5.814095 | -5.807990 |
| ARIMA(1,1,5) | **0.020507** | **8556.843** | **-5.814985** | -5.808880 |
| ARIMA(3,1,1) | 0.019623 | 8549.153 | -5.813710 | -5.807601 |
| ARIMA(3,1,5) | 0.005583 | 8523.252 | -5.799491 | -5.793383 |
| ARIMA(3,1,11) | 0.004119 | 8526.029 | -5.798020 | -5.791911 |
| ARIMA(4,1,11) | 0.003469 | 8521.794 | -5.797070 | -5.790960 |
| ARIMA(5,1,1) | 0.020478 | 8543.716 | -5.813966 | -5.807854 |
| ARIMA(5,1,3) | 0.005558 | 8521.509 | -5.798849 | -5.792737 |
| ARIMA(5,1,11) | 0.004886 | 8520.516 | -5.798173 | -5.792061 |
| ARIMA(11,1,3) | 0.003981 | 8499.841 | -5.795925 | -5.789813 |
| ARIMA(11,1,4) | 0.003349 | 8498.911 | -5.795301 | -5.789178 |
| ARIMA(11,1,5) | 0.004968 | 8501.293 | -5.796926 | -5.790804 |
| ***Highest $R^2$ and Log-likelihood; Least AIC and SBC.*** | | | | |

Table 3 displays the candidates for ARIMA model for Philippine Stock Exchange Composite index. ARIMA(1, 1,5) has the highest $R^2$, log-likelihood and lowest Akaike Information Criterion while ARIMA(0,1,1) or IMA(1,1) has the least Schwarz Bayesian Criterion. Though ARIMA(1,1,5) did not acquire all the necessary conditions, it is safe to say that the best ARIMA model among the candidates is **ARIMA(1,1,5)**. This can written as

$$\text{dlog(psei)}_t = 0.000548 + 0.131441 \text{ dlog(psei)}_{t-1} - 0.054326\, \varepsilon_{t-5} + \varepsilon_t.$$

*4.2.3. Verification of the Model.* The model is diagnosed to check the goodness of fit of the model. Researchers used Autoregressive Conditional Heteroskedasticity (ARCH) test for model diagnostic. The residuals of the estimated model is checked using ARCH test to examine if the existing model suffers from ARCH effect, that is, the variance of the residuals is not constant through time.

**Table 4.** ARCH Test for ARIMA(1,1,5)

| **Heteroskedasticitiy Test: ARCH** | |
|---|---|
| F-statistic | Probability F(12,2917) |
| 35.01592 | 0.0000 |

Table 4 represents the Autoregressive Conditional Heteroskedasticity test for the chosen ARIMA model. Constant variance should have F-statistic probability value greater than 0.05. In this case, the p-value of the F-statistic is 0.00 which is less than the critical value 0.05. Hence, the variance of error term is not constant through time, that is, the model has the volatility clustering effect. This empirical result agreed the characteristic of a financial time series, heteroskedasticity or time-varying variance.

*4.2.4. Equation of Conditional Variance.* Due to variance inconstancy, it is necessary to include variance equation to capture the ARCH effect in our time series. Therefore, the researchers decided to use ARCH model and some of its extensions. In finance, conditional volatility models such as ARCH/GARCH models usually describe "risks" perceived by investors.

*4.3. Measures of Statistical Performance of the Models*

To determine the best model for Philippine Stock Exchange Composite index, the researchers used statistical performance measures such as Mean Absolute Error (MAE), Mean Absolute Percentage Error (MAPE), Root Mean Squared Error (RMSE) and Theil Inequality Coefficient.

The table below summarizes the measures of statistical performance of mixture of ARIMA model with ARCH model and some of its extensions.

**Table 5.** Results of Statistical Performance of Models

| **Model** | **Statistical Performance Measurement** | | | |
|---|---|---|---|---|
| | **MAE** | **MAPE** | **RMSE** | **Theil-U** |
| **ARIMA(1,1,5) – ARCH(1)** | 28.10048 | 0.947536 | 43.75942 | 0.006432 |
| **ARIMA(1,1,5) – GARCH(1,1)** | 28.14948 | 0.948263 | 43.78427 | 0.006435 |
| **ARIMA(1,1,5) – EGARCH(1,1)** | 28.14739 | 0.947853 | 43.76435 | 0.006433 |
| **ARIMA(1,1,5) – TARCH(1,1,1)** | 28.16023 | 0.948130 | 43.77616 | 0.006435 |
| **ARIMA(1,1,5) – PARCH(1,1,1)** | 28.15703 | 0.948067 | 43.77259 | 0.006434 |

The model having the least MAE, MAPE, RMSE and Theil Inequality Coefficient will be chosen. Table 5 shows the summary results of statistical performance of the forecasting models. It can be seen that ARIMA(1,1,5) – ARCH(1) outperforms the other models having the lowest Mean Absolute Error, Mean Absolute Percentage Error, Root Mean Squared Error. Therefore, it is safe to say that among the mixture of ARIMA model with ARCH and some of its extensions, **ARIMA(1,1,5) – ARCH(1)** is the best model.

### *4.4. Causation Between PSEi and Economic Variables*

The study also aims to find out what is/are the economic variable(s) that can be helpful in predicting future Philippine Stock Exchange Composite index. Monthly closing price of PSEi from December 2001 to August 2013 is used to fit the frequency of the economic variables considered.

*4.4.1. Co-integration.* The Engle-Granger Methodology was used to test co-integration. This method is a two-step process; first, estimate an OLS regression on I(1) data and then a stationarity test (such as the ADF test) to the residual from this regression.

**Table 6.** Summary of Unit Root Test for Residuals

| **Variable** | **Include in Test Equation** | **ADF Test Statistics** | **Mackinon Critical Values** | | |
|---|---|---|---|---|---|
| | | | **1% level** | **5% level** | **10% level** |
| **COP** | Intercept | -11.974920 | -3.477835 | -2.882279 | -2.577908 |
| | Trend and Intercept | -11.974470 | -4.025426 | -3.442474 | -3.145882 |
| | None | -11.736730 | -2.581705 | -1.943140 | -1.615189 |
| **CPI** | Intercept | -11.335040 | -3.477835 | -2.882279 | -2.577908 |
| | Trend and Intercept | -11.316990 | -4.025426 | -3.442474 | -3.145882 |
| | None | -11.214980 | -2.581705 | -1.943140 | -1.615189 |
| **FE** | Intercept | -11.420170 | -3.477835 | -2.882279 | -2.577908 |
| | Trend and Intercept | -11.396290 | -4.025426 | -3.442474 | -3.145882 |
| | None | -11.171580 | -2.581705 | -1.943140 | -1.615189 |
| **GP** | Intercept | -11.453950 | -3.477835 | -2.882279 | -2.577908 |
| | Trend and Intercept | -11.440260 | -4.025426 | -3.442474 | -3.145882 |
| | None | -11.209050 | -2.581705 | -1.943140 | -1.615189 |
| **IR** | Intercept | -11.468920 | -3.477835 | -2.882279 | -2.577908 |
| | Trend and Intercept | -11.445660 | -4.025426 | -3.442474 | -3.145882 |
| | None | -11.222960 | -2.581705 | -1.943140 | -1.615189 |
| **MS** | Intercept | -11.350670 | -3.477835 | -2.882279 | -2.577908 |

| | | | | | |
|---|---|---|---|---|---|
| | Trend and Intercept | -11.326900 | -4.025426 | -3.442474 | -3.145882 |
| | None | -11.167000 | -2.581705 | -1.943140 | -1.615189 |
| **PE** | Intercept | -11.491930 | -3.477835 | -2.882279 | -2.577908 |
| | Trend and Intercept | -11.457230 | -4.025426 | -3.442474 | -3.145882 |
| | None | -6.408769 | -2.581827 | -1.943157 | -1.615178 |
| **PPI** | Intercept | -11.401780 | -3.477835 | -2.882279 | -2.577908 |
| | Trend and Intercept | -11.359220 | -4.025426 | -3.442474 | -3.145882 |
| | None | -11.018630 | -2.581705 | -1.943140 | -1.615189 |
| **TOT** | Intercept | -11.072920 | -3.477835 | -2.882279 | -2.577908 |
| | Trend and Intercept | -11.048760 | -4.025426 | -3.442474 | -3.145882 |
| | None | -10.771390 | -2.581705 | -1.943140 | -1.615189 |

Table 6 elaborates that the residuals of the nine (9) factors are stationary since the absolute value of ADF test statistics are greater than the absolute value of Mackinnon critical values for any test equations. Therefore, we say that PSEi is co-integrated for every economic variable considered.

*4.4.2. Lag Selection.*In selecting the lags, the researchers looked on the lag having the lowest Akaike Information Criterion (AIC). The table below displays the lag selection of each variable.

**Table 7.** Akaike Information Criterion of Each Variable at Different Lags

| Lags | Variables | | | | | | | | |
|---|---|---|---|---|---|---|---|---|---|
| | COP | CPI | FE | GP | IR | MS | PE | PPI | TOT |
| 1 | -4.902598 | -11.20667 | -8.017319 | -5.967508 | -7.570037 | -7.367106 | -5.529872 | -8.099914 | -4.342265 |
| 2 | -5.005374 | -11.23508 | -8.086729 | -6.088249 | -7.565244 | -7.503416 | -5.676516 | -8.255589 | -4.506782 |
| 3 | -5.052836 | -11.23830 | -8.207831 | **-6.125979** | -7.540696 | -7.587544 | -5.726543 | -8.290817 | -4.646528 |
| 4 | -5.017172 | -11.25496 | -8.211598 | -6.094422 | -7.522821 | -7.545751 | -5.732615 | -8.276141 | -4.682489 |
| 5 | -5.134964 | -11.24597 | -8.204656 | -6.113052 | -7.518117 | **-7.728846** | **-5.811397** | **-8.335168** | -4.723413 |
| 6 | -5.143216 | -11.23490 | -8.166524 | **-6.125979** | -7.483084 | -7.692001 | -5.768616 | -8.304754 | -4.757902 |
| 7 | **-5.147194** | -11.26475 | -8.141997 | -6.010105 | **-7.571015** | -7.636403 | -5.706559 | -8.250155 | -4.757331 |
| 8 | -5.127551 | -11.27886 | -8.137790 | -6.021403 | -7.527866 | -7.594387 | -5.652933 | -8.258144 | -4.691601 |
| 9 | -5.091543 | **-11.27990** | -8.092662 | -5.995569 | -7.520726 | -7.535082 | -5.661233 | -8.222991 | **-4.762973** |
| 10 | -5.041277 | -11.25164 | -8.254575 | -5.982922 | -7.465968 | -7.535675 | -5.636058 | -8.154406 | -4.718249 |
| 11 | -5.058537 | -11.25746 | **-8.271445** | -5.925706 | -7.410102 | -7.566829 | -5.606427 | -8.224659 | -4.696104 |
| 12 | -5.016723 | -11.23722 | -8.225671 | -5.913210 | -7.434513 | -7.518472 | -5.565515 | -8.203059 | -4.664895 |

Based on Table 7, COP has lowest AIC at lag 7; CPI has lowest at lag 9; FE at lag 11; GP at lags 3 and 6; IR at lag 7; MS, PE and PPI having the lowest AIC at lag 5; and DTOT at lag 9.

*4.4.3. Granger Causality Test.* After determining at which lag will each variable has the lowest AIC, we can now able to identify among the factors are Granger-cause to the PSEi. Granger Causality is used to investigate whether a time series $x_t$ can help forecast another series $y_t$..

**Table 8.** Economic Factors Granger Cause PSEi

| Null Hypothesis | F-Statistic | Probability |
|---|---|---|
| COP does Granger Cause PSEI | 3.00051 | 0.0062 |
| CPI does Granger Cause PSEI | 2.11531 | 0.0338 |

| | | |
|---|---|---|
| FE does Granger Cause PSEI | 3.16663 | 0.0010 |
| GP does not Granger Cause PSEI | 2.08365 | 0.1055 |
| IR does not Granger Cause PSEI | 1.49110 | 0.1769 |
| MS does not Granger Cause PSEI | 1.68209 | 0.1438 |
| PE does not Granger Cause PSEI | 0.49951 | 0.7761 |
| PPI does not Granger Cause PSEI | 0.47528 | 0.7941 |
| TOT does not Granger Cause PSEI | 1.13146 | 0.3467 |

Among the nine (9) factors, Consumers Price Index, Crude Oil Price, and Foreign Exchange obtained the probability lower than 0.05 which only means that they influenced the fluctuation of Philippine Stock Exchange index. In addition to that, the shown table below 9 gave supplementary information which of the factors can PSEi affect.

**Table 9.** PSEi Granger Cause Economic Factors

| Null Hypothesis | F-Statistic | Probability |
|---|---|---|
| PSEI does Granger Cause COP | 2.94884 | 0.0070 |
| PSEI does not Granger Cause CPI | 1.47139 | 0.1671 |
| PSEI does Granger Cause FE | 2.02478 | 0.0328 |
| PSEI does not Granger Cause GP | 0.76229 | 0.5172 |
| PSEI does not Granger Cause IR | 0.36254 | 0.9222 |
| PSEI does not Granger Cause MS | 0.48218 | 0.7890 |
| PSEI does not Granger Cause PE | 0.65190 | 0.6606 |
| PSEI does not Granger Cause PPI | 1.06175 | 0.3849 |
| PSEI does not Granger Cause TOT | 1.39967 | 0.1967 |

Table 9 shows that the Crude Oil Price and Foreign Exchange are the factors influenced or caused by PSEi because they attained probability lower than 0.05. Since Crude Oil Price and Foreign Exchange also Granger cause the PSEi, it indicates that there is bi-directional causality between Crude Oil Price and PSEi, and Foreign Exchange and PSEi.

## 5. Conclusion

The study was conducted to construct a time series model for daily closing price of PSEi and to find out some variables that can influence the behavior of PSEi. The daily closing price of PSEi from January 2, 2002 to December 27, 2013 is used to estimate the time series model. Meanwhile, Consumer Price Index (CPI), crude oil price, foreign exchange rate (U.S Dollar versus Philippine Peso), gold price, interest rate, money supply(M4), price-earnings ratio, Producers' Price Index (PPI) and terms of trade were the economic variables considered by the researchers. The mixture of ARIMA models and conditional variance equation, ARCH models and some of its extensions such as GARCH, EGARCH, TARCH and PARCH, were the generated time series model for PSEi. Also, Granger Causality test is used to examine which of the variables causes PSEi.

The researchers found out that time series data tend to be leptokurtic and mostly to have a non-normal distribution. The data were examined for unit root test and the researchers concluded that all of the data are non-stationary series. Difference of logarithm is used as data transformation for all of the variables. Correlogram of ACF and PACF were utilized to specify the number of lags in ARIMA model. Also, the model was diagnosed to check whether the model suffers from ARCH effect. The researchers discovered that the variance is not constant over time and hence heteroskedastic. ARCH and some of its extensions were used to capture the ARCH effect. According to ARCH(1) and GARCH(1,1), the current volatility was affected by past residuals. Also, the presence of leverage effects were noted by EGARCH(1,1), TARCH(1,1,1) and PARCH(1,1,1). The best fit model was chosen based on the statistical performance of the model. The researchers concluded that the best fit time series model for Philippine Stock Exchange Composite Index was ARIMA(1,1,5) – ARCH(1).

Granger causality test was applied by researchers in order to determine which of the considered economic variables could influence the future values of PSEi. Engle – Granger Methodology for testing co-integration was conducted. Afterwards, lags were determined based on Akaike Information Criterion. Granger causality test with the appropriate lags was executed. The researchers found out that out of nine (9) factors only Consumer Price Index, crude oil price and foreign exchange Granger cause PSEi. Additionally, it was also found out that PSEi can be used as a parameter for projecting future prices of crude oil prices and foreign exchange rate.

## 6. Recommendation

Upon the conclusion of the study, Philippine Stock Exchange Inc., players of the stock market, as well as investors may use this study as their basis in decision making. Also, they could seek the movement of Consumer Price Index, crude oil price and foreign exchange rate in order to project future stock prices.

One of the interests of the study was to develop a time series model for Philippine Stock Exchange Composite index using the mixture of ARIMA with some conditional variance equations such as ARCH, GARCH, EGARCH, TARCH, and PARCH. Thus, future researchers may use other conditional volatility models under the extensions of ARCH model.

The researchers concluded that using Granger causality test, Consumer Price index, crude oil price and foreign exchange rate can influence future values of the Philippine Stock Exchange Composite index. Hence, future researchers can utilize this finding in construction of factor models such as regression. In addition to that, other economic indicators may be used.

## 7. References

[1] Adam, Anokye M., George Tweneboah. *"Macroeconomic Factors and Stock Market Movement: Evidence from Ghana"*. School of Management, University of Leicester, UK. http://ssrn.com/abstract=1289842

[2] Agrawal, G. et. al. (December 2010). *"A Study of Exchange Rates Movement and Stock Market Volatility"*. International Journal of Business and Management, vol. 5, no. 12, pp 62-73. www.ccsenet.org/ijbm

[3] Alam, Zahangir, et. al. (April, 2013). *"Forecasting Volatility of Stock Indices with ARCH Model"*. International Journal of Financial Research. Vol. 4, No. 2, pp. 126-143. www.sciedu.ca/ijfr

[4] Ali, Mohammad Bayezid (2011). *"Stock Prices and Microeconomic Variables: T-Y Granger Causal Evidence From Dhaka Stock Exchange (DSE)"*. Research Journal of Finance and Accounting. Vol 2, No 6. www.iiste.org

[5] Arowolo, W.B. (May, 2013). *"Predicting Stock Prices Returns Using Garch Model"*. The International Journal of Engineering and Science. Vol. 2, issue 5, pp 32-37. www.theijes.com

[6] Cariño, Camille Joy F., et. al. (September, 2013). *"Philippine: Time Series Analysis of Stock Market Returns and Their Macroeconomic Impacts for Years 2002 Through 2011"*. De La Salle University, Manila.

# Topological tangle modeling of difference topology experiments: tangle analysis of DNA-protein complexes

**Soojeong Kim**

Assistant Professor, Yonsei University, International Campus,
85 Songdogwahak-ro, Yeonsu-gu, Incheon 406-840, Republic of Korea

E-mail: soojkim@yonsei.ac.kr

**Abstract.** An n-string tangle is a three dimensional ball with n strings which are properly embedded in it. Topological tangle analysis of DNA-protein complexes based on difference topology experiments are studied for last 30 years. In this paper, A 2-string tangle analysis of DNA-Tn3 complex [5, 6] and a 3-string tangle analysis of DNA-Mu complex [3, 13] and a biologically relevant 4-string tangle analysis [8, 10, 11] is reviewed. And a biologically relevant n-string tangle analysis is introduced.

## 1. Introduction

Mathematical $n$-string tangle is a three dimensional ball $B$ with $n$-strings which are properly embedded in $B$. I.e, each string's two end points are on the boundary of the ball $B$. In the early 1990's, C. Ernst and D. Sumners used 2-string tangle to analyze the conformation of DNA segments within the Tn3 and Phase$\lambda$ proteins based on N. Cozzarelli's experiments [4, 15, 18, 19]

In late 2000's, a 3-string tangle analysis of DNA topology within Mu-protein is introduced by I. Darcy et. al.[3]. Their work is motivated by S. Pathania et. al.'s difference experiment of Mu-transpososome [13]. I. Darcy et. al. figured out the 3-branched and 5 crossings conformation of DNA within Mu-proteins.

Recently, the author and I. Darcy developed 4-string tangle analysis of DNA-protein complexes [10] and this theory is generalized to n-string tangle analysis which is introduced in this paper.

Some basic concept of difference topology experience of DNA-protein complexes is explained in section 2. In section 3, each of 2, 3 and 4-string tangle modeling of DNA topology and analysis of difference topology experiment is explained. A generalized $n$-string tangle analysis is introduced at the end of this section 3.

## 2. Difference topology experiments of DNA-protein complexes

As widely known, every organism has its genetic information in DNA. In 1953, J. Watson and F. Crick discovered that the structure of DNA is double helical [20]. Here, 'double' means that one DNA string consists of two strands and they are connected to each other. 'Helical' means that the double stranded DNA is coiled around the helical axis at regular interval. Simply speaking, a three dimensional topology of DNA is twisted ladder as in Figure 1(a). Sometimes, to investigate

Journal of Physics: Conference Series **622** (2015) 012023 doi:10.1088/1742-6596/622/1/012023

a three dimensional shape(topology) of DNA, a double stranded DNA is represented by a string (see Figure 1(b)).

In nature, DNA is involved in many biological process including replication, transposition, recombination etc. Due to the double helical structure of it, a long DNA molecule contort into another shape. This is a supercoiling. A simple example of supercoiled DNA is shown in Figure 1(c).

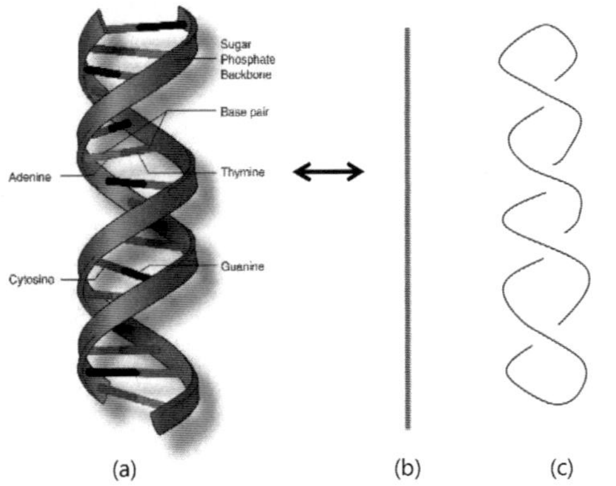


**Figure 1.** (a) The double helical structure of DNA. Figure Courtesy: National Human Genome Research Institute. (b) A double stranded DNA is simply represented by a string. (c) Supercolied DNA.

For performing a biological process, DNA used to bind with some proteins or enzymes. In this process, it is not easy to observe the topological shape of DNA which is bound by proteins. Scientists uses another protein to figure out the conformation of DNA within the proteins. For example, one can use a protein Cre. Cre is a protein which binds to DNA at two specific sites and cut both two sites and rejoin them in different way as in Figure 2. This type of action is called DNA recombination. In Figure 2, initially there is a supercoiled circular DNA with two Cre binding sites on it. There are two different type of DNA recombination reactions are occurred. First, the supercoiled DNA is unwounded and cre binds to its two binding sites. After Cre recombination, DNA conformation is turned out to unlink(two separated rings). Second, a unknown protein binds to the supercoiled DNA , and then Cre binds to its two binding sites. After Cre recombination, the conformation of DNA is changed to Hopf link (two rings with two crossings which cannot be removed by moving each component). Since two different types of Cre recombination experiments results in different DNA conformations, one can notice that the unknown protein used in second experiment wrapped some crossings of DNA. Furthermore, one can guess the number of crossings wrapped is two. This result is from the difference of two DNA topology between two experiments, so it is called 'difference topology experiment'.

## 3. Tangle analysis of difference topology experiments

As I mentioned in section 2, an n-string tangle is a three dimensional ball with $n$-strings properly embedded in it. Especially, if $n$-strings are connected as in Figure 3(a), it is called a 0-tangle(or zero-tangle). Examples of 2-string tangle and 3-string tangle are shown in Figure 3(b) and (c), respectively. Those two tangles in Figure 3(b) and (c) can be isotoped to a 2-string or 3-string zero-tangle if the boundary of the strings are allowed to move along the boundary of the ball. In other words, those tangles are *freely isotopic* to zero-tangle. In general, a *rational tangle* is a tangle which is freely isotopic to the zero-tangle.

C. Ernst and D. Sumners first used 2-string tangle model to analyze DNA-protein complexes based on difference topology experiments [5]. In that model, proteins are represented by a three dimensional ball and DNA within the proteins is represented by strings. They predicted

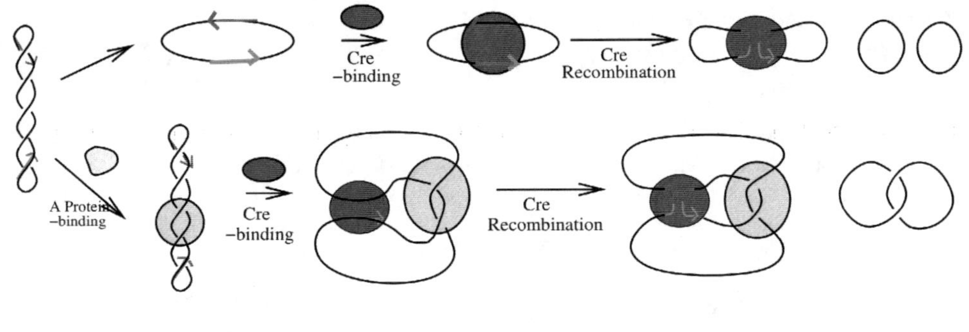


**Figure 2.** Example of difference topology experiments. Figure from [9]

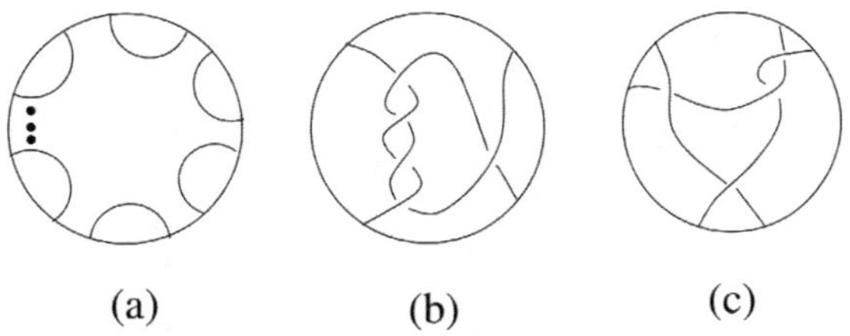


**Figure 3.** (a) $n$-string zero tangle (b) Example of 2-string tangle (c) Example of 3-string tangle

a conformation of DNA within a protein Phase$\lambda$ as in Theorem3.1. In this theorem, R is a two string tangle which represent DNA-Phase$\lambda$ complex. See [5, 6] for more detail.

**Theorem 3.1** *[5, 6] Suppose that 2-string tangles $O_b, R$ satisfy the three tangle equations in Figure 4, then the solution tangles for $O_b, R$ are as in Figure 5*

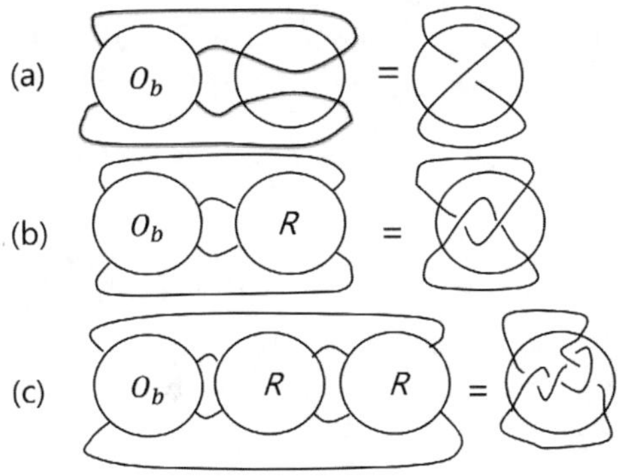


**Figure 4.**

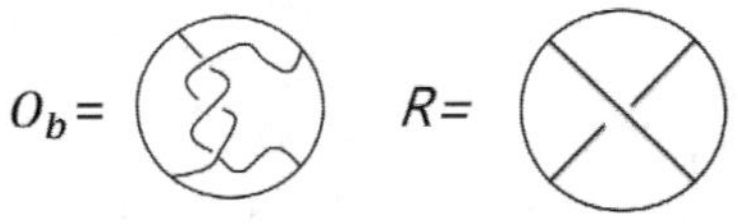


**Figure 5.**

Note that the tangle equations in Figure 4 imply processive Tn3 recombination experiments. For more detail, see [5, 6].

2-string tangle theories are well studied by many mathematicians. However, 3-string tangle analysis is much complicate. For example, a most biologically relevant 2-string tangle model is rational tangle and all 2-string rational tangle is classified. On the other hand, 3-string rational tangles have not been classified yet.

In 2002, Pathania et. al. performed difference topology experiments on DNA-Mu complexes [13]. Mu is a protein which binds to DNA at three specific sites. When Mu binds to a circular DNA with three Mu binding sites, DNA-Mu complex can be modeled by 3-string tangle with three outside loops as in Figure 6(a). Darcy et. al. used 3-string tangle model to support Pathania et. al.'s difference topology experiments of Mu-transpososome as in Theorem 3.2:

**Theorem 3.2** *[3] Let $T$ be a 3-string tangle that satisfies the system of tangle equations in Figure 6(a). If $T$ is freely isotopic to a projection with less than 8 crossings, then $T$ is the tangle in the Figure 6(b).*

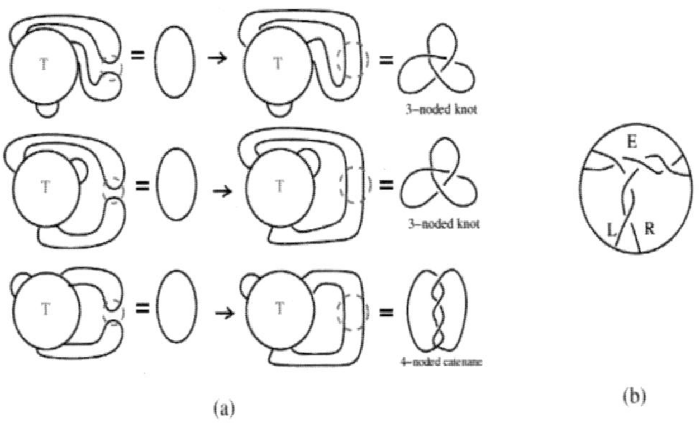


**Figure 6.**

Note that the tangle equations in Figure 6(a) imply difference experiments of Mu-transpososome with Cre recombination, and Figure 6(b) is a 3-string tangle model of DNA within Mu. For more detail, see [13, 3].

Mativated by 3-string tangle analysis, the author and I. Darcy developed a biologically relevant 4-string tangle analysis:

**Theorem 3.3** *[8, 10, 11] Suppose $T$ is 4-string tangle which has less than 8 crossings up to free isotopy. If $T$ is a tangle which satisfies all tangle equations in Figure 7, then $T$ is an R-standard tangle as in Figure 8 after allowing each pair of boundary points to move.*

Recently, the author extended a biologically relevant 3 or 4-string tangle analysis in Theorem 3.3 to $n$-string tangle analysis. The main theory is stated in Theorem 3.4:

**Theorem 3.4** *Suppose $T$ is an $n$-string rational tangle where $n \geq 5$. Let $G$ be a graph with edges $e_i, c_j$ and strings of $T$, as in Figure 9. If $T \cup_{1 \leq k \leq n} c_k$ is unknot and $T \cup_{k \neq i,j} c_k$ is $(2,p)$-torus knot/link, then $G$ is planar. I.e, a loop formed by all $c_i$'s and strings of $T$ bounds a disk.*

In this Theorem 3.4, two assumptions about $T(T \cup_{1 \leq k \leq n} c_k$ is unknot and $T \cup_{k \neq i,j} c_k$ is $(2,p)$-torus knot/link) imply difference topology experiments of a DNA-protein complex when the protein binds to DNA at $n$ sites. Even though there is no biological experimental data

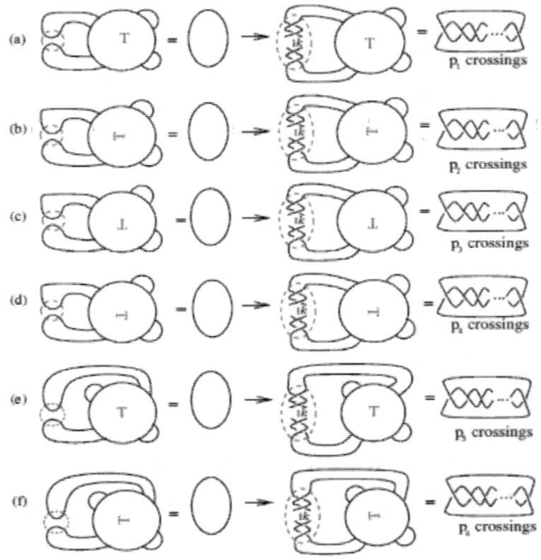


**Figure 7.**

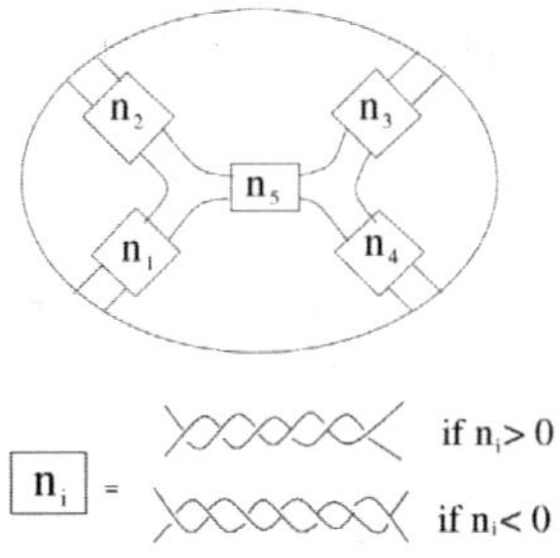


**Figure 8.**

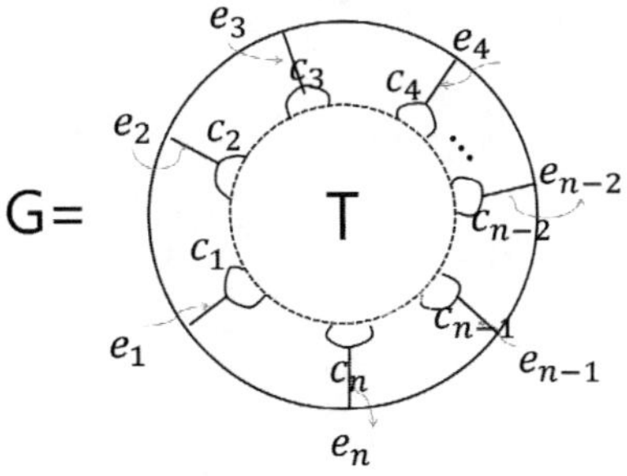


**Figure 9.**

for $n \geq 4$ yet, lots of biological processes (such as replication, transposition, translation, etc.) involve DNA-protein reaction at multiple sites. Hence developing $n$-string tangle analysis is greatly worth.

**Acknowledgments**
The author was supported by Basic Science Research Program through the National Research Foundation of Korea(NRF) funded by the Ministry of Education, Science and Technology(NRF-2012R1A1A3015842).

## References

[1] Bates A and Maxwell A 1993 DNA topology *IRL press oxford*
[2] Bauer W, Crick F and White J 1980 *Scientific american* **243** 118–133.
[3] Darcy I, Luecke J and Vazquez M 2009 *Algebraic and geometric topology,* , **9** 2247–2309.
[4] Dean F, Stasiak A, Koller T and Cozzarelli N 1985 *J. Biol. Chem.* **260** 4795–4983.
[5] Ernst C and Sumners D 1990 *Math. Proc. Camb. Phil. Soc.* **108** 489–515.
[6] Ernst C and Sumners D 1990 *Math. Proc. Camb. Phil. Soc.* **126** 23–36.
[7] Hirasawa M and Shimokawa K 2000 *Proc. Amer. Math. Soc.* **128** 3445–3451.
[8] Kim S 2011 *J. Ksiam* **15(3)** 161–175.
[9] Kim S 2013 *J. Ksiam* **17(2)** 87–102.
[10] Kim S and Darcy I 2009 *Mathematics of DNA structure, function and interacctions, the IMA volumes in nathematics and its applications, springer science + business media, LLC, New York*
[11] Kim S 2009 *Ph.D thesis* ,
[12] Lewin B 1994 Genes *Oxford University Press, 5th ed., New York*
[13] Pathania S, Jayaram M and Harshey R 2002 *Cell* **109(4)** 425–436.
[14] Saka Y and Vazquez M 2002 *Bioinformatics* **18** 1011–1012.
[15] Spengler S, Stasiak A and Cozzarelli N 1985 *Cell* **42** 325–334.
[16] Sumners D, Ernst C, Cozzarelli N and Spengler S 1995 *Quarterly Reviews of Biophysics* **28**
[17] Vetcher A, Lushnikov A, Navarra-Madsen J, Scharein R, Lyubchenko Y, Darcy I and Levene S 2006 *J. Mol. Biol.*, **357**, 1089–1104.
[18] Wasserman S and Cozzarelli N 1985 *Proc. Nat. Acad. Sci. U.S.A* **82** 1079–1083.
[19] Wasserman S, Dungan J and Cozzarelli N 1985 *Science* **229** 171–174.
[20] Watson J and Crick F 1953 *Nature* **171** 737-738

Journal of Physics: Conference Series **622** (2015) 012024 doi:10.1088/1742-6596/622/1/012024

# A GIS-based Computational Tool for Multidimensional Flow Velocity by Acoustic Doppler Current Profilers

**D Kim[1], M Winkler[2], M Muste[3]**
[1]Assistant Professor, Civil and Environmental Engineering, Dankook University Yongin, Geyonggi-do, South Korea, E-mail: dongsu-kim@dankook.ac.kr
[2]Program Manager, U.S. Army Corps of Engineers, Gulf Region District Baghdad, Iraq. E-mail:Michael.F.Winkler@usace.army.mil Phone: 1-540-722-2711
[3]Research Engineer, IIHR-Hydroscience & Engineering, The University of Iowa, Iowa City, IA 52242. Email: marian-muste@uiowa.edu. Phone: 1-319-594-0624

**Abstract.** Acoustic Doppler Current Profilers (ADCPs) provide efficient and reliable flow measurements compared to other tools for characteristics of the riverine environments. In addition to originally targeted discharge measurements, ADCPs are increasingly utilized to assess river flow characteristics. The newly developed VMS (Velocity Mapping Software) aims at providing an efficient process for quality assurance, mapping velocity vectors for visualization and facilitating comparison with physical and numerical model results. VMS was designed to provide efficient and smooth work flows for processing groups of transects. The software allows the user to select group of files and subsequently to conduct statistical and graphical quality assurance on the files as a group or individually as appropriate. VMS also enables spatial averaging in horizontal and vertical plane for ADCP data in a single or multiple transects over the same or consecutive cross sections. The analysis results are displayed in numerical and graphical formats.

## 1.Introduction

The Acoustic Doppler Current Profilers (ADCPs) has revolutionarily improved our ability to represent the hydrodynamics of riverine environments. ADCPs measure non-intrusively velocities along verticals and bathymetry with a great deal of spatial and temporal resolutions relatively enhanced with previous mechanical instrumentations. ADCPs provide three velocity components over flow depths and stream cross sections hence there is a considerable potential for supporting a wide range of scientific and practical analyses related to river investigations. Researchers in water communities have continuously attempted to utilize the detailed information provided by ADCPs based upon additional data processing software modules. Among the new innovative applications of ADCPs in support of riverine research are: (*i*) characterization of mean flow characteristics (e.g., Muste et al., 2004); (*ii*) estimation of turbulence quantities (e.g., Stacey et al., 1999, Lu and Lueck, 1999; Schemper and Admiraal, 2002; Nystrom et al., 2002; Howarth, 2002; and Kostaschuk et al., 2004); (*iii*) sediment transport (Rennie et al, 2002), (*iv*) characterization of the spatial distribution of velocities in riverine habitats and bed survey (Shields et al., 2003, Gaeuman and Jacobson, 2005), and (*v*) estimation of the longitudinal dispersion coefficient for river flow (Carr and Rehmann, 2007). However, the current ADCP companion software

1 Corresponding author.

(WinRiver and RiverSurveyor for RD Instruments and SonTek, respectively) only focused on calculating streamflow discharge beyond these additional processing capabilities. Presented in this paper, therefore, is a standalone software that assembles in one package visualization and post-processing tools using an advanced computational environment for the benefit of the river monitoring and research communities. The software, labeled 'VMS' (ADCP Velocity Mapping Software), provide many additional processing capabilities driven by research needs and specific requests from the ADCP user community. The VMS brings stronger data filtering and quality control features as well as enhanced visualization capabilities. The VMS software tool is freely-downloadable online at: http://chl.erdc.usace.army.mil/vms. The paper describes the descriptions of main features along with the many other processing and/or visualization capabilities. The VMS mainly provides five categories: general features, quality assurance, spatial averaging, geo-referencing, detailed single transect view (see Table 1).

**Table 1**. Main features of VMS

| Category | Features |
|---|---|
| General Features | • Read ASCII output format from WinRiver I and II (Teledyne RDI) and Matalab output format from RiverSurveyor (SonTek/YSI)<br>• Processing multiple ADCP files<br>• Incorporating ADCP data with GIS files (shapefile)<br>• Exporting processed data into various formats (picture, csv, shapefile, kml)<br>• Supporting 2D or 3D static or dynamic geographical display<br>• Supporting both English/SI unit system<br>• Supporting UTM coordinate (NAD 1983/1927)<br>• Computing 3D geo-referenced positions of beam depth<br>• Interpolating GPS locations using GGA and ADCP time<br>• Layer extraction |
| Quality Assurance | • Error velocity filter<br>• Beam depth filter<br>• Boat speed filter<br>• GPS location filter |
| Spatial Averaging | • Single group averaging along curvilinear path<br>• Multiple group averaging along curvilinear path<br>• Finding mean curvilinear path<br>• 2D/3D averaged velocity data representation<br>• 2D/3D averaging for user-supplied points (grids)<br>• Supporting uncertainty (random error) of averaging |
| Geo-referencing | • Vertical referencing – water surface elevation<br>• Horizontal referencing – no GPS data<br>• BT / GPS tracking |
| Single Transect View | • ASCII viewer (1,2 and 3D info.)<br>• Horizontal / Vertical viewer<br>• Contour plot<br>• BT/GPS track |

## 2. GENERAL FEATURES

The VMS was designed to efficiently process multiple ADCP files for both SonTek and TRDI ADCP formats. When the loaded files are geo-taggged with GPS data, they can be mapped, and the users can easily identify where their measurements are located. Also, the ADCP measurements can be synchronized with GIS shape files, which enables ADCP measurements to be more geographically recognized (see Figure 1). The ADCP files can be exported into GIS format (e.g., shapefile or kml) for further processing.

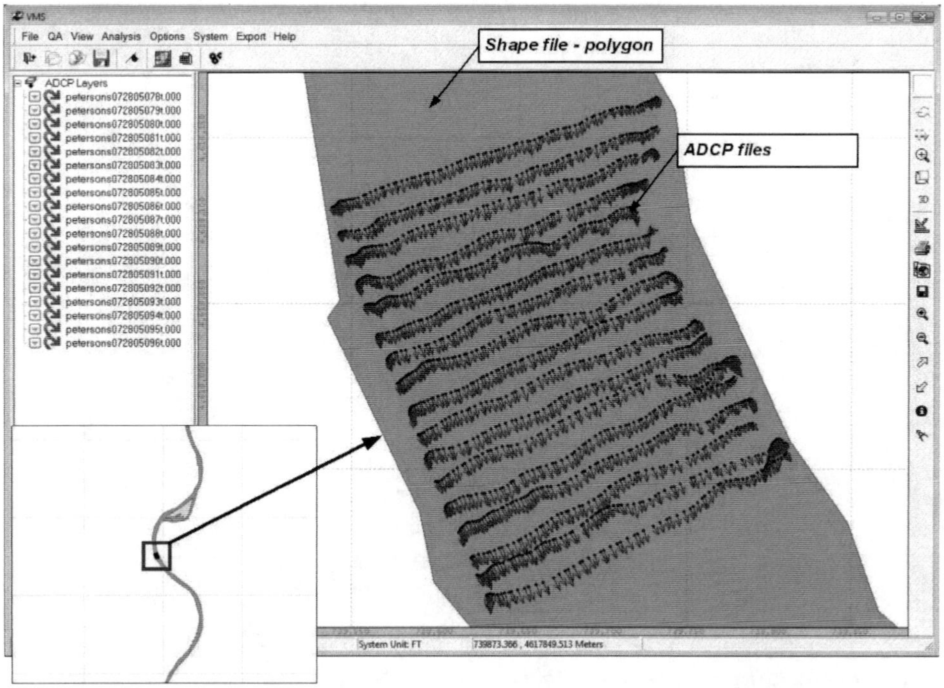


Figure 1. ADCP measurements back-grounded with a GIS shapefile of river boundary.

## 3. QUALITY ASSURANCE

The quality assurance in the VMS software is accomplished by applying various filters such as velocity filter, depth filter and boat speed filter. This quality assurance in VMS can be applied on a transect by transect. Moreover, the many of the data filters can be automated to provide a more efficient and consistent approach to quality assurance. The velocity filter uses the given error velocity, which represents error indicator of flow homogeneity checked by individual velocity measurement. When the error velocity is higher than the given criteria, the measured velocity is considered as an error, and filtered out of the dataset. The depth filter evaluates beam depth measurements which are usually averaged to generate mean depth. The spike or zero depth measurements are filtered out by applying Super Smoother (Friedman, 1984), where a smoothed line based on individual beam depths is generated and any spikes located above or below a given buffer of the smoothed line are filtered out and replaced. Figure 2 shows an application of the implemented filters (depth filter). Boat speed filter works similarly with the depth filter, where the abrupt speed change of the boat speed is filtered. After applying various filters, discharge is corrected.

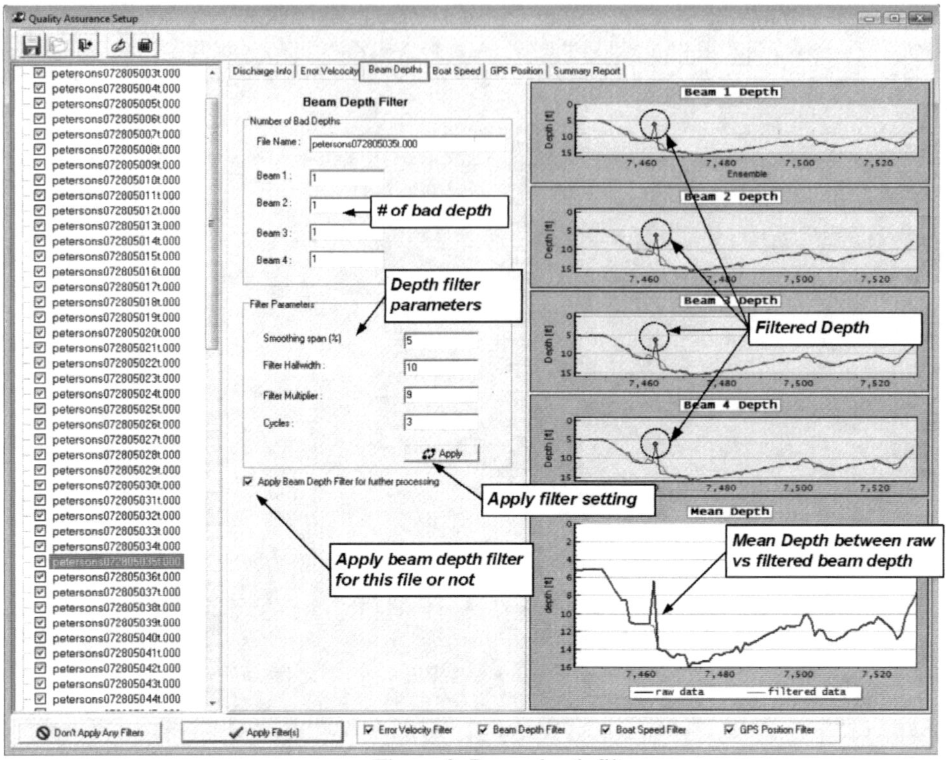


Figure 2. Beam depth filter.

## 4. SINGLE TRANSECT VIEW

VMS provides two different types of velocity profile: vertical and horizontal. First, the vertical profile at each ensemble can be examined in the Vertical tab (see Figure 3). Along the ship track, the user is able to move forward and backward to see velocity data in tabular and 3D graphical format. When the user starts the animation, dynamic variation of the 3D velocity in the transect can be visualized. Different from the conventional contour plot for the cross-section, dynamic three-dimensional plot of velocity vector with the two-dimensional location, the user is easily able to check out any erroneous velocity measurements or trend of the velocity direction. Second, the horizontal profile can be extracted for in the depth-averaged, nearest surface, nearest bottom, and user-defined layers.

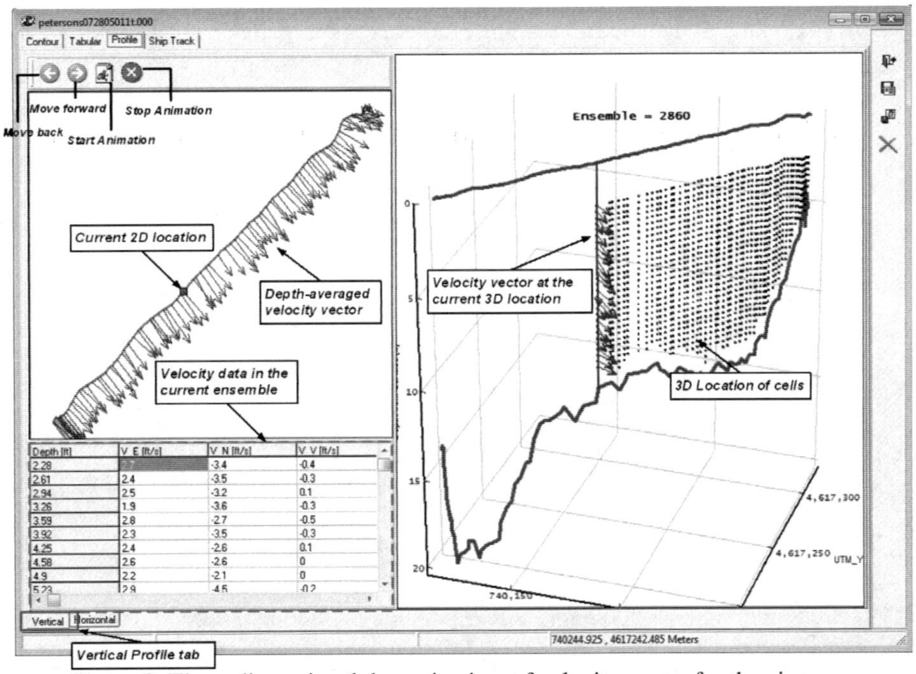


Figure 3. Three-dimensional dynamic view of velocity vector for the given cross-section.

## 5. SPATIAL AVERAGING

VMS supports horizontal averaging along the transect with specified averaging step (or spacing). The horizontal spacing of the averages is applied over the curvilinear path of the transect (see Figure 4). VMS provides averaging among multiple transects collected at the same cross section, which is expected to reduce the uncertainty associated with velocity averages. In general, the users collect multiple transects at the same cross section and look for horizontally and/or vertically averaged data using all of the transects in the group. The average path travelled for a group of transects is computed first to provide a method to compute a mean path. The spacing for horizontal averaging is then applied along that mean path and the points to be averaged will be those that fall within a user defined diameter. The depths along the mean path are averaged and smoothed, and depth at each spaced point will be taken from these depths. The user is able to process or reprocess the data for different purposes by changing options and averaging criteria without having to reselect the data files. VMS supports an automated approach to horizontal averaging that could avoid averaging across changes in flow distribution. This involves an algorithm to identify mean flow direction within the given search areas, where velocity vectors having less than a specified angle criteria are used in horizontal averaging. This approach only considers the flow distributions that are reasonably homogeneous and ensures data included in the average do not cross zones where flow directions change dramatically.

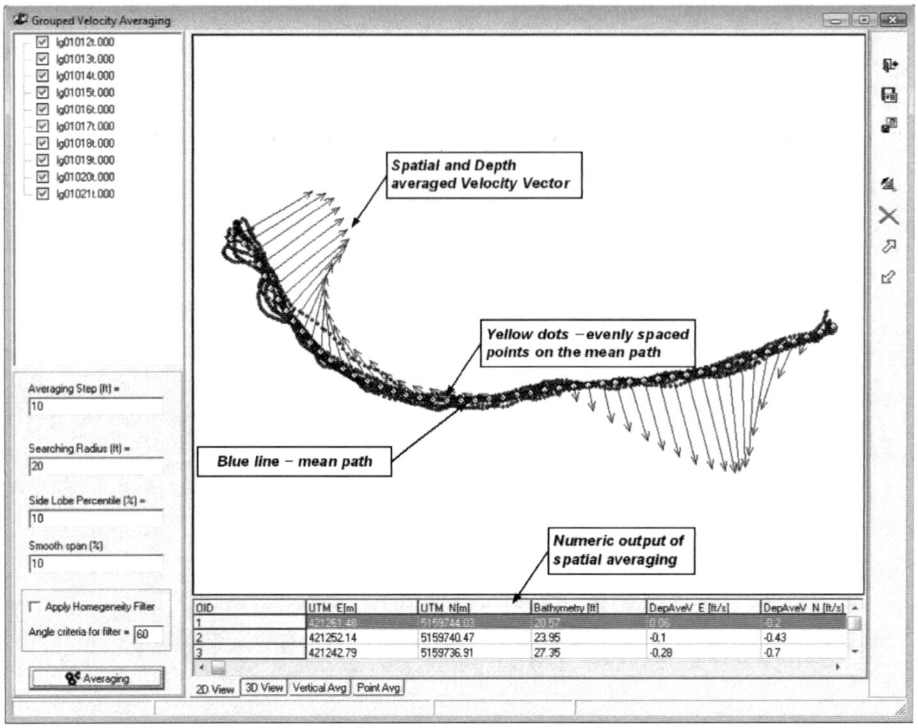


Figure 4. Spatial averaging of velocity vectors along the mean curvilinear path.

## 6. CONCLUSIONS AND FUTURE WORKS

The VMS currently contains several useful features that allow users to easily post-process multiple ADCP dataset to derive riverine information for further analysis. The key aspect of the VMS is to provide a community-based freely available ADCP post-processing tool, where the source algorithms and software are opened to the community and any contributors can update the code and interface of the VMS. The authors welcome any feedbacks from the community to eventually enhance the software.

## ACKNOWLEDGEMENTS
This work was supported by the National Research Foundation of Korea (NRF) grant funded by the Korea government (MEST) ((NRF-2012R1A2A2A02047549).

## REFERENCES

[1] P Biron, C Robson, M Lapointe, and S Gaskin 2004 Comparing Different Methods of Bed Shear Stress Estimates in Simple and Complex Flow Fields *Earth Surface Processes and Landforms*, B **29**, 1403-1415.

[2] H Chanson 2000 Boundary shear stress measurements in undular flows: Application to standing wave bed forms *Water Resources Research*, B **36** 10, 3063-3076

[3] R Dinehart and J Burau 2005 Repeated surveys by acoustic Doppler current profiler for flow and sediment dynamics in a tidal river *J of Hydrology* 1-21

[4] H Fischer, E List, R Koh, J Imberger and N Brooks 1979 *Mixing in Inland and Coastal Waters*, Academic Press, New York

[5] J Friedman 1984 A variable span smoother, Department of Statistics, Stanford University, Technical Report LCS5

[6] R Kostaschuk, P Villard and J Best 2004 Measuring Velocity and Shear Stress over Dunes with Acoustic Doppler Profiler *J. of Hydr. Engrg.*, ASCE, B 130 9, 932-936

[7] Y Lu and R Lueck 1999 Using a Broadband ADCP in a Tidal Channel. Part II: Turbulence *J of Atmospheric and Oceanic Technology*, B 16, 1568-1579

[8] M Muste, K Yu, T Pratt and D Abraham 2004 Practical Aspects of ADCP Data Use for Quantification of Mean River Flow Characteristics: Part II: Fixed-Vessel Measurements *J. of Flow Meas. and Instr.*, B 15 1, 17-28

[9] E Nystrom, K Oberg and C Rehman 2002 Measurement of Turbulence with Acoustic Doppler Current Profilers – Sources of Error and Laboratory Results *Proceedings of the ASCE-IAHR Hydraulic Measurements & Experimental Methods Conference*, Estes Park, CO

[10] RD Instruments 1996 Principles of Operation for Broadband ADCP, RD Instruments, San Diego, CA

[11] L Sime, C Ferguson and M Church 2007 Estimating shear stress from moving boat acoustic Doppler velocity measurements in a large gravel bed river *Water Resources Research*, B 43, W03418

[12] SonTek 2000 Doppler Velocity Log for ROV/AUV Applications *SonTek*, San Diego, CA

[13] M Stacey, S Monismith and J Burau 1999 Measurements of Reynolds Stress Profiles in Unstratified tidal flow *J. of Geophysical Research*, B 104, C5, 10933-10949

[14] R Szupiany, M Amsler, J Best and D Parsons 2007 A Comparison of Fixed- and Moving-vessel flow measurements with an acoustic Doppler profiler (aDp) in a large river *J of Hydraulic Engineering*, B 133 12, 1299-1309

[15] G Taylor 1954 The Dispersion of Matter in Turbulent Flow through a Pipe *Proceeding of the Royal Society of London, Series A, Mathematical and Physical Science*, 446-468

[16] F White 2003 *Fluid Mechanics - fifth edition*, McGraw-Hill, New York, NY

[17] P Wilcock 1996 Estimating local bed shear stress from velocity observations." *Water Resources Research*, B 32 11, 3361-3366

# Thermal Stability of Oil Palm Empty Fruit Bunch (OPEFB) Nanocrystalline Cellulose: Effects of post-treatment of oven drying and solvent exchange techniques

**E Indarti[1,2], Marwan[3] and W D Wanrosli[1]**

[1]Bioresources, Paper and Coating Division, School of Industrial Technology, Universiti Sains Malaysia, 11800 USM, Penang, Malaysia
[2]Agricultural Product Technology Dept., Faculty of Agriculture, University of Syiah Kuala, Banda Aceh 23111, Indonesia.
[3]Chemical Engineering Dept., Faculty of Engineering, University of Syiah Kuala, Banda Aceh 23111, Indonesia.

Corresponding author: eti_indarti@yahoo.com

**Abstract.** Nanocrystalline cellulose (NCC) from biomass is a promising material with huge potentials in various applications. A big challenge in its utilization is the agglomeration of the NCC's during processing due to hydrogen bonding among the cellulose chains when in close proximity to each other. Obtaining NCC's in a non-agglomerated and non-aqueous condition is challenging. In the present work NCC's was isolated from oil palm empty fruit bunch (OPEFB) using TEMPO-oxidation reaction method. To obtain non-agglomerated and non-aqueous products, the NCC's underwent post-treatment using oven drying (OD) and solvent exchanged (SE) techniques. The thermal stability of all samples was determined from TGA and DTG profiles whilst FTIR was used to analyze the chemical modifications that occurred under these conditions. NCC-SE has better thermal stability than the NCC-OD and its on-set degradation temperature and residue are also higher. FTIR analysis shows that NCC-SE has a slightly different chemical composition whereby the absorption band at 1300 $cm^{-1}$ (due to C-O symmetric stretching) is absent as compared to NCC-OD indicating that in NCC-SE the carboxylate group is in acid form which contribute to its thermal stability. (*Keyword: Nanocrystalline cellulose, oil palm empty fruit bunch, thermal stability, oven drying, solvent exchange*)

## 1. Introduction

Cellulose is the world most abundant natural renewable biodegradable polymer and has been predicted that globally around $10^{12}$ tons are synthesized and destroyed every year [1]. The use of non biomaterial will cause a serious environmental problem since they are not easily degraded in the environment after use. Many studies have shown that cellulose nanoparticles could be used as filler to improve mechanical and barrier properties of biocomposites. Growing interest in nanocomposites has been related to several reasons, among them are possible to design and create new materials and structures with exceptional flexibility and physical properties, almost free of defects when used as nanoscale fillers, and presenting a large volume of interfacing matrix material due to high specific surface area. For reinforcement applications, however, cellulose nano or micofibres present some disadvantages, for instance, high moisture absorption, poor wettability, incompability with most of polymeric matrices, and limitation of processing temperature.

Cellulose can be extracted from natural fibers and through systematic approach it can be converted to produce cellulose nanoparticles. There are various mechanical and chemical treatments offered depending on the source of lignocellulosic and properties to be achieved. Properties of cellulosic fibers are strongly influenced by different sources of plant, part of plant [2]. Agro-based material such as plant fibers is an example that is abundantly available in nature and more often are in the form of agro waste. Malaysia and Indonesia have abundance of this agro waste which is obtained from the oil palm (OP) industry of which the empty fruit bunch (EFB) is the largest available. Oil palm biomass comprise of 45-70% complex carbohydrate cellulose and hemicellulose. Therefore, this cheap and abundant biomass could compete with other plant biomass to produce cellulose microfibers or microfibriles at competitive price.

One of the most important fields of cellulose application in composite material is as reinforcement of biopolymer matrix [3]. However, certain drawbacks of the application are incompability with the hydrophobic biopolymer matrix, especially during processing it tend to form aggregates and water-swellable nature of cellulose [4]. Some research has recently stated that the addition of renewable and high biodegradable fillers like cellulose or cellulose derivatives could optimize cost performance balance and improve the mechanical and thermal behaviors [5]. More recently, there has been an increase use of cellulose nanowishkers or nanocrystals as the load bearing constituent in developments in new and inexpensive biodegradable materials due to their high aspect ratio, good mechanical properties [6] and fully degradable and renewable character. Some of biopolymer or biocomposite has low thermal stability, and also still limited and high production cost, therefore the use of cellulose, especially in nano size and crystalline form has a good potential to become the reinforcement [7]

The thermal properties are important to investigate in order to know the material application for biocomposite processing at high temperature [8]. Most of biocomposite was in the nonpolar form and hydrophobic that very good in water resistant properties, but weak in the thermal resistance, heat distortion temperature and rate of crystallization. Therefore the use of nano filler is potential to overcome the weakness of 'those' properties [9]. Some of researches have been reported one of biopolymer such as PLA low thermal properties, toughness, water vapor than gas barrier are inferior to those non biopolymers [10-12]. The challenge of crystalline nano fiber (CNF) used in industrial is to overcome the weak capability in hydrophobic material in order to enhance the dispersion and to increase the stress transfer towards the matrix. Most of the bio-polymeric or biocomposite are non-polar that cannot directly use NCC suspension in very polar or hydrophilic properties. Some research related to drying of NCC has been reported [13-16]. But NCC mostly in the form in water suspension and the use of the NCC as reinforcing is in solid state and suspension form also inconvenient during transportation. The problem is NCC tend to agglomeration during drying process and difficult to redisperse. Meanwhile the drying of NCC also has potential for next process and easier handling and reuse as a filler. Therefore the aim of this work is to study the thermal behavior of dried NCC as a reinforcement, heat barrier potential to overcome the weak thermal properties in most biocomposites. The NCC is prepared from OPEFB microcrystalline cellulose via the TEMPO-oxidation reaction technique and later underwent post-treatments using the oven drying and solvent exchange methods.

## 2. Experimental

*2.1 Material*

Oil Palm Empty Fruit Bunch (OPEFB) was obtained from SABUTEK Company, Perak, Malaysia while 4-acetamido-TEMPO (2,2,6,6-tetramethylpiperidin-1-oxyl) procured from Sigma Aldrich (USA). All other chemicals in the experiment were purchased from QReC (Malaysia).

*2.2 Preparation of Totally Chlorine Free Bleached OPEFB Pulp and OPEFB Microcrystalline Cellulose*

OPEFB fiber as raw material was cut into 4-5 cm lengths, then washed and dried and kept in polyethylene bags. Before pulping, OPEFB was first pre-hydrolyzed with water and later pulped using 26% NaOH at a temperature of 170°C for 2 h [17]. The obtained pulp is further bleached using a totally chlorine free (TCF) sequences of oxygen (O), ozone (Z) and peroxide (P) [18]. The final TCF OPEFB pulp has a Kappa no. of 1.4. Microcrystalline cellulose (MCC) was prepared according to Wanrosli et al. (2011). Briefly, the TCF OPEFB pulp was hydrolyzed with 2.5M HCl at a solid liquor ratio of 1:20 and refluxing at 105±2°C for 15 min. The MCC was thoroughly washed with distilled water before air drying. MCC powder was obtained by grinding the sample in a ball mill and subsequently kept in a desiccator.

*2.3 Isolation of NCC*

Nanocrystalline cellulose (NCC) was prepared using TEMPO- oxidation reaction [20]. 5 g of MCC was reacted with 4-acetamido-TEMPO (2,2,6,6-tetramethylpiperidin -1-oxyl) 98% (0.11 mmol/g) and sodium bromide (0.617 mmol/g) in 500 ml distilled water in an ultrasonic bath (Model 8510, Branson) for 4 hours. In the post-oxidation step, the mixture was reacted $NaClO_2$ at 70°C in water bath for 1 hour. The mixture was then washed with distilled water and unfibrillated with Sonicator (Sonifier 450, Branson). The suspension was centrifuged at 3500 rpm and the upper phase was separated as the NCC product.

*2.4 Drying and solvent exchange of NCC*

The Oven Drying (OD) process was conducted at 50°C for 24 hours in an air circulated oven dryer. Dried NCC was gently crushed with mortar and sieved to obtain uniform size. For solvent exchange, the procedure of Fujisawa et al (2013) was followed. NCC suspension was mixed with 1 M HCl to obtain pH 2, then stirred at slow speed at room temperature for 30 minutes. Then the mixture was washed with water and solvent exchanged to ethanol.

*2.5 Characterization*

*2.5.1 TGA*

The thermal stability was carried out using A Mettler TG 50 Module equipped to Mettler TC 11 4000 thermal Analyzer (USA). The specimen was analyzed in a nitrogen atmosphere at a flow rate 50 ml/min. The temperature was set from 30 to 800°C at a rate of 20°C per minute.

*2.5.2 Fourier Transform Infrared (FTIR) Spectroscopy Analysis*

FTIR analysis was carried out using a FTIR spectrometer (Nicolet iS10, Thermo Scientific). The films were cut around 3 cm and directly loaded for analyzing the transmittal spectra within the range of 4000–400 $cm^{-1}$.

## 3. Results and Discussion

*3.1 Thermogravimetric analysis*

Thermal properties of NCC-OD, NCC-SE and MCC, with the later as a control, were studied by thermogravimetric (TGA) and derivative thermogram (DTG). The TGA and DTG curves for all samples are shown in Figure 1. The thermal condition of degradation at $T_{on}$, T (50%) and the residual weight at

400°C and 700°C are presented in Table 1. A small weight loss for all samples occurs between 40-70°C which is attributed to the removal of absorbed water in cellulose [22]. As depicted in 1, the NCC-OD sample initiates a slightly more pronounced degradation process at around 255°C while for NCC-SE a more pronounced degradation process occurs at 290°C. The extremely high slope for all samples in the range of 240°C to 350°C is due to the high degradation or decomposition step of the cellulose. In this stage the cleavage of the glycosidic linkages of cellulose reduces the polymerization degree leading to the formation of $CO_2$, $H_2O$ and a variety of hydrocarbon derivatives [23]. It can also be seen that based on weight percentage as a function of temperature, the first initial mass loss portion was found between 25°C to about 125°C. No thermal degradation occurred in this region. The loss of weight in this region is about 5-10%, which is mainly caused by evaporation of loosely bound moisture on the surface of the NCC's [15, 24, 25].

a

b

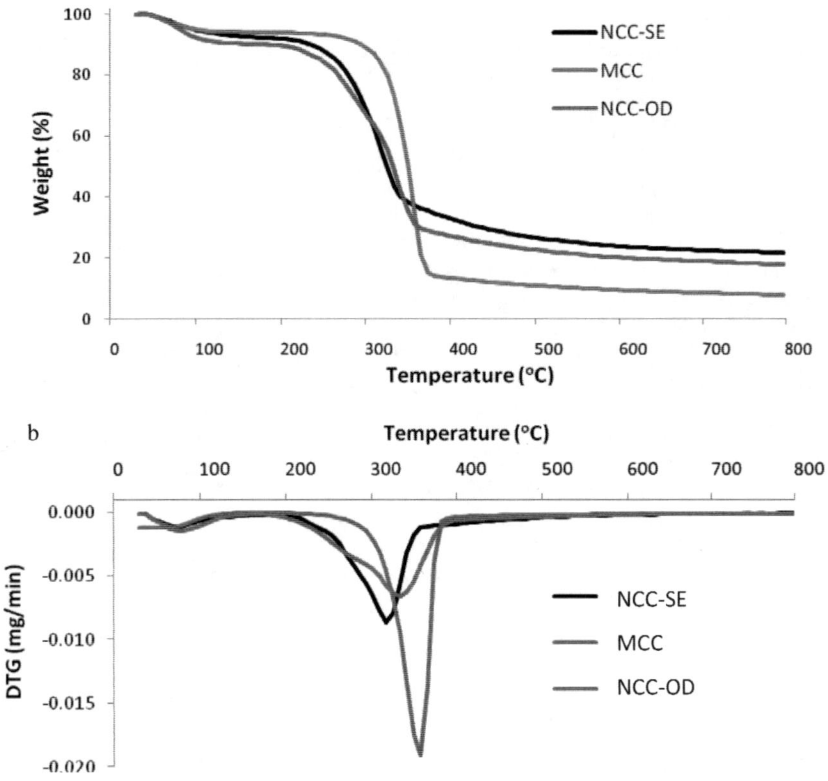


**Figure 1.** Typical TGA (a) and DTG (b) of MCC, NCC-OD and NCC-SE from palm oil empty fruit bunch

It can be seen that both processes of oven drying and solvent exchange treatment affected the thermal stability of NCC. NCC-OD and NCC-SE both has a lower degradation temperature ($T_{on}$) compare to MCC as a control. However, the NCC - SE has better thermal stability than the NCC-OD. This higher decomposition temperature is caused by the presence of free carboxylic group that is formed during the solvent exchange process. During the isolation of NCC using TEMPO-oxidation, the hydroxyl groups were replaced by the carboxylate groups. In the solvent exchange process (because of the acid condition where the pH was set to 2 by the addition of 1 M HCl at room temperature) these carboxylate groups were converted to free carboxyl groups [16, 21] leading to a viscous and flowable gel as compared to only a dispersable state for unprocessed NCC. On the other hand, NCC that is processed by oven drying is still in the form of sodium carboxylate salt. Thus, this different content of carboxylic group is responsible for the thermal behavior differences of both NCC.

It was also reported that the carbonation of the sodium carboxylic group degrade at a lower temperature than that of cellulose. Similar study also found for carboxymethyl cellulose that contain of sodium carboxyl groups [26]. The pathway of thermal decomposition is influenced by many chemical and physical factors such as temperature, type of atmosphere, size and texture of cellulose, and crystallinity [27].

Comparatively, MCC has a higher degradation behavior temperature than both NCC-OD and NCC-SE. The maximum of the DTG peak of MCC (Tmax = 358°C) is higher than the NCC-OD and NCC-SE of 334°C and 318°C respectively. This is probably due to the difference of the outer surface structure of each MCC and NCC resulting from the different processes that were used to prepare the materials. The NCC in this research was produced directly from OPEFB TCF pulp whereas MCC was produced by acid hydrolysis from the pulp with HCl. It has been reported that the chloride anions contributed to these differences because they affected the swelling of cellulose chains and breakage of hydrogen bonds in cellulose [28]. Hydrolysis process did not only dissolve the amorphous region, but also some crystalline region that caused more vulnerable degradation as temperature increased [29, 30]. There are two different degradation mechanism in cellulose. First is dehydration of cellulose by an endothermic process to form hehydrocellulose at 200-280°C and second is depolymerization of cellulose in competition with dehydration between 280-340°C that yield the volatile matter [15]. The mechanism of depolymerization occurs in very fast rate as shown very big slope on this stage. Decomposition of dehydrocelulose formed into gases and char residue becomes dominant at about 320°C. At temperature above 500°C, the levoglucosan breakdown to produce varying low molecular weight, including hydrocarbons and hydrogen as well as CO, $CO_2$ and $H_2O$ [15, 32].

**Table 1.** Thermal properties of MCC, NCC-OD and NCC-SE

| Samples | Degradation temperature (°C) | | DTG Peak Temp (°C) | Residue weight (wt %) | |
|---|---|---|---|---|---|
| | $T_{on}$ | $T_{(50\%)}$ | $T_{max}$ | $T_{400}$ | $T_{700}$ |
| MCC | 327.96 | 350.64 | 358 | 13.44 | 8.49 |
| NCC/OD | 254.40 | 332.83 | 334 | 27.20 | 18.85 |
| NCC/SE | 289.05 | 324.69 | 318 | 32.88 | 22.42 |

The NNC-SE and NCC-OD also has a higher residue at $T_{400}$ and $T_{700}$ compare to MCC. Similar trends were also reported from OPEFB nanocellulose whisker [32]. Moreover the NCC-SE has a higher residue at $T_{400}$ and $T_{700}$ (32.88% and 22.42%) compared to NCC-OD (22.42% and 18.85%). Again the probable reason for this observation is due to the presence of carboxylic form in NCC-SE as compared to the carboxylate form in NCC-OD. Other factors that can also affect the thermal degradation behavior are sonication time and temperature [8], both of which will be dealt in subsequent publications. It is interesting to note that in spite of the fact that both NCC-SE and NCC-OD has a lower thermal stability as compared to NCC-OD as seen in Fig.1 and Table 1, the char fraction (residue that remains after 700 °C) of the former is much higher (23 % & 19 % and 8% respectively). It is possible that the incorporation of the carboxylate functionality resulting from the TEMPO-oxidation has the flame-retardant effect which can have positive implications as a potential as a filler in lower thermal retardants in biopolimeric matrix [7].

*3.2 FTIR Analysis*

The FTIR spectroscopy is an essential technique for tracking chemical changes of the cellulose before and after treatment. The FITR spectra of MCC and NCC were presented in Figure 2. The vibrational assignments are summarized in Table 2.

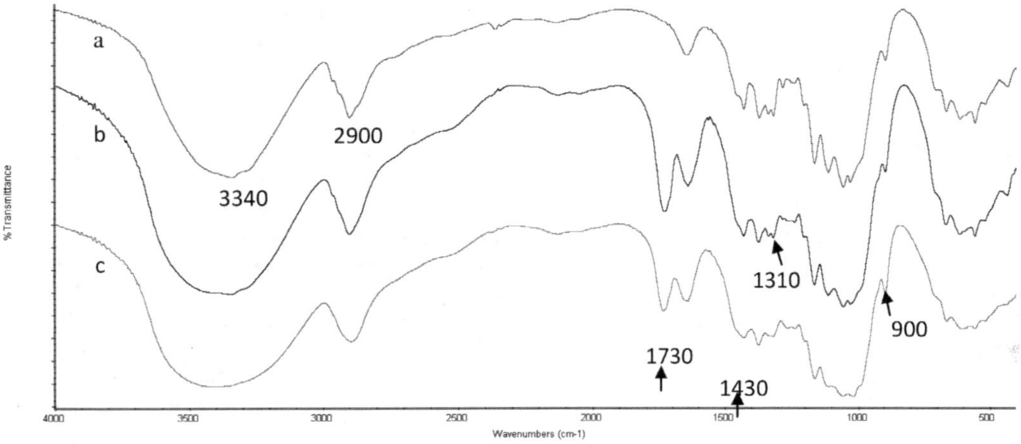


**Figure 2.** Typical FTIR spectra obtained from (a) MCC, (b) NCC-OD and (c) NCC-SE

In Figure 2, the broad peak for all samples observed in the region 3300-3400 $cm^{-1}$ is due to stretching vibration of O-H (Hydroxyl) groups [34]. The absorption around 2900 $cm^{-1}$ is attributed to stretching of C-H groups of cellulose [25, 35, 36]. When spectra of MCC are compared with those of NCC-OD and NCC-SE, the most different is a new appearance of the vibration peak observed at 1730 $cm^{-1}$ from spectra on NCC-OD and NCC-SE. The appearance of this peak is indicated by the C=O stretching of carboxylic group of TEMPO-oxidation nanocrystalline cellulose [37]. Meanwhile the MCC from OPEFB using HCl showed no C=O bond formation. Similar result were also reported by Wanrosli et al. (2011) that compares OPEFB MCC with pure cellulose before converting to cellulose phosphate. Comparatively, this C=O peak is lower in NCC-SE than NCC-OD which is a a consequence of pH reduction causing the presence of the free carboxilic group during solvent technique. This observation suggests that the higher content of carboxyl content in NCC-OD contributed higher thermal retardant in NCC-SE and the absence C=O bond MCC will cause the lower thermal retardant compare to both NCC post oxidation technique.

**Table 2.** FITR Spectral peak assignment for (a) MCC, (b) NCC-OD, (c) NCC-SE

| Peak frequency (cm-1) | | | |
|---|---|---|---|
| Peak assignment | MCC | NCC/OD | NCC/SE |
| OH Bending | 3343 | 3345 | 3413 |
| $CH_2$ Group Bending | 2901 | 2901 | 2898 |
| C=O stretching | - | 1732 | 1735 |
| O-H stretching | 1640 | 1639 | 1641 |
| $CH_2$ Bending | 1430 | 1430 | 1430 |
| C-O symmetric stretching | 1318 | 1318 | - |
| C-O-C Stretching | 1113 | 1164 | 1163 |
| C-H | 897 | 898 | 897 |

## 4. Conclusion

Nancocrystalline cellulose (NCC) was obtained from oil palm empty fruit bunch (OPEFB) by TEMPO oxidation. The resulting suspension was then dried or solvent exchanged as the way to prepare NCC for the next processing. In this work, the effect of drying and solvent exchange method on thermal stability of the OPEFB-NCC was studied. It reveals that the choice of between drying process and solvent exchange treatment of NCC determined thermal stability of NCC. The NCC prepared by either method has a lower degradation temperature ($T_{on}$) compare to MCC as a control. However, their thermal resistance properties are better than MCC as revealed from lower thermal degradation rate and higher residues. The NCC-SE has better thermal stability compared to NCC-OD due to the presence of the free carboxylic group during solvent exchange process, resulted from a reduction in the carboxyl group in NCC. FTIR spectra revealed such chemical change of the cellulose before and after different treatment. In addition to its function as filler, the result suggests NCC could contribute to better thermal properties of biocomposite. Future works will include development of the NCC as a filler of biocomposite by improving its interaction with the polymeric matrix without losing its thermal resistance properties.

## Acknowledgement

The authors express their gratitude to Universiti Sains Malaysia for the research university, grant scheme 1001/PTEKIND/814122 and Directorate General of Higher Education of Indonesia for sponsoring postgraduate studies of Eti Indarti.

## Refferences

[1] Hon D N S 1994 *Cellulose* **1** 1-25
[2] Siquera G, Bras J and Dufresne A 2010 *Polymer* **2** (4) 728-65
[3] Biagiotti J, Puglia D and Kenny J M 2004 *Journal of Natural Fibers* **1** (3) 23-65
[4] Hubbe M A, Rojas O J, Lucia L A and Sain M 2008 *Bioresources* **3** 929-80
[5] Oksman K, Skrifvars M and Selin J F 2003 *Composites science and technology* **63** (9) 1317-24
[6] Sturcova A, Davies G R and Eichhorn S J 2005 *Biomacromolecules* **6** (2) 1055-61
[7] Frone A N , Berlioz S, Chailan J F and Panaitescu D M 2013 *Carbohydrate Polymers* **91** 377–384
[8] Lavoine N, Desloges I, Dufresne A and Bras J 2012 *Carbohydrate polymer* **90** 735-764
[9] Raquez J M, Habibi Y, Murariu M and Dubois P 2013 *Progress in Polymer Science* **38** 1504- 42

[10] Pantani R, Gorrasi G, Vigliotta G, Murariu M and Dubois P 2013 *European Polymer Journal* **49** 3471-82
[11] Bonilla J, Fortunati E, Vargas M, Chiralt A and J M Kenny 2013 *Journal of Food Engineering* **119** 236-43
[12] Petersen K, Nielsen P and Olsen M 2001 *Starch* **53** 356-61
[13] Peng Y, Han Y and Gardner D J 2012 *Wood Fiber Sci* **44** (4) 448-61
[14] Peng Y, Gardner D J and Han Y 2012 *Cellulose* **19** (1), 91-102
[15] Peng Y, Gardner D J, Han Y, Kiziltas A, Cai Z, Tshabalala MA 2013 *Cellulose* **20** (50) 2379-92
[16] Fukuzumi H, Saito T, Okita Y and Isogai A 2010 *Polymer Degradation and Stability* **95** 1502-08
[17] Wanrosli W D, Leh C P, Zainuddin Z and Tanaka R 2003 *Holsforschung* **57** 106-13
[18] Leh C P, Wanrosli W D, Zainuddin Z, and Tanaka R 2008 *Industrial Crops and Products* **28** 260-7
[19] Wanrosli W D, Rohaizu R and Ghazali A 2011 Carbohydrate *Polymers* **84** 262-267
[20] Rohaizu R and Wanrosli W D 2014 submitted to *Carbohydrate Journal.*
[21] Fujisawa S, Saito T, Kimura S, Iwata T and Isogai A 2013 *Biomacromolecules* **14** (5), 1541-1546
[22] Bourbigot S, Chlebicki S and Mamleev V 2002 *Degradation and Stability* **78** 57-62
[23] Poletto M, Dettenborn J, Pistor V, Zeni M and Zattera A J 2010 *Material Research* **13** (3) 375-9
[24] Abraham E, Deepa B, Pothan L A, Jacob M, Thomas S, Cvelbar U and Anandjiwala R 2011 *Carbohydrate Polymers* **86** 1468-75
[25] Rosa S M L, Rehman N, De Miranda M I G, Nachtigali S M B and Bica C I D 2012 *Carbohydrate Polymers* **87** 1131-38
[26] Britto D and Assis O B G 2009 *Thermochim Acta* **494** 115 – 22
[27] Shafizadeh 1984 *Pyrolysis and combustion*. In R M Rowell, ed. *The chemistry of solid wood Advances in Chemistry Series* **207** Chapter 13 American Chemical Society Washington DC
[28] Zhang P P, Tong D S, Lin C X, Yang H M, Zhong Z K, Yu W H, Wang H and Zhou C H 2014 *Asia-Pac. J. Chem. Eng*. Published online in Wiley Online Library. DOI: 10.1002/apj.1812
[29] Bras J, Hassan M L, Bruzesse C, Hassan E A, El-Wakil N A M and Dufresne A 2010 *Industrial Crops and Products* **32** 627-33
[30] Mandal A and Chakrabarty D 2011*Carbohydrate polymer* **86** 1291-9
[31] Kilzer FJ and Broido A 1965 *Pyrodynamics* **2** 151-163
[32] Haafiz M K M, Hassan A, Zakaria Z and Inuwa I M 2014 *Carbohydrate Polymers* **103** 119–25
[33] Johnson R K, Zink-Sharp A, Renneckar S H and Glasser W G 2008 *Cellulose* **16**(2), 227-38
[34] Johar N, Ahmad I and Dufresne A 2012 *Industrial Crops and Products* **37** 93-95
[35] Jonoobi M, Khazaeian A, Tahir P M, Azry S S and Oksman K 2011 *Cellulose* **18** 1085-1095.
[36] Jain P and Vineshwaran N 2012 *Boresources Technology* **103** 219-26.
[37] Fujisawa S, Okita Y, Fukuzumi H, Saito T and Isogai A 2011 *Carbohydrate Polymers* **84** 579-83
[38] Kemala T, Budianto E and Soegiyono B 2012 *Arabian Journal of Chemistry* **5**, 103-108
[39] Qu P, Goa Y, Wu G F and Zhang L P 2010 *BioResouces* **5** (3) 1811-23

# Carboxymethyl Cellulose (CMC) from Oil Palm Empty Fruit Bunch (OPEFB) in the new solvent Dimethyl Sulfoxide (DMSO)/Tetrabutylammonium Fluoride (TBAF)

**M Y Eliza[1], M Shahruddin[1], J Noormaziah[1] and W D Wan Rosli[2]**

**[1]Faculty of Mechanical and Manufacturing Engineering, Universiti Tun Hussein Onn Malaysia, 86400 Johor, Malaysia**
**[2]School of Industrial Technology, Universiti Sains Malaysia, 11800 Minden, Pulau Pinang, Malaysia**

E-mail: [1]{elizay, sharudin, normazia}@uthm.edu.my and [2]wanrosli@usm.my

**Abstract.** The surplus of Oil Palm is the most galore wastes in Malaysia because it produced about half of the world palm oil production, which contributes a major disposal problem. Synthesis from an empty fruit bunch produced products such as Carboxymethyl Cellulose (CMC), could apply in diverse application such as for paper coating, food packaging and most recently, the potential as biomaterials has been revealed. In this study, CMC was prepared by firstly dissolved the bleached pulp from OPEFB in mixture solution of dimethyl sulfoxide (DMSO)/tetrabutylammonium fluoride (TBAF) without any prior chemical modification. It took only 30 minutes to fully dissolve at temperature $60^0C$ before sodium hydroxide (NaOH) were added for activation and monochloroacetate as terrifying agent. The final product is appeared in white powder, which is then will be analyzed by FTIR analysis. FTIR results show peaks appeared at wavenumber between 1609 $cm^{-1}$ to 1614 $cm^{-1}$ proved the existence of carboxymethyl groups which substitute OH groups at anhydroglucose (AGU) unit. As a conclusion, mixture solution of DMSO/TBAF is the suitable solvent used for dissolved cellulose before modifying it into CMC with higher Degree of Substitution (DS). Furthermore, the dissolution of the OPEFB bleached pulp was easy, simple and at a faster rate without prior chemical modification at temperature as low as $60^0C$.

## 1. Introduction

World palm oil production in 1990 doubled to 11.0 million tonnes from 5.0 million tonnes in 1980, and by the year 2000, the production doubled to 21.8 million tones [1]. Malaysia produced about half of the world palm oil production (10.8 million tonnes), thus, making Malaysia as the world's largest producer and exporter of palm oil during this period [2]. Palm oil surplus such as fronds and empty fruit bunches can be used in many process industries until recently Cellulose Phosphate (CP) one of the product synthesized from empty fruit bunch was used as material for biomedical applications because it has the potential due to its biocompatibility and ability to induce the formation of apatite nuclei [3]. CP has high Ca binding capacity, associated with lack of toxicity and indigestibility, has been used for decades in the treatment of Ca metabolism-related diseases [4 – 6]. Due to the reason, CP was then investigated for orthopaedic application to improve the osseointegration of cellulose [7].

Carboxymethyl Cellulose (CMC) is rarely used as a biomaterials. Conventionally, it was used in wide application of food, pharmaceuticals, toothpaste, detergents, oil drilling mud, paper coating and others [8]. Nowadays, CMC is used to enhance the viscosity, to control rheology of a solution, to avoid separation of water from a suspension and to improve surface or barrier properties [9]. Of late, it has been also investigated for its potential use as biomaterials [10].

The usage of CMC as a biomaterials base for the reason that the product is from polysaccharide group containing cellulose structural. Cellulose is a linear macromolecule in which anhydroglucose unit (AGU) are linked by β-1,4-glucosidic bonds [8]. AGU are presented from three reactive hydroxyl groups: the primary OH on C(6) and the two secondary OHs on C(2) and C(3) [8, 11]. Fig. 1 shows the general chemical structure of the cellulose presenting AGU unit and next the chemical structure of CMC (Fig. 2) after etherification process where OH groups are etherified with a carboxymethyl group.

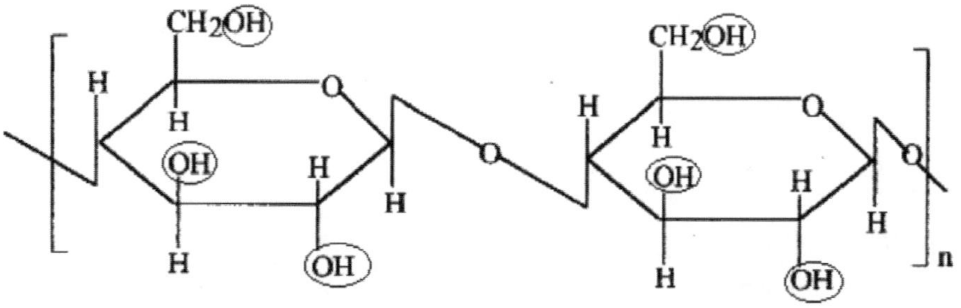

**Figure 1:** Chemical structure of cellulose. AGU units are circled at carbon number 2, 3 and 6.

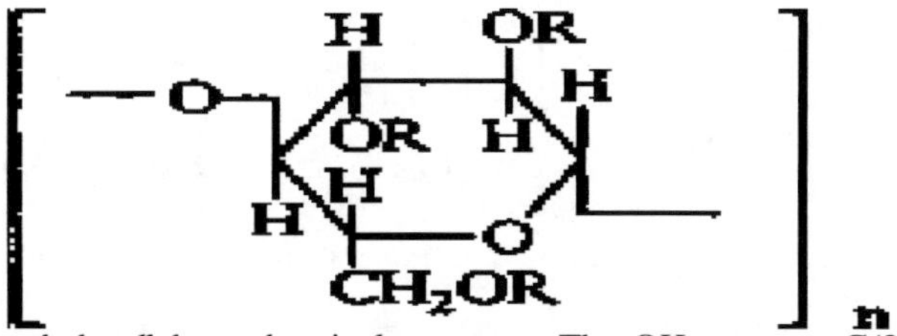

**Figure 2:** Carboxymethyl cellulose chemical structure. The OH group at C(2), C(3) and C(6) are substituted with Sodium Carboxymethyl groups where R = $CH_2CO_2Na$.

## 2. Carboxymethylated from various natural sources

CMC is produced by conversion of alkali cellulose in an organic liquid/aqeous NaOH with monochloroacetic acid or its sodium salt as etherifying agent [12] as illustrated in fig. 3. The sources of CMC were a lot such as from sisal [12], cotton linters [12, 13, 14] viscose staple fibre [13], spruce sulfite pulp [13], rayon grade wood pulp [14], durian rind [15], banana pseudo stem [16] and most recently from oil palm waste, empty fruit bunch [17].

Synthesization of CMC is simple and easy that commonly followed conventional way to produce it. The bleached pulp of OPEFB is treated with NaOH at a certain temperature, and the alcohol usually ethanol or isopropanol. This treatment is required as a swelling and impregnation stage and facilitates penetration of the NaOH into the cellulose structure. The product produced at this stage is called alkali cellulose (Na-cellulose) which is highly reactive towards monochloroacetic acid (MCA), or its sodium salt that normally carried out at about $50^0$C to $70^0$C, were added in the following etherification stage. The method has been practically used since 1920's [8] and already used frequently to produce CMC from various sources [8, 9, 11], [18].

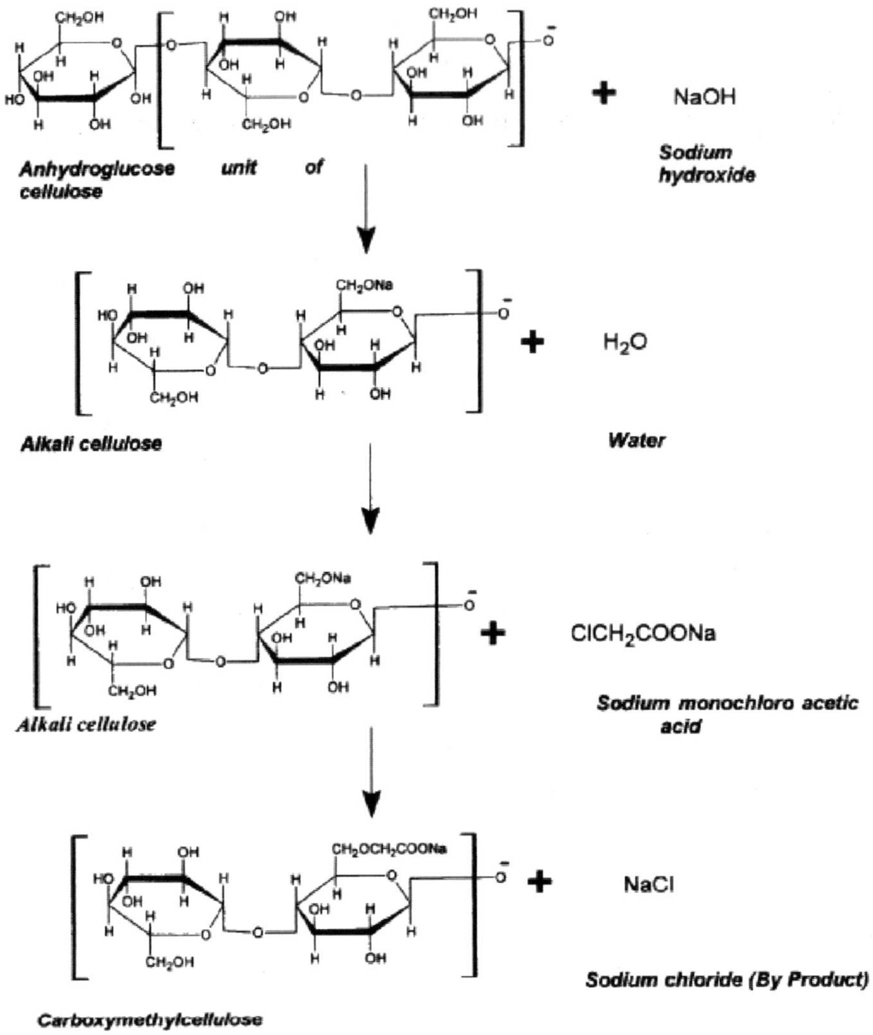


**Figure 3:** The reaction for the synthesis sodium carboxymethylcellulose [14] which shows reaction occur at Carbon number 6 [C(6)].

### 3. Cellulose dissolution in Tetrabutylammonium fluoride (TBAF)/Dimethyl sulfoxide (DMSO)

TBAF/DMSO is a powerful solvent for cellulose with a degree of polymerization as high as 650 can be dissolved without any pre-treatment within 15 minutes [19]. It is well known that cellulose is not soluble in conventional solvents, due to its supramolecular structure formed by inter- and intramolecular hydrogen bonds [20] – [26]. The cellulose dissolution power of the TBAF/DMSO system is essentially due to the strong ion-dipole interaction between fluoride ions from TBAF and the hydroxyl groups of cellulose [26]. The highly electronegative fluoride ion from hydrogen bonds with the cellulose hydroxyl groups deconstructing the hydrogen-bonded cellulose network [27]. Fig. 4 illustrates mixture solvents of TBAF/DMSO react with cellulose in disrupting the hydroxyl groups.

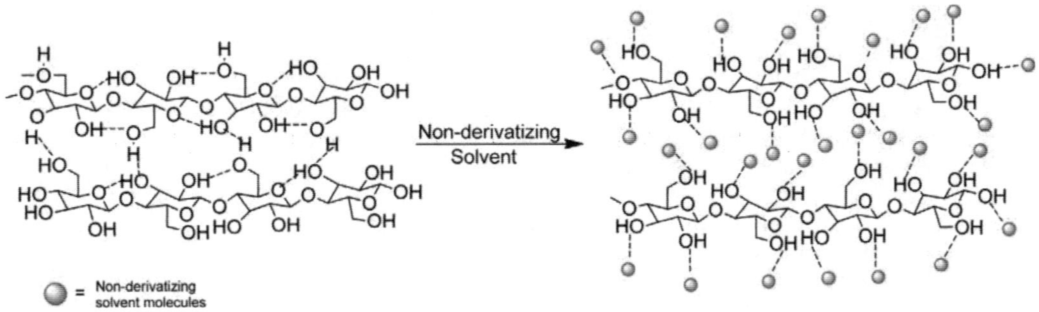


**Figure 4:** Non-derivatizing solvent of TBAF/DMSO reacts in disrupting the hydroxyl groups of cellulose [27].

The preparation of CMC from OPEFB is slightly different in this research where the bleached pulp was firstly dissolved in TBAF/DMSO mixture solution. This is due to the observation that CMC produced without prior dissolution, will not appear in powder form like CMC commercial. The product required to be appeared in powder form in order to suit with the requirement for biomaterial application. Fig. 5 shows different physical appearance between CMC without prior dissolution and CMC with prior dissolution.

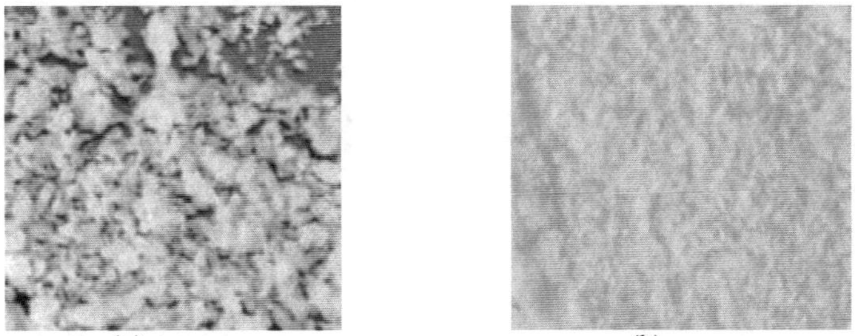

(a) (b)

**Figure 5:** CMC produced (a) without prior dissolution and (b) with prior dissolution

## 4. Materials and Methods

### 4.1 Materials

OPEFB was supplied by SABUTEK (M) Sdn. Bhd, and all the chemicals were bought from BGOilchem, Penang, Malaysia. All the experimental work was done at School of Industrial Technology, Universiti Sains Malaysia, Penang, Malaysia.

### 4.2 Extraction of bleached pulp from OPEFB

For this investigation, oil palm empty fruit bunch (OPEFB) cellulose was prepared using the environmentally benign process [28, 29]. This procedure involves water pre-hydrolysis of the OPEFB, followed by soda-anthraquinone pulping. Then the unbleached pulp was bleached using a totally chlorine-free (TCF) sequence of oxygen (O), ozone (Z) and peroxide (P) to a kappa number of 1.4.

### 4.3 Synthesis

#### 4.3.1 Dissolution of bleached pulp of OPEFB in TBAF/DMSO

Cellulose (bleached pulp of OPEFB) was immersed in a mixture solution of 60ml DMSO and 6.6g TBAF. Stirred evenly by magnetic stirrer for 30 minutes at temperature $60^0$C for 30 minutes in fume chamber, resulting in a clear solution of cellulose.

#### 4.3.2 Carboxymethylation of OPEFB bleached pulp in TBAF/DMSO

Following previous application [12], to the cellulose solution, 4.94g of NaOH which was suspended in 10ml of DMSO, and subsequently 10.78g of monochloroacetate were added under vigorous stirring using magnetic stirrer leading to gelation and partly precipitation of the cellulose. The temperature was raised to $70^0$C. After 4h, the mixture was cooled to room temperature and put into 300ml isopropanol. The solid product was filtered off, dissolved in 40ml of distilled water, neutralized with either NaOH if too acidic or with acetic acid if the alkali solution, and precipitated into 300ml of ethanol. After filtration, the product was washed with 90% (v/v) ethanol and dried in an oven at $50^0$C. Finally, to obtain fine white powder, the product was grind by using stone and mortar and kept in desiccators.

### 4.4 Characterization of Sodium Carboxymethylcellulose by FTIR analysis

Product of CMC was analyzed by using FTIR analysis with OMNIC software.

## 5. Results and Discussions

FTIR analysis using OMNIC software was done on the bleached pulp of OPEFB, CMC commercial, CMC with prior dissolution in TBAF/DMSO and CMC without prior dissolution in TBAF/DMSO, wavenumber between $4000cm^{-1}$ to $400cm^{-1}$. Fig. 6 is attached here as the comparison for FTIR result obtain for CMC synthesized from OZP-pulp of OPEFB. From fig. 6 and fig. 7, it shows obvious different pattern of graph and the existence peaks. It proves the attendance of new group when the pulp was synthesized to CMC, the water dissolved cellulose.

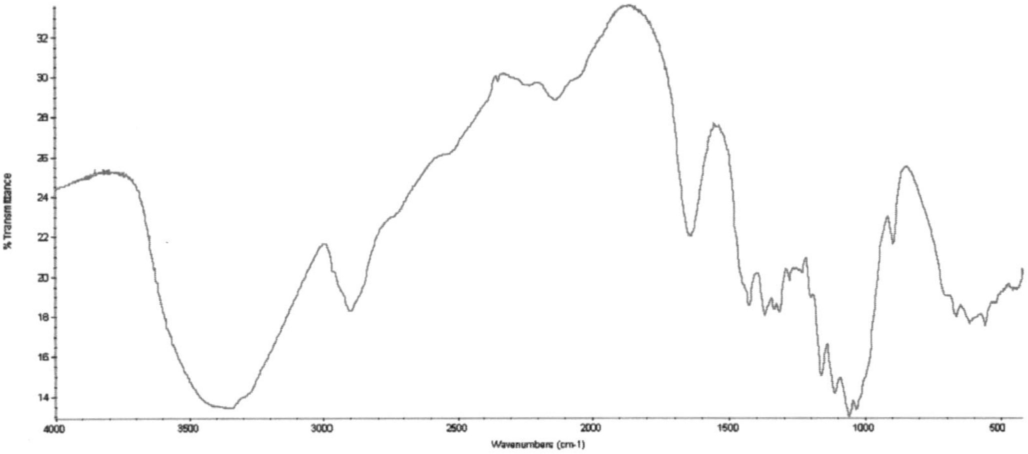


**Figure 6:** OZP-pulp

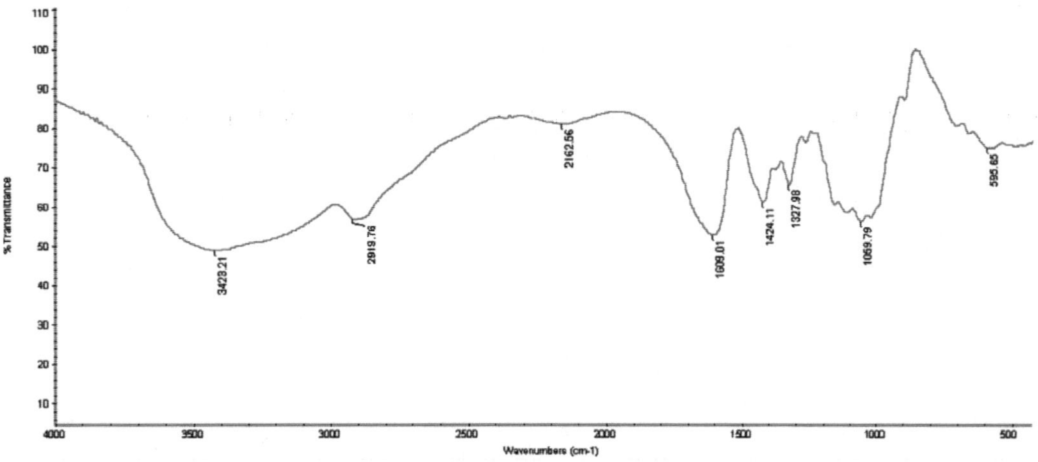


**Figure 7:** CMC commercial

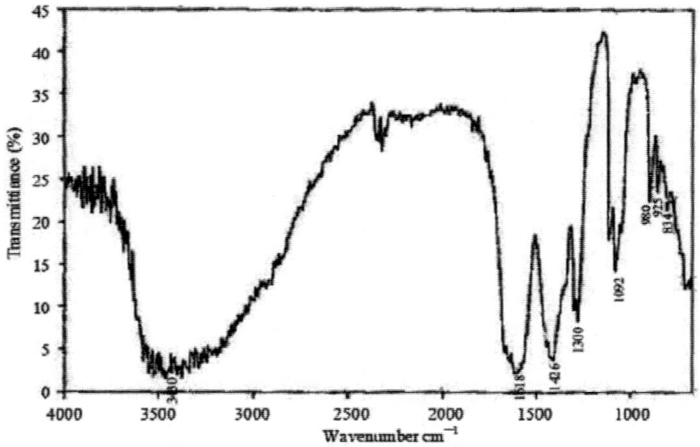


**Figure 8:** FTIR spectrum of synthesized CMC sample [30]

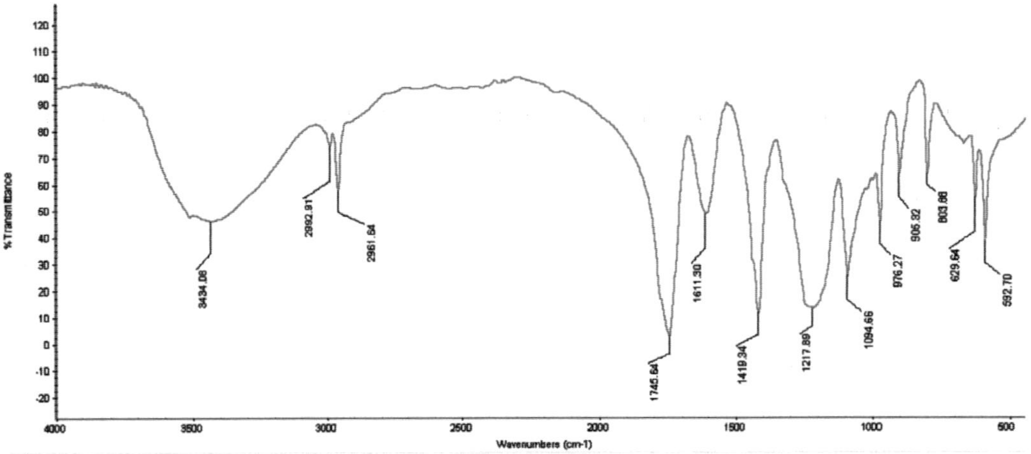


**Figure 9:** CMC from OPEFB with bleached pulp prior dissolution

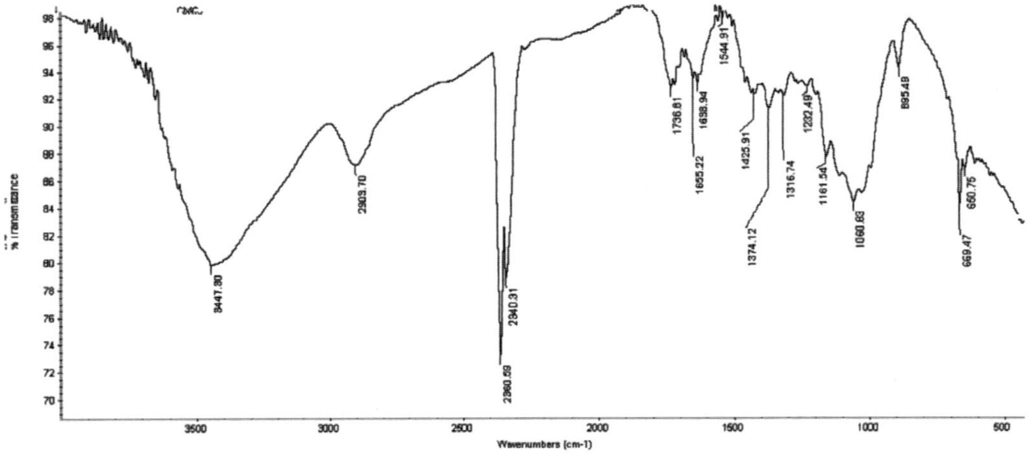


**Figure 10:** CMC from OPEFB without bleached pulp prior dissolution

Fig. 8 shows the FTIR spectrum of the sample with a rising peak of carboxyl, methyl and hydroxyl functional groups which are found at wavelength of 1618, 1426 and 1300 $cm^{-1}$ [30]. Comparing with fig. 9, the CMC from OPEFB with bleached pulp prior dissolution shows almost the same wavenumber as shown in fig. 7, CMC commercial and also in fig. 8, CMC from other publication. From the traced peak, the product is assumed to have higher degree of substitution (DS) because it contain sharp and long peak at wavenumber of carboxyl and methyl groups. Fig. 10 is CMC from OPEFB without bleached pulp prior dissolution shows peaks which also having almost identical wavenumber with fig. 8 but with larger value differentiation. The product also having the possibility to have lower DS due to the weak peak appears at carboxyl and methyl groups.

According to reported data, the peaks at wavelength of 1620 and 1423 $cm^{-1}$ represented two different functional groups in CMC [16, 31]. But, when referring to fig. 10, eventhough the peaks are appeared but not as sharp as wavelengths showed in fig. 9 at wavelength 1638$cm^{-1}$ and 1425$cm^{-1}$. Fig. 11 shows the additional peaks at wavelength of 2152 and 2376 $cm^{-1}$ might be done to the existence of the contamination from impurities or combination band with water [16]. It is obvious that broad absorption band at 3432$cm^{-1}$ is due to the stretching frequency of the hydroxyl group (–OH). The band at 2909$cm^{-1}$ is due to carbon-hydrogen bond (C–H) stretching vibration. The presence of a strong absorption band at 1603$cm^{-1}$ confirmed the presence of carboxyl group (–COO). The bands around 1423 and 1325 $cm^{-1}$ are assigned to $-CH_2$ scissoring and hydroxyl group (–OH) bending vibration, respectively. The band at 1061$cm^{-1}$ is due to $>CH-O-CH_2$ stretching [32]. All the data presented from previous publication supported the CMC attributed produced from OPEFB either with or without prior bleached pulp dissolution.

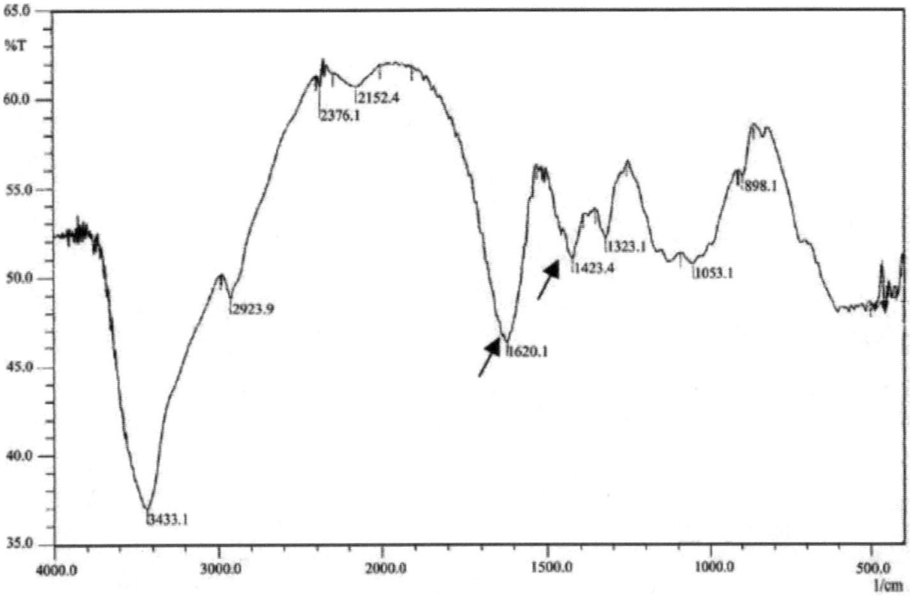


**Figure 11:** FTIR spectra of CMC made from cavendish banana pseudo stem cellulose with DS of 0.75 which was synthesized using 15% NaOH and 6 g of NaMCA (carboxyl groups substituent are indicates by arrows) [16].

## 6. Conclusion

The synthesization of CMC from OPEFB could easily done in laboratory at low temperature and in a short period. Fabrication method is conventional mostly practical from many researchers before and not dangerous. Commonly, one fabricate CMC without prior dissolution the sources, which usually in pulp form. From the experimental work done, CMC without prior dissolution is not suit to be used as biomaterials for all application. The simplest way in dissolving OPEFB bleached pulp was by the mixture solution of TBAF/DMSO where the process only took about 30 minutes at temperature 60$^{0}$C to be finished, resulting in clear solution of cellulose. FTIR analysis shown better result comparing with CMC without prior dissolution and the result was also discussed with comparing with other works to strengthen the obtaining FTIR data analysis. As a brief conclusion, CMC produced with prior dissolution of bleached pulp of OPEFB is better in term of its appearance, DS and also its application could be more diverse especially in biomaterials field.

## 7. References

[1] N. Abdullah and F. Sulaiman 2013 The Oil Palm Wastes in Malaysia *Intech* chapter 3.
[2] World Growth Palm Oil Green Development Campaign: Palm Oil — The Sustainable Oil a Report by World Growth September 2009. Available on line at <http://www.worldgrowth.org/assets/files/Palm_Oil.pdf>, (accessed 1st October 2014).
[3] Wan Rosli W D, Mohamad Haafiz M K and Azman S 2011 Cellulose phosphate from oil palm biomass as potential biomaterials *BioResources* 6 (2) pp 1719–1740.
[4] Mizusawa Y and Burke J R 1996 Prednisolone and cellulose phosphate treatment in idiopathic infantile hypercalcaemia with nephrocalcinosis *J Pediatr Child Health* 32 pp 350–352.
[5] Parfitt A M 1975 Effect of cellulose phosphate on calcium and magnesium homeostasis: studies in normal subjects and patients with latent hypoparathyroidism *Clin Sci Mol Med* 49 pp 83–90.
[6] Pak C Y C, Delea C S and Bartter F C 1974 Successful treatment of recurrent nephrolithiasis (calcium stones) with cellulose phosphate. *N Engl J Med* 290 pp 175–180.
[7] Barbosa M A, Granja P L, Barrias C C and Amaral I F 2005 Polysaccharides as scaffolds for bone regeneration *ITBM-RBM* 26 pp 212–217.
[8] Ambjornsson H A, Schenzel K and Germgard U 2013 Carboxymethyl Cellulose produced at different mercerization conditions and characterized by NIR FT Raman Spectroscopy in combination with multivariate analytical methods *BioResources* 8(2) pp 1918–1932.
[9] Stigsson V, Kloow G and Germgard U Historical overview of CMC production on an industrial scale *Paper Asia* 17(10) pp 16–21.
[10] Liuyun J, Yubao L and Chengdong X 2009 Preparation and biological properties of a novel composite scaffold of nano-hydroxyapatite/chitosan/carboxymethyl cellulose for bone tissue engineering *Journal of Biomedical Science* 16 (65) pp 1–10.
[11] Klemm D, Phillipp B, Heinze T, Heinze U and Wagenknecht W 2001a Comprehensive cellulose chemistry. Fundamentals and analytical methods *Wiley-VCH Verlag GmbH* (Weinheim, Germany) 1 (9) pp 135, ISBN 3-527-29413-9.
[12] Ramos L A, Frollini E and Heinze Th 2005 Carboxymethylation of cellulose in the new solvent dimethyl sulfoxide/tetrabutylammonium fluoride *Carbohydrate Polymers* 60 pp 259–267.
[13] Heinze T, Erler U, Nehls I and Klemm D 1994 Determination of the substituent pattern of heterogeneously and homogenously synthesized Carboxymethyl Celllulose by using high performance liquid chromatography *Die Angewandte Makromolekulare Chemie* 215 pp 93–106.
[14] Latif A, Anwar T and Noor S 2007 Two-step synthesis and characterization of Carboxymethylcellulose from rayon grade wood pulp and cotton linters *Jour. Chem. Soc. Pak.* 29 (2) pp 143–150.
[15] Rachtanapun P, Luangkamin S, Tanprasert K and Suriyatem R 2012 Carboxymethyl cellulose film from durian rind *LWT-Food Science & Technology* 48 pp 52–58.
[16] Adinugraha M P, Marseno D W and Haryadi 2005 Synthesis and characterization of sodium carboxymethylcellulose from Cavendish banana pseudo stem (Musa cavendishii LAMBERT), *Carbohyrdate Polymers* 62 pp 164–169.
[17] Amin M C I, Soom R M, Ahmad I and Lian H H 2007 Carboxymethyl cellulose from palm oil empty fruit bunch – their properties and use as a film coating agent *Jurnal Sains Kesihatan Malaysia* 4 (2) pp 53–62.
[18] Togrul H and Arslan N 2003 Production of Carboxymethyl Cellulose from sugar beet pulp cellulose and rheological behavior of carboxymethyl cellulose *Carbohydrate Polymers* 54 pp 73–82.
[19] Heinze T, Dicke R, Koschella A, Kull A H and Koch W 2000 Effective preparation of cellulose derivatives in a new simple cellulose solvent *Macromolecular Chemistry and Physics* 201 pp 627–631.
[20] Köhler S and Heinze T New solvents for cellulose: Dimethyl Sulfoxide/Ammonium Fluorides, *Macromol. Biosci.* 7 pp 307–314.
[21] Pinkert A, Marsh K N and Pang S 2010 Reflections on the solubility of cellulose *Ind. Eng. Chem. Res.* 49(22) pp 11121-11130.

[22] Pinkert A, Marsh K N, Pang S and Staiger M P 2009 Ionic liquids and their interaction with cellulose*Chem. Rev*. 109(12) pp 6712-6728.
[23] Lindman B, Karlstrom G, and Stigsson L 2010 On the mechanism of dissolution of cellulose *J. Mol. Liq*. 156(1) pp 76-81.
[24] Liebert T 2009 Cellulose solvents–remarkable history, bright future. In cellulose solvents: for analysis, shaping and chemical modification T. Liebert, T. J. Heinze, K. J. Edgar, Eds. *ACS Symposium Series 1033; American Chemical Society* (Washington, DC) ch. 1 pp 3-54.
[25] Klemm D, Heublein B, Fink H P and Bohn A 2005 Cellulose: Fascinating biopolymer and sustainable raw material *Angew. Chem., Int. Ed*. 44(22) pp 3358-3393.
[26] Eliza M Y, Shahruddin M, Noormaziah J and Chung W B 2015 The effectiveness of TBAF/DMSO in dissolving oil palm empty fruit bunch-cellulose phosphate, *Journal of Medical and Bioengineering* 4 (2) pp. 165–169.
[27] Sanghamitra S, James D M and Dimitris S A 2013 Review of cellulose non-derivatizing solvent interactions with emphasis on activity in inorganic molten salt hydrates *Journal of Sustainable Chemistry and Engineering* 1 pp 858-870.
[28] Wan Rosli W D, Leh C P, Zainuddin Z and Tanaka R 2003 Optimisation of Soda Pulping Variables for Preparation of Dissolving Pulps from Oil Palm Fiber *Holzforschung* 57 pp 106-113.
[29] Leh C P, Wan Rosli W D, Zainuddin Z and Tanaka R 2008 Optimization of oxygen delignification in production of totally chlorine free cellulose pulps from oil palm empty fruit bunch fiber *Industrial Crops and Products* 28 pp 260-267.
[30] Heydarzadeh H D, Najafour G D and Nazari-Moghaddam A A 2009 Catalyst-free conversion of alkali cellulose to fine Carboxymethyl Cellulose at mild conditions *World Applied Sciences Journal* 6 (4) pp 564–569.
[31] Pecsok R L, Shields L D, Cairns T and McWilliam I G 1976 Modern method of chemical analysis *John-Wiley New York*.
[32] Biswal D R and Singh R P 2004 Characterisation of Carboxymethyl cellulose and polyacrylamide graft copolymers *Carbohydrated Polymers* 57 pp 379–387.

**Acknowledgement**
The author would like to thank Universiti Tun Hussein Onn Malaysia (UTHM) Johor who support this research through grant funds of RAGS 021.

# Follicle Detection on the USG Images to Support Determination of Polycystic Ovary Syndrome

**Adiwijaya[1], B. Purnama[2], A. Hasyim[3], M. D. Septiani[4], U. N.Wisesty[5], W. Astuti[6]**

School of Computing Telkom University, Bandung

[1]adiwijaya@telkomuniversity.ac.id, [2]bedy@telkomuniversity.ac.id, [5]untarinw@telkomuniversity.ac.id, [6]astutiwidi@telkomuniversity.ac.id

**Abstract.** Polycystic Ovary Syndrome (PCOS) is the most common endocrine disorders affected to female in their reproductive cycle. This has gained the attention from married couple which affected by infertility. One of the diagnostic criteria considereded by the doctor is analysing manually the ovary USG image to detect the number and size of ovary's follicle. This analysis may affect low varibilites, reproducibility, and efficiency. To overcome this problems, automatic scheme is suggested to detect the follicle on USG image in supporting PCOS diagnosis. The first scheme is determining the initial homogeneous region which will be segmented into real follicle form The next scheme is selecting the appropriate regions to follicle criteria, then measuring the segmented region attribute as the follicle. The measurement remains the number and size that aimed at categorizing the image into the PCOS or non-PCOS. The method used is region growing which includes region-based and seed-based. To measure the follicle diameter, there will be the different method including stereology and euclidean distance. The most optimum system plan to detect PCO is by using region growing and by using euclidean distance on quantification of follicle.

**Keywords : polycystic ovary syndrome, region growing, stereology, euclidean distance, follicle segmentation, follicle detection.**

## 1. Introduction

Polycystic Ovary Syndrome (PCOS) is the most common endocrine disorders affected to female in their reproductive cycle. According to the criteria of National Institutes of Health (NIH), 6% - 10% female who has PCOS and Rotterdam can widely define as 15% prevalency of PCOS. PCOS has gained the attention from married couple when they acquire infertility. Yet, there have been 4 conferences in terms of identifying the definition and criteria of PCOS dignosis. The last conference was sponsored by the experts from National Institutes of Health (NIH) which suggested the existence of diagnostic criteria produced by Rotterdam conference [2]. One of the Rotterdam criteria in diagnosing PCOS remains the ovary polycystic (PCO) [3]. Female can be said that they have PCO syndrome when there has been 12 or more a 2-9mm follicle or the increasing in ovary volume to more than 10 cm3 [3]. For now, to recognize that female is affected by polycystic ovary syndrome, doctor can manually calculate number of follicle on ovary and identifies its size on ultrasonographic (USG) image. In this research, doctor needs the thoroughness and high accuracy so there will not be low variability, reproducibility, and efficiency. Thus, there will be the system built automatically which can detect PCOS. Based on the ultrasonographic images as the input, this research can idetify PCOS considering to follicle morphology on ovary using NIH criteria so it is expected to assist the doctor in analyzing PCO accurately on PCOS patient using USG image.

The method used in identifying follicle object on USG image is algorithm approach of Region Growing. This aims at segmenting both structural characteristic follicle and geometric follicle comprehensively [4]. Moreover, Region Growing is known as follicle segmentation algorithm which has the highest follicle recognition about 78% and has a better accuracy than its competitor [4]. After segmentation process of follicle structure and existence, there will be stereology method and euclidean distance to identify geometric structures quantitatively to the follicle on ovary [5], so it can be used to detect PCO on ultrasonographic image. These approaches will be used on medical images for mobility and authentication[1]. In this paper, detection of a follicle is done automatically. Hence, it will be faster then conventional ones when the measurement carried out offline.

## 2. PCO detection system

The primary four processes in creating PCO detection system are identification of homogeneous regions, region growing, follicle extraction, and follicle quantification. The process can be described as follows:

1. The identification process of homogeneous region is begun when the user selects template 1 and template 2. The cropping size on template 1 is 330x389 in which the whole ovary is filled up. The template 2 has the cropped image which remains variety of size. The next process goes to determining threshold value (T1) from the cropped image by calculating the average value of greylevel in each pixel. This T1 threshold value will be the algorithm limitation of adaptive neighborhood median filter in filtering process. Lower intensity pixel of threshold T1 will be filtered using a 11x11 neighborhood, while higher one will be using 5x5 neighborhood. The 11x11 neighborhood is selected because a 50-60 pixel on USG image is approximately 10mm in a real.
   The twice-filtered images by adaptive neighborhood filter will determine threshold T2 value by calculating the average pixel intensity on smoothed image then it is subtructed by its pixel deviation standard. In threshold T2 value will be used by binary segmentation to create binary image in which homogeneous regions remain. It is conducted by determining lower pixel intensity of T2 as the white pixel while higher one as black pixel. The white pixel or the pixel which has white neighborhood is stated as the region.
   The previous procedure has produced number of unexpected homogeneous region since it is too small to become follicle region. This region will be deleted by considering to the criteria where the area of the region is less than threshold T3. It is stated by 50 which is predicted as well as a quarter of the detected minimum wide-region of follicle [4]. T1, T2, and T3 value are obtained from the earlier research in refference [4]. Furthermore, the region passed through the limit on the image, will be deleted considering to the artificial region in the edge of image. Therefore, this process can produce some homogeneous region.
2. The identified homogeneous region of the previous step can represent the prediction of initial follicle region and can determine the seed of region. Region growing step in this paper aims to compare both the identified seed and homogeneous region to enlarge its region for reaching the real

Journal of Physics: Conference Series **622** (2015) 012027 doi:10.1088/1742-6596/622/1/012027

follicle limits. Each of homogeneous region will enlarge its region repeatedly using region's pixel outer limits.

a. Regarding to reffrence [4], there will be some criteria used to the region growing based on the region as follows:
   1) Grey-level criteria, by equation:
   $$|I_k(p)-\mu(R_0)| \geq \alpha\sigma R_0 \tag{1}$$
   2) Edge criteria and texture information, by equation:
   $$grad(p_i)=\|\nabla I_k(p)\|(\frac{c}{e^{tex(p)}-1}) \tag{2}$$
   3) Grey-level criteria using weighted gradient, by equation:
   $$|grad(p)-\mu(grad(R_0))| \leq |grad(p)-\mu(grad(R_0))| \leq \alpha\sigma\mu(grad(R_0)) \tag{3}$$

   The refference [4] describes that the pixel's outer limits of region filled up the two criteria (i and iii) is marked as the potential candidate which may be gathered with its region. It will be conducted either when there has been four neighboring pixels of connected environment in their region which fulfilled those criteria or the pixel marked as potential candidate. The connected environment is observed using 2 type including 4-neighborhood and 8-neighborhood. However, if the pixel is not marked by potential candidate or does not comply the integrated criteria, so it can not enlarge its region. The repeated process from the outer pixels of region will be stopped since there is no one which fills up the criteria of integrated pixel and region. Therefore, it will be added to segmented region image.

b. Seed-based region growing uses six criteria to enlarge homogeneous region as follows:

$$\text{K1: } if\ S_{q(x)}+i>0\ and\ S_{q(x)}+i \leq Row \tag{4}$$
$$\text{K2: } if\ S_{q(y)}+j>0\ and\ S_{q()}+j \leq Col \tag{5}$$
$$\text{K3: } if\ \sqrt{S_{q(x)}+i-(S_{q(x)})^2} < MaxDist \tag{6}$$
$$\text{K4: } if\ \sqrt{S_{q(y)}+j-(S_{q(y)})^2} < MaxDist \tag{7}$$
$$\text{K5: } if\ I(S_{q(x)}+i, S_{q(y)}+j) \leq (\text{Re}\, gVal+ThresVal) \tag{8}$$
$$\text{K6: } if\ I(S_{q(x)}+i, S_{q(y)}+j) \geq (\text{Re}\, gVal-ThresVal) \tag{9}$$

Based on the reference [7], Row is the number of the original image lines, Col is the number of the original image coloums, MaxDist is the maximum distance to the initial *seed* with infinite value, RegVal is the value of graylevel pixel which is the initial *seed (centroid),* and an empirical value that will be observed. If it meets those six criteria, the pixel will be marked as follicle region.

3. Follicle extraction stage is the stage of final decision taken on each *region* which has been expanding its territory, whether it is follicle or not. Each *region* in the segmented image has area, which is the number of pixels bounded by a line that forms a homogeneous space. *Region* that has area of less than 220 pixels will be deleted. The *threshold* is obtained from a minimum size of follicle (2-3mm dimeter ) [4]. Each *region* will calculate its ratio between the width of *the bounding box* and the area. *Region* that has a ratio of more than 0.5 will be marked as a *region* in accordance with the follicle, but if the ratio is less than 0.5, the *region* will be deleted and not marked as the actual follicle [5].

4. In the process of quantifying the quantification of follicles will be conducted to obtain the feature vector from the image that has been recognized as the follicles in the previous stage. The information will be used to classify follicles on ovary whether it is PCOS or not. Information of structural and geometric attributes taken from the follicles in the image is the number and the size of follicles. In this paper, the process of follicle quantification uses two approaches, namely stereology and *euclidean distance.* Both of these methods will later be compared to find out which one is better in measuring the diameter of the detected follicles so later it can be determined whether a follicle belongs to PCO or not. In process of stereology, each recognized follicle in the image will be determined its pixel area. Then, the recognized pixel area will be calculated in order to determine the number and size of each follicle in the image using the equation (10):

$$\frac{\sqrt{4\times\frac{Area}{\pi}}}{treshold} \qquad (10)$$

where *Area* is the number of pixels that forms the follicle region *(region). Threshold* is obtained from the size of pixels per millimeter, usually 50-60 pixels in the image of USG ovary has a distance of 1cm in its original size. It means that every millimeter has 5 to 6 pixels, so the value of *threshold* that will be used in order to measure the follicles per millimeter is 5,5piksel / mm [4].

Meanwhile, *euclidean distance* method is used to measure the diameter of follicles. *Euclidean distance* is a calculation method that is most often used to calculate the similarity of two vectors [6]. *Euclidean distance* calculating the squared root of the difference of two vectors. In this research, a vector is a follicle coordinate. The formula for calculating the diameter of follicles and *euclidean distance* uses the equation (11).

$$D=\sqrt{\sum_{t=1}^{n}(x_{it}-x_{jt})^2+(y_{it}-y_{jt})^2} \qquad (11)$$

where,

$D$ = diameter of the follicle (pixel),

$x_i, x_j$ = coordinate of axis $x$ follicle,

$y_i, y_j$ = coordinate of axis $y$ follicle,

$t$ = the number of points.

In this test, the coordinate of axis $x$ follicle and axis $y$ follicle is farthest pixel of the axis x and y of a follicle. The farthest pixel can be found by using the function of MajorAxisLength and MinorAxisLength which are already contained in the Matlab program.

## 3. Testing and Analysis

In this testing, a total of 12 images of USG Ovary is obtained from Clinic of Permata Bunda Syari'ah, Cirebon (Indonesia). Each image contains two images, the left ovary and the right ovary. We use a total of 12 images of USG Ovary on the testing process. It was felt sufficient the amount of data for testing in this research, because this research is focused on the size and number of follicles were detected, not a person identification PCO syndrome or not. The number and size of follicles of all images have been validated by Dr. Zulkifli Ahmad, SpOG., K.FER., M.Kes as Female Fertility Consultant in Clinic of Permata Bunda Syari'ah. Testing the system is conducted by performing simulations on some pre-defined scenarios. Testing the success of estimating the size and number of follicles is conducted by calculating the accurate success rate of follicle number detection (RR), the error rate of follicle size (MR), and the error rate of detection (FR).

3.1. Scenario Parameter Values of Alpha and Thresval on Region Growing
For area-based region growing, it is used the empirical values of variable alpha in equation (1), (2) and (3). Experiment A is conducted by changing the value of variable (alpha = 1, or alpha = 1.5, or alpha = 2). The value of observed threshold is assumed that the distribution of pixel intensity and weighted gradient in the image is classified as normal distribution. Therefore, the observed value of alpha is derived from the fixed normal distribution curve by selecting the interval 66% - 95% of the data.
Meanwhile, for the seed-based region growing it is used the empirical value on variable ThresVal in equation (8) and (9). Experiment B is conducted by changing the value of variable (Thresval = 0.025, or Thresval = 0.050, or Thresval = 0075). Empirical values of alpha and ThresVal provide different performance measured from the average success of detection (RR).

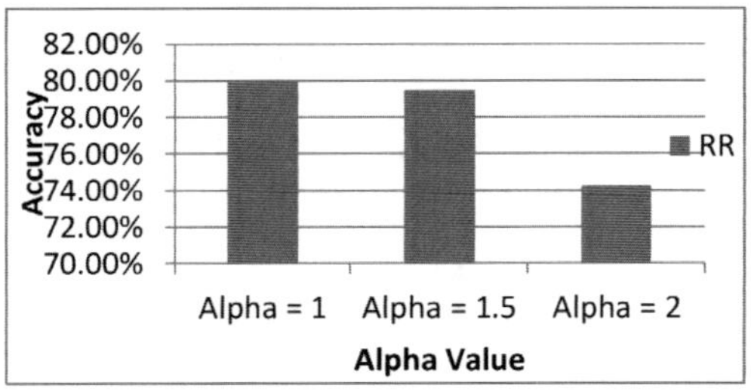


**Figure 1**. Success Rate Detection Accuracy Testing Follicles (RR) in Experiment A

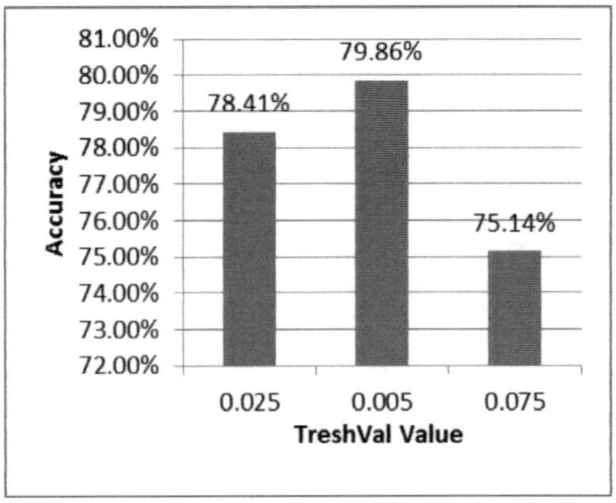


**Figure 2.** Succes Rate Detection Accuracy Testing Follicles (RR) in Experiment B

Figure 1 shows the average success rate of the detection on alpha equals 1 is 80%, alpha equals 1.5 is 79.5%, and alpha equals 2 is 74.25%. While in Figure 2, it can be seen that the average success rate of detection on thresVal Value equals 0.025 is 78.41%, thresVal Value equals 0050 is 79.86%, and thresVal Value equals 0075 is 75.14%.

### 3.2. Scenario Size of Template Image

In this scenario, the test is conducted with the configuration on the size of the template image that is used. Test A is combined with the region growing parameter (alpha = 1, or alpha = 1.5, or alpha = 2 and connectivity pixel 4- neighborhood or 8 -neighborhood). However, test B is combined with the parameter region growing (ThresVal = 0.025, or ThresVal = 0.050, or ThresVal = 0.075 and connectivity pixel 4- neighborhood or 8 -neighborhood).

Testing in this scenario aims to determine the initial estimation of the region of optimal follicles that can affect the region growing method in determining the actual follicles. Based on the testing that has been conducted, the performance in this scenario is affected by the size of the template image that is used. The size of the template used is the template 1 and template 2 that give different performance measured from the average error rate of detection (FR).

Based on Figure 3, it can be seen that the error level rate on template 1 is 17.67% and template 2 is 6.17%. Meanwhile, from figure 4, it can be seen that the error rate of template 1 is 16.36% and template 2 is 15.93%.

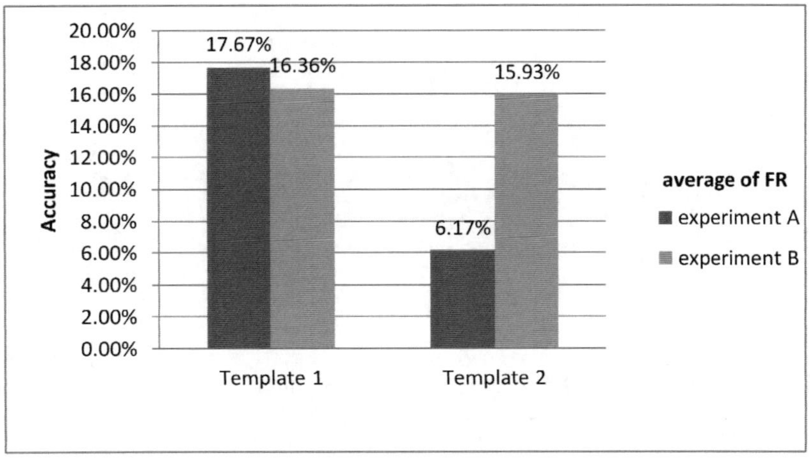


**Figure 3.** Detection Error Rate (FR) based on Template of Experiment A and B

Based on the experiment A and B, it can be seen that the error rate (FR) of template 2 is lower than FR of template 1. It is because template 2 only takes the most essential image part. The most essential image part is the part that contains follicles reffered by specialists that makes speckle noise at template 2 lesser than template 1 that captures the complete image of a single ovarium that results in some findings of the other structures at ovarium, such as blood vessel, lymph node and dendrite.

### 3.3. Scenario of Pixel Connectivity Type

This scenario is conducted by comparing the region of pixel connectivity type of 4-neighborhood and 8-neighborhood. The pixel used 4-neighborhood connectivity is classified as a part of an object if the neighborhood along the horizontal and vertical line fits with the criteria, while 8-neighborhood adds neighborhood on the diagonal line. Observation is conducted by changing the pixel connectivity type. On this scenario, the observation is conducted towards the most appropriate pixel connectivity size on the region growing method as to extend the region. Type of connectivity is proportional to the similarity between pixels in each template. But inversely proportional to the long process of computation.

Experiment A is combined with the parameter of region growing (alpha =1, or alpha=1.5, or alpha =2, and 4-neighborhood or 8-neighborhood pixel connectivity) and the parameter of image template size (template 1 or template 2). Meanwhile, experiment B is combined with the parameter of region growing (ThresVal =0.025, or ThresVal =0.050, or ThresVal =0.075 and pixel connectivity of 4-neighborhood or 8-neighborhood). Each experiment is conducted by using two quantification methods, where the measuring methods used are stereologi and euclidean distance.

The experiment in this scenario is aimed for determining the pixel connectivity region in the most optimum region growing method. The determination of pixel connectivity region can influence the region growing method used to determine the outest region of pixel, whether the pixel is connected to the region or not so that it can influence the stereology method in the measurement of the follicle region. The second aim of the scenario of experiment is to find the follicle quantification method that has the lowest error level of follicle measurement.

Based on the experiment, the performance in the scenario is influenced by the pixel connectivity type and the quantification method used. The pixel connectivity types used are 4-neighborhood and 8 neighborhood and the follicle quantification methods used are stereology and euclidean distance that give different performance measured from the error level of follicle detection (MR).

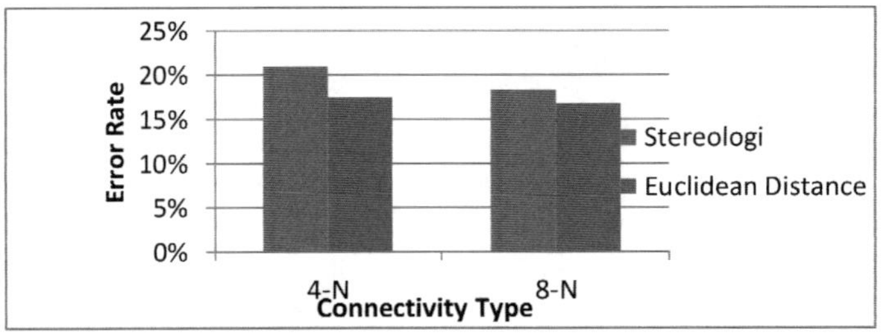


**Figure 4.** Follicle Size Error Rate Testing (MR) Based on the Type of Pixels Connectivity using Stereologi Method and Euclidean Distance in Experiment A

Based on figure 4, it can be seen that the error rate of follicle detection (MR) of the 4-neighborhood connectity using stereology is 21% and the one of 8-neighborhood is 18.33%, meanwhile that of the 4-neighborhood connectivity using euclidean distance is 17.50% and that of 8-neighborhood is 16.83%.

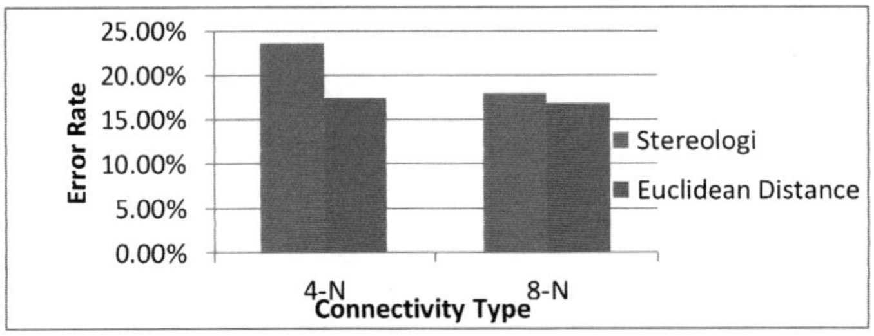


**Figure 5.** Follicle Size error Rate Testing (MR) Based on the Type of Pixels Connectivity using Stereologi Method and Euclidean Distance in Experiment B

Figure 5 provide the error rate of follicle detection (MR) on the 4-neighborhood connectivity using stereology is 23.62% and the one of 8-neighborhood is 17.59%; while that of 4-neighborhood connectivity using euclidean distance method is 17.50% and that of 8-neighborhood is 16.83%. Based on test A and B, it can be seen that the rate of MR shows that 8-neighborhood connectivity is the one that has lower error level on follicle detection compared to stereology. It is because 8-neighborhood has more neighbors that touch each pixels, so that it influences the number of pixels found on the follicle border that causes the follicle region wider resembling the real shape of the follicle. Euclidean distance method has a better performance compared to the stereology since euclidean distance conducts the follicle quantification referred to the farthest diameter. While the stereology conducts the follicle quantification through diameter equivalen. Euclidean distance method is in accordance with the validation conducted by specialists so that the performance of euclidean distance is better than that of stereology.

## 4. Conclusion

Based on the analysis towards the test on the PCO detection system of ultrasonographic image, it can be concluded that the region growing and euclidean distance methods can be implemented for detecting PCO. In the experiment A, the system configuration scenario that generates the most optimum performance is resulted from template 2, 8-neighborhood connectivity, and the threshold of alpha of 2 while the most optimum quantification method is euclidean distance. The testing scenario, when the performance is compared to the validation of specialists, results in success level of follicle detection (RR) of 78%, error level of follicle size (MR) of 17%, and error level of follicle detection (FR) of 12%. In the experiment B, the system configuration scenario that generates the most optimum performance is resulted from template size 2, 8-neighborhood connectivity, and the threshold of thresVal of 0.050 while the most optimum quantification method is euclidean distance. The testing scenario, when the performance is compared to the validation of specialists, results in success level of follicle detection (RR) of 76%, error level of follicle size (MR) of 20%, and error level of follicle detection 9FR) of 9%. Based on the second experiment, it can be concluded that the most optimum system to detect PCO is by using region growing and by using euclidean distance on quantification of follicle. It can be seen from the success level of follicle detection (RR), the error level of follicle size 9MR) and error level of follicle detection (FR).

## References

[1] Adiwijaya, T.A.B. Wirayuda, S.D. Winanjuar, U. Muslimah, *The Multiple Watermarking on Digital Medical Image for Mobility and Authenticity*, Operations Research Proceedings 2012, 457-462

[2] National Institutes of Health, *Evidence-based Methodology*, Final report of Workshop on Polycystic Ovary Syndrome. 2012 December 3–5.

[3] M.J. Lawrence, M.G. Eramian, R.A. Pierson, E. Neufeld, *Computer Assisted Detection of Polycystic Ovarian Morphology in Ultrasound Images*, Proceedings of the 4th Canadian Conference on Computer and Robot Vision. 2007, IEEE Computer Society.

[4] Rotterdam ESHRE/ASRM-Sponsored PCOS Consensus Workshop Group. *Revised 2003 consensus on diagnostic criteria and long-term health risks related to polycystic ovary syndrome*. Fertil Steril. 2004 Jan; 81(1) : 19-25.

[5] B. Potoˇcnik and D. Zazula. *Automated analysis of a sequence of ovarian ultrasound images. PartI : segmentation of single 2D images*. Image and Vision Computing. 2002, 20(3) : 217–225.

[6] E. Weibel, *Stereological Methods. Vol. 1: Practical Methods for Biological Morphometry*. 1979, Academic Press, London, U.K.

# Data Hiding Scheme on Medical Image using Graph Coloring

**Widi Astuti[1], Adiwijaya[2], Untari Novia Wisety[3]**

School of Computing, Telkom University, Bandung

[1]astutiwidi@telkomuniversity.ac.id; [2]adiwijaya@telkomuniversity.ac.id;
[3]untarinw@telkomuniversity.ac.id;

**Abstract.** The utilization of digital medical images is now widely spread[4]. The medical images is supposed to get protection since it has probability to pass through unsecure network. Several watermarking techniques have been developed so that the digital medical images can be guaranteed in terms of its originality. In watermarking, the medical images becomes a protected object. Nevertheless, the medical images can actually be a medium of hiding secret data such as patient medical record. The data hiding is done by inserting data into image - usually called steganography in images. Because the medical images can influence the diagnose change, steganography will only be applied to non-interest region. Vector Quantization (VQ) is one of lossy data compression technique which is sufficiently prominent and frequently used. Generally, the VQ based steganography scheme still has limitation in terms of the data capacity which can be inserted. This research is aimed to make a Vector Quantization-based steganography scheme and graph coloring. The test result shows that the scheme can insert 28768 byte data which equals to 10077 characters for images area of 3696 pixels.

## 1. Introduction

Nowadays, more digital medical images are transmitted between hospitals which are located differently via computer network [5]. Nevertheless, it is hard to find the network which is secure. Therefore, several methods to protect medical images are developed. The research to protect medical images is mainly classified into two purpose categories : to ensure the images originality or to conceal the patient medical record in medical image[5]. Embedding secret data into an image to conceal the secret data is called steganography. The image will change but the most important thing is that no one can read the secret data.

Vector Quantization (VQ) is one of data compression techniques which is lossy, prominent and frequently used. VQ makes use of the fact that many blocks in images have similarities so that an image can be divided into several blocks, and then each block is quantified based on the similarity with a dictionary called codebook. Some data experts attempt to utilize the Vector Quantization in steganography by adopting its principle. VQ and steganography have a similiarity, their process is basically change an image but keep the change visually tolerable. The researches divides the codebook into some clusters then manipulating the quantization process. The clustering process has a very important role since it decided the resulted images quality and data capacity which can be concealed. This clustering process is rarely based on graph coloring. Steganography scheme based on VQ without graph coloring still has limitation in terms of data capacity which can be inserted.

Yue et al. [1] proposed a steganography scheme based on VQ and graph coloring which has good performance with regard to capacity or security. This research attempts to utilize the proposed scheme by Yue et al. for the case of patient medical record data hiding in medical images. Because data hiding in the medical image is very risky due to changes in medical image could lead to changes in diagnosis, this scheme will only be done in its region of non interest, a specific region that is not so sensitive to the change. It is expected that the scheme will be a secure scheme of steganography which will not destroy the important part of medical images but has large capacity of insertion.

## 2. Data Hiding Scheme in Medical Image

The medical image that will be used in this research is ultrasonography(USG) image since USG is widely used. The medical image can be separated into two region i.e region of interest (ROI) which become the focus in diagnosis making and region of non-interest (RONI) as the figures below shows.

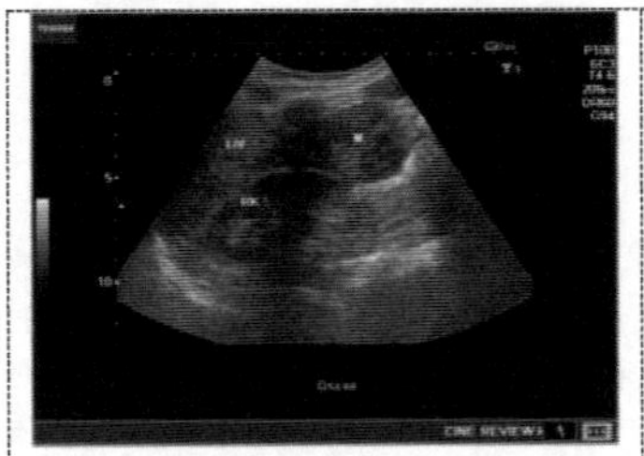

**Figure 1.** a medical image

**Figure 2.** RONI of medical image from figure 1

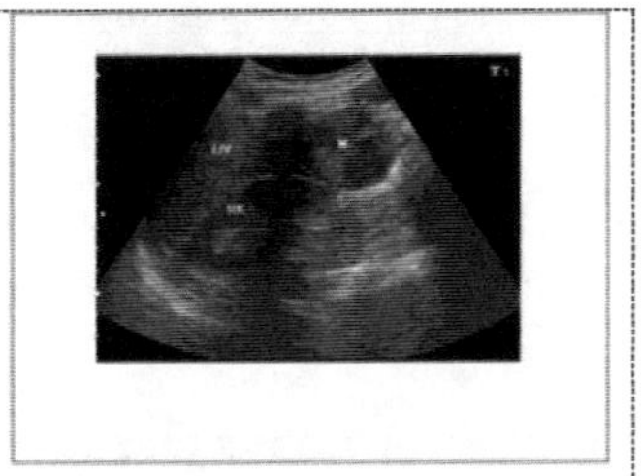

**Figure 3.** ROI of medical image from figure 1

This paper is described generally described in this flowchart.

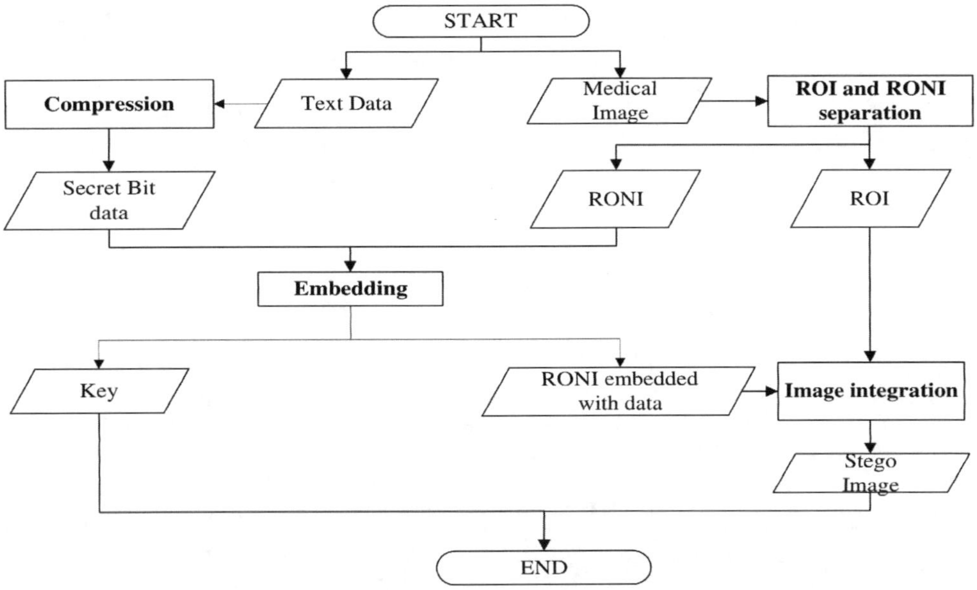


**Figure 4.** General Scheme

Journal of Physics: Conference Series **622** (2015) 012028 doi:10.1088/1742-6596/622/1/012028

Based on figure 4, secret text data that will be inserted in the image need to be compressed first in order to enlarge the hiding capacity. This research use Huffman Algorithm as the compression method. In the next step, secret data that have been transformed into bit data will be embedded in the image's RONI which had been separated from the ROI, an area that is selected manually. There are two outputs form the embedding process i.e. stego image and key. Stego image is an image that contain secret data. Key is something to protect the secret data. Without the key, the secret data can not be retrieved. Embedding procedure which is modified from [1] is described in figure 5.

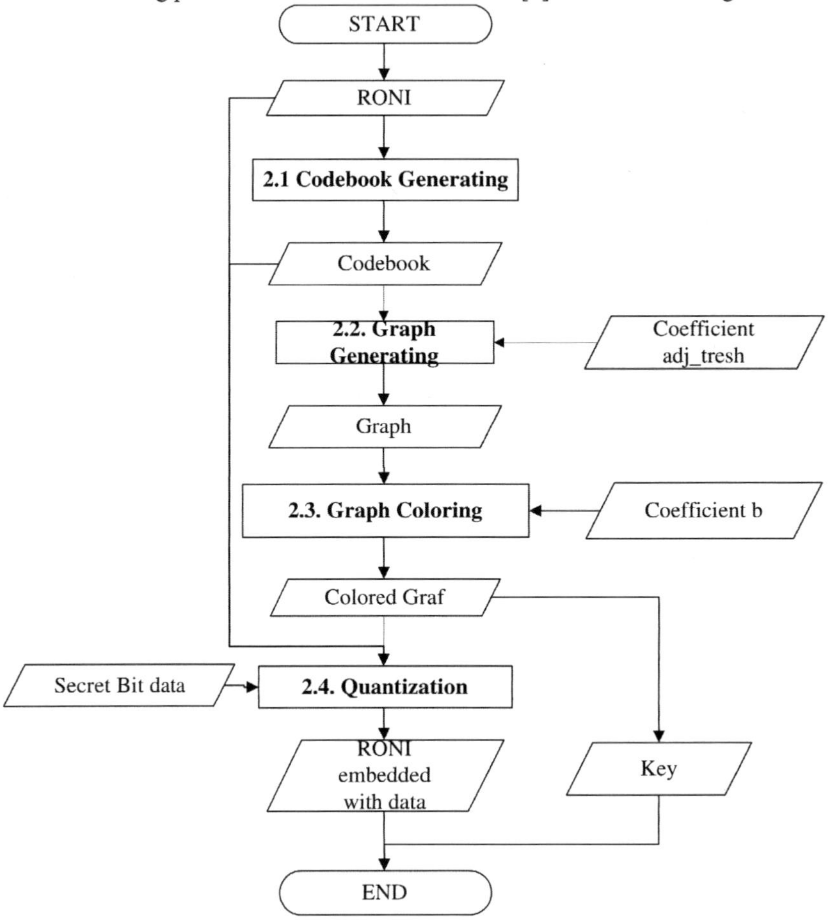


**Figure 5.** Embedding Procedure

Embedding procedure actully adopt the VQ procedure. First step in VQ is codebook generating, the system construct a codebook that consists of codewords. Each codeword represent a block of pixels which usually the size of one block is 4x4 pixels. Next step in VQ is quantization. In quantization, VQ will process each block by compare it with codebook and find the most similiar codeword. The most similiar codeword is determined by smallest Euclidian distance between the block and each codeword *cw*. The equation for the Euclidian distance is shown in Equation (1).

$$D(blok, cw) = \sqrt{\sum_{i=0}^{15}(block[i] - cw[i])^2)} \quad (1)$$

Where,
*D(block,cw)* = Euclidian Distance between block and codeword cw
block[i] = pixel value in block[i],
cw = pixel value in cw[i],

Then, VQ outputs the index of the most similar codeword as the current compression code in the index table, a table for storing each block compression code. After processing every block in the image, VQ will keep the index table and the codebook.

In this research secheme, before the quantization, scheme do graph coloring in order to clustering the codebook.

### 2.1. Codebook generating

There are some codebook generating algorithms in VQ. Lu [3] explained that in codebook generating, an image will be divided into some training vectors and the training vectors will undergo a clustering process. This research uses one of the codebook generating algorithm that easily to be implemented i.e. Linde Buzo Gray (LBG) algorithm.

### 2.2. Graph generating

A graph is constructed by adding a vertex for each codeword in the codebook so that every vertex is representating the codebook. For every vertex v[i] and v[j] in graph representating cw[i] and cw[j], scheme will add an edge between v[i] and v[j] only and if only the values of cw[i] and c[j] are close enough. The values of cw[i] and cw[j] are considered as close if Euclidian Distance between cw[i] and cw[j] is not exceed adj_tresh, a constant coefficient that determined empirically.

### 2.3. Graph Coloring

The graph is colored using a number of color called color_number. Each color can be represented using an integer so that color ={ 0,1,2,3,...,color_number-1}. Color_number value must be satisfied $2^x$ for an integer x. Each color is coresponding with x bit of secret message. For example, if color_number=8, the colors are: = {0,1,2,3,4,5,6,7} ={000,001,010,011,100,101,110,111}. Since graph coloring is a NP complete problem, this research will utilize Particle Swarm Optimization to perform it. Graph Coloring in this scheme is adopting Cui[2] scheme. The difference is Cui's scheme is only used to do graph coloring using 4 colors, but in this scheme, Cui's scheme is modified so that it can be used to do graph coloring using $2^n$ colors.

### 2.4. Quantization

The last step on embedding process is a quantization. The quantization is described as follows.
Inputs : RONI, colored graph, codebook, and bit data
Outputs: RONI embedded with data

1. The scheme divides RONI into some blocks
2. The scheme reads the first block
3. The scheme searchs codeword c that is most similiar with current block using Euclidian Distance
4. The scheme searchs vertex v that is coresponding with codeword c.
5. The scheme retrieves x bit from secret bit data where $2^x$=color_number.
6. The scheme searchs using Breadth First Search for vertex v' that either directly or indirectly connected with v and satisfied color(v')=x; If v' found, go to step 7, If v' can not be found go to step 8
7. Codeword c' that is coresponding with v' will be the output. x bit from secret bit are deleted from secret bit data. Go to step 9.
8. Codeword c will be the output. Go to step 9.
9. The scheme processes next block using step 3-8 until every blocks are processed.

Journal of Physics: Conference Series **622** (2015) 012028 doi:10.1088/1742-6596/622/1/012028

## 3. Testing and Analysis

In this testing, a same RONI of USG image will be embedded by some secret text using 8 scenarios. Two parameters will be used to measure the scheme performance i.e. hiding capacity and Peak Signal to noise Ratio (PSNR). The scenarios are created by changing three parameters i.e *codebook_size*, *adj_tresh*, and *color_number*.

### 3.1. Codebook size

Generally, the size of codebook used in VQ is 256 [1]. But, [1] shows that in a data hiding scheme , codebook size is affecting the scheme performance. Testing in this research changing the codebook size between 64 and 128 because only some region from the image will be embedded by secret data and RONI of medical image tends to be homogenous so there is no need codebook as big as common image.

### 3.2. Value of adjacency treshold

Adj_tresh is a coffecient that has important role in graph generating phase since the changing of adj_tresh will cause the change of graph generated. Test is conducted by changing the value of variable (adj_tresh = 60, or adj_tresh = 70).

### 3.3. Color number

Color_number is a parameter that is determinating in graph coloring and quantization. Test is conducted by changing the value of variable (color_number = 8, or color_number = 16).

After conducting the test, the result is shown on the table below.

| Scenario | *codebook_size* | *adj_tresh* | *color_number* | PSNR | Capacity (bit) |
|---|---|---|---|---|---|
| 1 | 64 | 8 | 60 | 17,35% | 22 371 |
| 2 | 64 | 8 | 70 | 16,94% | 28 768 |
| 3 | 64 | 16 | 60 | 17,21% | 22 650 |
| 4 | 64 | 16 | 70 | 17,78% | 94 |
| 5 | 128 | 8 | 60 | 17,46% | 22368 |
| 6 | 128 | 8 | 70 | 18,20% | 4836 |
| 7 | 128 | 16 | 60 | 16,93% | 22668 |
| 8 | 128 | 16 | 70 | 18,10% | 1060 |

The test is aimed at determining the best variable at the scheme. Best result is determined by highest PSNR and hiding capacity. Based on the test results above, it can be seen PSNR will increase as codebook_size, adj_tresh and color_number increase. In term of hiding capacity, as adj_tresh and color_number increase, the hiding capacity will decrease. It can not be decided yet how codebook size can affect the hiding capacity. It can be conclude that the most optimum parameter is codebook_size=64 , color_number = 16 , and adj_tresh=60 with byte capacity 28768 equivalent to 10077 characters.

## 4. Conclusion

Based on the analysis towards the test on the data hiding scheme, it can be concluded that the Vector Quantization and Graph Coloring can be implemented for hiding the data in medical image. The best performance is produced by the parameter codebook_size=64 , color_number = 16 , and adj_tresh=60 with byte capacity 28768 equivalent to 10077 characters. In term of security, the data which is concealed on medical images using this scheme can not be retrieved without knowing the key, ie the colored graph.

## References

[1] Yue, Shuai *et al.*, 2013, " An Image Data Hiding Scheme Based On Vector Quantization and Graph Coloring". Springerlink: Recent Advances in Information Hiding and Applictions , 1-17. [1] Cheddad, Abbas (dkk), 2010, "Digital Image Steganography: Survey and analysis of current methods". Signal Processing , 90 (3), 727-752.

[2] Cui, Guangzhao, *et al.*, 2007, "Modified PSO algorithm for solving planar graph coloring problem". Progress in Natural Science 18 (2008), 353-357.

[3] Lu, Tzu-Chuen, 2010, "A Survey of VQ Codebook Generation". Journal of Information Hiding and Multimedia Signal Processing , 1 (3), 190-203.

[4] Memon, Nisar Ahmed, 2010, " Watermarking of Medical Images for Content". Phd Thesis, Faculty of Computer Science and Engineering, GIK Institute of Engineering Sciences and Technology, Khyber Pakhtunkhwa

[5] Adiwijaya, T.A.B. Wirayuda, S.D. Winanjuar, U. Muslimah, "The Multiple Watermarking on Digital Medical Image for Mobility and Authenticity", Operations Research Proceedings 2012, 457-462

# Development of Ozone Technology Rice Storage Systems (OTRISS) for Quality Improvement of Rice Production

**M.Nur[1], E.Kusdiyantini[2], W.Wuryanti[3], T.A. Winarni[4], S.A. Widyanto[5], and H. Muharam[6]**

[1]Department of Physics, [2]Department of Biology, [3]Department of Chemistry, Faculty of Science and Mathematics, Diponegoro University, Semarang, Indonesia
[4]Department of Fishery Product Technology, Faculty of Fisheries and Marine Science, Diponegoro University, Semarang, Indonesia
[5]Department of Machanical Engineering, Faculty of Engineering, Diponegoro University, Semarang, Indonesia
[6]Department of Management, Faculty of Business and Economics, Diponegoro University, Semarang, Indonesia
E-mail: m.nur@undip.ac.id

**Abstract.** This research has been carried out by using ozone to address the rapidly declining quality of rice in storage. In the first year, research has focused on the rice storage with ozone technology for small capacity (e.g., household) and the medium capacity (e.g., dormitories, hospitals). Ozone was produced by an ozone generator with Dielectric Barrier Discharge Plasma (DBDP). Ozone technology rice storage system (OTRISS) is using ozone charateristic which is a strong oxidizer. Ozone have a short endurance of existence and then decompose, as a result produce oxygen and radicals of oxygen. These characteristics could kill micro-organisms and pests, reduce air humidity and enrich oxygen. All components used in SPBTO assembled using raw materials available in the big cities in Indonesia. Provider of high voltage (High Voltage Power Supply, 40-70 kV, 23 KH, AC) is one of components that have been assembled and tested. Ozone generator is assembled with 7 reactors of Dielectric Barrier Discharge Plasma (DBDP). Rice container that have been prepared for OTRISS have adjusted so can be integrated with generator, power supply and blower to blow air. OTRISS with a capacity of 75 kg and 100 kg have been made and tested. The ability of ozone to eliminate bacteria and fungi have been tested and resulted in a decrease of microorganisms at 3 log CFU/g. Testing in food chemistry showed that ozone treatment of rice had not changed the chemical content that still meet the standard of chemical content and nutritional applicable to ISO standard milled rice. The results of this study are very likely to be used as an alternative to rice storage systems in warehouse. Test and scale-up is being carried out in a mini warehouse whose condition is mimicked to rice in National Rice Storage of Indonesia (Bulog) to ensure quality. Next adaptations would be installed in the rice storage system in the Bulog.

## 1. Introduction

Technology of rice storage in Indonesia has not been able to keep the quality of rice in a long time. Often, distribution of rice in large quantities had to be canceled by the damage of rice. due to the quality of the rice that was not fit for consumption. It is very worrying the government in this case Bulog can not replace the rice in a large amount in a short time. On the other side the needs of the

Journal of Physics: Conference Series **622** (2015) 012029 doi:10.1088/1742-6596/622/1/012029

community may not be delayed. In remote regions that are not producing rice this problem may be more fatal, such as famine. The rapid of decay of rice occurs due to high water levels and breeding of microorganisms contained in the rice storage container. Rice that has been reduced quality is also easily become breeding places of which often called mite pests of rice.

So far, in rice storehouses owned Bulog, fumigation and spraying insecticides have been done as efforts for maintaining the quality of rice. Fumigation is a method of insect pest control with specific gases generated by the fumigant. With this method, pest insects that were in or at surface of a sack of rice can be eradicated. To improve the effectiveness of the insect pests eradication on rice and sacks, fumigation collaborated with insecticide spraying method. Insecticides was sprayed on surface of the sack, sack piles sidelines, warehouse floors, and the surrounding environment that serves to eradicate insects that exist in the area. So the insect pest population can be reduced and also can prevent the insects return to the rice barn.

Technology of rice storage has not been able to keep the quality of rice. The new technology is needed to resolve the issue. The alternative is ozone technology. Based on research, this technology proved can maintain dan even increase the quality of cereals (especially wheat). Utilization of appropriate ozone generating technology and proper dosage of ozone is believed to be used for rice storage technology.

Ozone can also be generated by using a plasma reactor for a particular purpose. Ozone consists of three oxygen atoms and has chemical formula O3. Ozone is an unstable molecule with a very short life span (20-30 minutes) before return to be oxygen, therefore ozone will always seek 'target' to able release one oxygen atom by means of oxidation, so it can be turned into oxygen molecules stable (O2). The precise dose of ozone is determined by configuration of the ozone reactor and the generator. [1].

Ozone as an oxidant has many potential applications in food industry due to the advantages of ozone technology compared to traditional food preservation techniques. Ozone has been known to be extensive utilization for microbial inactivation. Ozone has been reported to eliminate micro-organisms including bacteria, fungi, viruses, protozoa, fungi and bacteria spores. Another advantage of the use of ozone for food is that ozone compounds rapidly decompose and produce oxygen so it does not leave residues in food [2]. Ozone treatments for wheat and corn with ozone concentration of 50 ppm for (3-5) days treatment can kill insect pests reach 70% - 100% [2]. The factors that determine the ability of ozone to kill insects, eliminate fungus and decontamination of toxins, depends on: ozone concentration, exposure time, pH and moisture content of the grains (e.g., wheat, and corn) [3] [4] [5]. In recent years, Diponegoro University has developed research of applications of plasma technology including ozone produced by the plasma technology [6] [7]. Utilization of ozone to maintain the shelf life of rice [8]. In 1995, ozone expressed by Generalize Recognized as Safe (GRAS) or is generally recognized as safe by FDA (U.S. Food and Drug Administration) for the treatment of drinking water bottles. These applications are then developed as "GRAS" for processing food by some experts in last few years [9].

Disinfectant properties of ozone have been tested by several researchers. For example Tiwari et al. [2], which examined the effectiveness of ozone against grain pests such as Tribolium castaneum and Ephestia elutella and the coliform bacteria, Micrococcus, Bacillus spores, fungi and the like. In his journal Tiwari et al. [2] write that the respiratory system be the entrance of ozone to the inside the body of the insect. Ozone causes oxidative tissue damage (even at low concentrations), causing DNA damage, changes in pulmonary system, bronchial sensitivity, membrane oxidation or mutation in vivo. Increased respiratory system with an increase in temperature can result excessive gas exchange due to the increase in metabolism and respiration rate. [10].

In a different study, Ran et al. [11] examined the effectiveness of ozone against Criptosporidium (intestinal protozoa, one of the causes of diarrhea) and protozoan parasite Giardia. In addition, Ran also investigated the mechanism of cell destruction Cryptosporidium by ozone using scanning electron microscopy (SEM) at treatment of time 0, 60, 120, and 480 seconds. As a result, the cell Cryptosporidium suffered the greatest damage in 480 seconds.

Research on microbial inactivation of bacteria and molds (fungi) using ozone has been done by several researchers. For example, Zorlugenc et al. [12] who observed the effectiveness of ozone gas and liquid against the bacteria E. coli and coliform bacteria as well as fungi. The tests proved, disinfectant properties (against E. coli and coliform bacteria) ozone gas is lower than the liquid ozone. Similarly, when testing against molds (fungi), which perfectly reduced the mold in 15 minutes using liquid ozone while in ozone gas still leaving approximately 1 log CFU/g. Types of mold that identified having degraded populations include Aspergillus flavus, Aspergillus niger, Aspergillus parasiticus [12]. Test effectiveness of ozone against E. coli was also conducted by Patil et al. [13], where they use ozone to sterilize orange juice from bacterial contamination.
The bacteria can be symbiotic mutualism and/or parasitism. In many ways, the bacteria associated with causing disease (parasites and pathogens), accelerating the decay and the like. Overcome the things that adverse caused by bacteria, it is used various types of disinfecting agents. Disinfectant agent that is quite popular in use today is chlorine. Chlorine can kill bacteria with a low concentration, in quick time, and little risk. Chlorine kills bacteria in a way, destabilize cell membranes and inhibit biochemical reactions. Besides chlorine, ozone and ultraviolet is also used. Disinfecting ability of ozone has been studied by several experts in recent years. Ozone, in some cases can kill the bacteria in which chlorine is quite difficult to use. Effectiveness of the disinfectant agent is different, depending on the type of microbes, concentration, and the like [14].
The ability of ozone to kill bacteria, fungi, and insects used as basis for utilizing ozone in rice storage system. In addition, due to ozone within 40 minutes will decompose and possible to bind water molecules, the water content in rice will also be reduced. This paper discusses the use of ozone for rice storage. The discussion is about ozone generation systems, the ability to hold the growth of microorganisms and the quality of rice that has undergone ozonized. In this paper also discussed adaptation of ozonized system into the storage system in Bulog warehouse.

## 2. Method

Ozone generators are made in the laboratory can be adapted for rice storage system which is based on techniques Dielectric Barrier Discharge Plasma (DBDP) and raised with an AC voltage source; 40-70 kV, frequency of 23 kHz.

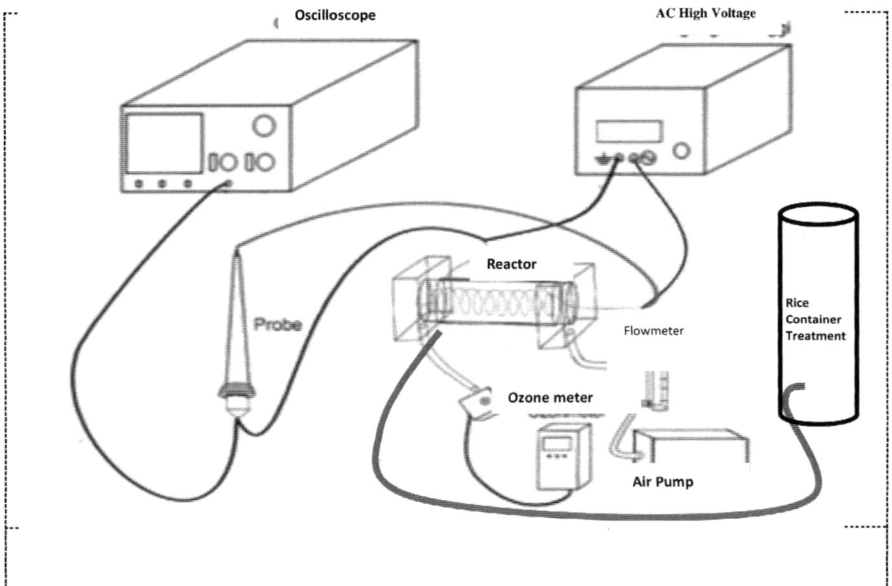


**Figure 1.** Experimental set up

Active electrode is made of spiral wire copper and passive electrodes made from tube-shaped copper plate. Between the two electrodes is given dielectric made of pyrex glass with varying diameter (2 cm, 3 cm and 5 cm). Air pump (Amara AA-250) was used to deliver the air into the reactor, and ozone from recombination process, into treatment media. Probe (Ser. No. 20048737) was used to divide the power supply voltage to 1000 times smaller before being connected to an oscilloscope. Oscilloscope (Instek, GOS-620 20Mhz) was used to monitor the voltage, and operating frequency of the power supply. Ozone detector (Model OS-4, serial # 1208) was used to test the generated ozone levels for treatment (figure 1). Ozone generator was integrated of DBDP reactor with a spiral configuration of copper and copper-dielectric cylindrical pyrex pipes (diameter 3 cm and length 15 cm for research on rice with small and medium capacity). Adjustments to the requirements of adaptation to SPBTO containers, using one generator unit with 7 reactor DBDP. Rice samples were placed in a cylindrical media. Ozone is supplied using an air pipe through the bottom of rice samples. Variation of ozone exposure time was 30, 60, 90, 120, 150 and 180 minutes. Each treatment was performed 4 times, so to be obtained at least 20 data from each object, where the objects of this study are bacteria and mold or fungi which contaminate rice. Testing of samples for chemical analysis (proximate), biological analysis (bacteria and mold/fungi), physical analysis (amilografi, hardness, viscosity and whiteness) and SEM analysis (SEM morphology and SEM EDS).

## 3. Results and Discussion

### *3.1. Characterization of Current-Voltage and Ozone concentration*

The influence of voltage to electrical current average is presented in Figure 2 for the integrated generator which consists of 7 pieces reactor Spiral Cylinder (SC) with 3 cm of diameter pyrex tube.
Total electric current has two components: capacitive electric current from the capacitor barrier and capacitive electrical current from the capacitor gas. Therefore, the electric current that actually being measured is the average capacitive electrical current. Influence of electric voltage to ozone concentrations are presented in figure 3 for spiral cylinder reactor of pyrex tube with a diameter of 3 cm.

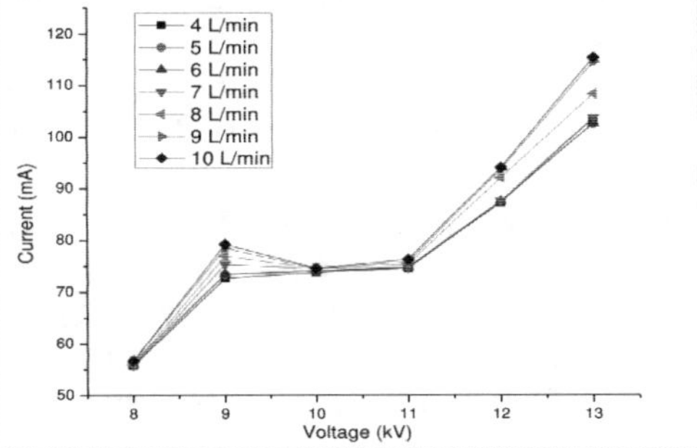


**Figure 2**. Electric current as a function of applied voltage for DBDP spiral-cylinder reactor, to 3 cm diameter pyrex tube.

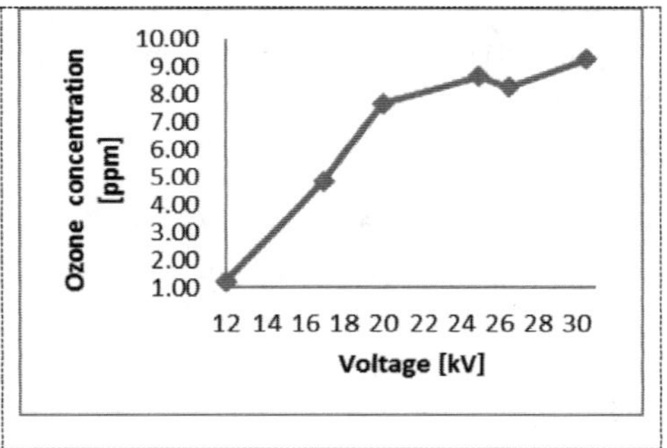


**Figure 3.** Ozone concentration as function of applied voltage, for DBDP reactor spiral-cylinder, for diameter of Pyrex tube of 3 cm

### *3.2. Inactivation of Bacteria and Fungi*

In Figure 4, we find that the effect of ozone on the amount of bacteria that contaminate rice showed a negative correlation. Bacterial growth is inversely related to ozone exposure time. The average number of bacteria decreased 0.41 log CFU / g every 30 minutes during treatment lasts beyond. This condition is caused by the amount of contaminating bacteria do not survive to breeding process. Resulting slower in population growth of bacteria. The curve above shows, the optimum point decline in bacteria occurs on treatment for 180 minutes.

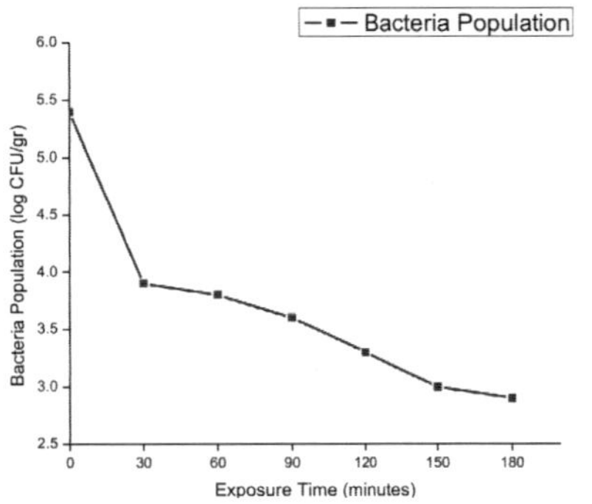


**Figure 4.** Bacteria colonies as function of time of treatment by ozone

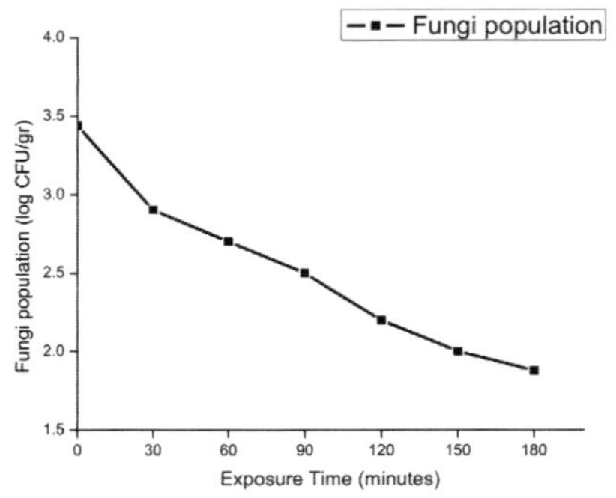


**Figure 5.** Fungi colonies as function of time of treatment by ozone

The number of bacteria decreased to 2.48 log CFU / g from the initial state (5.39 log CFU/g). Bacterial growth at 180 minutes, only reached 2.91 log CFU/g indicate that the highest number of bacteria which died before the proliferation (in vitro incubator) done. Remaining bacteria after treatment for 180 minutes is 54.02% of the total number of bacteria before treatment, or can be said to be 45.98% of the bacteria died before breeding. The most significant decrease in the yield curve occurs in treatment between 0 to 30 minutes, in which the number of bacteria decreased 1.44 log CFU/g compared to the transition time of 30-60 minutes. The influence of ozone to the number of fungi is presented in Figure 5. Research of ozone treatment for inactivation of fungi indicate decline from an average of 3.44 log CFU/g to an average of 1.88 log CFU/g. The curve above shows the optimal reduction in the number of fungi occurred at 180 minutes. Where the number of fungi population decreased 1.56 log CFU/g compared with no treatment. The number of fungi surviving 180 minutes after treatment was 1.88 log CFU/g (approximately 54.56% than without treatment). The most significant decrease in the curve occurs at the transition 0 to 30 minutes, in which the number of fungi fell 0.52 log CFU/g. The curve is a significant decrease in the number of fungi in the first 30 minutes also occur in bacteria. This decrease may be due to some fungi can not survive environmental changes abruptly.

Journal of Physics: Conference Series **622** (2015) 012029 doi:10.1088/1742-6596/622/1/012029

### *3.3. Content of Carbohydrate, Amylose and Amylopectin*

Chemical content in rice are being subjected to ozone should be studied seriously Application of ozone at sufficient doses for effective decontamination of rice can affect various quality parameters. Utilization of ozone are not all beneficial and in some cases can lead to oxidative degradation of the chemicals present in rice. Surface oxidation, discoloration or change unwanted odors can be occur from excessive use of ozone [5].

The main part of rice is carbohydrate, then the rice used to meet nutritional needs as a source of carbohydrate. Carbohydrate content in rice, corn, and potatoes are 78.9; 72.4; 19.1% respectively [16]. In different exposure time tends to decrease carbohydrate content, but in relatively small amounts. Carbohydrate in rice consists of amylose and amylopectin. The content both of the compounds is commonly used as a standard of rice quality (figure 6, figure 7).

The results shows that there is no change in significant chemical content in rice for conditions in this study. Table 1 shows chemical composition Nutritional value per 100 grams rice.

**Table 1.** Nutritional value per 100 grams rice [15]

| No | Chemical composition | Great contains |
|---|---|---|
| 1 | Carbohydrate | 79 g |
| 2 | Glucose | 0,12 g |
| 3 | Dietary fiber | 1,3 g |
| 4 | Fat | 0,66 g |
| 5 | Protein | 7,13 g |
| 6 | Water content | 14 g |
| 7 | Thiamine (Vit.B1) | 0,070 mg (5 % AKG) |
| 8 | Riboflavin (Vit B2) | 0,049 mg (3 % AKG) |
| 9 | Niacin (Vit B3) | 1,6 mg mg (11 % AKG) |
| 10 | Pantotenik acid (B5) | 1,014 mg (20 % AKG) |
| 11 | Vitamin B6 | 0,164 mg (13 % AKG) |
| 12 | Folik Acid Vit B9 | 8 μg (2 % AKG) |
| 13 | Forum | 0,8 mg (6 % AKG) |
| 14 | Phosphorus | 115 mg (16 % AKG) |
| 15 | Potassium | 115 mg (2 % AKG) |
| 16 | Calcium | 28 mg (3 % AKG) |
| 17 | Magnesium | 25 mg (7 % AKG) |
| 18 | Zinc | 1,09 mg (11% AKG) |
| 19 | Calories | 354 kcal |

There are 4 classes of rice [17], the first class is rice with high amylose content about 25-33%, the second class is rice with intermediate amylose content about 20-25%, the third class is rice with low amylose content about 9-20%; the fourth class is rice with very low amylose content about < 9%.

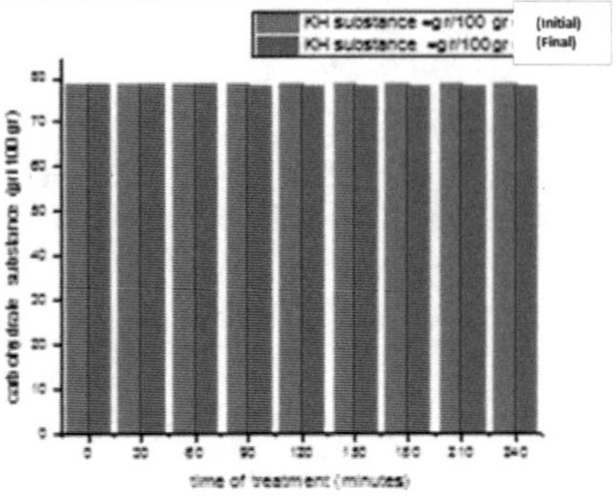


**Figure 6.** The test result of carbohydrate contain on rice with variation of time of treatment by ozone, compared to control, for the same rice condition.

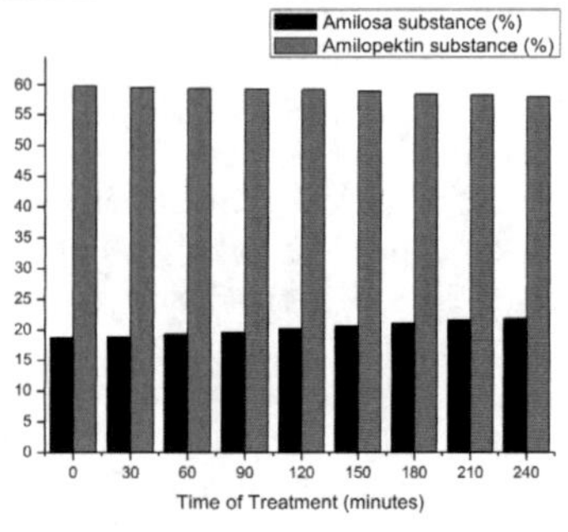


**Figure 7.** The test result of amylose and amylopectin content in rice with variation of time of treatment by ozone, compared to control, for the same rice condition

The analysis result of amylose and amylopectin content in rice known that generally, amylose content increased with increasing of exposure time by ozone. The treated rice by ozone content 16.26 % of amylose. As previously described, based on the amylose content, the quality of rice is classified into 4 classes [17], the first class is rice with high amylose content about 25-33%, the second class is rice with intermediate amylose content about 20-25%, the third class is rice with low amylose content about 9-20%; the fourth class is rice with very low amylose content about < 9%.

The samples that used in this research are rice with low amylose and high amylopectin. Therefore, when it cooked into cooked rice, it has high adhesion level. Content of amylose and amylopectin in rice are one one of the factors that determine the quality of rice. Glutinous rice has a low amylose content of less than 2%, it caused the glutinous rice more sticky than rice.

### *3.4 Content of Fat and Protein*

From the data of test result known that the fat content tended to decrease after treat by ozone irradiation. Changes in fat content in rice were being treatment by ozone are still in a very small amount such that there is no significant change after rice being treat by ozone (Figure 8). Fat content is generally below the analytical results of the fat content of milled rice [16]. This is not being problem because the rice used as a source of carbohydrate.

Raw white rice contains about 6.7% protein, while if it is cooked, it contain about 2%. Rice contains glutelin protein group that is orizenin and in the protein contained 24.1% of glutamate amino acid and slightly lysine. Physical properties of protein of cereal grains is very important, although the protein content is not so high. The size of protein content depends on: species, type of soil, fertilizers, and climate. In general, changes in protein content is very small due to the irradiation is done at room temperature about 29 °C, less than 50 °C. While the character of protein is easy to denatured at temperatures about 50 °C and having surface denaturation in range of pH of 4.5 to 6.0. In the Figure 9 shown that there is an slightly increase of protein on 60 minutes exposure by ozone. The increase can not provide conclusion, as to the provision of higher ozone, protein levels return as in the control

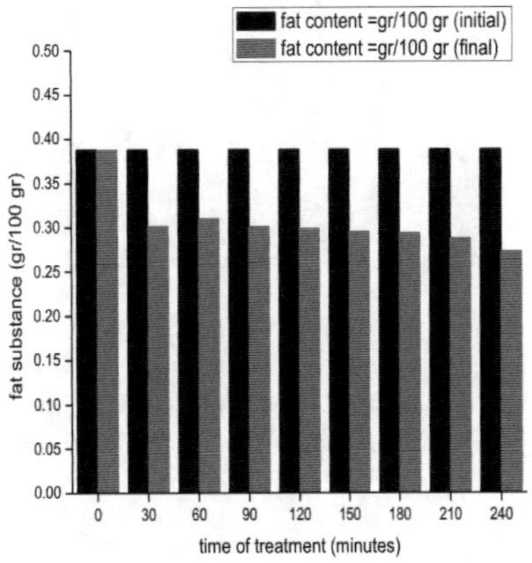


**Figure 8.** The test result of fat content on rice with variation of time of treatment by ozone, compared to control, for the same rice condition.

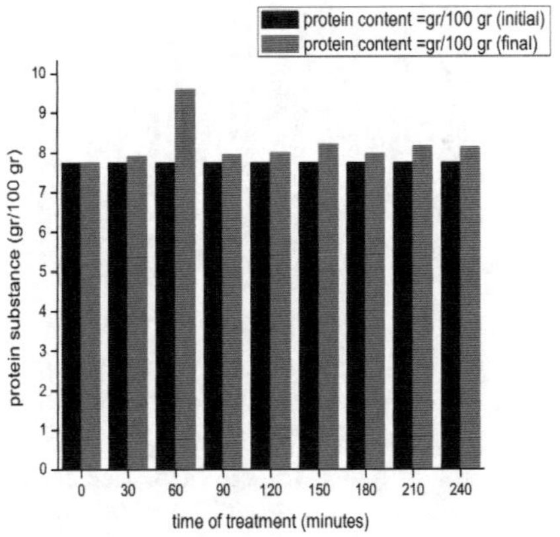


**Figure 9**. The test result of protein contain on rice with variation of time of treatment by ozone, compared to control, for the same rice condition

### *3.5 Content of Ash and Water*

The analysis result of ash content on treated rice by ozone is shown in figure 10. It shows that there is no significant change of ash content, and still in the range of milled rice standard. From the data analysis it appears that the water content before and after irradiation are not much different (figure 11). Rice with minimum water contain will be more resistant to microbial attack due to the properties of microbial that tend to live in high humid condition. In addition, rice with minimum water contain is also will be more resistant to insect attack. The level of water contain reduced by 12.4%, and according to the experience on Bulog, it can be stored for more than 3 years.

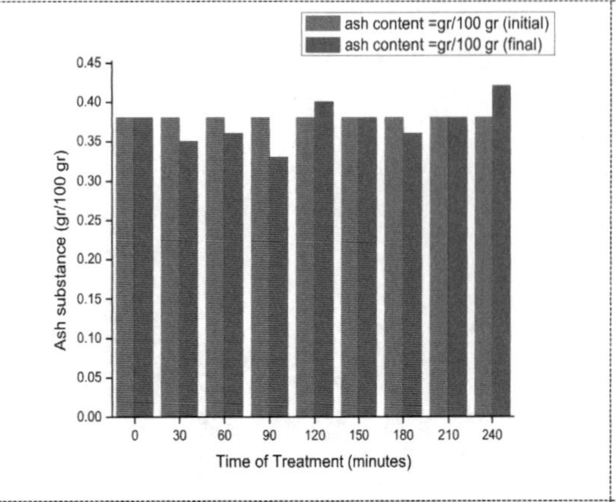


**Figure 10**. The test result of ash content on rice with variation of time of treatment by ozone, compared to control, for the same rice condition

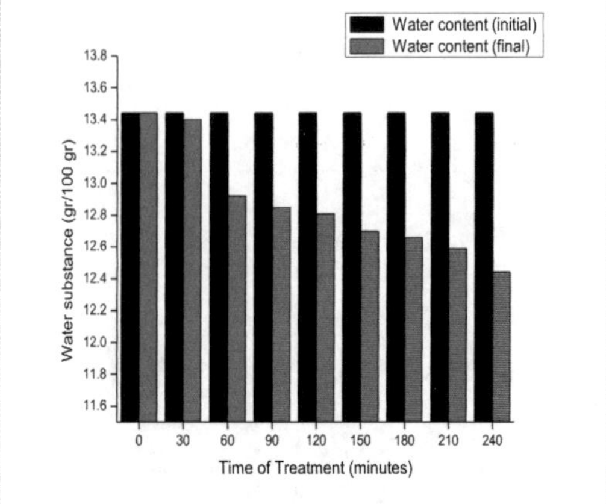


**Figure 11.** The test result of water content on rice with variation of time of treatment by ozone, compared to control, for the same rice condition

## 4. Conclusion

In this study has successfully developed ozone technology rice storage system (SPBTO). Ozone was generated with application of plasma technology by Dielectric Barrier Discharge Plasma (DBDP) reactor. Development of the product begins with the electro mechanical testing of DBDP plasma reactor and ozone generator. The results indicate that the DBDP reactor can produce ozone properly and can be adjusted to the desired concentration to maintain the quality of rice. The test results on treated rice by ozone indicate that the bacteria in rice greatly reduced over a periode time, and the most effective decrease is in 180 minutes.
Utilization of ozone have also been able to reduce the growth of fungi, it also obtained that the optimal reduction in the number of fungi occurred at 180 minutes. This time will be used for each repetition period of ozonation on rice. There are no insects on treated rice by ozone although it has been stored for 12 months. The analysis of fat, water, and protein, ash, and carbohydrate that content in rice before and after treatment by ozone results that there is no significant change.

## Acknowledgment

All members of group of research thanks to High Education Directorate (DIKTI) for financial of this research in the schame Flagship Research for National Strategic (Penelitian Unggulan Trategis Nasional) of the year 2014.

## References

[1] M.Nur, 2011, *Fisika Plasma dan Aplikasinya*, Undip Press, Semarang,

[2] Tiwari, B.K., Brennan, C.S., Curran, T., Gallagher, E., Cullen, P.J., & O'Donnell, C.P., 2010. *Application of ozone in grain processing*. Journal of Cereal Science, 51, 248-255.

[3] J.A. Hardin, C.L. Jones, E.L. Bonjour, R.T. Noyes, R.L. Beeby, D.A. Eltiste, S. Decker, 2010, Ozone fumigation of stored grain; closed-loop recirculation and the rate of ozone consumption. Journal of Stored Products Research 46, pp. 149-154

[4] M. X. McDonough , A. Carlos, Campabadal, L.J. Mason, D.E. Maier, A. Denvir and C. Woloshuk, (2011), Ozone application in a modified screw conveyor to treat grain for insect pests fungal contaminants, and mycotoxins,Journal of Stored Product Research, vol 47. Pp. 249-254

[5] F.Mendez, D.E. Maier, L.J. Mason, C.P. Woloshuk,(2003), Penetration of ozone into columns of stored grains and effects on chemical composition and processing performance, Journal of Stored Products Research vol. 39 pp. 33–44.

[6] M. Nur, Supriati, A. Setyaningrum, D.H. Gunawan, Munir, M. Sumariyah, (2009), Ozone Generator by Using Dielectric Barrier Discharge Plasma Technology With Spiral-Cylinder Configuration: Comparison Between Oxygen and Air As Sources, Berkala Fisika, ISSN : 1410-962, Vol.2 pp.69-76

[7] M.Nur, T.A. Winarni, M.Restiwijaya,(2014), Dielectric Barrier Discharge Plasma Reactor Analysis as Ozone Generator, Proceeding ISTMET 2014, IEEE Explore, pp 129 – 132

[8] M.Nur, A.Solichin, E. Kusdiayantini, T.A. Winarni, D. A. Rahman, R.Maryam, S. Teke, H. Muharam (2013), Ozone production by Dielectric Barrier Discharge Plasma for microbial inactvation in rice, Proceeding ICBME2013, IEEE Explore, pp 221 – 225

[9] Voidarou, C., Tzora, A., Skoufos, I., Vassos, D., Galogiannis, G., Alexopoulos, A., Bezirtzoglou, E., 2007. Experimental Effect of Ozone upon Some Indicator Bacteria for Preservation of an Ecologically Protected Watery System. Water Air Soil Pollut, 181, 161-171.

[10] Cullen, P.J., Tiwari, B.K., O'Donnell, C.P., Muthukumarappan, K., 2009. Modelling Approaches to Ozone Processing of Liquid Foods. *Trends in Food Science & Technology*, 20, 125-136.

[11] Ran, Z., Li, S., Huang, J., Yuan, Y., Cui, C., Williams, C. D., 2010. Inactivation of Cryptosporidium by Ozone and Cell Ultrastructures. Journal of Environmental Sciense, 22(12), 11954-1959

[12] Zorlugenc, B.Keroglu, Zorlugenc, F., Oztekin S., Evliya I 2008, The Influence of Gaseous Ozone and Ozonated Water Microbial Flora and decradal alfaloxm B1 in dried figs, *Food Chem. Texica*l 46.3593-359

[13] Patil, S., Bourke, P., Frias, J.M., Tiwari, B.K., & Cullen, P.J., 2009. Inactivation of Escherichia coli in orange juice using ozone. *Innovative Food Science and Emerging Technologies*, 10, 551-557.

[14] Narins, B.(editor), 2003. *World of Microbiology and Immunology*, Vol. 1 - (A-L)(366s). The gale group, inc: USA.

[15] Juliano, B.O. and Bechtel, D.B. (1985). The grain and its gross composition. In: Rice: *Chemistry and Technology* p. 17–57. Los Banos, Laguna: IRRI

[16] Anonim, 1979. *Daftar Komposisi Bahan Makanan*. Direktorat Gizi Departemen Kesehatan Republik Indonesia Bharata Karya Aksara. Jakarta.

[17] Winarno, FG, 1992. *Biokimia Pangan dan Gizi*; PT. Gramedia Pustaka Utama. Jakarta.

# Soil Identification using Field Electrical Resistivity Method

**Z A M Hazreek[1,8], S Rosli[2], W D Chitral, A Fauziah[4,9], A T S Azhar[5], M Aziman[6] and B Ismail[7]**
[1,5,6] Faculty of Civil and Environmental Engineering, Universiti Tun Hussein Onn Malaysia, 86400 Batu Pahat Johor, MALAYSIA
[2,8] School of Physics, Universiti Sains Malaysia, 11800 USM Penang, MALAYSIA
[3,7] Research Center for Soft Soil, Universiti Tun Hussein Onn Malaysia, 86400 Batu Pahat Johor, MALAYSIA
[4] School of Civil Engineering, Universiti Sains Malaysia, 14300 Nibong Tebal Penang, MALAYSIA
[9] Institute of Postgraduate Studies, Universiti Sains Malaysia, 11800 USM Penang, MALAYSIA

E-mail: hazreek@uthm.edu.my

**Abstract.** Geotechnical site investigation with particular reference to soil identification was important in civil engineering works since it reports the soil condition in order to relate the design and construction of the proposed works. In the past, electrical resistivity method (ERM) has widely being used in soil characterization but experienced several black boxes which related to its results and interpretations. Hence, this study performed a field electrical resistivity method (ERM) using ABEM SAS (4000) at two different types of soils (Gravelly SAND and Silty SAND) in order to discover the behavior of electrical resistivity values (ERV) with type of soils studied. Soil basic physical properties was determine thru density (ρ), moisture content ($w$) and particle size distribution (d) in order to verify the ERV obtained from each type of soil investigated. It was found that the ERV of Gravelly SAND (278 Ωm & 285 Ωm) was slightly higher than Silty SAND (223 Ωm & 199 Ωm) due to the uncertainties nature of soils. This finding has showed that the results obtained from ERM need to be interpreted based on strong supported findings such as using direct test from soil laboratory data. Furthermore, this study was able to prove that the ERM can be established as an alternative tool in soil identification provided it was being verified thru other relevance information such as using geotechnical properties.

## 1. Introduction

Conventional soil identification commonly involved the process of soil sampling and laboratory experiment. Soil sampling was obtained either from disturbed or undisturbed condition during the site investigation works. Then, several testing related to index properties determination was performed using particle size distribution (sieve test), density, water content, Atterberg limits (liquid limit and plastic limit test), etc. Those classical soil identification processes may faced limitation due to the cost and time particularly when it deals with large and difficult construction area. Moreover, conventional site investigation tools such as drilling methods experience difficulties in steep and hilly terrain, swampy areas, coastal regions and complex geomaterial areas which need to be investigated [1]. According to [2], high number of soil sampling was required in order to achieve high accuracy of soil properties assessment and as reported by [3], this objective was hard to achieve by conventional sampling which generally time-consuming and costly. Conventional laboratory experimental testing also can be very costly and took a longer time when dealing upon large number of soil.

Nowadays, geophysical method involving electrical resistivity, seismic, gravity, ground penetrating radar, magnetic, etc has increasingly applied in site investigation. Geophysical method applies the principles of physics properties in studying the earth such as resistivity, velocity, density, magnetic susceptibility, etc. General utilization of geophysical method was related to engineering, environment and archeological studies. The main purposed of geophysical method adoption in those areas was mainly focused on mapping or detection purposes. From past experienced, geophysical method has been recognized as an alternative tool in engineering, environmental and archeological studies which able to produced good efficiency due to the cost, time and data coverage. As reported by [1] and [4] – [6], geophysical techniques contributes several advantages such as it can be performed fast and low cost and has the ability to cover greater areas more thoroughly. However, the standard performance of individual geophysical method always depends on fundamental physical limitation (e.g. penetration, resolution, and signal to-noise ratio) [7].

Electrical resistivity method has widely used in soil profiling. Currently, the main purpose of ERM utilization was heavily used as detection and mapping tools related to locate boulder, bedrock and overburden materials [8] and [9]. However, several considerations related to results reliability are still been discussed and argued between those related parties (geophysicists, engineers, geologists, etc). Traditionally, the entire process of electrical resistivity method (field measurement (data acquisition), data analysis (processing) and results interpretation) has been performed by geophysicists due to its field of expertise. However, lack of cooperation with other related expert may cause some unreliable outcome due to the weak result interpretation and justification. In the past, geophysicists are too obsessed with the qualitative anomaly changes during the interpretation stage. That scenario has create some difficulty to the geotechnical engineers since most of the electrical resistivity results and justification was too abstract which always changes relative to the individual interpreter. Major problem in geophysical data interpretation was often a result arising from ambiguous data since the process of field measurement (data acquisition), data processing and interpretation was always exposed to noise which may influenced by devices, earth and environmental uncertainties or human factor (e.g. equipment operator) culminating in the geophysical data ambiguity [10]. Moreover, electrical resistivity value result interpretation was always exposed to the limitation with particular reference to the geomaterial resistivity overlapping properties. According to [11], previous ERV geomaterial reference charts and tables are not easily adaptable due to its wide range of variation in the parametric values and the overlapping values. Moreover, [12] has reported that the electrical resistivity value used to characterize subsurface profile material is necessarily subjected to local ground conditions, and the characterization occurs within overlapping classifications. In current geotechnical site assessment, the engineer desires strong verification from the geophysicist since the electrical resistivity method was a non-destructive method which being performed thru surface measurement to determine the subsurface profile. Geophysicists still possess only limited appreciation of the engineer's point of view due to the lack knowledge of the mechanics of soils. Geophysical methods are unable to stand alone in order to provide solutions to any particular problems [13] and [14].

Hence, strong verification from direct test (field test, experimental test, etc.) was important to support the electrical resistivity anomaly interpretation. Geophysical method has a good prospect in order to solve some of the problems related to the conventional site investigation methods [15]. As a result, this study performed a field electrical resistivity test for soil identification purposes together with in situ and laboratory soil geotechnical properties verification in order to reduced electrical resistivity anomaly ambiguity interpretation thus increased confidence level of the results interpretation and conclusion produced.

## 2. Materials and Methods

This study was divided into three (3) stages via fieldwork, laboratory test and data processing. Firstly, field 2-D electrical resistivity survey (2DERS) was conducted using ABEM SAS 4000 equipment at two different sites particularly in Kuala Kangsar, Perak Malaysia and Lenggong, Perak Malaysia representing two different types of soil investigated. Testing configuration was based on Schlumberger array using two (2) resistivity land cables, forty one (41) numbers of electrode and forty two (42) numbers of jumper cable. Equal electrode spacing of 0.05 meter was used for all 41 electrodes producing total electrical resistivity survey length of 2.0 meter. Immediately after the 2DERS finished, two different point of the field density test was performed using sand replacement method in line within 2DERS as referred to [16]. Fieldwork schematic diagram representing soil sampling and 2DERS line alignment was given in Figure 1. Soil sample obtained from field was taken to the geotechnical laboratory for water content, particle size distribution (dry and wet sieve) and specific gravity tests as referred to [16]. Finally, raw data from field 2DERS (Terrameter SAS 4000 data logger) was transferred to the computer for processing. During the processing stage, smooth constraint least square method was used in order to produce a smooth boundary representing soil material underneath. According to [11], smooth constraint least square method will produce a smooth boundary change which was considered more suitable for representing the soil material in contrast with rock and fractured material which preferred to analyzed using a robust method due to its sharp geomaterial boundary. RES2DINV commercialize software [17] was used to process and analyze the raw data thus producing an inverse model that estimate the actual subsurface structure. The inversion algorithm of RES2DINV was used to process the data, as proposed by [18] in order to obtain the 2-D resistivity section.

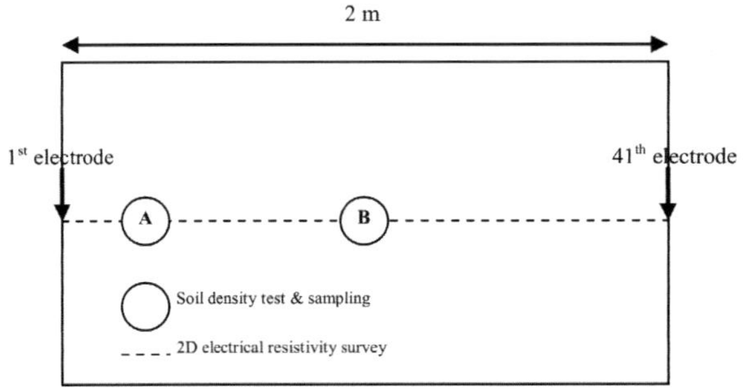


**Figure 1.** Schematic diagram of the soil sampling and testing position together with resistivity line alignment (drawing not to scale).

## 3. Results and Discussions

All results presented and discussed are based on field electrical resistivity value (ERV) which being verified using basic geotechnical properties as given in Figure 2 – 3 and Table 1 – 2.

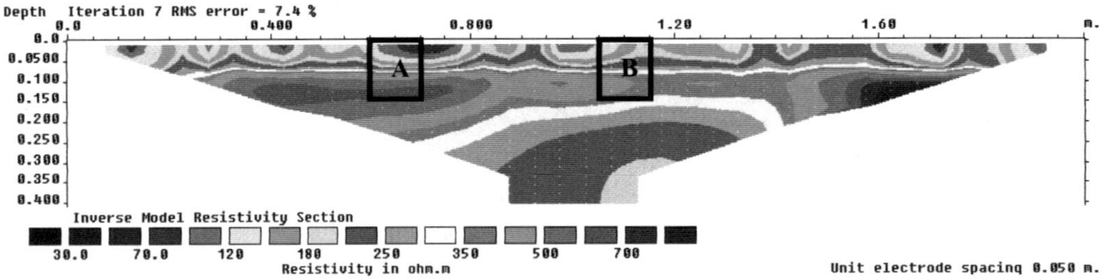


**Figure 2.** 2D electrical resistivity section and localize selected point of ERV for Gravelly SAND.

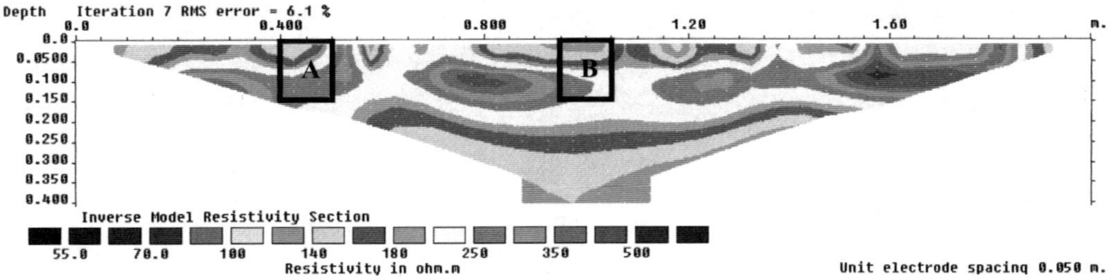


**Figure 3.** 2D electrical resistivity section and localize selected point of ERV for Silty SAND.

Electrical resistivity value (ERV) was determined by measuring the potential difference at points on the ground surface which caused the propagation of direct current through the subsurface [19]. Based on Figure 2 and 3, it was found that the ERV for Gravelly SAND (A = 278 Ωm & B = 285 Ωm) was greater than Silty SAND (A = 223 Ωm & B = 199 Ωm). As reported by [20] and [21], the ERV value for sand and wet silty sand is in the range of 10 – 800 Ωm and 100 – 250 Ωm respectively. Hence, this outcome has proved that the electrical resistivity method was able to mapped different type of soils thus contributed in assisting geotechnical engineer for field soil identification. However, the reliability of this finding was difficult to accept due to its qualitative anomaly perspectives. Electrical resistivity value is highly influenced by water concentration and lithology variations [22]. Hence, those electrical resistivity anomaly need to be verified using basic soil properties obtained from geotechnical field and laboratory tests as given in Table 1 and 2.

Soil classification test using particle size distribution test (dry and wet sieve) showed that soil at Kuala Kangsar was classed as Gravelly SAND while soil at Lenggong was classed as Silty SAND. It was found that the soil at Kuala Kangsar was dominated by coarse grained soil (sand and gravel) while soil at Lenggong consist of a mixture from fine (clay and silt) and coarse (sand and gravel) grained soil. For both sites, individual soil particle quantity obtained from the particle distribution test was analyzed and presented in Table 1 and 2. It was found that, soil moisture content ($w$) for Gravelly SAND (A = 6.62 % & B = 5.50 %) was lower than Silty SAND (A = 13.03 % & B = 14.27 %) as shown in Table 1 and 2. As stated by [23], electrical current may propagate in geomaterial via the process of electrolysis where the current was carried by ions at a comparatively slow rate. The application of field electrical resistivity has theoretically stated that the water content in subsurface materials has a close positive correlation with the electrical conductivity [24]. The variation of soil moisture content for Gravelly SAND and Silty SAND may due to the variation of porosity derived from different particle grained fraction between both soils thus influencing the variation of ERV. Silty SAND has great ability to absorb and retained more water due to its high composition of fine grained soil (clay and silt). Moreover, high composition of fine grained soil may produce low permeability thus has higher capacity to retained more water in order to produced high soil moisture content. In contrast, Gravelly SAND has lower ability to absorbed more water due to its high coarse grained particle (sand and gravel) which easily permit the water to seep thru (higher permeability) underneath. This study found that Gravelly SAND (A = 0.319 & B = 0.362) has more soil porosity compared to the Silty SAND (A = 0.271 & B = 0.256). This condition may due to the soil composition such as Gravelly SAND composed more coarse grained soil while Silty SAND consists of a mixture from fine and coarse grain soil. Soil porosity will decreased due to the high existing of fine grained soil since more void was filled by small grain soil as experienced by Silty SAND. This study has revealed that the ERV of Gravelly SAND was greater than Silty SAND due to the low soil moisture content and high air filled void. As reported by [25] – [32], air filled void posses a higher resistivity value compared with the water filled void. Higher soil moisture content and fine grained soil (clay and silt) has influenced the lower ERV of Silty SAND compared to the Gravelly SAND ERV. Soil's electrical resistivity value generally varies inversely proportional to the water content and dissolved ion concentration as clayey soil exhibit high dissolved ion composition, lowest resistivity of all soil materials was a wet clayey soils while dry sand, coarse and gravel deposits and massive bedded and hard bedrocks have the highest ERV [4]. Based on Table 1 and 2, it was found that soil density of Silty SAND was higher than Gravelly SAND. This condition may due to the high soil moisture content within Silty SAND compared to the Gravelly SAND. Hence, this finding has revealed that high soil density cannot easily interpreted as granular soil since a mixture of fine and granular soil will also produced greater density with the existing of high quantity of soil moisture content.

This study has demonstrated that the electrical resistivity method proves to have good prospects in soil identification determination. Prediction of soil types can easily be determined at field which enables the engineers in fast, economically and able to obtain more data. This study also can contribute to the enhancement of anomaly interpretation which is traditionally too much relative based on stand-alone qualitative approach. According to [14] and [33] methods are insufficient to stand alone in order to provide solutions to particular problems. This technique was applicable to assist and improve the confidence level of conventional electrical resistivity anomaly interpretation due to its ability to verify type of soils interpreted based on quantification of geotechnical properties. Geotechnical property quantification is an important factor for geophysical method used in engineering application [34]. Hence, the reliability of traditional anomaly interpretation and conclusion can be increased using supported additional geotechnical data with particular reference to soil moisture content, grain size fraction, void ratio and porosity, density, etc.

**Table 1.** Extracted ERV, soil grain fraction and moisture content properties of Gravelly SAND.

| Soil Sample | A | | | | B | | | |
|---|---|---|---|---|---|---|---|---|
| Electrical resistivity, $\rho$ ($\Omega$m) | 278 | | | | 285 | | | |
| Moisture content, *w* (%) | 6.62 | | | | 5.50 | | | |
| Bulk Density, $\rho_{Bulk}$ (Mg/m$^3$) | 1.77 | | | | 1.67 | | | |
| Dry Density, $P_{Dry}$ (Mg/m$^3$) | 1.66 | | | | 1.58 | | | |
| Void ratio, e | 0.468 | | | | 0.566 | | | |
| Porosity, $\eta$ | 0.319 | | | | 0.362 | | | |
| Particle size analysis, d (µm – mm, %) | Clay | Silt | Sand | Gravel | Clay | Silt | Sand | Gravel |
| | 0.00 | 4.24 | 67.85 | 27.10 | 0.00 | 3.35 | 65.36 | 30.41 |

**Table 2.** Extracted ERV, soil grain fraction and moisture content properties of Silty SAND.

| Soil Sample | A | | | | B | | | |
|---|---|---|---|---|---|---|---|---|
| Electrical resistivity, $\rho$ ($\Omega$m) | 223 | | | | 199 | | | |
| Moisture content, *w* (%) | 13.03 | | | | 14.27 | | | |
| Bulk Density, $\rho_{Bulk}$ (Mg/m$^3$) | 1.90 | | | | 1.96 | | | |
| Dry Density, $P_{Dry}$ (Mg/m$^3$) | 1.68 | | | | 1.72 | | | |
| Void ratio, e | 0.372 | | | | 0.345 | | | |
| Porosity, $\eta$ | 0.271 | | | | 0.256 | | | |
| Particle size analysis, d (µm – mm, %) | Clay | Silt | Sand | Gravel | Clay | Silt | Sand | Gravel |
| | 7.95 | 24.25 | 43.78 | 24.02 | 7.05 | 21.20 | 51.42 | 20.33 |

## 4. Conclusion

Electrical resistivity method was successfully performed for field soil identification. Conventional electrical resistivity anomaly interpretation for field soil identification purpose will increasingly meaningful due to the supported additional basic soil properties verification. Electrical resistivity value was relatively influenced by the variation of basic geotechnical properties related to soil moisture content, grain size fraction, porosity, density, etc. Hence, this study has shown that the application of field electrical resistivity method together with supported basic soil properties data was applicable to identify type of soils which efficient in term of cost, time and more data coverage compared to the previous conventional method. Furthermore, electrical resistivity technique used surface method which able to eliminate site damageability thus contributes to the sustainable and green environment in construction industry.

**Acknowledgment**
This research was funded by Universiti Tun Hussein Onn Malaysia (E040) and Universiti Sains Malaysia. First author wish to acknowledge gratefully to supervisors and research members for their tremendous guidance, work and cooperation.

**References**

[1] Khatri R, Shrivastava V K and Chandak R 2011 Correlation between vertical electric sounding and conventional methods of geotechnical site investigation *Int. Journal of Advanced Engineering Sciences and Technologies*. **4** 042-053

[2] Pozdnyakova A and Pozdnyakova L 2002 Electrical fields and soil properties. *Proceedings of 17th World Congress of Soil Science*. **53** 1-11

[3] Osman F I S and Osman, S B A S 2012 Integrating geo-electrical and geotechnical data for soil characterization. *International Journal of Applied Physics and Mathematics*. **2** 104-106

[4] Liu C and Evett J B 2008 *Soils and Foundation* (New Jersey: Pearson International)

[5] Godio A, Strobbia C and De Bacco G 2005 Geophysical characterisation of a rockslide in an alpine region *Engineering Geology*. **83** 273-286

[6] Cosenza P, Marmet E, Rejiba F, Jun Cui Y, Tabbagh A and Charlery Y 2006 Correlations between geotechnical and electrical data: A case study at Garchy in France *Journal of Applied Geophysics*. **60** 165-178

[7] Mauritsch H J, Seiberl W, Arndt R, Romer A, Schneiderbauer K and Sendlhofer G P 2000 Geophysical investigations of large landslides in the Carnic Region of Southern Austria *Engineering Geology*. **56** 373-388

[8] Saad R, Muztaza N M and Mohamad E T 2011 The 2D Electrical Resistivity Tomography (ERT) Study for Civil and Geotechnical Engineering Purposes *Electronic Journal of Geotechnical Engineering*. **16** 1537-1545

[9] Chik Z and Islam S M T 2012 Finding Soil Particle Size through Electrical Resistivity in Soil Site Investigations *Electronic Journal of Geotechnical Engineering*. **17** 1867-1876

[10] Abidin M H Z, Wijeyesekera D C, Saad R and Ahmad F 2013 The Influence of Soil Moisture Content and Grain Size Characteristics on its Field Electrical Resistivity *Electronic Journal of Geotechnical Engineering*. **18/D** 699-705

[11] Abidin M H Z, Saad R, Ahmad F, Wijeyesekera D C and Baharuddin M F T 2014. Correlation Analysis between Field Electrical Resistivity Value and Basic Geotechnical Properties. *Soil Mechanics and Foundation Engineering*. **51** 117-125

[12] Solberg I L, Hansen L, Rønning J S, Haugen E D, Dalsegg E and Tønnesen J 2011 Combined Geophysical and Geotechnical Approach to Ground Investigations and Hazard Zonation of a Quick Clay Area, Mid Norway *Bulletin of Engineering Geology and the Environment*. **71** 119-133

[13] Fraiha S G C and Silva J B C 1994 Factor analysis of ambiguity in geophysics, *Geophysics*. **59** 1083-1091

[14] Benson R C, Yuhr L and Kaufmann R D 2003 Some Considerations for Selection and

Successful Application of Surface Geophysical Methods *Proc. The 3rd Int. Conf. on Applied Geophysics (Orlando, Florida. 2003)*
[15] Clayton C R I, Matthews M C and Simons N E 1995 *Site Investigation* (UK: Blackwell Science Ltd)
[16] British Standard 1377 1990 *Methods of test for Soils for Civil Engineering Purposes*
[17] Loke M H, Acworth I and Dahlin T 2003 A comparison of smooth and blocky inversion methods 2-D electrical imaging surveys. *Exploration Geophysysics.* **34** 182–187
[18] Loke M H and Barker R D 1996 Rapid least squares inversion of apparent resistivity pseudosection using a quasi-Newton method. *Geophysical Prospecting* **44(3)** 131–152
[19] Burger H R, Sheehan A F and Jones C H 2006 *Introduction to Applied Geophysics* (New York: W.W. Norton & Company)
[20] Telford W M, Geldart L P and Sheriff R E 1976 *Applied Geophysics* (Cambridge: Cambridge University Press)
[21] Lee T S 2002. *Slope Stability and Stabilization Methods* (New York: John Wiley & Sons, Inc.)
[22] Griffiths D H and King R F 1981 *Applied Geophysics for Geologist and Engineers-The Element of Geophysical Prospecting* (Oxford: Pergamon Press)
[23] Telford W M, Geldart L P and Sheriff R E 1976 *Applied Geophysics* (Cambridge: Cambridge University Press)
[24] Abidin M H Z, Saad R, Ahmad F, Wijeyesekera D C and Baharuddin M F T 2012 Integral Analysis of Geoelectrical (Resistivity) and Geotechnical (SPT) Data in Slope Stability Assessment *Academic Journal of Science.* **1** 305-316
[25] Jusoh Z 2010 Application of 2-D resistivity imaging and seismic refraction technique in subsurface investigation for civil engineering, *M.S. thesis*, Sch. Physics, Science Univ., Penang, Malaysia
[26] Abidin M H Z, Ahmad F, Wijeyesekera D C, Saad R and Baharuddin M F T 2013 Soil resistivity measurements to predict moisture content and density in loose and dense soil *Applied Mechanics and Materials.* **353-356** 911-917
[27] Abidin M H Z, Wijeyesekera D C Ahmad F, Saad R and Baharuddin M F T 2013 Correlation of soil box resistivity teston a different moisture content and density for silty sand and gravelly sand *Information Engineering Latters.* **3(2)** 1-10
[28] Abidin M H Z, Saad R, Ahmad F, Wijeyesekera D C and Yahya A S 2013. Soil moisture content and density prediction using laboratory resistivity experiment. *International Journal of Engineering and Technology*. **5(6)** 731-735
[29] Abidin M H Z, Saad R, Ahmad F, Wijeyesekera D C and Yahya A S 2013. The influence of soil moisture content and density on its small embankment electrical resistivity *Recent Advanced in Civil and Mining Engineering*: *Proc. of 4th European Conference of Civil Engineering (Antalya, Turkey, 8-10 October 2013)* pp 166-174
[30] Abidin M H Z, Saad R, Wijeyesekera D C and Ahmad F 2013 Soil resistivity influence due to the different utilization of electrical resistivity array *Electronic Journal of Geotechnical Engineering.* **18/X** 5643-5654
[31] Abidin M H Z, Ahmad F, Wijeyesekera D C, Saad R and Baharuddin M F T 2014 Small soil embankment electrical resistivity value on its array, moisture content and density influences *International Journal of Geology.* **8** 9-18
[32] Abidin M H Z, Ahmad F, Wijeyesekera D C and Saad R 2014 The influence of basic physical properties of soil on its soil electrical resistivity valueunder loose and dense condition *Journal of Physics: Conference Series.* **49** 1-13
[33] Fraiha S G C and Silva J B C 1994 Factor analysis of ambiguity in geophysics. *Geophysics.* **59** 1083 - 1091
[34] Kibria G and Hossain M 2012 Investigation of Geotechnical Parameters Affecting Electrical Resistivity of Compacted Clays. *Journal of Geotechnical and Geoenvironmental Engineering*, **452** 1520-1529

# Efficient evaluation of the sample variance of an interval-valued dataset

**Michal Černý**
Department of Econometrics, University of Economics Prague, Winston Churchill Square 4, 13067 Prague, Czech Republic

E-mail: cernym@vse.cz

**Abstract.** Given a set of interval-valued data, a general problem is to compute bounds for a particular statistic, such as sample mean or variance, variation coefficient or entropy. It is well known that computation of the upper bound of sample variance is an NP-hard problem. Here we consider a variant of an algorithm by Ferson et al., which is exponential in the worst case, and investigate its behavior under a natural probabilistic model. A simulation study shows that the undesirable case, which forces the algorithm to work in exponential time (and which appears in the proof of NP-hardness), occurs very rarely in an environment when the interval data are generated by probabilistic processes which are natural from a statistical viewpoint. The main finding is that the the algorithm is practically very efficient and that the NP-hardness result usually "does not matter too much".

## 1. Introduction

Let $x_1, \ldots, x_n$ denote a set of one-dimensional data. Often we face the problem that the dataset cannot be observed exactly. What we observe instead of $x_1, \ldots, x_n$ is a family of intervals $[\underline{x}_i, \overline{x}_i]$, $i = 1, \ldots, n$, such that it is guaranteed that

$$\underline{x}_i \leq x_i \leq \overline{x}_i, \quad i = 1, \ldots, n. \tag{1}$$

We often meet interval datasets in practice. Example include:

- A real-valued data point $x$ is affected by rounding to the nearest integer $\widetilde{x}$; then we cannot observe the true value $x$, but only the interval $[\widetilde{x} - \frac{1}{2}, \widetilde{x} + \frac{1}{2}] \ni x$. The same problem occurs in scientific computing whenever we represent rational numbers by data types of fixed size.
- Another situation where we need to handle datasets of the form (1) is when data suffer from imprecision. Measurement devices sometimes guarantee the measurement precision in the form $\widetilde{x} \pm \Delta$; that is, when the device reports the value $\widetilde{x}$, then we only know that the true value $x$ lies in the interval $[\widetilde{x} - \Delta, \widetilde{x} + \Delta]$.
- Further examples are encountered in econometrics, where we often need to work with interval predictions of future values of economic quantities (such as inflation or interest rates) or expert estimates, which are also often of interval nature.
- Another example is when we observe only daily mins/maxs of a continuous random process (which is the case of financial data). Let $x(t)$ be a trajectory of a random process with time $t \geq 0$, representing the price of an asset, say. We are often reported only daily minimum $\underline{x}$ and maximum $\overline{x}$. Then we only know that the value $x(t)$ lies in $[\underline{x}, \overline{x}]$.

Journal of Physics: Conference Series **622** (2015) 012031 doi:10.1088/1742-6596/622/1/012031

- In general, data subject to categorization, censoring or dicrestization can be usually interpreted as intervals of possible values.
- Another example is natural in physics or chemistry: "constants" are rarely constant, e.g. the gravity acceleration constant often should be treated as an interval of possible values (since the true value is not constant, depending on the position on the Earth).

## 2. The main question

We are interested in computation of a particular characteristic of the dataset $x_1, \ldots, x_n$, such as sample mean or variance. But this cannot be done directly since we observe only the intervals satisfying (1). Let $S(x_1, \ldots, x_n)$ be a statistic (or, generally, a continuous function). We need to replace $S$ by another statistic $S^*(\underline{x}_1, \overline{x}_1, \ldots, \underline{x}_n, \overline{x}_n)$ giving us similar information to $S$.

**Example.** Let $x_1, \ldots, x_n$ be sampled from $N(\mu, \sigma^2)$, where the parameters $\mu$ and $\sigma^2$ are unknown and are to be estimated. The assumption of a particular distribution allows us to write down the likelihood function and construct the max-likelihood estimator

$$(\widehat{\mu}, \widehat{\sigma}) = \arg\max_{\substack{\mu\in\mathbb{R}\\ \sigma\geq 0}} \prod_{i=1}^{n} \left[\Phi\left(\frac{\overline{x}_i - \mu}{\sigma}\right) - \Phi\left(\frac{\underline{x}_i - \mu}{\sigma}\right)\right],$$

where $\Phi$ stands for the cumulative distribution function of $N(0, 1)$. □

Another approach, complementary to the max-likelihood construction of the Example, is the possibilistic approach; for details see e.g. [1]. This approach does not rely on the assumption of a particular form of the underlying distribution. It simply considers an interval of *all possible* values of the statistic $S$, regardless of the underlying distribution (which is often unknown). So, the possibilistic version of the statistic $S$ is the interval $[\underline{S}, \overline{S}]$, where

$$\overline{S} = \max\{S(x_1, \ldots, x_n) : \underline{x}_i \leq x_i \leq \overline{x}_i,\ i = 1, \ldots, n\}, \tag{2}$$
$$\underline{S} = \min\{S(x_1, \ldots, x_n) : \underline{x}_i \leq x_i \leq \overline{x}_i,\ i = 1, \ldots, n\}. \tag{3}$$

Now the assumption (1) implies that $S \in [\underline{S}, \overline{S}]$. For example, when $S$ is the sample mean $\widehat{\mu} = \frac{1}{n}\sum_{i=1}^{n} x_i$, then we get the interval

$$[\underline{\widehat{\mu}}, \overline{\widehat{\mu}}] = \left[\frac{1}{n}\sum_{i=1}^{n} \underline{x}_i, \frac{1}{n}\sum_{i=1}^{n} \overline{x}_i\right].$$

In this text we will restrict ourselves to the case when $S$ is the sample variance:

$$S(x_1, \ldots, x_n) = \frac{1}{n-1}\sum_{i=1}^{n}\left(x_i - \frac{1}{n}\sum_{j=1}^{n} x_j\right)^2.$$

It is easily seen that computation of $\underline{S}$ is easy, since the optimization problem (3) reduces to convex quadratic minimization, which is an efficiently solvable problem (see e.g. Vavasis [7]). But now we arrive at the main problem, which has been proved in [2], [4]:

**Theorem 1.** *For every fixed $\delta \geq 0$ it is NP-hard to compute a value $\Sigma$ such that $|\overline{S} - \Sigma| \leq \delta$.* □

Theorem 1 tells us that the value $\overline{S}$ cannot be computed efficiently, even if we do not insist on an exact value but we are allowed to compute it only approximately with a prescribed absolute error $\delta$. So we can expect only algorithms working in time $\approx 2^n$ or worse. The good news is that we can indeed do it in time $2^n$. The geometry of the optimization problem (2) shows that we are to maximize a convex function over the box $[\underline{x}_1, \overline{x}_1] \times \cdots \times [\underline{x}_n, \overline{x}_n]$. It follows that a maximizer is in a vertex. Thus the $2^n$ algorithm works as follows:

$$\overline{S} = \max\{S(x_1, \dots, x_n) : x_i \in \{\underline{x}_i, \overline{x}_i\},\ i = 1, \dots, n\}. \tag{4}$$

However, when $n$ is larger than 50 (say), the approach (4) is useless. So the main question is: what to do then?

## 3. A practically useful approach

The situation with computation of $\overline{S} = \max\{\frac{1}{n-1}\sum_{i=1}^{n}\left(x_i - \frac{1}{n}\sum_{j=1}^{n} x_j\right)^2 : \underline{x}_i \le x_i \le \overline{x}_i,\ i = 1, \dots, n\}$ is not as bad as it might look from Theorem 1.

For an interval $\boldsymbol{x} = [\underline{x}_i, \overline{x}_i]$ we define its center point and radius

$$x^C = \tfrac{1}{2}(\overline{x} + \underline{x}), \quad x^\Delta = \tfrac{1}{2}(\overline{x} - \underline{x}).$$

and denote $\boldsymbol{x} = [x^C \pm x^\Delta]$.

Now the interval dataset $[\underline{x}_1, \overline{x}_1], \dots, [\underline{x}_1, \overline{x}_1]$, can be denoted as $\boldsymbol{x}_1 = [x_1^C \pm x_1^\Delta], \dots, \boldsymbol{x}_n = [x_n^C \pm x_n^\Delta]$.

Ferson et al. [4] proved the following result:

**Theorem 2.** *Let $k \in \{2, \dots, n\}$ be a number such that for every $I \subseteq \{1, \dots, n\}$ such that $|I| = k$ we have*

$$\bigcap_{i \in I} [x_i^C \pm \tfrac{1}{n} x_i^\Delta] = \emptyset. \tag{5}$$

*Then $\overline{S}$ can be computed in time $O(n^2 2^k)$.* □

The computation time $O(n^2 2^k)$ can be expected to be much better than $2^n$ in (4), since for datasets appearing in practice it will often be $k \ll n$. However, there are "extremal" examples: when all intervals $\boldsymbol{x}_1, \dots, \boldsymbol{x}_n$ share a common point (which can happen, for example, when $x_1^C = x_2^C = \cdots = x_n^C$), then the condition (5) is satisfied with no $k$.

So, when analyzing a particular dataset, a good approach is to find

$$k^* = \min\left\{k : \bigcap_{i \in I} [x_i^C \pm \tfrac{1}{n} x_i^\Delta] = \emptyset,\ I \subseteq \{1, \dots, n\},\ |I| = k\right\}$$

and assess whether the computation time $n^2 2^{k^*}$ is acceptable; often it will be the case.

**Remark.** Observe that computation of $k^*$ is an easily solvable problem: it suffices to sort the $(2n)$-tuple of numbers $a_1 := x_1^C - \frac{1}{n}x^\Delta, a_2 := x_1^C + \frac{1}{n}x^\Delta, \dots, a_{2n-1} := x_n^C - \frac{1}{n}x^\Delta, a_{2n} := x_n^C + \frac{1}{n}x^\Delta$ into the form $a_{\pi(1)} \le a_{\pi(2)} \le \cdots \le a_{\pi(2n)}$, where $\pi$ is a permutation, and for each interval $[a_{\pi(i)}, a_{\pi(i+1)}]$ to check in how many intervals $[x_i^C \pm \frac{1}{n}x_i^\Delta]$ it belongs.

The crucial parameter affecting computational complexity is $k^*$. The main message of this paper is: though the algorithm of Theorem 2 is exponential in the worst case, in practice we usually find out that $k^*$ is reasonably low, so that the factor $2^{k^*}$ "does not matter too much". This is what we will do in the next section.

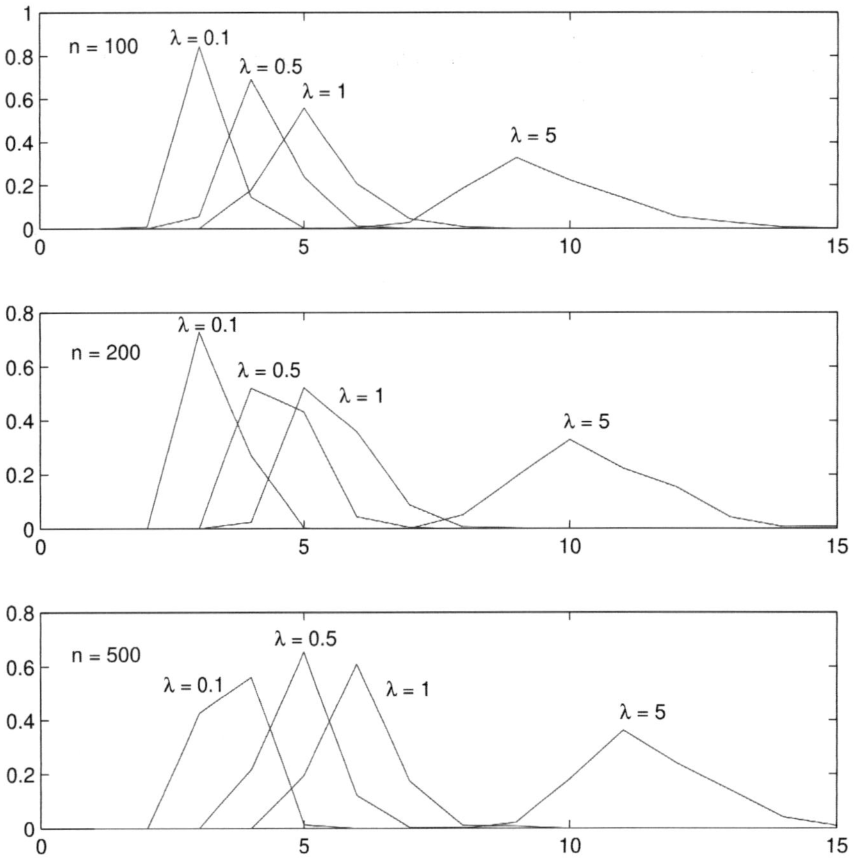


**Figure 1.** Distribution of $k^*$ for $n \in \{100, 200, 500\}$ and $\lambda \in \{0.1, 0.5, 1, 5\}$.

## 4. A simulation

We will illustrate how $k^*$ behaves in a quite natural model: we select

$$x_i^C \sim N(0, \sigma^2), \quad x_i^\Delta \sim \mathrm{Exp}(\kappa), \quad i = 1, \ldots, n. \tag{6}$$

It seems reasonable to measure $k^*$ as a function of $n$ and

$$\lambda = \frac{\mathrm{var}(x_i^\Delta)}{\mathrm{var}(x_i^C)} = \frac{\kappa}{\sigma}.$$

**Remark.** Though we select the data generating process in the particular form (6), further simulations (not presented here) show similar behavior also for other distributions with bounded variance.

The simulated distributions of $k^*$ for $n = 100, 200, 500$ and $\lambda = 0.1, 0.5, 1, 5$ are depicted in Figure 1. It is not surprising that the average value of $k^*$ grows with both $n$ and $\lambda$. But the

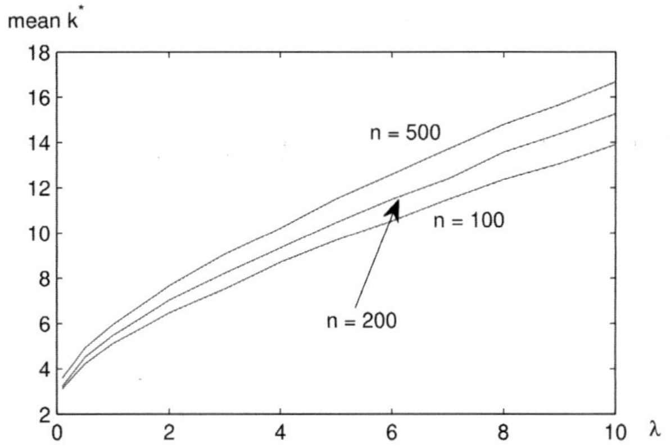


**Figure 2.** Estimated mean value of $k^*$ for $n \in \{100, 200, 500\}$ and $\lambda \in [0.1, 10]$.

main message is that in the selected setup we encounter values of $k^*$ at most 13 (say), and $2^{13} = 8\,192$, which is still an acceptable computation time.

We also plot the (simulated) mean value $k^*$ as a function of $\lambda$ in Figure 2.

## 5. Further results

As far the author is aware, the algorithm of Theorem 2 is the best practically useful tool for computation of $\overline{S}$. However, there are two more results which are complementary: the algorithm by Dantsin et al. [3], which computes $\overline{S}$ in polynomial time under the condition that no interval $[x_i^C \pm \frac{1}{n} x_i^\Delta]$ is a proper subinterval of another interval $[x_j^C \pm \frac{1}{n} x_j^\Delta]$. Many datasets appearing in practice fulfill this condition.

And finally we should mention the pseudopolynomial algorithm of Černý and Hladík [2], which is applicable under the condition that all numbers $\underline{x}_1, \overline{x}_1, \ldots, \underline{x}_n, \overline{x}_n$ are integers. Then, the pseudopolynomial algorithm works in time $O(n^3 M^3)$, where

$$M = \max\{|\underline{x}_1|, |\overline{x}_1|, \ldots, |\underline{x}_n|, |\overline{x}_n|\}.$$

So, this algorithm is polynomial when $M$ can be bounded by a polynomial in $n$. From a practical viewpoint, the pseudopolynomial algorithm is fast when the dataset does not contain excessively large numbers.

## 6. Conclusions

Though the problem of computation of $\overline{S}$ is NP-hard and inapproximable in general, we have seen that with a very natural probabilistic data-generating model (6) we almost do not encounter a hard instance which would require an excessively high computation time. This is good news for practice since the problem of computation of $\overline{S}$ seems to be practically feasible, at least in many cases. The most tempting question for further research is: given a probabilistic data-generating model, how to estimate the probability that $k^*$ attains a high value? This question makes a bridge to graph theory. It can be reformulated as follows. Let $\boldsymbol{x}_i$, $i = 1, \ldots, n$, be random intervals. Consider the interval graph $G$ determined by the dataset; that is, let $\boldsymbol{x}_i$, $i = 1, \ldots, n$, be the vertices, and $\boldsymbol{x}_i$ and $\boldsymbol{x}_j$ $(i \neq j)$ are connected with an edge iff $\boldsymbol{x}_i \cap \boldsymbol{x}_j \neq \emptyset$. How could we bound the probability that $G$ has a large clique (under some reasonable assumptions, say that the random centers and random radii of the intervals are iid with finite variance)? If an estimate would show that the probability is reasonably small, we would have a proof that the algorithm of Theorem 2 works "fast" on average, even though in the worst case it still can be exponential.

## Acknowledgment
The work was supported by the Czech Science Foundation under Project No. P402/12/G097.

## References
[1] Černý M, Antoch J and Hladík M 2013 *Inform. Sci.* **244** 23–47
[2] Černý M and Hladík M 2014 *Comp. Stat. Data Anal.* **80** 26–43
[3] Dantsin E, Kreinovich V, Wolper A and Xiang G 2006 *Reliab. Comput.* **12** 273–80
[4] Ferson S, Ginzburg L, Kreinovich V, Longpré L and Aviles M 2005 *Reliab. Comput.* **11** 207–33
[5] Horowitz J L, Manski C F, Ponomareva C F and Stoye J 2003 *Reliab. Comput.* **9** 419–40
[6] Kreinovich V, Longpré L, Patangay P, Ferson S and Ginzburg L 2005 *Reliab. Comput.* **11** 59–76
[7] Vavasis S A 1991 *Nonlinear Optimization: Complexity Issues* (Oxford: Oxford University Press)

# Mathematical analysis of the computational complexity of integer sub-decomposition algorithm

**Ruma Kareem K. Ajeena**[1] **and Hailiza Kamarulhaili**[2]

[1,2] School of Mathematical Sciences, Universiti Sains Malaysia, 11800, Penang, Malaysia

E-mail: ruma.usm@gmail.com & hailiza@cs.usm.my

**Abstract.** In this paper, the computational complexity to compute a scalar multiplication on classes of elliptic curve over a prime field that have efficiently-computable endomorphisms is analyzed mathematically. This scalar multiplication is called integer sub-decomposition (ISD) method, which is based on the GLV method of Gallant, Lambert and Vanstone that was initially proposed in the year 2001. In this work, the mathematical proof of the computational cost of ISD implementation is presented. The computational cost of ISD algorithm has been proven based on the operations counting. Two types of the operations are used, namely, elliptic curve operations represented by elliptic curve point addition $A$ and elliptic curve point doubling $D$ and finite field operations represented by field inversion $I$, field multiplication $M$ and field squaring $S$ that design the computation of the running time of ISD scalar multiplication.

**Key words and phrases:** Elliptic curves, scalar multiplication, efficiently-computable endomorphisms, integer sub-decomposition, computational complexity.

## 1. Introduction

The attractive features of elliptic curves history awarded it studying by mathematicians over a hundred of years to solve a variety of problems. The entry of these curves into cryptography independently by Neal Koblitz [1] and Victor Miller [2] in 1985 who suggested elliptic curve public key cryptosystems. The elliptic curves performance has active importance in the security level as a traditional asymmetric cryptosystem, such as RSA [3],[4]. The fundamental step of elliptic curve cryptosystems is to compute elliptic curve scalar multiplication $kP$ for a point $P$ which has a large prime order $n$. To accomplish this end, various methods have been innovated, adopting on elliptic curves $E$ over finite fields. A group of methods cleverly employs a distinguished endomorphism $\psi \in End(E)$ to split a large computation into a sequence of cheaper ones, so that the overall computational cost will be lowered [3].

Recently, Gallant, Lambert and Vanstone [5],[6],[7] used such a technique that, contrary to the previous ones, also applied to special curves defined over large prime fields. Their method uses an efficiently computable endomorphism $\psi \in End(E)$ to rewrite $kP$ as

$$kP = k_1P + k_2\psi(P), \text{with } max\{|k_1|, |k_2|\} = O(\sqrt{n}). \tag{1.1}$$

Their key point is an algorithm, that will be called the GLV method, which inputs integers $n$ and $\lambda \in [1, n-1]$ and produces for any $k$ $(mod\ n)$, two residues $k_1$ and $k_2$ $(mod\ n)$ such that

$$k = k_1 + \lambda k_2\ (mod\ n). \tag{1.2}$$

Starting with analyzing the GLV method of Gallant, Lambert and Vanstone, our study uses two fast endomorphisms with minimal polynomials $p_j(X) = X - \lambda_j \ (mod\ n)$, for $j = 1, 2$ to compute any multiple $kP$ of a point $P$ of order $n$ lying on an elliptic curve. When both values or one of them is not bounded by $\pm\sqrt{n}$, the value $k$ is then decomposed into the values $k_1$ and $k_2$. The sub-decomposition from $k = k_1 + k_2\lambda \ (mod\ n)$ is shown clearly as follows:

$$k_1 = k_{11} + k_{12}\lambda_1 \ (mod\ n) \quad \text{and} \quad k_2 = k_{21} + k_{22}\lambda_2 \ (mod\ n). \tag{1.3}$$

We calculate, in particular, the integer sub-decomposition (ISD) as follows:

$$\begin{aligned} kP &= k_{11}P + k_{12}[\lambda_1]P + k_{21}P + k_{22}[\lambda_2]P \\ &= k_{11}P + k_{12}\psi_1(P) + k_{21}P + k_{22}\psi_2(P). \end{aligned} \tag{1.4}$$

where $-C\ \sqrt{n} < k_{11}, k_{12}, k_{21}, k_{22} < C\ \sqrt{n}$, where $C > 1$ as proven in [9].

The mathematical proofs of the computational complexity to compute a scalar multiplication $kP$ on elliptic curve over a finite field $F_p$ are determined in this paper. This determination depends on the elliptic curve operations and finite field operations that design the computation of the running time of scalar multiplication. The fast computation of addition and doubling on an elliptic curve indicates an efficient scalar multiplication algorithm that is performed based on these sped-up computations. On the other hand, using endomorphism applications can be applied to improve performance in elliptic curve scalar multiplication, such as in the GLV decomposition and ISD sub-decomposition of a scalar $k$ in a scalar multiplication $kP$.

The elliptic curve model and coordinate model used as well as the algorithm applied for computing a scalar multiplication $kP$ consider as the three principal factors to determine the computational cost of a scalar multiplication $kP$. The outline of this paper shows: Section 2 gives a summary of the mathematical background to clarify elliptic curve $E$ over prime field and endomorphisms on it. Section 3 reviews the procedure of ISD scalar multiplication method. Section 4 presents the mathematical proof of the computational complexity of ISD algorithm that includes the mathematical proofs of all sub-algorithms which form ISD method through two types of operations, namely, elliptic curve operations and finite field operations. Finally, Section 5 draws the concluding remarks.

## 2. Mathematical Background

### *2.1. Elliptic Curves over Prime Fields*

**Definition 2.1.** *Let $p \neq 2, 3$. An elliptic curve $E(F_p)$ over $F_p$, is defined by an equation of the form [8]:*

$$E : Y^2 = X^3 + AX + B \ (mod\ p), \tag{2.5}$$

*where $A, B \in F_p$. The curve $E$ is said to be non-singular if it has no double zeroes, that means the discriminant $D_E = 4A^3 + 27B^2 \neq 0 \ (mod\ p)$.*

### *2.2. Endomorphisms of Elliptic Curve over Prime fields*

Assume that $E$ is an elliptic curve defined over the finite field $F_p$. The point at infinity is denoted by $O_E$. The set of $F_p-$rational points on $E$ forms the group $E(F_p)$. A rational map $\psi : E \rightarrow E$ satisfies $\psi(O_E) = O_E$ dubbed an endomorphism of $E$. The endomorphism $\psi$ will be defined over $F_q$ where $q = p^n$, if the rational map is defined over $F_q$. Therefore, clearly, for any $n \geq 1$, $\psi$ is a group homomorphism of $E(F_p)$ and also of $E(F_q)$ [3] and [8].

**Definition 2.2.** *The endomorphism of elliptic curve $E$ defined over $F_q$ is the $m-$ multiplication map $[m] : E \rightarrow E$ defined by*

$$P \rightarrow mP \tag{2.6}$$

*for each $m \in Z$. The negation map $[-1] : E \rightarrow E$ defined by $P \rightarrow -P$ is a special case from $m-$multiplication map [3].*

**Theorem 2.3.** *(Hasse Theorem). Let $E$ be an elliptic curve over a finite field $F_p$ [3]. Then, the order of $E(F_p)$ satisfies*

$$|p + 1 - \#E(F_p)| \leq 2\sqrt{p}. \quad (2.7)$$

**Definition 2.4.** *The rectangle norm [4] of $(x, y)$ is defined by $max\{|x|, |y|\}$. We denote it by $|(x, y)|$.*

## 3. Integer sub-decomposition (ISD) method

The idea of GLV method is the main source on which the ISD method depends to obtain an efficient scalar multiplication on an ordinary elliptic curve $E$ defined in equation (2.5). This method primarily aims to sub-decompose the values $k_1$ and $k_2$ when one or both values are not bounded by $\pm\sqrt{n}$. Decomposing a scalar $k \in [1, n-1]$ in a scalar multiplication $kP$ gives two new scalars $k_1$ and $k_2$. According to GLV idea the values of $k_1$ and $k_2$ satisfy the main GLV condition, that is $max\{|k_1|, |k_2|\} \leq \sqrt{n}$ on the interval $[1, n-1]$. However, generically this decomposition produces also new scalars $k_1$ and $k_2$ lie outside the range $\sqrt{n}$ on the same interval $[1, n-1]$. In other words, there are other scalars $k_1$ and $k_2$ on the same interval $[1, n-1]$ satisfy $max\{|k_1|, |k_2|\} > \sqrt{n}$. But with the last condition, the GLV idea to compute $kP$ cannot work. So, this gap considers as a main reason to sub-decompose $k_1$ and $k_2$ that lie outside the range $\sqrt{n}$ through our proposed method called Integer sub-decomposition (ISD) that is based on new sub-scalars $k_{11}, k_{12}$ and $k_{21}, k_{22}$ of scalars $k_1$ and $k_2$ respectively. The sub-decomposition is expressed by these formulas:

$$k_1 = k_{11} + k_{12}\lambda_1 (mod\ n) \text{ and } k_2 = k_{21} + k_{22}\lambda_2 (mod\ n),$$

as shown in Equation (1.3). To accomplish sub-decomposition, one should first find a GLV generator $\{v_1, v_2\}$ by using a GLV generator algorithm in [7] for a given $n$ and $\lambda$, where $n$ is a large prime order of elliptic curve point $P$ and $\lambda$ is a root of the characteristic polynomial of endomorphism $\psi$ of $E$. Consequently, $k \in [1, n-1]$ is decomposed into $k_1$ and $k_2$. This decomposition can be performed using the balanced length-two representation of a multiplier $k$ algorithm in [3]. Our proposed algorithm can then be used to generate the ISD generators $\{v_3, v_4\}$ and $\{v_5, v_6\}$ such that each component of $v_3, v_4$ and $v_5, v_6$ is bounded by $\sqrt{n}$ and relatively prime to each other. These generators can be easily computed by solving the shortest vector problem in a lattice that is involved in using an extended Euclidean algorithm in [5]. $k_1$ and $k_2$ can be decomposed again into integers $k_{11}, k_{12}$ and $k_{21}, k_{22}$ which means that the sub-decomposition of $k$ is as follows:

$$k = k_{11} + k_{12}\lambda_1 + k_{21} + k_{22}\lambda_2 (mod\ n) \quad (3.8)$$

with $-C\sqrt{n} < k_{11}, k_{12}, k_{21}, k_{22} < C\sqrt{n}$ from any ISD generators $\{v_3, v_4\}$ and $\{v_5, v_6\}$. Finally, the scalar multiplication $kP$ can be computed as follows.

$$\begin{aligned} kP &= k_{11}P + k_{12}[\lambda_1]P + k_{21}P + k_{22}[\lambda_2]P \\ &= k_{11}P + k_{12}\psi_1(P) + k_{21}P + k_{22}\psi_2(P). \end{aligned}$$

as shown in Equation (1.4). This formula demonstrates our proposal which includes computing two endomorphisms, namely, $\psi_1(P) = \lambda_1 P$ and $\psi_2(P) = \lambda_2 P$, where $P \in E(F_q)$, $\lambda_1, \lambda_2 \in [1, n-1]$ and $\lambda_1 \neq \pm\lambda_2$. The implementation results for computing ISD scalar multiplication $kP$ are provided using Algorithm (1). This algorithm first finds the shortest integer vectors $v_1$ and $v_2$ using GLV algorithm (1) in [7] that consists from extended Euclidean algorithm (EEA) in the first part or necessary condition in the second part. These vectors consider as the reduced integer lattice points. The next step of ISD algorithm is to implement balanced length-two representation of a multiplier algorithm (3.74) in [3] to decompose $k$ into $k_1$ and $k_2$

depending on the coordinates of the vectors $v_1$ and $v_2$ and computing $c_1 = \lfloor -t_{m+2}k/n \rceil$ and $c_2 = \lfloor t_{m+1}k/n \rceil$ with $v_1 = (r_{m+1}, -t_{m+1})$ and $v_2 = (r_{m+2}, -t_{m+2})$ are the shortest vector that help us to compute $k_1 = k - c_1 r_{m+1} - c_2 r_{m+2}$ and $k_2 = -c_1(-t_{m+1}) - c_2(-t_{m+2})$.

**Algorithm 1** Computing ISD Scalar Multiplication Through Interleaving Based on wNAF Expansions.

**Input:** The integers $p$, $n$, $\lambda$, $w_j$ with $j = 1, 2, 3, 4$ and a point $P = (x, y)$ on $E$.
**Output:** $kP$.
**Computation stage:**
Run GLV generator Algorithm (1) that is given in [7] to find the generator $\{v_1, v_2\}$ such that $v_1 \leftarrow (r_{m+1}, -t_{m+1})$ and $v_2 \leftarrow (r_m, -t_m)$ or $v_2 \leftarrow (r_{m+2}, -t_{m+2})$.
Choose $k \in [1, n-1]$.
Run balanced length-two representation of a multiplier algorithm (3.74) that given in [3] to decompose $k$ into $k_1$ and $k_2$.
**if** $max\{|k_1|, |k_2|\} > \sqrt{n}$ **then**
  Set $\lambda_1, \lambda_2 \in [1, n-1]$
  **Precomputation stage:**
  The precomputation of the endomorphism $\psi_1(P) = \lambda_1 P$ and $\psi_2(P) = \lambda_2 P$.
  **Computation stage:**
  Run ISD generators Algorithm (1) given in [9], [10] with input $(n, \lambda_1, \lambda_2)$ to find $\{v_3, v_4\}$ and $\{v_5, v_6\}$ such that $v_3 \leftarrow (r_{(m+1)_1}, -t_{(m+1)_1})$, $v_4 \leftarrow (r_{m_1}, -t_{m_1})$ or $v_4 \leftarrow (r_{(m+2)_1}, -t_{(m+2)_1})$ and $v_5 \leftarrow (r_{(m+1)_2}, -t_{(m+1)_2})$, $v_6 \leftarrow (r_{m_2}, -t_{m_2})$ or $v_6 \leftarrow (r_{(m+2)_2}, -t_{(m+2)_2})$, respectively.
  Use Algorithm (2) in [9], [10] to sub-decompose $k_1$ and $k_2$ into $k_1 \equiv k_{11} + k_{12}\lambda_1 \ (mod\ n)$ and $k_2 \equiv k_{21} + k_{22}\lambda_2 \ (mod\ n)$, respectively.
  **return** $k_{11}, k_{12}$ and $k_{21}, k_{22}$.
  Use generalized computing wNAF Algorithm (1) or (2) in [11] to compute $w_j NAF$ expansions for $j$ from 1 to 4 of integers $k_{11}, k_{12}, k_{21}$ and $k_{22}$.
  Use extended interleaving Algorithm (3) or (4) in [11] to compute $kP = k_{11}P +_E k_{12}\psi_1(P) + k_{21}P +_E k_{22}\psi_2(P)$.
**else**
  Stop and go to choose another value of $k$.
**end if**

## 4. Mathematical analysis of the computational complexity ISD method

The computational cost of the ISD method is discussed in this section mathematically based on presenting the proof of the computational costs of all sub-algorithms that form the ISD algorithm.

### *4.1. Computational complexity for GLV generator algorithm*

Generating the GLV generator $\{v_1, v_2\}$, where $v_1$ and $v_2$ are linearly independent vectors, can be performed through the GLV generator algorithm (1) provided in [7]. GLV generator algorithm consists of two parts, namely, the extended Euclidean algorithm (EEA) shown in [5] to generate shortest vectors in two dimensions, $v_1 = (a_1, b_1)$ and $v_2 = (a_2, b_2)$ or the second part that is the necessary condition part (NCP). The following sections discuss the computational complexity for the said parts.

- **Computational Complexity of the Extended Euclidean Algorithm.**

An extended Euclidean algorithm (EEA) is a useful way to compute the reduction lattice in two dimensions to determine short vectors $v_i$ for $i = 1, 2$ for a given $n$ and $\lambda$. The computational cost of extended Euclidean algorithm $C_{EEA}$ can be determined by computing the field operations that provide the following result:

**Lemma 4.1.** *For a prime number $n$ and positive integer $\lambda$ such that $n \geq \lambda$ and $gcd(n, \lambda) = 1$, a tuple of Equations $s_i n + t_i \lambda = r_i$ for $i = 0, 1, 2, ..., m, m+1, m+2, ..., z-1$ that results from EEA exists. The computational complexity of extended Euclidean algorithm $C_{EEA}$ to determine the tuple of variables $r_i, t_i$ and $s_i$ for $i = 0, 1, 2, ..., m, m+1, m+2, ..., z-1$ with $z > m$ is*

$$C_{EEA} = Z(I + 3M), \tag{4.9}$$

*where $Z$ is the dimension of the returned variables $r_i$, $t_i$ and $s_i$; $M$ is a field multiplication; and $I$ is a field inversion.*

*Proof.* A reasonable way to determine the computational cost of extended Euclidean algorithm $C_{EEA}$ is by implementing the EEA provided as follows:

$u \leftarrow \lambda, v \leftarrow n.$
$t_1 \leftarrow 1, s_1 \leftarrow 0, t_0 \leftarrow 0, s_0 \leftarrow 1, r_0 \leftarrow n, r_1 \leftarrow \lambda.$
While $u \neq 0$
$q \leftarrow \lfloor v/u \rfloor, r \leftarrow v - qu, t \leftarrow t_0 - qt_1, s \leftarrow s_0 - qs_1.$
$v \leftarrow u, u \leftarrow r, t_0 \leftarrow t_1, t_1 \leftarrow t, s_0 \leftarrow s_1, s_1 \leftarrow s.$
Endwhile

The implementation of this algorithm for a loop operation shows the running time in field operations that has the cost $1I + 3M$. Repeating the implementation for another loop operation yields another cost $1I + 3M$. Thus, the cost of two loops is

$$1I + 3M + 1I + 3M = 2I + 6M = 2(1I + 3M). \tag{4.10}$$

The third iteration costs

$$1I + 3M + 1I + 3M + 1I + 3M = 3I + 9M = 3(1I + 3M).$$

The iteration of the loops continues until $r = 0$ and also obtains $u = 1$. If the last loop reaches $Z$ times, then the total cost of this algorithm is

$$\underbrace{1I + 3M + ... + 1I + 3M}_{Z-times} = ZI + 3ZM = Z(I + 3M) \tag{4.11}$$

□

- **Computational Complexity of the Necessary Condition Part of a GLV generator algorithm.**

The study of the necessary condition part (NCP) of the algorithm GLV generator provided in algorithm (1) in [7]. Determining the step cost can finally identify the total cost of the NCP of GLV algorithm by

$$ZI + 3ZM + 1I + 2M + 1I + 3M + 4M = Z(I + 3M) + 2I + 9M, \tag{4.12}$$

where $M$ is a field multiplication, $I$ is a field inversion and $Z$ is the demission of sequence of the variables $r_i, t_i$ and $s_i$ for $i = 0, 1, 2, ..., m, m+1, m+2, ..., z-1$ with $z > m$.

### *4.2. Computational Complexity of Decomposing a Scalar k*

The following result can be used to show the computational cost of decomposing a scalar $k$.

**Lemma 4.2.** *Let $E$ be a prime curve and $P$ be a point on $E$ that has prime order $n$. The scalar $k \in [1, n-1]$ of scalar multiplication $kP$ decomposes into new scalars $k_1$ and $k_2$ such that $k \equiv k_1 + k_2\lambda \ (mod\ n)$, where $\lambda$ is a root of characteristic polynomial $p(X) = X - \lambda$ of endomorphism $\psi$. Thus, the computational cost of this decomposition is*

$$C_{Decomposing\ k} = \begin{cases} Z(I+3M) + 4S + 6M + 2I, & \textit{with using EEA}, \\ Z(I+3M) + 4S + 4I + 15M, & \textit{with using NCP}. \end{cases} \quad (4.13)$$

*Proof.* Let $n$ be a prime order of $P$ which is a point on elliptic curve $E$. Assume that $\psi$ is a nontrivial endomorphism that has characteristic polynomial $p(X) = X - \lambda \ (mod\ n)$. The computational cost of decomposition $k$ that was written by sum formula $k \equiv k_1 + k_2\lambda \ (mod\ n)$ as defined in Equation (1.2) requires the exact computational cost of GLV generator contraction. This contraction is performed by applying the EEA to determine two vectors, $v_1$ and $v_2$. The components of each vector are determined based on a tuple of variables $r_i$ and $t_i$ for $i = 0, 1, 2, ..., m, m+1, m+2, ..., z-1$ with $z > m$ . The cost to find a tuple of variables $r_i$ and $t_i$ is $Z(I + 3M)$ as shown in Lemma (4.1). Identifying the largest index $r_m \geq \sqrt{n}$ provides a possibility to write $v_1 = (r_{m+1}, -t_{m+1})$ and select $v_2$ between the shortest vectors $(r_m, -t_m)$ and $(r_{m+2}, -t_{m+2})$. Selecting a vector $v_2$ can be conducted by verifying the condition defined in Algorithm (1) in [7], Step 3. Computing this condition requires $4S$ operations.

Furthermore, the cost of decomposing a scalar $k$ depends on computing $c_1 = \lfloor -t_{m+2}k/n \rceil$ and $c_2 = \lfloor t_{m+1}k/n \rceil$ that both have a cost of $2M + 2I$. Moreover, determining $k_1 = k - c_1 r_{m+1} - c_2 r_{m+2}$ and $k_2 = c_1 t_{m+1} + c_2 t_{m+2}$ costs $2M$ for each one. Therefore, the total computational complexity of decomposing $k$ is

$$C_{Decomposing\ k} = \begin{cases} Z(I+3M) + 4S + 6M + 2I, & \text{with using EEA}, \\ Z(I+3M) + 4S + 4I + 15M, & \text{with using NCP}. \end{cases}$$

□

### *4.3. The Computational Complexity of the ISD Generators*

In this section, the computational complexity of the ISD generators is provided. This cost is determined based on one of the costs, the cost of the generalization of the extended Euclidean algorithm GEEA because further sub-decomposition in ISD method needs further extension of the EEA that was earlier used in GLV method. For that we have generated the EEA and develop a modified algorithm to perform the sub-decomposition process. On the other hand, cost of the necessary condition part (NCP) of ISD generators algorithm (1) given in [9], [10] can be used to determine the cost of the ISD generators.

- **The Computational Complexity of the ISD Generators Based on the Generalization of the Extended Euclidean Algorithm GEEA.**

  Employing the generalization of the extended Euclidean algorithm GEEA to determine two tuples of variables used to form new generators called ISD generators is possible. These computations include identifying linearly independent vectors $v_3, v_4$ and $v_5, v_6$ in the $kerT$, which considers the integer lattice points in two dimensions. The components in each vector must satisfy the conditions represented by the components of each vector being less than $\sqrt{n}$ and the relation between the components in each vector is a relative prime to form ISD

generators $\{v_3, v_4\}$ and $\{v_5, v_6\}$.
When the output vectors $v_3, v_4$ and $v_5, v_6$ from implementation of the GEEA cannot satisfy these conditions, applying the NCP shown in the ISD generators Algorithm (1) in [9], [10] to generate these vectors that satisfy these conditions is necessary, especially the vectors $v_4$ and $v_6$. Therefore, identifying the computational cost of ISD generators is conducted by recognizing the cost of the ISD generators algorithm that consist of the GEEA in the first part or computing the NCP in the second part.
The computational cost to compute ISD generators is provided by the following result based on the GEEA.

**Theorem 4.3.** *Let $n$ be a prime number, the positive integers $\lambda_j \in [1, n-1]$ for $j = 1, 2$ with $gcd(n, \lambda_j) = 1$ and $\lambda_1 \neq \pm\lambda_2$. The generalization of the extended Euclidean algorithm (GEEA) for $(n, \lambda_1)$ and $(n, \lambda_2)$ produces two tuples of variables $r_{i_1}, t_{i_1}, s_{i_1}$ for $i = 0, 1, ..., m_1, (m+1)_1, (m+2)_1, ..., (z-1)_1$ and $r_{i_2}, t_{i_2}, s_{i_2}$ for $i = 0, 1, ..., m_2, (m+1)_2, (m+2)_2, ..., (z-1)_2$ to find linearly independent vectors $v_3, v_4$ and $v_5, v_6$ which generate ISD generators $\{v_3, v_4\}$ and $\{v_5, v_6\}$. Then the computational cost to form ISD generators, when the algorithm uses GEEA that is given in the first part of ISD generators algorithm (1) given in [9], [10], is*

$$C_{ISD\ generators} = \begin{cases} Z(I+3M)+4S, & \text{if } Z = Z_1 = Z_2 \text{ when use GEEA}, \\ max(Z_1, Z_2)(I+3M)+4S, & \text{if } Z_1 \neq Z_2 \text{ when use GEEA}. \end{cases} \tag{4.14}$$

*where $Z$ is a dimension of two tuples of variables $r_{i_1}, t_{i_1}, s_{i_1}$ which have the same dimension $Z_1$ and $r_{i_2}, t_{i_2}, s_{i_2}$ that have dimension $Z_2$, namely when $Z_1 = Z_2 = Z$, $I$ is a field inversion, $M$ is a field multiplication and $S$ is a field squaring.*

*Proof.* The proof is presented in [12]. □

- **The Computational Complexity of the ISD Generators Algorithm Implemented Based on NCP.**

Computing the ISD generator algorithms (1) given in [9], [10] can be implemented with the necessary condition defined in the second part of this algorithm. The computational cost of ISD generators can be determined as follows.

$$C_{NCP} = Z(I+3M)+2I+9M \tag{4.15}$$

when two tuples of the variables $r_{i_j}, t_{i_j}$ and $s_{i_j}$ for $j = 1, 2$ $i = 0, 1, 2, ..., m_j, (m+1)_j, (m+2)_j, ..., (z-1)_j$, with $z_j > m_j$ have same dimension $Z_1 = Z_2$, $M$ is a field multiplication, and $I$ is a field inversion. Otherwise, when these two tuples of the variables have different dimensions $Z_1$ and $Z_2$ such that $Z_1 \neq Z_2$, then the computational cost of the NCP is

$$C_{NCP} = max(Z_1, Z_2)(I+3M)+2I+9M. \tag{4.16}$$

Thus, from Equations (4.15) and (4.16), the final computational complexity of the ISD generators can be rewritten by formula

$$C_{ISD\ generators} = \begin{cases} Z(I+3M)+2I+9M+4S, & \text{if } Z = Z_1 = Z_2, \text{ when use NCP} \\ max(Z_1, Z_2)(I+3M)+2I & \\ +9M+4S, & \text{if } Z_1 \neq Z_2, \text{when use NCP}. \end{cases} \tag{4.17}$$

where $4S$, in Equation (4.17) is the cost to determine the shorter vectors $v_4$ and $v_6$.

The computational complexity of ISD generators algorithm can be rewritten by

$$C_{ISD\ generators} = \begin{cases} Z(I+3M)+4S, & \text{if } Z=Z_1=Z_2 \text{ and use GEEA,} \\ max(Z_1,Z_2)(I+3M)+4S, & \text{if } Z_1 \neq Z_2 \text{ and use GEEA,} \\ Z(I+3M)+2I+9M+4S, & \text{if } Z=Z_1=Z_2 \text{ and use NCP,} \\ max(Z_1,Z_2)(I+3M)+2I+9M+4S, & \text{if } Z_1 \neq Z_2 \text{ and use NCP.} \end{cases} \tag{4.18}$$

### *4.4. The Computational Complexity of Sub-Decomposition of Scalars $k_1$ and $k_2$*

The computational complexity to sub-decompose scalars $k_1$ and $k_2$ is given by the following result.

**Theorem 4.4.** *Let $E$ be an prime curve and $P$ be a point on $E$ which has a prime order $n$. The scalar $k \in [1, n-1]$ of a scalar multiplication $kP$ was decomposed into $k_1$ and $k_2$ with $max(|k_1|, |k_2|) > \sqrt{n}$. The ISD sub-decomposition decomposes $k_1$ and $k_2$ into new sub-scalars $k_{11}, k_{12}, k_{21}$ and $k_{22}$ such that $k \equiv k_{11}+k_{12}\lambda_1+k_{21}+k_{22}\lambda_2 \ (mod\ n)$ with $max(|k_{11}|, |k_{12}|) < C\sqrt{n}$ and with $max(|k_{21}|, |k_{22}|) < C\sqrt{n}$. Then the computational complexity to find $k_{11}, k_{12}, k_{21}$ and $k_{22}$ is*

$$C_{Sub-Decomposition\ k_1\ and\ k_2} = \begin{cases} Z(I+3M)+4S+6M+2I, & \text{if } Z_1 = Z_2, \\ max(Z_1,Z_2)(I+3M)+4S+6M+2I, & \text{if } Z_1 \neq Z_2, \end{cases} \tag{4.19}$$

*when the generation of ISD generators uses GEEA. Or*

$$C_{Sub-Decomposition\ k_1\ and\ k_2} = \begin{cases} Z(I+3M)+4S+4I+15M, & \text{if } Z_1 = Z_2, \\ max(Z_1,Z_2)(I+3M)+4S+4I+15M, & \text{if } Z_1 \neq Z_2, \end{cases} \tag{4.20}$$

*when the generation of ISD generators uses NCP.*

*Proof.* The proof is presented in [12]. □

### *4.5. The Computational Complexity for Parallel Pre-computation of Two Efficient Computable Endomorphisms*

The computational complexity to pre-compute two endomorphisms $\psi_1$ and $\psi_2$ which are defined by $\psi_1(P) = \lambda_1 P$ and $\psi_2(P) = \lambda_2 P$ that act on the subgroup $\langle P \rangle$ generated by a point $P$ that lies on $E$ is given by the following result:

**Theorem 4.5.** *Suppose $\psi_1$ and $\psi_2$ are two nontrivial separable endomorphisms on a prime curve $E$ which are defined by $\psi_1(P) = \lambda_1 P$ and $\psi_2(P) = \lambda_2 P$ for an elliptic curve point $P$ and $\lambda_j \in [1, n-1]$ is a root of its characteristic polynomial $p_j(X) = X - \lambda_j$, $j = 1, 2$ and $\lambda_1 \neq \pm\lambda_2$. Then the cost to compute such endomorphisms, $C_{\psi_1(P)\ and\ \psi_2(P)}$, is:*

$$C_{\psi_1(P)\ \&\psi_2(P)} = (max(\lambda_1, \lambda_2) - 1)\, I + 2\,(max(\lambda_1, \lambda_2) - 1)\, M + (max(\lambda_1, \lambda_2) - 1)\, S. \tag{4.21}$$

*Proof.* The proof is presented in [12]. □

*4.6. The Computational Complexity for computing the $w_j$NAF expansions of Positive Integers.*

In this section, the computational cost to represent four positive integers $a_j$ for $j = 1, 2, 3, 4$ using $w_j$NAF expansions is provided by the following result:

**Theorem 4.6.** *Let $w_j$ for $j = 1, 2, 3, 4$ be width windows. Suppose the w-NAF expansions of positive integers of scalar $a_j$ that form a scalar $k$ of ISD scalar multiplication $kP$ with $P$ is a point on elliptic curve $E$, are represented. Then the computational complexity for computing w-NAF expansions of $a_j$ simultaneously is*

$$C_{w_jNAF_{a_j}} = 4D + \sum_{j=1}^{4}(2^{w_j-2} - 1)A, \tag{4.22}$$

*where $D$ and $A$ indicate to point doubling and point addition on elliptic curve $E$ over $F_p$.*

*Proof.* For the proof, see [12]. □

*4.7. The Computational Complexity of the Interleaving Method Based on $w_j$NAF Expansions.*

The computational cost of the interleaving method based on $w_j$NAF expansions for $j = 1, 2, 3, 4$ is provided by the following result.

**Theorem 4.7.** *Let $w_j$, for $j = 1, 2, 3, 4$, be width windows that are positive integers $\geq 2$ and $a_j$ be any positive integers. Then the computational cost of the interleaving method that based on $w_j$NAF expansions is*

$$C_{Interleaving} = \left[4D + \sum_{j=1}^{4}(2^{w_j-2} - 1)A\right] + \left[max_{j=1:4}\ l_j\ D + \sum_{j=1}^{4}\frac{l_j}{w_j+1}A\right]. \tag{4.23}$$

*where $A$ indicates to elliptic curve point addition, $D$ denotes an elliptic curve point doubling and $l_j$ indicates to the length of $NAF_{w_j}(a_j)$.*

*Proof.* For the proof, see [12]. □

Finally, the computational complexity of ISD method is defined by this result.

**Theorem 4.8.** *Let $E$ be an elliptic curve over finite field $F_p$ and $P$ be a point on $E$ which has prime order $n$. Let $\psi$ be a nontrivial endomorphism defines by $\psi_j(P) = \lambda_j P$ where $\lambda_j$ are roots of characteristic polynomials $p_j(X) = X - \lambda_j \pmod n$ of $\psi_j$ for $j = 1, 2$. The scalar $k \in [1, n-1]$ of scalar multiplication $kP$ that has ISD sub-decomposition such that $kP = k_{11}P +_E k_{12}\psi_1(P) +_E k_{21}P +_E k_{22}\psi_2(P)$ with $max\{|k_{11}|, |k_{12}|\} < C\sqrt{n}$ and $max\{|k_{21}|, |k_{22}|\} < C\sqrt{n}$. Then, generally, the computational complexity of ISD method is. In particular, the computational complexity of ISD method take one of the following costs.*

**I.** *The computational cost of ISD algorithm through the implementation GEEA to generate ISD generators and with using one interleaving method [11] is*

$$C_{ISD\ Method} = \left\{ \begin{array}{ll} 2Z(I+3M)+8S+12M+4I, & if\ Z_1 = Z_2 \\ Z(I+3M)+max(Z_1,Z_2)(I+3M)+8S+12M+4I, & if\ Z_1 \neq Z_2 \end{array} \right\}$$
$$+4D+\sum_{j=1}^{4}(2^{w_j-2}-1)A+\ max_{j=1,2,3,4}\ l_j D$$
$$+\sum_{j=1}^{4}\frac{l_j}{w_j+1}A. \tag{4.24}$$

**II.** *The computational cost of ISD algorithm through the implementation NCP to generate ISD generators and with using one interleaving method [11] is*

$$C_{ISD\ Method} = \left\{ \begin{array}{ll} 2Z(I+3M)+8S+30M+8I, & if\ Z_1 = Z_2 \\ Z(I+3M)+max(Z_1,Z_2)(I+3M)+8S+30M+8I, & if\ Z_1 \neq Z_2 \end{array} \right\}$$
$$+4D+\sum_{j=1}^{4}(2^{w_j-2}-1)A+\ max_{j=1,2,3,4}\ l_j D$$
$$+\sum_{j=1}^{4}\frac{l_j}{w_j+1}A. \tag{4.25}$$

**III.** *The computational cost of ISD algorithm through the implementation GEEA to generate ISD generators and with using two interleaving method [11] is*

$$C_{ISD\ Method} = \left\{ \begin{array}{ll} 2Z(I+3M)+8S+12M+4I, & if\ Z_1 = Z_2 \\ Z(I+3M)+max(Z_1,Z_2)(I+3M)+8S+12M+4I, & if\ Z_1 \neq Z_2 \end{array} \right\}$$
$$+max\left[\left(2D+\sum_{j=1}^{2}(2^{w_j-2}-1)A\right)+\left(max_{j=1:2}\ l_j\ D\right.\right.$$
$$\left.+\sum_{j=1}^{2}\frac{l_j}{w_j+1}A\right),\left(2D+\sum_{j=3}^{4}(2^{w_j-2}-1)A\right)$$
$$\left.+\left(max_{j=3:4}\ l_j\ D+\sum_{j=3}^{4}\frac{l_j}{w_j+1}A\right)\right]. \tag{4.26}$$

**IV.** *The computational cost of ISD algorithm through the implementation NCP to generate ISD generators and with using two interleaving method [11] is*

$$C_{ISD\ Method} = \left\{ \begin{array}{ll} 2Z(I+3M)+8S+30M+8I, & if\ Z_1 = Z_2 \\ Z(I+3M)+max(Z_1,Z_2)(I+3M)+8S+30M+8I, & if\ Z_1 \neq Z_2 \end{array} \right\}$$
$$+max\left[\left(2D+\sum_{j=1}^{2}(2^{w_j-2}-1)A\right)+\left(max_{j=1:2}\ l_j\ D\right.\right.$$
$$\left.+\sum_{j=1}^{2}\frac{l_j}{w_j+1}A\right),\left(2D+\sum_{j=3}^{4}(2^{w_j-2}-1)A\right)$$
$$\left.+\left(max_{j=3:4}\ l_j\ D+\sum_{j=3}^{4}\frac{l_j}{w_j+1}A\right)\right]. \tag{4.27}$$

*Proof.* The proof is presented in [12]. □

## 5. Conclusion

This paper presents the computational complexity of the sub-decomposition method (that is, ISD) of a scalar $k$ in a scalar multiplication $kP$ on the elliptic curve $E$ over a prime finite field $F_p$. The novelty of this work relies on the theoretical part of the computational cost of the ISD method which has been proven mathematically by computing the elliptic curve operations and finite field operations that give the executing time for implementation of this method. Determining the computational cost in the ISD method depends on computing the cost of all sub-algorithms used in this method.

In particular, the type of implemented sub-algorithm determines the cost of this method. For instance, the computational cost of the ISD method was determined by four formulas, namely, the GEEA, NCP, one, and two interleaving methods as shown in Table (1).

**Table 1.** The cost types of ISD method.

| Implementation Type | Cost of ISD Method |
|---|---|
| GEEA with one interleaving | Type I |
| NCP with one interleaving | Type II |
| GEEA with two interleaving | Type III |
| NCP with two interleaving | Type IV |

## 6. Acknowledgments

The authors would like to express their gratitude to the Fundamental Research Grant (FRGS), 203/PMATHS/6711320, funded by the Ministry of Higher Education and the School of Mathematical Sciences, Universiti Sains Malaysia.

## References

[1] N. Koblitz, Elliptic curve cryptosystems, *Mathematics of computation*, **48**, (1987), pp. 203-209.
[2] V. Miller, Use of elliptic curves in cryptography, in *Advances in Cryptology-CRYPTO'85 Proceedings*, (1986), pp. 417-426.
[3] D. Hankerson, A. J. Menezes, and S. Vanstone, *Guide to elliptic curve cryptography*, Springer Verlag, USA, (2004).
[4] F. Sica, M. Ciet, and J.-J. Quisquater, Analysis of the Gallant-Lambert-Vanstone method based on efficient endomorphisms: Elliptic and hyperelliptic curves, in *Selected areas in cryptography*, (2003), pp.21-36.
[5] R. Gallant, R. Lambert, and L. Vanstone, Faster point multiplication on elliptic curves with efficient endomorphisms, in *Advances in Cryptology-CRYPTO 2001*, (2001), pp.190-201.
[6] M. Ciet, J.-J. Quisquater, and F. Sica, Preventing differential analysis in GLV elliptic curve scalar multiplication, *Cryptographic Hardware and Embedded Systems-CHES 2002*, (2003), pp. 1-13.
[7] D. Kim, S. Lim, Integer decomposition for fast scalar multiplication on elliptic curves, in *Selected Areas in Cryptography*, Springer, (2003), pp.13-20.
[8] L. C. Washington, *Elliptic curves: number theory and cryptography*, Chapman & Hall/CRC, USA, (2008).
[9] R.K.K. Ajeena, H. Kamarulhaili, 2013. Analysis on The Elliptic Scalar Multiplication Using Integer Sub-Decomposition Method. International Journal of Pure and Applied Mathematics, 87(1): 95-114.
[10] R.K.K. Ajeena, H. Kamarulhaili, 2014. Point Multiplication using Integer Sub-Decomposition for Elliptic Curve Cryptography. Journal of Applied Mathematics & Information Sciences, 8(2): 517-525.
[11] R.K.K. Ajeena, H. Kamarulhaili. Accelerating Integer Sub-Decomposition for Elliptic Scalar Multiplication using $w$NAF Expansion Method. Malaysian Journal of Mathematical Sciences.
[12] Ruma Kareem K. Ajeena. Integer Sub-Decomposition Method for Elliptic Curve Scalar Multiplication. PhD thesis, November, 2014, School of Mathematical sciences, Universiti Sains Malaysia, Penang, Malaysia.

Journal of Physics: Conference Series **622** (2015) 012033 doi:10.1088/1742-6596/622/1/012033

# Analysis of selected prioritization methods in the analytic hierarchy process

**J Jablonsky**
Department of Econometrics, University of Economics, Prague
W. Churchill Sq. 4, 13067 Praha 3, Czech Republic

E-mail: jablon@vse.cz

**Abstract.** A crucial problem in the analytic hierarchy process (or its extension analytic network process) deriving priorities form pairwise comparison matrices. The most popular methods for deriving priorities are eigenvector method, which is originally proposed by Saaty [5], logarithmic least square method, and least square method. The paper deals with other alternative approaches using goal programming methodology - one of them is based on minimization of sum of absolute or relative deviations and the other one on minimization of maximum deviation. The results of methods are compared on a set of randomly generated matrices of different sizes and consistency levels.

## 1. Introduction

The Analytic Hierarchy Process (AHP) was formulated by T. Saaty in 1977 – more detailed information about it can be found e.g. in Saaty [6] – and this method has become one of the most popular tools for analysis of complex decision making problems. Even though the AHP was proposed many years ago it is still subject for research and a powerful tool for applications in various fields of human being. The Analytic Network Process (ANP) is a generalization of the AHP methodology that was introduced by Saaty [7]. ANP splits the decision making problem into clusters and allows interdependencies among clusters and within clusters.

AHP (and ANP) is still theoretically discussed and developed, and used as a main or secondary modelling tool in variety of real-world case studies. Among others, in the theoretical area the review paper [4] contains main developments of the AHP and ANP in the last decade. A great attention is paid to prioritization methods in the AHP/ANP. Their recent review is published in [2]. An interesting approach to this problem based on combination of current methods is presented in [9]. A wide review of the AHP/ANP applications can be found in [8].

One of the main features of both methods consists in expression of priorities by decision maker among evaluated elements using pairwise comparisons and then deriving priorities from pairwise comparison matrices. The priorities can be derived using several methodological approaches. Saaty [6] proposes eigenvalue method which has several advantages but is quite computationally uncomfortable. Other commonly used methods for deriving priorities are least square method, logarithmic least square method, modified least square method and others. We propose method that are based on goal programming methodology and either minimizes the sum of deviations or maximum deviation of theoretical and empirical pairwise comparisons.

The paper is organized as follows. Section 2 summarizes main principles of AHP and ANP. Next section formulates models for deriving priorities from pairwise comparison matrices. Comparison of results on the set of randomly generated matrices of different sizes and consistency levels is given in Section 4. The results and possible directions for a future research are discussed in the final section of the paper.

## 2. Analytic hierarchy/network process

AHP is a powerful and popular tool for analysis of general decision making problems. This approach organizes the decision problem as a hierarchical structure containing always several levels. The first level defines the main goal of the decision problem and the last one usually describes the decision alternatives or scenarios. The levels in between can contain secondary goals, criteria and subcriteria of the decision problem. The number of the levels is not limited, but in the typical case it does not exceed four or five.

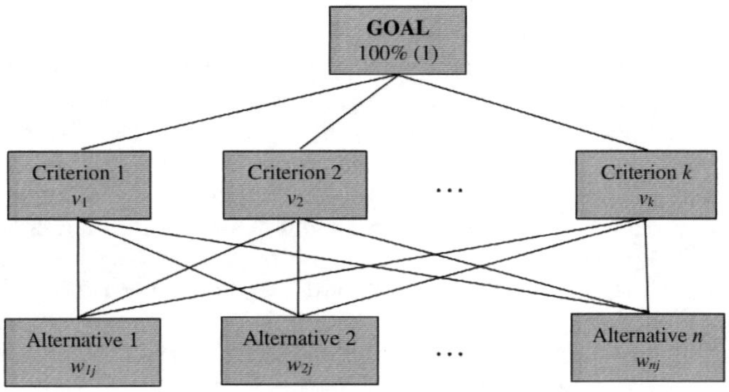


**Figure 1.** AHP hierarchy with three levels.

Let us consider a simple three-level hierarchy that represents a standard decision problem with a finite set of alternatives - $n$ alternatives $X_1, X_2, \ldots, X_n$, are evaluated by $k$-criteria $Y_1, Y_2, \ldots, Y_k$. The hierarchy corresponding to this problem is presented on Figure 1 – this figure is a modified version of the one presented in Jablonsky [5]. The decision maker expresses his preferences by comparing the importance of the elements on the given level with respect to an element of the preceding level. The decision maker's judgements with respect to a given element are synthesised into the local priorities. They can express, e.g. the relative importance of criteria (weight coefficients - in Figure 1 denoted by $v_j$, $i = 1, 2,\ldots, k$) or preference indices of the units with respect to the given criterion ($w_{ij}$, $i = 1, 2\ldots, n$, $j = 1, 2,\ldots, k$). In a conventional AHP model the decision maker's judgements on elements in a given level with respect to an element in the level above are organised into pairwise comparison matrices. The judgements are estimates of the preference between two elements of the lower level with respect to the element in the level above. Let us denote the pairwise comparison matrix $\mathbf{A} = \{a_{ij} \mid a_{ji} = 1/a_{ij}, a_{ij} > 0, i, j = 1, 2,\ldots, k\}$, where $k$ is the number of elements in the particular comparison set of the lower level. Elements of pairwise comparison matrix $a_{ij}$ express how many times is the $i$-th criterion (alternative) more (or less) important than the $j$-th criterion (alternative). Saaty [6] proposes to use $a_{ij}$ integers in the range 1 through 9 to express preference, where 1 means that the $i$-th and the $j$-th element are equally important and 9 means that the $i$-th element is absolutely more important than the $j$-th element. The local priorities from matrix $\mathbf{A}$ can be derived either by Saaty's original method, which is solving the following eigenvector problem (1), or by several alternative approaches that are formulated in the next section of the paper:

$$\mathbf{A}.\mathbf{w} = \lambda_{\max}\mathbf{w},$$
$$\sum_{i=1}^{n} w_i = 1, \tag{1}$$

where $\lambda_{max}$ is the largest eigenvalue of matrix **A** and **w** is the normalised right eigenvector belonging to $\lambda_{max}$. The vector **w** can be computed using various algorithms. One of them is based on the following relation:

$$c\mathbf{w} = \lim_{p\to\infty} \frac{\mathbf{A}^p\mathbf{e}}{\mathbf{e}^{\mathrm{T}}\mathbf{A}^p\mathbf{e}}, \tag{2}$$

where $\mathbf{e}^{\mathrm{T}} = (1, 1,\ldots, 1)$ and $c$ is a scalar. Calculation of the right eigenvector w using formula (2) is not complex but it is quite time demanding and that is why other approximation methods are used as proposed in Saaty [6].

In order to get appropriate results the inconsistency level of matrix **A** must not exceed given threshold. Measuring of consistency (or inconsistency) is based on the property of positive reciprocal matrices. For fully consistent positive reciprocal $(n, n)$ matrices $\lambda_{max} = n$, otherwise $\lambda_{max} > n$. Saaty defines so called consistency index C.I. = $(\lambda_{max} - n)/(n - 1)$ and pairwise comparison matrix is considered to be consistent if C.I. < 0.1.

Taking into account the hierarchy given on Figure 1, the decision maker must build one pairwise comparison matrix of order $k$ on second level of hierarchy (comparison of criteria with respect to the main goal). Priorities derived from this matrix are weights of the criteria $v_j$, $j = 1, 2,\ldots, k$. Then, he/she must build $k$ comparison matrices of order $n$ on the third level of hierarchy (comparison of alternatives with respect to particular criteria). Priorities from these matrices ($w_{ij}$, $i = 1, 2,\ldots, n$, $j = 1, 2,\ldots, k$) can be explained as preference indices of alternative $i$ with respect to given criterion $j$. Global priorities of alternatives are given as a simple weighted sum, i.e.

$$u(\mathrm{X}_i) = \sum_{j=1}^{k} v_j w_{ij}, \quad i = 1,2,\ldots,n. \tag{3}$$

The aim of this paper is not discuss in detail the theory of the AHP or ANP. ANP is a generalization of the AHP that takes into account dependencies among the elements of the hierarchy. The motivation for working with ANP models rather than AHP models consists in fact that many decision making problems cannot be structured hierarchically because they contain dependencies that can be expressed within the hierarchy but must be constructed as a network with clusters and their elements. Detailed description of this methodology can be found in Saaty's original book [7].

## 3. Prioritization methods in the AHP

Since formulation of principles of the AHP (and later ANP) several prioritization methods for deriving priorities from pairwise comparison matrices have been proposed. The original Saaty's procedure, as presented above, computes the prioritization vector as the right eigenvector **w** belonging to the largest eigenvalue $\lambda_{max}$ of the pairwise comparison matrix **A**. Due to computational problems (solving using (2) may be quite time demanding) with solving of problem (1) some other prioritization methods have been formulated by Saaty and later by other researchers. All of them are based on minimization of a metric (a deviation function) between elements of pairwise comparison matrices $a_{ij}$ on one side and ratios of estimated priorities $w_i/w_j$ on the other side.

*Least square method* (*LSM*) constructs the deviation function as the sum of squares of deviations between elements $a_{ij}$ and ratios $w_i/w_j$, i.e. the model is as follows:

Minimize $$\sum_{i=1}^{n}\sum_{j=1}^{n}\left(a_{ij}-\frac{w_i}{w_j}\right)^2 \tag{4}$$

subject to $$\sum_{i=1}^{n} w_i = 1, \tag{5}$$
$$w_i \geq 0,\ i = 1, 2, \ldots, n.$$

The problem (4)-(5) is a difficult non-linear problem with non-unique solutions that are hardly computable. That is why the *LSM* cannot be used for practical purposes. A modification of the *LSM* is the *logarithmic least square method* (*LLSM*) that minimizes the objective function

$$\sum_{i=1}^{n}\sum_{j=1}^{n}\left(\ln a_{ij}-\ln\left(\frac{w_i}{w_j}\right)\right)^2 \tag{6}$$

with respect to constraints (5). The solution of the problem (5)-(6) can be simply given as the geometric mean of the elements of each row of matrix A that is normalized to unit sum. That is why this method (originally proposed by Saaty) is often called *geometric mean method*. Solution of this problem is identical to the eigenvector problem (1) in case the matrix **A** is fully consistent and it is close to this solution when the consistency measure is on a satisfactory level.

Because of computational problems with *LSM* method, its modification that minimizes the following metric has been proposed

$$\sum_{i=1}^{n}\sum_{j=1}^{n}\left(a_{ij}w_j - v_i\right)^2 \tag{7}$$

The objective function (7) is not linear but it can be easily transformed into a system of linear equations – see e.g. Bozoki [1] or Gao [3]. Let us denote this method as *modified LSM* (*MLSM*).

The models with objective functions (4), (6) and (7) minimize the sum of squares. They need not be (and usually are not) easily solvable. Instead of the minimization of squares it is possible to apply a goal programming methodology that allows to minimize the sum of positive and negative deviations or to minimize the maximum deviation. In both cases the deviations can be measured either as their absolute values or as relative deviations (in %). The optimization problem for minimization of the sum of relative deviations can be written as follows – let us denote this problem as *RSUM*:

Minimize $$\sum_{i=1}^{n}\sum_{j=1}^{n}\frac{d_{ij}^{-}+d_{ij}^{+}}{a_{ij}},$$

subject to $$a_{ij}+d_{ij}^{-}-d_{ij}^{+}=\frac{w_i}{w_j},\qquad i=1,2,\ldots,n,\ j=1,2,\ldots,n, \tag{8}$$

and constraints (4).

A solution that minimizes the maximum relative deviation can be given by solving the following optimization problem – let us denote this problem as *RMAX*:

Minimize $D$,

subject to $$a_{ij} + d_{ij}^- - d_{ij}^+ = \frac{w_i}{w_j}, \quad i = 1,2,\ldots,n,\ j = 1,2,\ldots,n, \tag{9}$$

$$\frac{d_{ij}^- - d_{ij}^+}{a_{ij}} \le D, \quad i = 1,2,\ldots,n,\ j = 1,2,\ldots,n,$$

and constraints (4).

A significant part of constraints in both problems (8) and (9) are not linear but their solution can be given quite easily by any non-linear solver, e.g. included in modeling and optimization system LINGO (www.lindo.com).
To avoid non-linearity in models (7) and (8) their simplified version can be formulated and solved. The model for minimization of deviations is as follows:

Minimize $$\sum_{i=1}^{n}\sum_{j=1}^{n}\left|a_{ij}w_j - w_i\right| \tag{10}$$

subject to constraints (4). The model that minimizes maximum deviation is

Minimize $$\max_{i,j}\left|a_{ij}w_j - w_i\right| \tag{11}$$

subject to constraints (4). The models (10) and (11) are not linear but it is possible to re-formulate them using deviational variables into linear models very easily. The solution of models (8) and (9) on one hand and models (10) and (11) on the other hand is identical only for fully consistent matrices.

Formulation of all models in this section assumes that all elements of matrix **A** are taken into account either in constraints or the objective function. Due to the reciprocal nature of the pairwise comparison matrix **A** it is questionable whether to consider all elements or the elements greater or equal 1 only, i.e. $a_{ij} \ge 1$. All the models presented here can be modified accordingly.

## 4. Computational experiments

The models for deriving priorities presented in the previous section were tested on randomly generated matrices of different sizes and different consistency levels. We are going to present the results for (4, 4) pairwise comparison matrices. Consistency indices (C.I.) of generated matrices are from very small values (less than 0.01) until values that indicate inconsistent matrices (more than 0.1; the largest value was approx. 0.2). Table 1 presents priorities derived by six methods – eigenvector method, *LLSM*, minimization of the sum of absolute and relative deviations (*ASUM* and *RSUM*) and minimization of the maximum deviation (absolute *AMAX* and relative *RMAX*) – for one of the almost consistent matrices (C.I. approx. 0.015).

**Table 1**. Results for an almost fully consistent matrix (C.I. = 0.015)

| | *Eigenval* | *LLSM* | *ASUM* | *RSUM* | *AMAX* | *RMAX* |
|---|---|---|---|---|---|---|
| $w_1$ | 0.5052 | 0.5047 | 0.4706 | 0.4706 | 0.4807 | 0.5068 |
| $w_2$ | 0.3328 | 0.3322 | 0.3529 | 0.3529 | 0.3503 | 0.3245 |
| $w_3$ | 0.1057 | 0.1061 | 0.1177 | 0.1177 | 0.1038 | 0.1039 |
| $w_4$ | 0.0563 | 0.0570 | 0.0588 | 0.0588 | 0.0652 | 0.0648 |
| *ASUM* | 3.30 | 3.27 | **1.67** | **1.67** | 3.55 | 3.75 |
| *RSUM* [%] | 84.62 | 85.19 | **58.33** | **58.33** | 101.43 | 104.19 |
| *AMAX* | 0.97 | 0.87 | 1.00 | 1.00 | **0.63** | 0.99 |
| *RMAX* [%] | 24.08 | 24.02 | 33.33 | 33.33 | 31.39 | **21.92** |

All optimization problems are solved for elements of the generated matrix greater or equal 1 only and each method is described in Table 1 by four characteristics – the best (optimal) values for all characteristics are given in bold:

- Sum of absolute deviations of original elements $a_{ij}$ and ratios $w_i/w_j$ (*ASUM*).
- Sum of relative deviations (*RSUM*), i.e. the objective function of model (8).
- Maximum absolute deviation (*AMAX*).
- Maximum relative deviation (*RMAX*).

The results show that the original eigenvector procedure has average values of all four characteristics and is not the best in any of them. Differences in priorities are quite high with regard to quite low C.I. value. It is clear that application of other methods than the first two are can lead to quite different final results in evaluation of alternatives, scenarios, etc.

Table 2 contains the same information as Table 1 but for a matrix that is inconsistent and its C.I. slightly exceeds the recommended threshold 0.1. It is clear and understandable that all optimization criteria are much worse for matrices with higher C.I. The derived priorities are very different for all methods except the first two ones. It is quite surprising that eigenvector method and *LLSM* is worse in all four criteria than almost all other methods. Maximum difference in *LLSM* is nearly 7 and it is really the extremely high value. This fact leads to question whether the commonly used methods are acceptable for inconsistent or nearly inconsistent matrices. This question is difficult to answer but all computational experiments show that the other methods are much better. On the other hand they are much computationally demanding and that is why their real applications depend on availability of appropriate powerful solvers.

**Table 2**. Results for an inconsistent matrix (C.I. = 0.108)

| | *Eigenval* | *LLSM* | *ASUM* | *RSUM* | *AMAX* | *RMAX* |
|---|---|---|---|---|---|---|
| $w_1$ | 0.5634 | 0.5604 | 0.4038 | 0.5526 | 0.4480 | 0.5085 |
| $w_2$ | 0.2636 | 0.2721 | 0.4038 | 0.2763 | 0.3639 | 0.2832 |
| $w_3$ | 0.1309 | 0.1271 | 0.1346 | 0.0921 | 0.1427 | 0.1587 |
| $w_4$ | 0.0421 | 0.0404 | 0.0578 | 0.0790 | 0.0454 | 0.0496 |
| ***ASUM*** | 13.55 | 13.38 | **8.67** | 9.33 | 11.81 | 12.56 |
| ***RSUM* [%]** | 239.74 | 236.03 | 177.78 | **163.89** | 224.55 | 238.84 |
| ***AMAX*** | 6.37 | 6.87 | 3.67 | 4.83 | **2.86** | 3.26 |
| ***RMAX* [%]** | 90.98 | 98.19 | 66.67 | 80.56 | 58.97 | **46.61** |

## 5. Conclusions

AHP (and its generalization ANP) belongs to one of the most popular methods for structuring and analysis of complex decision making problems. Both the methods are based on deriving priorities for the elements on each level of hierarchy by pairwise comparisons. The paper presents a survey of possible methods for deriving priorities. The standard methods as *eigenvector method* or *LLSM* are both computationally simple but they does not reach acceptable values of the optimization criteria as maximum deviation, sum of deviations, etc. The alternative procedures seem to be better with respect to all optimization criteria but they are computationally more demanding and their wider using is questionable for users without advanced background in optimization. Future research can be focused on analysis of rank reversals in case different prioritization methods are used or are combined in solving a decision problem.

**Acknowledgments**
The research is supported by the Czech Science Foundation, project no. P403/12/1387.

**References**

[1] Bozoki S 2008 Solution of the Least Squares Method problem of pairwise comparison matrices. *Central European Journal of Operations Research,* **16**, 345 – 358.

[2] El Hefnawy A and Mohammed AS 2014 Review of different methods for deriving weights in the analytic hierarchy process. *Int. J. of the Analytic Hierarchy Process*, **6**, 95 – 123.

[3] Gao S, Zhang Z and Cao C 2009 New methods of estimating weights in AHP. *Proceedings of the ISIP'09*, Huangshan, 201 – 204.

[4] Ishizaka A and Labib A 2011 Review of the main developments in the analytic hierarchy process. *Expert Systems with Applications*, **38**, 14336 – 14345.

[5] Jablonsky J 2007 Measuring the efficiency of production units by AHP models. *Mathematical And Computer Modelling*, **46**, 1091–1098.

[6] Saaty TL 1990 *The Analytic Hierarchy Process: Planning, Priority Setting, Resource Allocation*. RWS Publications, Pittsburgh.

[7] Saaty TL 1996 *Decision making with dependence and feedback: The Analytic Network Process*, Pittsburgh, RWS Publications.

[8] Vaidya OS and Kumar S 2006 Analytic hierarchy process: An overview of applications. *European Journal of Operational Research*, **169**, 1 – 29.

[9] Srdjevic B 2005 Combining different prioritization methods in the analytic hierarchy process synthesis. *Computers & Operations Research*, **32**, 1897 – 1919.

# Three types of Marine microalgae and *Nannocholoropsis oculata* cultivation for potential source of biomass production

**Vijendren Krishnan[1], Yoshimitsu Uemura[1]*, Nguyen Tien Thanh[1], Nadila Abdul Khalid[1], Noridah Osman[1], Nurlidia Mansor[2]**

[1]Centre for Biofuel and Biochemical Research, Universiti Teknologi PETRONAS, 31750 Tronoh, Perak, Malaysia,

[2]Department Of Chemical Engineering, Universiti Teknologi PETRONAS, 31750 Tronoh, Perak, Malaysia

yoshimitsu_uemura@petronas.com.my

**Abstract**. Microalgae have been vastly investigated throughout the world for possible replacement of fossil fuels, besides utilization in remediation of leachate, disposal of hypersaline effluent and also as feedstock for marine organisms. This research particularly has focused on locally available marine microalgae sample and *Nannochloropsis oculata* for potential mass production of microalgae biomass. Biomass produced by sample 1 and sample 2 is 0.6200 g/L and 0.6450 g/L respectively. Meanwhile, sample 3 and *N. oculata* has obtained maximum biomass concentration of 0.4917 g/L and 0.5183 g/L respectively. This shows that sample 1 and sample 2 has produced approximately 20% higher biomass concentration in comparison to sample 3 and *N. oculata*. Although sample 3 and *N. oculata* is slightly lower than other samples, the maximum biomass was achieved four days earlier. Hence, the specific growth rate of sample 3 and *N. oculata* is higher; meanwhile the specific growth rate of *N. oculata* is the highest. Optical density measurements of all the sample throughout the cultivation period also correlates well with the biomass concentration of microalgae. Therefore, *N. oculata* is finally selected for utilization in mass production of microalgae biomass.

## 1. Introduction

There is an intense research focus towards production of biofuel from microalgae as an alternative for fossil fuels. Unlike other energy crops, microalgae growth is extremely rapid besides having high photosynthetic efficiency and very high lipid content [1]. Although, the growth of microalgae is species dependent, they effortlessly double their biomass within 24 hour. Moreover, microalgae production provides a solution for the mitigation of carbon dioxide which causes climate changes. Whereby, one ton microalgae are estimated to consume 1.83 ton of atmospheric carbon dioxide [2]. Further, biofuel made from microalgae is non-toxic, biodegradable and renewable resource [3].

On the other hand, microalgae are also vastly being utilized in remediation of leachate from municipal waste and disposal of hypersaline effluent from desalination plants [4,5]). Also, microalgae rich in lipid especially in essential fatty acids are widely cultivated as diet for juvenile fish, crab, shellfish, and rotifers [6]. However, large scale fresh water microalgae production could endanger water availability. Therefore, marine microalgae species are more favourable for sustainable biofuel production [7]. Yet, there is no commercialization of biodiesel from microalgae. This is mainly due to high biomass production cost related to the microalgae cultivation.

A few plans have been strategized in order to reduce costs of microalgae biofuel which include usage of natural seawater or wastewater, reduce energy for cultivation and harvesting, and importantly increase the productivity and oil content [8]. Therefore, this research is undertaken to investigate potential source for mass production of microalgae biomass. This research investigates biomass production rate of locally available microalgae from Straits of Malacca Sea in comparison to pure strain *Nannochloropsis oculata.*

## 2. Material and Method

*Nannochloropsis oculata* purchased from UTAR Microalgae Sdn. Bhd. and three mixed cultures obtained from Malacca Straits Sea were used for all the experiments. The mixed cultures and their source location are listed in the table 1. Source of sample 1 and 2 are directly from seawater, meanwhile sample 3 is brakish water. These sources are chosen in order to compare the growth of microalgae from the seawater and brakish water. As there is higher diversity of microalgae in seawater, two seawater sources were selected.

Table 1: Sample location of wild microalgae obtained from Malacca Straits Sea

| **Sample** | **Location** | **Google GPS coordinate** |
|---|---|---|
| 1 | Teluk Batik Beach | N 4° 11' 20.4", E 100° 36' 20.8" |
| 2 | Marina Cove Resort | N 4° 12' 48.06", E 100° 36' 08.1" |
| 3 | Titi Panjang (Lumut Jetty) | N 4°13' 56.0", E 100° 38' 31.6" |

### 2.1. Cultivation

Small scale culture was conducted in 500 ml culture flask containing F2 medium for all the microalgae. During the exponential phase, the microalgae were then transferred into 5.0 L glass bottle containing F2 medium at 23 ± 2 °C. All the cultures were continuously illuminated with fluorescent lamps approximately at 5000 lux and aerated with air at 3.0 L/min.

### 2.2. Data Collection

2.2.1. *Optical density.* Microalgae cultures were sampled every day in order to determine its optical density. One ml of culture was diluted four times with distilled water prior to optical density measurement. Measurements were conducted on Shimadzu UV-Vis Spectrophotometer (UV-2600) at wavelength 688 nm.

2.2.2. *Biomass determination.* On daily basis, 20 ml of culture medium was sampled from the microalgae culture and transferred into three glass vials respectively. Culture mediums were then centrifuged at 4000 rpm for 15 min to produce biomass pellets. Supernatants were removed and the microalgae pellets were dried in oven at 105°C for 24 hours before the dry mass is weighed.

2.2.3. *Microalgae growth determination.* The growth rate of each microalgae was characterized based on daily biomass determination. The specific growth rate, μ of each microalgae sample was calculated from the slope of the linear regression of time and nature log of biomass concentration in exponential growth phase; as specified by Song et al., 2013.

$$\text{Specific growth rate, } \mu = (\text{Ln } M_0 - \text{Ln } M_1)/ (t_0 - t_1)$$

Where, $\mu$ is the specific growth rate in exponential phase, $M_0$ is the biomass concentration at the beginning of exponential phase ($t_0$) and $M_1$ represent the biomass concentration at the end of the exponential phase ($t_1$).

## 3. Result and Discussion

Generally, growth of microalgae can be segregated into four phases which are lag phase, log phase or exponential, stationary phase and finally death phase [9]. Lag phase is the initial phase of cultivation in which the microalgae adapts to the surrounding such as medium, pH, temperature and lighting. Subsequently, the microalgae begin to undergo active cell division and the biomass of the culture will increase usually in exponential order. Thereafter, stationary phase begins which ceases biomass increase. This is due to the equal rate of the cell division and cell death. This phase mainly occurs due to depletion of nutrients in the medium. Lastly, the microalgae death rate will be higher than the cell division rate, hence the graph shows decrease in biomass.

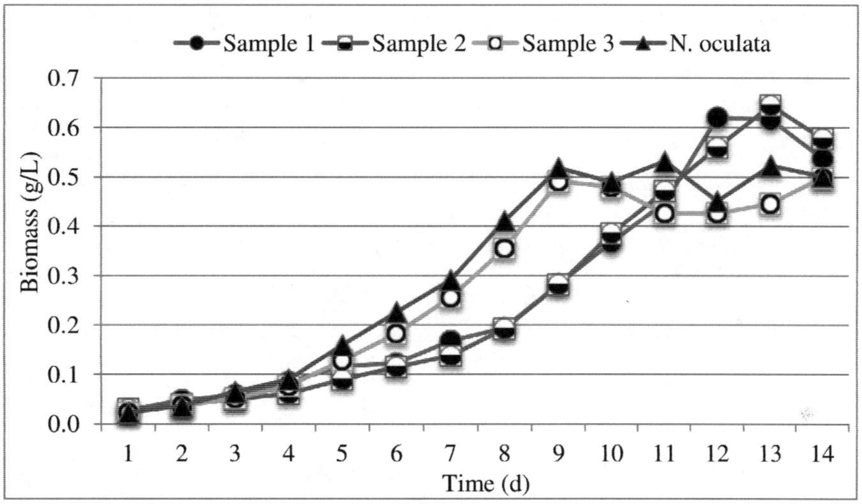


Figure 1: Biomass concentration of microalgae cultivated in F2 medium

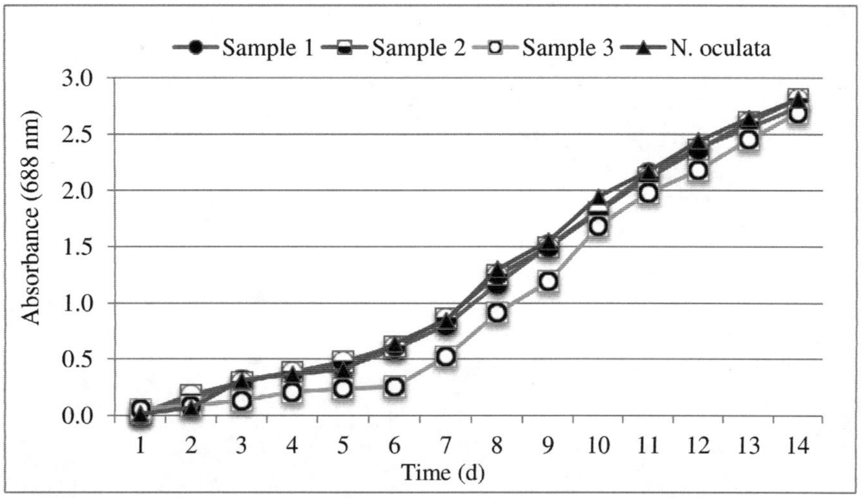


Figure 2: Absorbance measured at 688 nm for microalgae cultivated in F2 medium

Sample 3 and N. oculata reached its maximum biomass concentration on day 9 with 0.4917 g/L and 0.5183 g/L respectively. On the other hand, sample 1 and sample 2 reaches its maximum biomass concentration on day 12 and 13, later than sample 3 and N. oculata. Sample 1 and sample 2 have recorded biomass concentration of 0.6200 g/L and 0.6450 g/L respectively, which is higher than sample 3 and *N. oculata* approximately by 20%. Therefore, sample 3 and *N. oculata* is considered to have higher growth rate compared to other microalgae culture. Microalgae that produce high biomass concentration in short period of time is vital for high production of biodiesel or other valuable products. The trend of growth for sample 1 and sample 2 is similar, this may be due to the similar sampling location and therefore the microalgae biodiversity is also the comparable.

Occasionally, optical density is also used to measure the growth of microalgae. Optical density is used to measure the intensity of chlorophyll pigments in the microalgae cells. Highest absorbance of *N. oculata* was recorded at 688 nm, this is due to the fact that *N. oculata* contain chlorophyll α. Therefore the optical density was measured at wavelength of 688nm. However, it does not accurately represent the growth of microalgae. In figure 2, absorbance of all microalgae cultures were exhibited. In actual microalgae growth based on biomass determination, the exponential phase begins from day 5, whereby absorbance diagram shows exponential phase on day 7. This is mainly due to the colour intensity of the culture which begins to become intense on day 7 onwards. The dark green colour of the microalgae culture explains the rapid growth of the microalgae. In addition, all microalgae culture exhibit similar trend of absorbance. It is also important to note that, the gradient of the curves reduces on day 13 to day 14, indicating that the intensity of the green colour is becoming plateau.

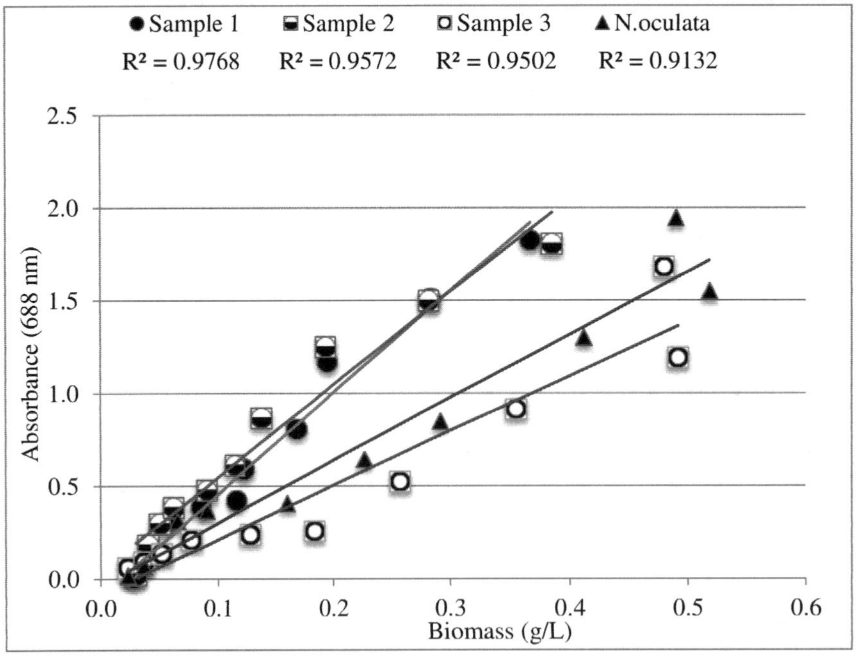


Figure 3: Correlation between absorbance measured at 688 nm and biomass concentration of microalgae

On the other hand, correlation between absorbance and biomass concentration were presented on figure 3. The absorbance correlates well with biomass concentration, especially during the exponential phase. As the microalgae reaches stationary phase, the curves bends upwards showing less correlation to the biomass concentration. Hence, absorbance measurement can be used to represent biomass concentration with a limitation, that it can only be used during the exponential phase. Therefore, absorbance measured at lag phase and stationary phase has been omitted from the diagram. The regression value obtained is reasonable, whereby the minimum $R^2$ value obtained is 0.9132 for all microalgae culture conducted. Figure 3 also shows that, at same biomass concentration of different samples, the absorbance measured was different. This is mainly due to the presence of high chlorophyll content per unit cell which result in high optical density of microalgae cells. In addition, optical density can be utilized for a rapid estimation of microalgae biomass.

Table 2: Specific growth rate of various microalgae sample cultivated in F2 medium

| Microalgae | Specific growth rate, μ ($d^{-1}$) |
|---|---|
| Sample 1 | 0.2558 |
| Sample 2 | 0.2488 |
| Sample 3 | 0.3386 |
| *N. oculata* | 0.3445 |

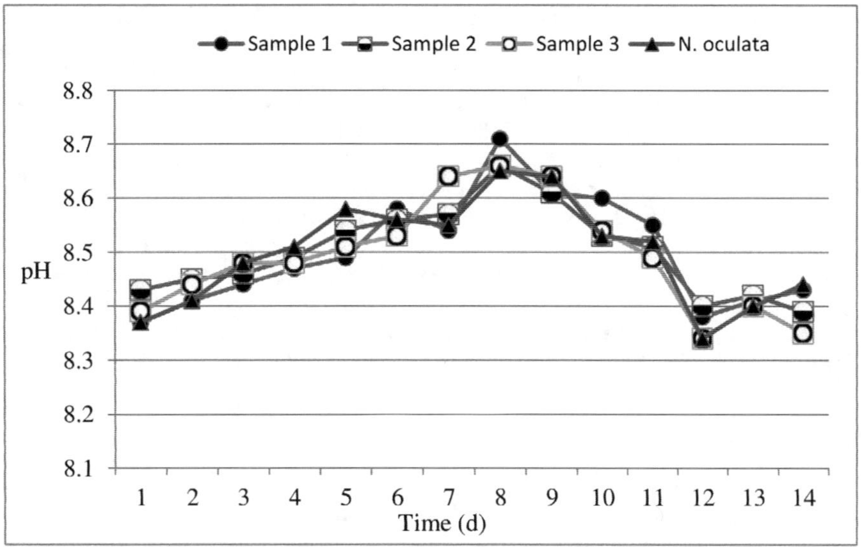


Figure 4: pH changes observed during microalgae cultivation in F2 medium.

Specific growth rate of microalgae were affected by cultivation of nutrient in culture medium and also the reproduction rate of the microalgae itself [9]. Sample 1 and sample 2 exhibited low specific growth rate that is 0.2558 and 0.2488 respectively (Table 2). In comparison, sample 3 and N. oculata exhibited significantly higher specific growth rate which is 0.3387 and 0.3445 respectively. Hence, it is obvious that N. oculata exhibit the highest growth rate among all samples investigated and it also reaches maximum biomass concentration earlier than other microalgae.

Microalgae growth are influenced by a few factors namely, nutrient concentration, carbon dioxide concentration, light intensity, aeration rate and pH. On a daily basis, pH of the microalgae culture was measured. Initial pH of nutrient medium was adjusted to 8.00 ± 0.05. Changes of pH in nutrient medium were presented in figure 4. All microalgae cultures exhibit similar trend that is increasing continuously until late exponential phase thereafter the pH gradually decreases. Maximum pH was observed on day 8 of cultivation which ranges from 8.66 to 8.71.

## 4. Conclusion

Three types of locally available marine microalgae and *N. oculata* were successfully cultivated in F2 medium supplemented with artificial seawater. Among all samples investigated, *N. oculata* exhibited the highest specific growth rate and high biomass concentration in a shorter duration of cultivation time. Therefore, high biomass productivity can supply demand for high biodiesel production or other value added products which on the other hand will reduce the production cost. Due to that reason, *N. oculata* is selected for mass production of microalgae biomass.

## 5. Acknowledgements

The authors would like to thank MOE LRGS ($CO_2$ rich natural gas value chain program from wells to wealth: A green approach) and the Mitsubishi Corporation Education Trust Fund

## 6. References

[1] Suzana W, Ani I, Sitti R, and Muhamad S 2013 *Bioresource Technology* **129** 7
[2] Chisti Y 2007 *Biotechnology Advances* **25** 294
[3] Song M, Haiyan P, Wenrong H, and Guixia M 2013 *Bioresource Technology* **141** 245
[4] Richards R G, and Mullins B J 2013 *Ecological Modelling* **249** 59
[5] Aravantinou A F, Marios A. Theodorakopoulos, Manariotis I D 2013 *Bioresource Technology* **147** 130
[6] Li S, Jilin X, Jiao C, Juanjuan C, Chengxu Z, XiaojunY 2014 Aquaculture
[7] San Pedro A, González-López C V, Acién F G, and Molina-Grima E 2013 *Bioresource Technology* **134** 353
[8] Bondioli P,Laura D B, Gabriele R, Graziella C Z, Niccolò B, Liliana R, David C, Matteo P, David C, and Mario R. T 2012 *Bioresource Technology* **114** 567
[9] Moazami N, Alireza A, Reza R, Mehrnoush T, Roghieh E, and Ali S N 2012 *Biomass and Bioenergy* **39** 449

Journal of Physics: Conference Series **622** (2015) 012035 doi:10.1088/1742-6596/622/1/012035

# Benzimidazole as corrosion inhibitor for heat treated 6061 Al-SiC$_p$ composite in acetic acid

**Melby Chacko and Jagannath Nayak**

Department of Metallurgical and Materials Engineering, National Institute of Technology Karnataka - 575025, India

melbychacko@gmail.com

**Abstract**. 6061 Al-SiCp composite was solutionized at 350 °C for 30 minutes and water quenched. It was then underaged at 140 °C (T6 treatment). The aging behaviour of the composite was studied using Rockwell B hardness measurement. Corrosion behaviour of the underaged sample was studied in different concentrations of acetic acid and at different temperatures. Benzimidazole at different concentrations was used for the inhibition studies. Inhibition efficiency of benzimidazole was calculated for different experimental conditions. Thermodynamic parameters were found out which suggested benzimidazole is an efficient inhibitor and it adsorbed on to the surface of composite by mixed adsorption where chemisorption is predominant.

## 1. Introduction

Aluminium alloys, particularly 6061, reinforced with SiC have been the focus of the attention because of their application potential in an extensive range of demanding applications, such as automobiles, aerospace and defense. There are many other applications which involve exposure of the composites to potential corrosive environment [1], [2]. Because of the duplex nature of the composites, they are prone to accelerated corrosion compared to their monolithic counterpart. Corrosion of these composites not only limit their service life but also lead to deterioration of their unique mechanical properties for which they are designed [3]. Studies reveal the role of micro structural changes and processing routes on the corrosion behaviour. Also, aging treatment was found to have influence on the corrosion rate as the heat treated samples showed higher corrosion rates as compared to the non-treated samples [4]-[6].

Corrosion studies in organic acid solutions are rare in comparison with similar studies in mineral acids. Acetic acid is a frequently used organic acid in many industrial processes [7]. At high temperatures, these acids dissociate, generating new aggressive ions which cause faster corrosion and they can provide sufficient protons to act as true acids [8], [9].

Adding inhibitors to the corrosion medium, is a general practice for the corrosion protection. Heterocyclic organic compounds which contain oxygen, sulphur, phosphorous, nitrogen and aromatic rings are considered to be the most active and resourceful inhibitors in acidic corrosive medium for the metals, alloys and composites [10], [11]. Latterly, benzimidazole have established considerable reputation on their corrosion inhibition properties for metals and alloys, owing to the existence of aromatic rings and nitrogen atom [12]. In the present study benzimidazole is used as inhibitor.

Journal of Physics: Conference Series **622** (2015) 012035 doi:10.1088/1742-6596/622/1/012035

## 2. Experimental work

### *2.1. Specimen*

6061 Al-$SiC_p$ composite material with 15 vol% SiC particles of size 23 μm and 99.9% purity is studied. The composites were cast in the form of cylinders each of 90 mm diameter and 240 mm length by stir casting technique. These cylinders were extruded at 430 °C – 480 °C with extrusion ratio of 30:1. The experiments were performed with composite in extruded rod form. The extruded samples were cut in cylinders of 5 cm length and diameter of 1 cm. The composition of base metal alloy is given in Table 1.

**Table 1**. Composition of 6061 aluminium alloy (matrix)

| Element | Cu | Mg | Si | Mn | Cr | Others | Al |
|---|---|---|---|---|---|---|---|
| Weight % | 0.15- 0.4 | 0.8-1.2 | 0.4-0.8 | 0.15 | 0.04-0.35 | 1.25 | Balance |

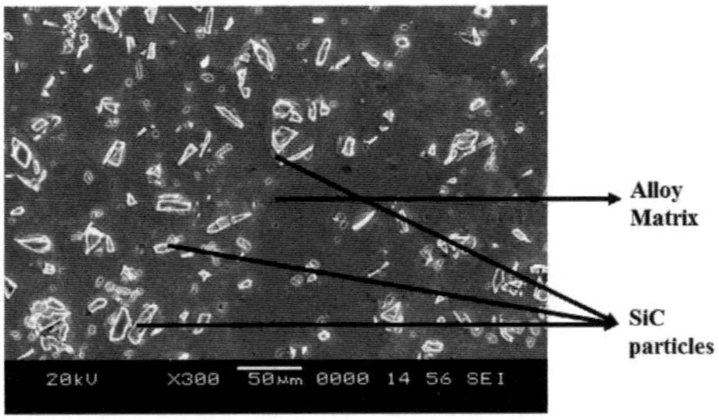


**Figure 1**. SEM picture of 6061 Al/ SiCp composite.

For corrosion tests, heat treated cylindrical coupons were mounted in epoxy material. The specimens were polished with various grades of emery paper and wet polished with diamond paste. Acetone was used to clean and dry the samples.

### *2.2. Electrolyte for Corrosion Tests*

Test solutions were prepared using distilled water and analytical grade acetic acid for all the experiments. Range of concentrations of acetic acid for corrosion tests was decided to be 10%, 20% and 30 volume%.

### *2.3. Methods*

The samples for heat treatment were polished and solution treated at a temperature of 350 °C for 30 minutes. They were then water quenched at room temperature. This was followed by aging the quenched composites at 140 °C (T6 treatment) for various durations of time. The aging behaviour of the composite was studied using Rockwell hardness measurement. Rockwell B Hardness (HRB) tests were performed immediately after aging. Each hardness value was the average of at least three measurements in two samples. Time taken for under aging the sample was found out as one hour, using aging curve and then corrosion tests were performed on under aged samples.

For corrosion tests, an area of 1 $cm^2$ of the polished, under aged 6061 Al- SiC specimens were exposed to 150 ml of acetic acid solution (10, 20 and 30 volume%) at different temperatures like 30 °C, 35 °C, 40 °C, 45 °C and 50 °C. Experiments were carried out in the presence (50, 100 and 200 ppm) and absence of inhibitor. Electrochemical measurements were performed with a potentiostat under software control, BIOLOGIC, SP-150. Polarization scans were initiated from -250 mV cathodically and scanned to approximately +250 mV anodically, all potentials being relative to the open-circuit corrosion potential. Potentiodynamic scan rates were 10 mV/s. A three electrode cell consisting of a large area platinum as counter electrode, saturated calomel electrode (SCE) as a reference electrode and sample as working electrode was used. All potentials are referenced to the SCE. The corrosion rate is obtained from the expression (1)

$$Corrosion\,Rate\,(mpy) = (0.129 * EW * icorr)/D \quad (1)$$

where, $i_{corr}$ is the corrosion current density in μA/ cm$^2$, EW is the equivalent weight of the corroding specimen in gm and D is the density of the corroding species in gm/cm$^3$.

*2.4. Inhibitor*

As obvious from literature, benzimidazole has established considerable responsiveness on its inhibition properties. So, for the present study benzimidazole at different concentrations such as 50 ppm, 100 ppm and 200 ppm, is chosen as inhibitor. The details of inhibitor such as molecular structure, abbreviation and molar mass are given in Table 2. Optimized structure of inhibitor is shown in Figure 2.

**Table 2**. Molecular structure, abbreviation and molar mass of inhibitor.

| Inhibitor | Structure | Abbreviation | Molar mass (g/mol) |
|---|---|---|---|
| Benzimidazole | 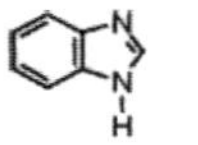 | BI | 118.14 |

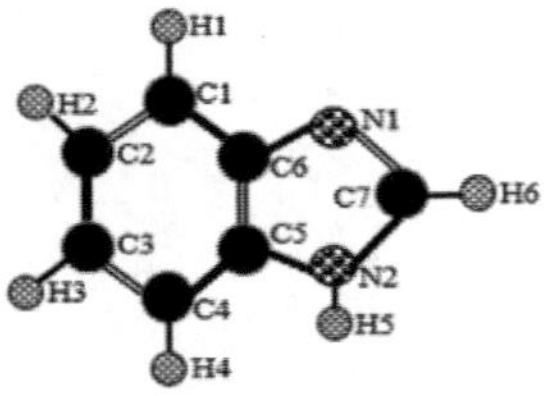


**Figure 2**. Optimized structure of benzimidazole (BI).

## 3. Results and discussion

Rockwell B hardness values of 6061 Al-SiCp composite, which are T6 treated at 140 °C are given in Table 3 and aging curve is displayed in Figure 3. The hardness measurements showed a sharp rise in hardness for the first few hours, it reaches a peak value and then decreases.

**Table 3**. Rockwell hardness values of composite aged at 140 °C.

| Time of Aging (hours) | 0 | 1 | 2 | 3 | 4 | 5 |
|---|---|---|---|---|---|---|
| Rockwell B hardness values | 62 | 68 | 73 | 78 | 88 | 79 |

From the hardness profile of Figure 3, the aged specimens can be categorized into three groups namely a) under aged b) peak aged and c) over aged. The variation in hardness is associated with the microstructural evolution. The formation of Mg and Si clusters takes place initially. This occurs in successive stages. The aging sequence for 6061 Al alloy and its composite are as follows: Super-saturated solid solution→ clusters of solute atoms and vacancies (Primitive Guinier-Preston [GP] zones) → needle shaped GP zones → rod shaped, metastable, semi-coherent β' phase → stable, incoherent, $Mg_2Si$ precipitate (β phase) [13]. Formation of primitive GP zones contribute considerably to the rise in hardness. Peak-aging is related to the dense population of β needle-shaped precipitates. Since these precipitates are coherent precipitates; large amount of lattice strain is developed because of lattice mismatch. This results in high hardness. Only a part of these precipitates remain in the microstructure during over-aging as the metastable phases like β' form. These precipitates are incoherent and there by the lattice strain involved is less, which results in the lowering of hardness values [14]. Corrosion rates (mpy) of under aged sample at 140 °C, in the presence and absence of inhibitor at different temperatures are shown in Table 4.

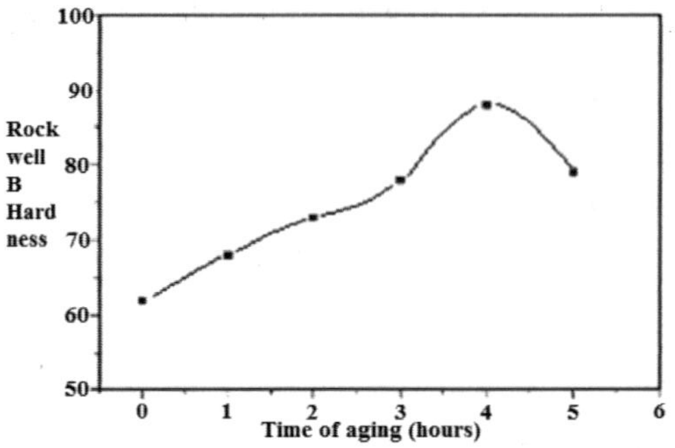


**Figure 3**. Hardness variation with time in T6 treatment at 140 °C.

**Table 4**. Corrosion rates of under aged sample with and without inhibitor, at different temperatures.

| Temperature | Concentrations of Acetic acid (vol %) | Corrosion rates (mpy) [Inhibitor: BI] | | | |
|---|---|---|---|---|---|
| | | 0 ppm | 50 ppm | 100 ppm | 200 ppm |
| 30 °C | 10 % | 4.5 | 3.9 | 3.5 | 3.2 |
| | 20 % | 5.7 | 4.7 | 3.9 | 3.5 |
| | 30 % | 6.2 | 5.5 | 4.8 | 4.0 |
| 35 °C | 10 % | 6.0 | 5.1 | 5.0 | 4.4 |
| | 20 % | 7.2 | 5.9 | 5.6 | 5.0 |
| | 30 % | 8.4 | 7.0 | 5.9 | 5.2 |
| 40 °C | 10 % | 10.2 | 6.7 | 5.2 | 4.6 |
| | 20 % | 15.0 | 7.7 | 5.8 | 5.3 |
| | 30 % | 17.8 | 12.9 | 8.8 | 8.0 |
| 45 °C | 10 % | 14.7 | 8.5 | 8.0 | 7.2 |
| | 20 % | 17.8 | 9.7 | 8.4 | 7.9 |
| | 30 % | 18.9 | 13.3 | 10.6 | 8.9 |
| 50 °C | 10 % | 24.1 | 13.5 | 11.8 | 10.0 |
| | 20 % | 25.3 | 19.6 | 15.9 | 11.8 |
| | 30 % | 30.6 | 21.3 | 17.8 | 12.6 |

The results indicate that under aged 6061 Al-SiC composite is reasonably susceptible to corrosion in lower concentrations of acetic acid. But its corrosion rate is comparatively higher at higher temperatures like 45 °C and 50 °C, in all the three selected concentrations of acetic acid. Corrosion rate of the 6061 Al-SiC increases with increase in temperature for all concentrations of acetic acid. It is possibly because of the increased kinetics of the reaction. As the acid concentration increases from 10 % to 30 %, corrosion rate shows an increase, as more acidic ions are present and so, more number of metal ion dissolution will take place leading to an increase in the corrosion rate. Tafel plots for the sample in 10% acetic acid at 30 °C in the presence and absence of inhibitor is presented in Figure. 4. Similar Tafel plots were obtained for all other experimental environments, from which corrosion rates were calculated.

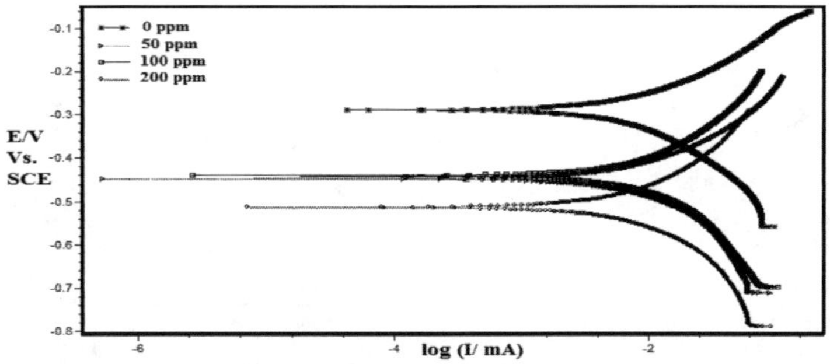


**Figure 4**. Tafel plots for underaged 6061 Al-SiCp composite in 10% acetic acid at 30 °C with and without inhibitor.

**Table 5**. Inhibition efficiency of BI

| Temperature (°C) | Inhibitor concentration (ppm) | Inhibitor efficiency (IE) (%), Acetic Acid (%) | | |
|---|---|---|---|---|
| | | 10 % | 20 % | 30 % |
| 30 °C | 50 | 13.29 | 17.05 | 10.29 |
| | 100 | 21.81 | 30.42 | 21.79 |
| | 200 | 28.53 | 37.98 | 34.19 |
| 35 °C | 50 | 14.41 | 17.88 | 16.28 |
| | 100 | 16.02 | 22.10 | 29.15 |
| | 200 | 25.73 | 29.90 | 37.05 |
| 40 °C | 50 | 33.71 | 48.15 | 27.61 |
| | 100 | 48.31 | 60.82 | 50.22 |
| | 200 | 54.08 | 64.58 | 53.14 |
| 45 °C | 50 | 41.91 | 45.21 | 29.46 |
| | 100 | 45.00 | 52.33 | 43.64 |
| | 200 | 50.46 | 55.60 | 52.89 |
| 50 °C | 50 | 44.02 | 22.34 | 30.50 |
| | 100 | 50.75 | 37.05 | 41.98 |
| | 200 | 58.56 | 53.25 | 58.90 |

Aluminum is an exceptionally reactive metal, but it has a good corrosion resistance due to the formation of passive oxide film. But, aluminum is susceptible to galvanic corrosion when it is coupled to a nobler material. This is a matter of great concern that, whether the reinforcement constituents will encourage galvanic corrosion of the metal or alloy matrix. The degree of galvanic corrosion depends on the environment to which the MMC is exposed and on the properties and amount of the reinforcement constituent. Corrosion in Al based MMCs is also influenced by microstructural changes during fabrication, caused by the reinforcement/matrix reactions, which may lead to the formation of new phase or intermetallics at the reinforcement/matrix interface [15].

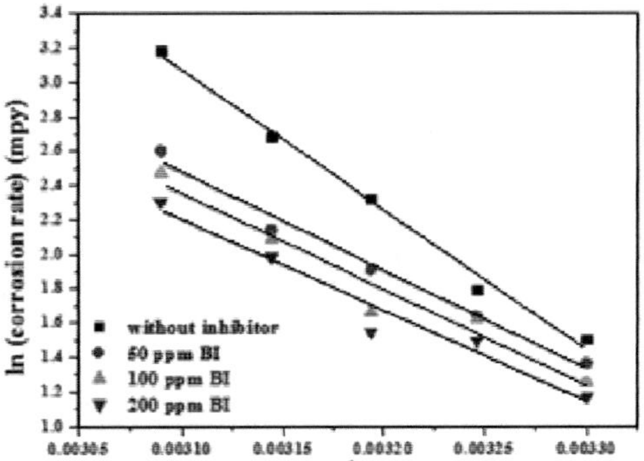


**Figure 5**. Arrhenius plots for 10 % acetic acid in the presence of BI.

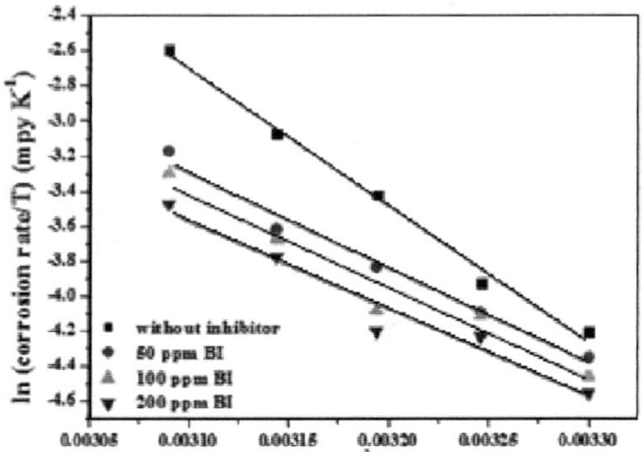


**Figure 6**. Plot of ln (CR/T) vs. (1/T) for 10% acetic acid in the presence of BI.

**Table 6**. Activation parameters for the corrosion of 6061 al-sic in acetic acid in the presence of BI

| Acetic acid concentration | Inhibitor concentration | Ea (KJ/mol) | ΔH# (KJ/mol) | ΔS# (J/mol K) |
|---|---|---|---|---|
| | 0ppm | 67.671 | 64.985 | -62.926 |
| 10 vol % | 50 ppm | 42.585 | 41.597 | -145.658 |
| | 100 ppm | 46.274 | 43.673 | -137.990 |
| | 200 ppm | 47.541 | 45.780 | -132.589 |
| | 0 ppm | 61.622 | 59.099 | -82.962 |
| 20 vol % | 50 ppm | 45.356 | 42.809 | -140.772 |
| | 100 ppm | 50.887 | 48.103 | -122.713 |
| | 200 ppm | 53.128 | 50.585 | -113.355 |
| | 0 ppm | 64.135 | 61.592 | -73.957 |
| 30 vol % | 50 ppm | 43.534 | 41.976 | -142.134 |
| | 100 ppm | 52.536 | 48.378 | -120.174 |
| | 200 ppm | 54.145 | 50.728 | -110.856 |

From Table 5, it is apparent that, BI is an effective inhibitor. The maximum inhibition efficiency obtained using BI was 64.58%. It was attained at 40 °C in 20 % acetic acid when 200 ppm of BI was added. Activation energy (Ea) for the corrosion processes of Al-SiC samples in varying concentrations of acetic acid at different temperatures was calculated from the Arrhenius equation and The enthalpy of activation (ΔH#) and entropy of activation (ΔS#) values for the corrosion process were calculated from transition state theory equation. Free energy of adsorption was calculated using the relation (2)

$$\Delta Gads = -RT\ln[55.5 * \frac{\theta}{C(1-\theta)}] \qquad (2)$$

where, ΔGads = Free energy of adsorption (KJ/mol), R is real gas constant, T is temperature (K), C is concentration of inhibitor (mol/dm$^3$), $\theta$ is surface coverage of inhibitor molecules. Figure 5 shows Arrhenius plots for 10% acetic acid in the presence of BI, at different temperatures. The plot of ln (CR/T) against (1/T) for 10% acetic acid are shown in Figure 6. Activation energy, enthalpy of activation, entropy of activation and free energy of adsorption for other experimental conditions were also calculated.

**Table 7**. Free energy of adsorption (ΔG ads) of inhibitor BI

| Acetic acid concentration | Temperature | Free energy of adsorption ($\Delta G_{ads}$) (KJ/mol) | | |
|---|---|---|---|---|
| | | 50 ppm | 100 ppm | 200 ppm |
| | 30 °C | -24.960 | -24.723 | -23.881 |
| | 35 °C | -25.613 | -24.158 | -23.912 |
| 10 % | 40 °C | -28.905 | -28.685 | -27.484 |
| | 45 °C | -30.292 | -28.792 | -27.540 |
| | 50 °C | -30.999 | -29.864 | -28.852 |
| | 30 °C | -25.700 | -25.855 | -24.959 |
| | 35 °C | -26.271 | -25.174 | -24.444 |
| 20 % | 40 °C | -30.472 | -30.006 | -28.622 |
| | 45 °C | -30.647 | -29.569 | -28.086 |
| | 50 °C | -31.994 | -28.360 | -28.273 |
| | 30 °C | -24.230 | -24.720 | -24.544 |
| | 35 °C | -25.982 | -26.126 | -25.269 |
| 30 % | 40 °C | -28.157 | -28.884 | -27.386 |
| | 45 °C | -28.846 | -28.646 | -27.797 |
| | 50 °C | -29.433 | -28.914 | -28.889 |

It is a common supposition that the adsorption of inhibitors on to the sample surface is the first step in the inhibition mechanism. Organic molecules could be adsorbed on the metal surface by one of the following mechanisms:

- Electrostatic interaction between charged surface of metal and the charge of inhibitor.
- Interaction of unshared electron pairs in the inhibitor molecule with the metal.
- Interaction of $\pi$-electron with metal.
- A combination of all the above types [16].

BI is a 5-membered N-containing heterocycle (imidazole), attached to a benzyl nucleus. It is very stable towards oxidation and reductive agents. It contains two nitrogen hetero-atoms: N1 or 'pyrrolic' nitrogen atom at the 1-position in the molecule, whose unshared electron pair take part in imidazole $\pi$-

electron sextet and N3 or 'pyridyne' nitrogen atom at the 3-position in the molecule, whose unshared electron pair is free and eases the reaction with acids to give salts. So it is apparent that 'pyridine' N3 is a donor of electrons in BI molecule. However BI is an organic base which protonize in an acid medium, predominantly affecting the nitrogen atom (N3) in the imidazole ring. Thus, they become cations. Also the presence of acetate ions in the solution should be considered too. They have a strong adsorbability on the metal surface which brings about a negative charge supporting the adsorption of cation type inhibitors [17]. The inhibitors are classified into three groups according to the temperature effects:

- Inhibitors whose IE decreases with temperature increase. The value of the activation energy Ea found is greater than that in the uninhibited solution.
- Inhibitors whose IE does not change with temperature variation. The activation energy does not change with the presence or absence of inhibitors.
- Inhibitors in whose presence the IE increases with temperature increase, while the value of Ea for the corrosion process is smaller than that obtained in the uninhibited solution.

Studies also indicate that the corrosion process taking place in the presence of powerful inhibitors is characterized by an activation energy whose value is smaller than that of the uninhibited process. The lower value of the activation energy of the process in the presence of an inhibitor when compared to that in its absence is because of chemisorption, while the opposite is the case with physical adsorption. It is considered that the increase of IE with temperature increase, is owing to the change in the nature of adsorptions. To be exact, the inhibitor is adsorbed physically at lower temperatures, while chemisorption is favored as temperature increases. [18].

It has been reported that the ΔGads value up to -20 KJ/mol or less negative specifies a physisorption, while values more negative than -40 KJ/mol involves chemisorption where sharing or transfer of electrons between inhibitor and metal to form a coordinate type bond takes place [11]. Values of ΔGads between this ranges indicates that adsorption may involve mixed interactions, i.e. chemical and physical adsorption. Recently studies has attributed values of ΔGads between -27.10 to -32.41 KJ/mol to chemisorption which involves direct adsorption in the presence of lone pairs of electrons and π-electrons [19].

It is obvious from Table 5, IE increases with temperature. Table 6 displays that, Ea values obtained for inhibited solutions are smaller than that obtained in the uninhibited solution. Values of ΔGads for BI, comes in the range of -23.88 KJ/mol to -31.99 KJ/mol. Taking into consideration the literature and results mentioned above, subsequent inferences could be drawn concerning the inhibitor:

Experimental nature of data confirmed that the inhibition effect attained is good with BI. This effect can be credited to a strong adsorption bond, which can be attributed to chemisorption. The lower value of the activation energy in the presence of BI when compared to that in its absence is attributed to its chemisorption. ΔGads values obtained proposes that the inhibitor is adsorbed physically at lower temperatures, while chemisorption is favored as temperature increases. In view of this, it can be established that the inhibitor get adsorbed on the metal surface by mixed adsorption, where chemisorption is significant.

## 4. Conclusions

Aging profile of T6 treated Al-$SiC_p$ was obtained using Rockwell B hardness. Under aged sample was subjected to corrosion in acetic acid. Highest corrosion rate obtained was 30.69 mpy at 50 °C in 30 volume % acetic acid. Corrosion was inhibited using BI and maximum efficiency of inhibition obtained was 64.58 %. Activation parameters and free energy of adsorption were found out for all experimental conditions. IE increases with temperature and Ea values obtained for inhibited solutions are smaller than that obtained in the uninhibited solution which suggested that BI gets adsorbed on the composite surface by mixed adsorption, where chemisorption is predominant.

## 5. References

[1] Lloyd D J 1994 *International Materials Reviews* **39** 1
[2] Paciej R C and Agarwala V S 1986 *Corrosion* **42** 718
[3] McCafferty E, Hubler G K and Natishan P P M 1987 *Mat. Sci. Eng.* **87** 1
[4] Burleigh T D, Rennick R C and Bovard F. S 1993 *Corrosion* **49** 683
[5] Nayak J and Hebbar K R 2008 *Trans. Indian Inst. Met.* **61** 221
[6] Rajasekaran S, Udayashankar N K, and Nayak J 2012 *International Scholarly Research Network* **2012** 1
[7] Singh S K, Mukherjee A K and M. M. Singh M M 2011 *Indian Journal of Chemical Technology* **18** 291
[8] Rafiquee M Z A, Khan S, Saxena N and Quraishi M A 2007 *Portugaliae Electrochimica Acta,* **25** 419
[9] Singh V B and Singh R N 1995 *Corrosion Science* **37** 1399
[10] Viswanathan S S 2010 *Recent Patents on Corrosion Science* **2** 6
[11] Wang H L, Fan H B and Zheng J S 2002 *Mater. Chem. Phys.* **77** 655
[12] Xiume W, Huaiyu Y and Fuhui W 2011 *Corrosion Science* **53** 113
[13] Appendino P, Badini C, Marino F and Tomasi A 1991 *Mat. Sci. Eng.***A135** 275
[14] Ozturk F, Sisman A, Toros S, Kilic S, and Picu R C 2010 *Materials and Design* **31** 972
[15] Aylor D M and Moran P J 1985 *Journal of Electrochemical Soc.* **132** 1277
[16] Aljourani J, Raeissi K and Golozar M A 2009 *Corrosion Science* **51** 1836
[17] Popova A, Sokolova E, Raicheva S and Christov M 2004 *Corrosion science* **46** 1333
[18] Popova A, Sokolova E, Raicheva S and Christov M 2003 *Corrosion Science* **45** 33
[19] Obot I B and Obi-Egbedi N O 2010 *Corrosion Science* **52** 657

# Ranking Schools' Academic Performance Using a Fuzzy VIKOR

**Suhaina Musani[1,2] and Abdul Aziz Jemain[3]**

[1]Faculty of Computing and Informatics, Universiti Malaysia Sabah, Labuan International Campus, Jalan Sg. Pagar, 87000 Wilayah Persekutuan Labuan, Malaysia.

[3]School of Mathematical Sciences, Universiti Kebangsaan Malaysia, 43600 Bangi, Selangor Darul Ehsan, Malaysia

**Abstract.** Determination rank is structuring alternatives in order of priority. It is based on the criteria determined for each alternative involved. Evaluation criteria are performed and then a composite index composed of each alternative for the purpose of arranging in order of preference alternatives. This practice is known as multiple criteria decision making (MCDM). There are several common approaches to MCDM, one of the practice is known as VIKOR (Multi-criteria Optimization and Compromise Solution). The objective of this study is to develop a rational method for school ranking based on linguistic information of a criterion. The school represents an alternative, while the results for a number of subjects as the criterion. The results of the examination for a course, is given according to the student percentage of each grade. Five grades of excellence, honours, average, pass and fail is used to indicate a level of achievement in linguistics. Linguistic variables are transformed to fuzzy numbers to form a composite index of school performance. Results showed that fuzzy set theory can solve the limitations of using MCDM when there is uncertainty problems exist in the data.

Keywords: Multi Criteria Decision Making; VIKOR; fuzzy numbers; composite index

## 1. Introduction

Since the last few years, the evaluation of the education system with excellent schools ranks has become increasingly popular as a benchmark of the education system. School ranking as an evaluating school performance will provide a direct influence on schools involved [1] [2]. According to Wu et al. [3], evaluating performance is important to the administration of educational institutions to determine the market which will influence the publics' perception to the institution. In addition, it also affects the expenses allocated for student recruitment and operations, as well as acting as a guide to strategic planning institution. On this basis, implementation of school ranking is not meant to punish but to identify schools in need of assistance in terms of infrastructure, financing, enhancement of teachers and also for the development of an environment conducive in teaching and learning process [4]. To achieve these goals, the study will focus on the standard school examination results. Assessment of

[2] Corresponding author. E-mail address: shonsue@yahoo.com.my (Suhaina Musani)

academic achievement on standardized examination results is the most practical way to rank the school in this study [5].

To choose an alternative by priority usually take into many factors to consider such as limited resources, organizational goals and requirements, risk and many others. Both qualitative and quantitative criteria may affect the assessment of each alternative will make the selection process more complex, challenging and unique. Then a more systematic method must be determined to tackle the problem of these criteria [6]. Often the criteria evaluated are determined by the subjective perception and personal considerations, then fuzzy MCDM approach can explain in more detail how decision makers can make an assessment of the alternatives, then choose the best solution. As stated in the assessment of different linguistic variables criteria, then the evaluation process should be carried out in a fuzzy environment. This study aims to develop an effective fuzzy MCDM approach to solve school ranking problems with application of fuzzy VIKOR method. The main idea of this method is to utilize compromise ranking method by comparing the measure of closeness to the ideal alternative to find the best solution. Recently, the usage of fuzzy VIKOR method has been increasing as a medium of decision making in fuzzy MCDM problem solving [7] [8] [9] [10] [11].

This article is organized as follows; the methodology fuzzy MCDM is presented briefly in the next section. An empirical study of a school ranking is presented in Section 3. Finally, conclusion is provided in the final section in this study.

## 2. Methodology

The Methodology of applying Fuzzy VIKOR for school ranking is presented in the following subsections.

### 2.1 Fuzzy MCDM

In the classic MCDM, evaluation of alternatives and weights measured in numbers or crisp and it depends on the consideration of the researcher. Usually the alternatives assessment and the important weights of the criteria cannot be measured reliably; in which case it may come from a variety of sources including the information cannot be quantified, imprecise, and uncertain with conflict of preferences involved in the selection process [12]. In this situation, fuzzy set theory is introduced into MCDM by Bellman and Zadeh [13] to model the uncertainty inherent in human judgment and is known as fuzzy MCDM.

In fuzzy MCDM, performance evaluation and weighting usually represented by fuzzy numbers. According to Liu et al. [14] triangular fuzzy numbers and trapezoidal fuzzy numbers (TzFN) are the most commonly used in the theory and practice of fuzzy number. In fact, the triangular fuzzy number is a special case of TzFN. When the two middle values are the same number, TzFN will become fuzzy triangular numbers. For the sake of simplicity and without loss of generality, TzFN prefer to represent linguistic variables in this study. For example, a positive TzFN $\tilde{A}$ marked as $(x_1 = 2, x_2 = 3, x_3 = 5, x_4 = 6)$ shown in Figure 1.

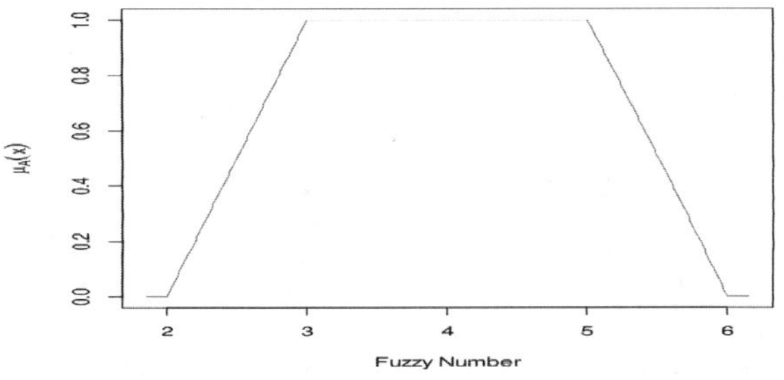


Figure 1: Trapezoidal fuzzy numbers $\tilde{A}$

Given any two positive TzFN $\tilde{A} = (a_1, a_2, a_3, a_4)$, $\tilde{B} = (b_1, b_2, b_3, b_4)$ and a positive real numbers $r$, the algebraic operations of the TzFN can be expressed as follows:

$$\tilde{A} \oplus \tilde{B} = [a_1 + b_1, a_2 + b_2, a_3 + b_3, a_4 + b_4], \quad \tilde{A} \Theta \tilde{B} = [a_1 - b_4, a_2 - b_3, a_3 - b_2, a_4 - b_1],$$

$$\tilde{A} \otimes \tilde{B} \cong [a_1 b_1, a_2 b_2, a_3 b_3, a_4 b_4], \qquad r \otimes \tilde{B} \cong [rb_1, rb_2, rb_3, rb_4]. \tag{1}$$

A linguistic variable is a variable whose value is specified in the form of non-numerical or in words [15]. The concept of a linguistic variable is very useful in dealing with situations which are too complex or too difficult to be described by the quantitative expression. The linguistic values represented by fuzzy numbers. Zadeh [16] provide a level of expertise that is more appropriate in fuzzy linguistic variables. For example, a student achieves academic level as excellence, honours, average, pass and fail depending on the subjective assessment by the assessor. Table 1 below gives the trapezoidal fuzzy numbers to five linguistic variables as previously stated, while Figure 2 shows the membership functions for the sake of visualization.

Table 1: Linguistic variable for each level of achievement

| Linguistic Variable | Trapezoidal Fuzzy Numbers (TzFN) |
|---|---|
| Excellent (g5) | (8,9,10,10) |
| Honours (g4) | (6,7,8,9) |
| Average (g3) | (3,4,5,7) |
| Pass (g2) | (1,2,3,4) |
| Fail (g1) | (0,0,0,2) |

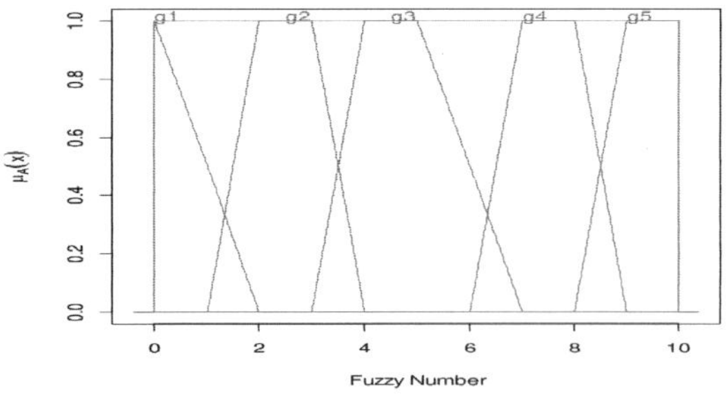


Figure 2: Membership functions for each level of academic achievement

## 2.2 The Fuzzy VIKOR Method

Multi Criteria Optimization and Compromise Solution methods (or VIKOR) has been developed for multi-criteria optimization in a complex system [17]. It determines the compromise solution and best solution from a set of alternatives. Compromise solution will be presented by comparing the degree of closeness to the ideal alternative and each alternative can be evaluated by each criterion function [18]. A systematic approach of a fuzzy VIKOR method for multi criteria in fuzziness environment is given in this section. According Tzeng et al. [19] this approach aims to find the best compromise solution between decision-makers to be consistent with the objectives of human cognition.

VIKOR algorithm based on modified fuzzy numbers stated as follows:

**Step 1**: Expressed multi criteria decision making problem in the matrix format.

There are $m$ alternatives can defined as $A_i$( $i$ = 1, 2, …, $m$) which will be evaluated based on the criteria selected that is $C_j$( $j$ = 1, 2, …, $n$). Each criteria has five grade achievement $g$ = 1, 2, …, 5. Subjective evaluation is done to determine the decision matrix $X = \{x_{ijg}, i = 1,2,...,m; j = 1,2,...,n; g = 1,2,...,5\}$ using linguistic variable as shown in Table 1.

Decision matrix can be expressed as follows:

$$X = \begin{array}{c} \\ A_1 \\ A_2 \\ \vdots \\ A_m \end{array} \begin{array}{c} \begin{array}{cccc} C_1 & C_2 & \cdots & C_n \end{array} \\ \begin{bmatrix} x_{11} & x_{12} & \cdots & x_{1n} \\ x_{21} & x_{22} & \cdots & x_{2n} \\ \vdots & \vdots & & \vdots \\ x_{m1} & x_{m2} & \cdots & x_{mn} \end{bmatrix} \end{array}, \quad i = 1, 2, \ldots, m; j = 1, 2, \ldots, n$$

$$W = [w_1, w_2, ..., w_n]$$

Where $A_1, A_2, ..., A_m$ are the alternatives to be chosen, $C_1, C_2, ..., C_m$ are the evaluation criteria, $x_{ij}$ is the rating of alternative $A_i$ with respect to $C_j$, $w_j$ is the importance weight of the $j$ th criterion holds.

**Step 2**: Construct a fuzzy decision matrix.
The aggregated fuzzy rating $x_{ijg}$ of alternatives with respect to trapezoidal fuzzy numbers is modification from the method of arithmetic weighted average [20] and calculated using the following equation:

$$\tilde{X} = \sum_{i=1}^{m}\sum_{j=1}^{n}\sum_{g=1}^{5} x_{ijg} \otimes TzFN = \left[\tilde{X}_{ij}\right]_{m\times n} = \begin{bmatrix} \tilde{x}_{11} & \tilde{x}_{12} & \cdots & \tilde{x}_{1n} \\ \tilde{x}_{21} & \tilde{x}_{22} & \cdots & \tilde{x}_{2n} \\ \vdots & \vdots & \vdots & \vdots \\ \tilde{x}_{m1} & \tilde{x}_{m2} & \cdots & \tilde{x}_{mn} \end{bmatrix} \quad (2)$$

$$\tilde{W} = [\tilde{w}_1, \tilde{w}_2, ..., \tilde{w}_n]$$

This method is most often used an aggregation process because of simple and flexible operations and fits well with the goals of the study. $\tilde{x}_{ij}$ and $\tilde{w}_j$ are linguistic variables denoted by trapezoidal fuzzy number where $\tilde{x}_{ij}$ is the rating of alternative $A_i$ with respect to $C_j$, $\tilde{w}_j$ is the importance weight of the $j$ th criterion. A trapezoidal fuzzy number can be defined as $\tilde{x}_{ij} = (\tilde{a}_{ij}, \tilde{b}_{ij}, \tilde{c}_{ij}, \tilde{d}_{ij})$.

**Step 3**: Evaluate the fuzzy importance weight of criteria.
The fuzzy weighted values for each criterion will be determined based on the importance of each criterion. Degree of importance of each criterion depends on the burden borne by each school. Relative value is directly proportional to the number of candidates sitting for specific subjects. This clearly shows the value of a higher weight should be given to the criteria that have more number of candidates because it brings an additional burden to ensure that each candidate can understand the subject well [21]. Therefore, if the number of candidates taking the subject $j$, then the fuzzy importance of subjects is given as specified by Diakoulaki et al. [22]:

$$\tilde{w}_j = \tilde{s}_j \Big/ \sum_{j=1}^{n} \tilde{s}_j \quad (3)$$

$\tilde{s}_j$ is the standard deviation value for the criterion $C_n$. Standard deviation $\tilde{s}_j$ is given as follows:

$$\tilde{s}_j = \sqrt{\frac{1}{M}\sum_{m=1}^{M}\left(\tilde{x}_{mn} - \bar{\tilde{x}}_n\right)^2} \tag{4}$$

with $\bar{\tilde{x}}_n = \frac{1}{M}\sum_{m=1}^{M}\tilde{x}_{mn}$ , $0 \le \tilde{w}_j \le 1$ and M = Total number of alternative.

**Step 4**: Determine the fuzzy best value ( $\tilde{x}_j^*$ ) and fuzzy worst value ( $\tilde{x}_j^-$ )

$$\tilde{x}_j^* = \max_i \tilde{x}_{ij},$$
$$\tilde{x}_j^- = \min_i \tilde{x}_{ij}. \tag{5}$$

**Step 5**: Compute the normalized fuzzy decision matrix. The normalized fuzzy decision matrix is calculated to ensure that each criterion value between 0 and 1, so that all the criteria are the standard and are comparable with each other. In this situation, VIKOR method using linear normalization to stabilize [17]. Linear normalization formula indicated by the score $\tilde{S}_i$ and $\tilde{R}_i$ as follows:

$$\tilde{S}_i = \sum_{j=1}^{n}\tilde{w}_j\left(\frac{\tilde{x}_j^* - \tilde{x}_{ij}}{\tilde{x}_j^* - \tilde{x}_j^-}\right) \quad \text{and} \quad \tilde{R}_i = \underset{j}{Max}\left[\tilde{w}_j\left(\frac{\tilde{x}_j^* - \tilde{x}_{ij}}{\tilde{x}_j^* - \tilde{x}_j^-}\right)\right] \tag{6}$$

**Step 6**: Compute the index VIKOR $\tilde{Q}_i$

$$\tilde{Q}_i = v\left(\frac{\tilde{S}_i - \tilde{S}^-}{\tilde{S}^+ - \tilde{S}^-}\right) + (1-v)\left(\frac{\tilde{R}_i - \tilde{R}^-}{\tilde{R}^+ - \tilde{R}^-}\right) \tag{7}$$

where

$$\tilde{S}^+ = \max_i \tilde{S}_i, \tilde{S}^- = \min_i \tilde{S}_i$$
$$R^+ = \max_i \tilde{R}_i, \tilde{R}^- = \min_i \tilde{R}_i$$

$v$ is introduced as the weight in strategy of the maximum group utility. From the literature, it has been inferred that the VIKOR index value is mostly taken as $v = 0.5$.

**Step 7**:

Sorting the value $\tilde{S}$ , $\tilde{R}$ and $\tilde{Q}$ in descending order. The best alternative in order of $\tilde{Q}$ is the maximum possible value of $\tilde{Q}$ based on merit points that was done in this study and symbolized $A^{(1)}$. With the second largest alternative referred to $A^{(2)}$ and so on until an alternative with the smallest value of $\tilde{Q}$ is expressed as $A^{(m)}$.

**Step 8:**

The alternatives $A^{(1)}$ that are in the best position with the maximum value of $\tilde{Q}$ will be proposed as the best alternatives in providing a compromise solution if and only if satisfy two conditions:

***C 1***: Acceptable advantage.

The alternative $A^{(1)}$ accepted as the best advantages when the difference index VIKOR $\tilde{Q}$ between alternative $A^{(2)}$ and $A^{(1)}$ must be greater than or equal to the value of *DQ*, or in other words

$$\tilde{Q}_{(A^{(2)})} - \tilde{Q}_{(A^{(1)})} \ge DQ \text{ with } DQ = \frac{1}{M-1}.$$

***C 2***: Acceptable stability in decision making.
Alternative $A^{(1)}$ must also be in the best ranked by $\tilde{S}$ or $\tilde{R}$.

When one of the conditions is not satisfied, a set of compromise solution will be proposed as follows:
i. If ***C 1*** is not satisfied:
Then the alternative set $A^{(1)}, A^{(2)}, ..., A^{(m)}$ considered together with its best $A^{(m)}$ determined by the relationship $\tilde{Q}_{(A^{(m)})} - \tilde{Q}_{(A^{(1)})} < DQ$.
ii. If ***C 2*** is not satisfied:
Thus, both alternative $A^{(1)}$ dan $A^{(2)}$ are recommended as the best option position [7] or in other words the two alternative can be described as the best alternative.
R software 2:15 was used to analyse data using fuzzy VIKOR method.

## 3 Empirical Study

This study used a sample data of academic achievement for the Sijil Pelajaran Malaysia (SPM) examination results from one of the state in Malaysia. It is aim to evaluate and rank the schools containing multiple conflicting criteria subject. The selected ten schools (School 1, School 2, ..., School 10) are to be evaluated by four major subjects (Subject 1, Subject 2, Subject 3, Subject 4) in five grades for each subject which are excellent (g5), honours (g4), average (g3), pass (g2) and fail (g1) as shown in Table 2:

Table 2: Data on the percentage of academic achievement, $x_{ijg}$

| | Subject | | | | | | | | | | | | | | | | | | | |
|---|---|---|---|---|---|---|---|---|---|---|---|---|---|---|---|---|---|---|---|---|
| | Subject 1 | | | | | Subject 2 | | | | | Subject 3 | | | | | Subject 4 | | | | |
| School | g1 | g2 | g3 | g4 | g5 | g1 | g2 | g3 | g4 | g5 | g1 | g2 | g3 | g4 | g5 | g1 | g2 | g3 | g4 | g5 |
| 1 | 0.017 | 0.238 | 0.213 | 0.336 | 0.196 | 0.174 | 0.234 | 0.191 | 0.204 | 0.196 | 0.106 | 0.264 | 0.140 | 0.162 | 0.328 | 0.120 | 0.295 | 0.115 | 0.150 | 0.321 |
| 2 | 0.110 | 0.294 | 0.298 | 0.212 | 0.086 | 0.209 | 0.402 | 0.176 | 0.148 | 0.066 | 0.275 | 0.275 | 0.199 | 0.144 | 0.106 | 0.206 | 0.324 | 0.139 | 0.080 | 0.252 |
| 3 | 0.099 | 0.167 | 0.167 | 0.257 | 0.311 | 0.186 | 0.190 | 0.072 | 0.176 | 0.376 | 0.323 | 0.173 | 0.123 | 0.114 | 0.268 | 0.241 | 0.159 | 0.027 | 0.100 | 0.473 |
| 4 | 0.000 | 0.000 | 0.025 | 0.364 | 0.610 | 0.000 | 0.076 | 0.254 | 0.415 | 0.254 | 0.000 | 0.085 | 0.186 | 0.356 | 0.373 | 0.025 | 0.136 | 0.110 | 0.178 | 0.551 |
| 5 | 0.052 | 0.278 | 0.268 | 0.309 | 0.093 | 0.224 | 0.316 | 0.235 | 0.143 | 0.082 | 0.469 | 0.250 | 0.146 | 0.094 | 0.042 | 0.306 | 0.337 | 0.082 | 0.092 | 0.184 |
| 6 | 0.061 | 0.287 | 0.243 | 0.291 | 0.117 | 0.143 | 0.313 | 0.165 | 0.187 | 0.191 | 0.367 | 0.218 | 0.162 | 0.144 | 0.109 | 0.223 | 0.266 | 0.135 | 0.135 | 0.240 |
| 7 | 0.031 | 0.336 | 0.288 | 0.205 | 0.140 | 0.276 | 0.443 | 0.110 | 0.123 | 0.048 | 0.333 | 0.228 | 0.259 | 0.114 | 0.066 | 0.368 | 0.342 | 0.075 | 0.066 | 0.149 |
| 8 | 0.070 | 0.140 | 0.193 | 0.333 | 0.263 | 0.14 | 0.333 | 0.123 | 0.246 | 0.158 | 0.228 | 0.228 | 0.333 | 0.140 | 0.070 | 0.246 | 0.351 | 0.105 | 0.123 | 0.175 |
| 9 | 0.207 | 0.414 | 0.207 | 0.135 | 0.036 | 0.366 | 0.429 | 0.071 | 0.054 | 0.080 | 0.509 | 0.287 | 0.130 | 0.065 | 0.009 | 0.291 | 0.400 | 0.100 | 0.073 | 0.136 |
| 10 | 0.143 | 0.388 | 0.240 | 0.158 | 0.071 | 0.235 | 0.281 | 0.143 | 0.189 | 0.153 | 0.51 | 0.201 | 0.139 | 0.093 | 0.057 | 0.297 | 0.25 | 0.073 | 0.078 | 0.302 |

The modified fuzzy VIKOR method is used to solve this multi criteria decision making problem and the computational procedures are stated as follow:

**Step 1**: The observations in decision matrix described the percentage of students who obtained the results of each subjects are shown in Table 2. For School 1, the percentage of students who obtained a fail grade for Subject 1 is 1.7 percent, percentage of students earned a pass grade is 23.8 percent, average grade is 21.3 percent, 19.6 percent for honours and 17.4 percent for excellent grade.

**Step 2**: Five grades for each subject use the linguistic variables. The corresponding fuzzy numbers of five linguistic variables are shown in Table 1 and membership functions for each linguistic variable are shown in Figure 2. According to equation (1) and (2), convert the linguistic variables into TzFN

$(\tilde{a}_{ij}, \tilde{b}_{ij}, \tilde{c}_{ij}, \tilde{d}_{ij})$ as well aggregate the skor (percentage $x_{ijg}$) with TzFN. Fuzzy decision matrix can be referred in Table 3. To be clearer, aggregate the skor and TzFN of School 1 with respect to Subject 1 is computed as:

$$\tilde{x}_{School1,Subject1,\tilde{a}} = (0.017\times 0)+(0.238\times 1)+(0.213\times 3)+(0.336\times 6)+(0.196\times 8)=4.460$$

$$\tilde{x}_{School1,Subject1,\tilde{b}} = (0.017\times 0)+(0.238\times 2)+(0.213\times 4)+(0.336\times 7)+(0.196\times 9)=5.443$$

$$\tilde{x}_{School1,Subject1,\tilde{c}} = (0.017\times 0)+(0.238\times 3)+(0.213\times 5)+(0.336\times 8)+(0.196\times 10)=6.426$$

$$\tilde{x}_{School1,Subject1,\tilde{d}} = (0.017\times 2)+(0.238\times 4)+(0.213\times 7)+(0.336\times 9)+(0.196\times 10)=7.460$$

Table 3: Aggregated trapezoid fuzzy number decision matrix

| | Subject | | | | | | | | | | | | | | | |
|---|---|---|---|---|---|---|---|---|---|---|---|---|---|---|---|---|
| | Subject 1 | | | | Subject 2 | | | | Subject 3 | | | | Subject 4 | | | |
| School | $\tilde{a}$ | $\tilde{b}$ | $\tilde{c}$ | $\tilde{d}$ | $\tilde{a}$ | $\tilde{b}$ | $\tilde{c}$ | $\tilde{d}$ | $\tilde{a}$ | $\tilde{b}$ | $\tilde{c}$ | $\tilde{d}$ | $\tilde{a}$ | $\tilde{b}$ | $\tilde{c}$ | $\tilde{d}$ |
| 1 | **4.460** | **5.443** | **6.426** | **7.460** | 3.600 | 4.426 | 5.251 | 6.421 | 4.277 | 5.170 | 6.064 | 6.983 | 4.103 | 4.983 | 5.863 | 6.778 |
| 2 | 3.147 | 4.037 | 4.927 | 6.249 | 2.340 | 3.131 | 3.922 | 5.242 | 2.585 | 3.309 | 4.034 | 5.403 | 3.235 | 4.029 | 4.824 | 5.916 |
| 3 | 4.694 | 5.595 | 6.495 | 7.450 | 4.471 | 5.285 | 6.100 | 6.982 | 3.368 | 4.045 | 4.723 | 5.900 | 4.623 | 5.382 | 6.141 | 6.936 |
| 4 | 7.144 | 8.144 | 9.144 | 9.559 | 5.364 | 6.364 | 7.364 | 8.364 | 5.763 | 6.763 | 7.763 | 8.576 | 5.941 | 6.915 | 7.890 | 8.475 |
| 5 | 3.680 | 4.629 | 5.577 | 6.804 | 2.531 | 3.306 | 4.082 | 5.459 | 1.583 | 2.115 | 2.646 | 4.219 | 2.602 | 3.296 | 3.990 | 5.194 |
| 6 | 3.704 | 4.643 | 5.583 | 6.770 | 3.461 | 4.317 | 5.174 | 6.291 | 2.441 | 3.074 | 3.707 | 5.127 | 3.406 | 4.183 | 4.961 | 6.079 |
| 7 | 3.550 | 4.520 | 5.489 | 6.668 | 1.895 | 2.618 | 3.342 | 4.680 | 2.215 | 2.882 | 3.548 | 5.075 | 2.154 | 2.785 | 3.417 | 4.711 |
| 8 | 4.825 | 5.754 | 6.684 | 7.684 | 3.439 | 4.298 | 5.158 | 6.263 | 2.632 | 3.404 | 4.175 | 5.667 | 2.807 | 3.561 | 4.316 | 5.491 |
| 9 | 2.135 | 2.928 | 3.721 | 5.099 | 1.607 | 2.241 | 2.875 | 4.232 | 1.139 | 1.630 | 2.120 | 3.750 | 2.227 | 2.936 | 3.645 | 4.900 |
| 10 | 2.628 | 3.485 | 4.342 | 5.653 | 3.066 | 3.832 | 4.597 | 5.821 | 1.629 | 2.119 | 2.608 | 4.201 | 3.354 | 4.057 | 4.760 | 5.828 |

**Step 3**: Fuzzy important weight of criteria are addressed in Table 4. To make it explicit, the fuzzy weight of Subject 1 with respect to $\tilde{a}$ ($\tilde{w}_{Subjek1,\tilde{a}}$) is calculated as:

$$\bar{\tilde{x}}_{\tilde{a}} = \frac{1}{10}\sum_{m=1}^{10}\tilde{x}_{m\tilde{a}} = \frac{1}{10}(39.967) = 3.997,$$

$$\tilde{s}_{Subject1,\tilde{a}} = \sqrt{\frac{1}{10}\sum_{m=1}^{10}\left(\tilde{x}_{m\tilde{a}} - \bar{\tilde{x}}_{\tilde{a}}\right)^2} = \sqrt{\frac{1}{10}\begin{pmatrix}(4.460-3.997)^2+(3.147-3.997)^2+(4.694-3.997)^2+\\(7.144-3.997)^2+(3.680-3.997)^2+(3.704-3.997)^2+\\(3.550-3.997)^2+(4.825-3.997)^2+(2.135-3.997)^2+\\(2.628-3.997)^2\end{pmatrix}}$$

$=1.404$

$\tilde{s}_{Subject1,\tilde{b}} = 1.451$, $\tilde{s}_{Subject1,\tilde{c}} = 1.499$, $\tilde{s}_{Subject1,\tilde{d}} = 1.229$,

$\tilde{s}_{Subject2,\tilde{a}} = 1.156$, $\tilde{s}_{Subject2,\tilde{b}} = 1.242$, $\tilde{s}_{Subject2,\tilde{c}} = 1.329$, $\tilde{s}_{Subject2,\tilde{d}} = 1.186$,

$\tilde{s}_{Subject3,\tilde{a}} = 1.391$, $\tilde{s}_{Subject3,\tilde{b}} = 1.551$, $\tilde{s}_{Subject3,\tilde{c}} = 1.712$, $\tilde{s}_{Subject3,\tilde{d}} = 1.436$,
$\tilde{s}_{Subject4,\tilde{a}} = 1.172$, $\tilde{s}_{Subject4,\tilde{b}} = 1.258$, $\tilde{s}_{Subject4,\tilde{c}} = 1.346$, $\tilde{s}_{Subject4,\tilde{d}} = 1.126$.

$$\tilde{w}_{Subject1,\tilde{a}} = \frac{1.404}{\begin{pmatrix}1.404+1.451+1.499+1.229+1.156+1.242+1.329+1.186+\\ 1.391+1.551+1.712+1.436+1.172+1.258+1.346+1.126\end{pmatrix}} = \frac{1.404}{21.489} = 0.065$$

Table 4: Fuzzy important weight of the criteria

| Subject 1 | | | | Subject 2 | | | | Subject 3 | | | | Subject 4 | | | |
|---|---|---|---|---|---|---|---|---|---|---|---|---|---|---|---|
| $\tilde{a}$ | $\tilde{b}$ | $\tilde{c}$ | $\tilde{d}$ | $\tilde{a}$ | $\tilde{b}$ | $\tilde{c}$ | $\tilde{d}$ | $\tilde{a}$ | $\tilde{b}$ | $\tilde{c}$ | $\tilde{d}$ | $\tilde{a}$ | $\tilde{b}$ | $\tilde{c}$ | $\tilde{d}$ |
| (**0.065**, | 0.068, | 0.070, | 0.057) | (0.054, | 0.058, | 0.062, | 0.055) | (0.065, | 0.072, | 0.080, | 0.067) | (0.055, | 0.059, | 0.063, | 0.052) |

**Step 4**: The fuzzy best value ($\tilde{x}_j^*$) and fuzzy worst value ($\tilde{x}_j^-$) are listed in Table 5.

Table 5: Fuzzy best value ($\tilde{x}_j^*$) and fuzzy worst value ($\tilde{x}_j^-$)

| | Subject 1 | Subject 2 | Subject 3 | Subject 4 |
|---|---|---|---|---|
| $\tilde{x}_j^*$ | (7.144, 8.144, 9.144, 9.559) | (5.364, 6.364, 7.364, 8.364) | (5.763, 6.763, 7.763 8.576) | (5.941, 6.915, 7.890, 8.475) |
| $\tilde{x}_j^-$ | (2.135, 2.928, 3.721, 5.099) | (1.607, 2.241, 2.875, 4.232) | (1.139, 1.630, 2.120, 3.750) | (2.154, 2.785, 3.417, 4.711) |

**Step 5**: Skor $\tilde{S}_i$ and $\tilde{R}_i$ are computed respectively in Table 6 as states in equation (6).

**Step 6**: According to equation (7) the score $\tilde{S}^+$, $\tilde{S}^-$, $\tilde{R}^+$ and $\tilde{R}^+$ are listed below:

$$\tilde{S}^+ = 0.902, \tilde{S}^- = 0.101$$

$$R^+ = 0.120, \tilde{R}^- = 0.050$$

By applying equation (7), the index VIKOR $\tilde{Q}_{School1}$ can be calculated as:

$$\tilde{Q}_{School1} = 0.5 \times \left(\frac{0.288-0.101}{0.902-0.101}\right) + (1-0.5) \times \left(\frac{0.050-0.050}{0.120-0.050}\right) = 0.117 \ .$$

Therefor the values $\tilde{Q}_i$ are shown in Table 6.

Table 6: Ranking of schools for $v = 0.5$

| Schools | $\tilde{S}_i$ | Ranking | $\tilde{R}_i$ | Ranking | $\tilde{Q}_i$ ( index VIKOR) | Ranking |
|---|---|---|---|---|---|---|
| 1 | 0.288 | 9 | 0.050 | 10 | 0.117 | 9 |
| 2 | 0.821 | 3 | 0.083 | 6 | 0.685 | 4 |
| 3 | 0.322 | 8 | 0.069 | 8 | 0.271 | 8 |
| 4 | 0.101 | 10 | 0.055 | 9 | 0.035 | 10 |
| 5 | 0.704 | 6 | 0.093 | 4 | 0.682 | 5 |
| 6 | 0.776 | 4 | 0.104 | 2 | 0.801 | 3 |
| 7 | 0.686 | 7 | 0.074 | 7 | 0.536 | 7 |
| 8 | 0.715 | 5 | 0.090 | 5 | 0.665 | 6 |
| 9 | 0.867 | 2 | 0.100 | 3 | 0.833 | 2 |
| 10 | 0.902 | 1 | 0.120 | 1 | 1.000 | 1 |

**Step 7**: The values of $\tilde{S}_i$, $\tilde{R}_i$ dan $\tilde{Q}_i$ are calculated for all schools by selecting $v = 0.5$ (consensus) as shown in Table 6. As stated in Table 6, the larger $\tilde{Q}_i$ implies the better performance of the schools. Hence $A^{(1)} = 1.000$, $A^{(2)} = 0.833$, …, $A^{(m)} = 0.035$.

**Step 8**:

Once the value $\tilde{S}_i$, $\tilde{R}_i$ and $\tilde{Q}_i$ are obtained by descending order, ***C1*** tested whether filled by the following equation:

$$\tilde{Q}_{(A^{(2)})} - \tilde{Q}_{(A^{(1)})} \geq DQ,$$

The difference value $Q$ for alternative $A^{(1)}$ and $A^{(2)}$ is $-0.167$. This means $\tilde{Q}_{(A^{(2)})} - \tilde{Q}_{(A^{(1)})}$ is less than threshold $DQ = 0.111$. Therefore, first condition ***C1*** is not satisfied then the ***C2*** is tested. The result shows that the second condition ***C2*** is satisfied. Since ***C2*** is satisfied, this research propose the school 10, 9, 6, 2, 5, 8, 7, 3, 1 and 4 as a set of compromise solutions by order preference such an analysis are presented in Table 6 or can be stated as follows:

$$School10 > School9 > School6 > School2 > School5 >$$
$$School8 > School7 > School3 > School1 > School4.$$

## 4 Conclusion

VIKOR method is an effective technique for analysing various types of criteria, and it has been widely used in the rank of a compromise in the field of management. The provided case study has demonstrated the capability of the proposed fuzzy MCDM model to effectively solve school ranking problem under a fuzzy environment. In this method, the ranking of schools are assessed in linguistic variable by trapezoidal fuzzy numbers and the importance weights of criteria are also in fuzzy number. Several studies have applied the method of performance analysis in educational institutions such as Chen and Tzeng [1], Wu et al. [2] and Wu et al. [3]. This study used fuzzy VIKOR method to determine the priority ranks of the performance for ten schools. By using the suggested approach, the ambiguities involved in the assessment of academic achievement on examination results data could be effectively represented and processed to assure a more effective evaluation process. Based on the result, this study can give management implication for the school administrators and Ministry of Education who wish to take countermeasures. In addition, the process and results of this study could provide a reference point for other schools and related educational institutions in their efforts to improve their performances, conduct academic evaluations and to legislate educational policies. To accurately reflect the real performance situations of schools in Malaysia, future research is recommended to take daily schools, boarding schools, religious schools and vocational/technical schools into consideration.

## References

[1] Chen C H & Tzeng G H 2011 Creating the aspired intelligent assessment systems for teaching materials *Expert Systems with Applications* 38(10): 12168-12179

[2] Bradley S, Johnes G & Millington J 2001 The effect of competition on the efficiency of secondary school in England *European Journal of Operational Research* 135:545-568

[3] Wu H Y, Chen J K, Chen I S & Zhuo H H 2012 Ranking universities based on performance evaluation by a hybrid MCDM model *Measurement* 45(5): 856-880

[4] Wu H Y, Lin Y K & Chang C H 2011 Performance evaluation of extension education centres in universities based on the balanced scorecard *Evaluation and Program Planning* 34(1): 37-50

[5] Borhan M & Jemain A A 2011 Assessing schools' academic performance using a belief structure *Social Indicators Research* 106:187–197
[6] Tzeng G H, Teng M H, Chen J J & Opricovic S 2002 Multicriteria selection for a restaurant location in Taipei *International Journal of Hospitality Management* 21:171-187
[7] Büyüközkan G & Ruan D 2008 Evaluation of software development projects using a fuzzy multi-criteria decision approach *Mathematics and Computers in Simulation* 77(5–6): 464-475
[8] Jeya Girubha R & Vinodh S 2012 Application of fuzzy VIKOR and environmental impact analysis for material selection of an automotive component *Materials & Design* 37(0): 478-486
[9] Sanayei A, Farid M S & Yazdankhah A 2010 Group decision making process for supplier selection with VIKOR under fuzzy environment *Expert Systems with Applications* 37(1): 24-30
[10] Shemshadi A, Shirazi H, Toreihi M & Tarokh M J 2011 A fuzzy VIKOR method for supplier selection based on entropy measure for objective weighting *Expert Systems with Applications* 38(10): 12160-12167
[11] Yücenur G N & Demirel N Ç 2012 Group decision making process for insurance company selection problem with extended VIKOR method under fuzzy environment *Expert Systems with Applications* 39(3): 3702-3707
[12] Chen L Y & Wang T C 2009 Optimizing partners' choice in IS/IT outsourcing projects: The strategic decision of fuzzy VIKOR *International Journal of Production Economics* 120(1): 233-242
[13] Bellman R E, Zadeh L A 1970 Decision making in a fuzzy environment. *Management Science* 17(4):141–164
[14] Liu H C, Liu L L, Nan & Mao L X 2012 Risk evaluation in failure mode and effects analysis with extended VIKOR method under fuzzy environment *Expert Systems with Applications* 39(17): 12926-12934
[15] Herrera F, Herrera-Viedma, E & verdegay J L 1996 A model of consensus in group decision making under linguistic assessments *Fuzzy Sets and Systems* 78(1): 73-87
[16] Zadeh L A 1975 The concept of a linguistic variable and its applications to approximate reasoning-Part I *Information Sciences* 8: 199–249
[17] Opricovic S & Tzeng G H 2004 Compromise Solution by MCDM Methods: A Comparative Analysis of VIKOR and TOPSIS *European Journal of Operational Research* 156(2): 445-455
[18] Opricovic S & Tzeng G H 2007 Extended VIKOR method in comparison with outranking methods *European Journal of Operational Research* 178(2): 514-529
[19] Tzeng G H, Lin C W & Opricovic S 2005 Multi-Criteria Analysis of Alternative-Fuel Buses for Public Transportation *Energy Policy* 33(11): 1373-1383
[20] Zarghami M & Szidarovszky F 2009 Revising the OWA operator for multi criteria decision making problems under uncertainty *European Journal of Operational Research* 198(1): 259-265
[21] Musani S & Jemain A A 2013 A fuzzy MCDM approach for evaluating school performance based on linguistic information *AIP Conference Proceeding: The 2013 UKM FST Postgraduate Colloquium* 1571: 1006:1012
[22] Diakoulaki D, Mavrotas G & Papayannakis L 1995 Determining objective weights in multiple criteria problems: The CRITIC method *Computers & Amp; Operations Research* 22(7): 763-770

# Linear representation of algebras with non-associative operations which are satisfy in the balanced functional equations

**Amir Ehsani**

Department of Mathematics, Mahshshar Branch, Islamic Azad University, Mahshahr, Iran.

E-mail: a.ehsani@mhriau.ac.ir

**Abstract.** Algebras with a pair of non-associative binary operations $(f, g)$ which are satisfy in the balanced quadratic functional equations with four object variables considered. First, we obtain a linear representation for the operations, of this kind of binary algebras $(\mathbf{A}, f, g)$, over an abelian group $(\mathbf{A}, +)$ and then we generalize the linear representation of operations, to an algebra $(\mathbf{A}, F)$ with non-associative binary operations which are satisfy in the balanced quadratic functional equations with four object variables.

## 1. Introduction

A binary algebra $\mathbf{A}$ is an ordered pair $(A, F)$, where $A$ is a nonempty set and $F$ is a family of binary operations $f : A^2 \to A$. The set $A$ is called the universe (base, underlying set) of the algebra $\mathbf{A} = (A, F)$. If $F$ is finite, say $F = \{f_1, \ldots, f_k\}$, we often write $(A, f_1, \ldots, f_k)$ for $(A, F)$. The algebra $\mathbf{A}$ is a groupoid if it has only one binary operation.

A quasigroup is a natural generalization of the concept of a group. Quasigroups differ from groups in that they need not be associative. A quasigroup is usually defined to be a groupoid $(A, f)$ such that for any $a, b \in A$ there are unique solutions $x, y$ of $A$, for the equations $f(a, x) = b$ and $f(y, a) = b$.

A loop is a quasigroup with unit $(e)$ such that $f(e, x) = f(x, e) = x$. Groups are associative quasigroups, i.e. they satisfy: $f(f(x, y), z) = f(x, f(y, z))$ and they necessarily contain a unit.

If $(A, f)$ is a quasigroup we say that $f$ is a quasigroup operation.

Quasigroups are important algebraic (combinatorial, geometric) structures which arise in various areas of mathematics and other disciplines. We mention just a few of their applications: in combinatorics (as latin squares, see [1]), in geometry (as nets/webs, see [1] and [2]), in statistics (see [3]), in special theory of relativity (see [4]), in coding theory and cryptography ([5]).

A triple $(\alpha, \beta, \gamma)$ of bijections from a set $G$ onto a set $H$ is called an isotopism of a groupoid $(G, \cdot)$ onto a groupoid $(H, \circ)$ provided $\gamma(x \cdot y) = \alpha x \circ \beta y$ for all $x, y \in G$. $(H, \circ)$ is then called an

Journal of Physics: Conference Series **622** (2015) 012037 doi:10.1088/1742-6596/622/1/012037

isotope of $(G, \cdot)$, and groupoids $(G, \cdot)$ and $(H, \circ)$ are called isotopic to each other. An isotopism of $(G, \cdot)$ onto $(G, \cdot)$ is called an autotopism of $(G, \cdot)$.

Isotopism is a generalization of isomorphism. Isotopic image of a quasigroup is again a quasigroup. A loop isotopic to a group is isomorphic to it. Every quasigroup is isotopic to some loop i.e., it is a loop isotope.

Let $\alpha$ and $\beta$ be permutations of $G$ and let $\iota$ denote the identity map on $G$. Then $(\alpha, \beta, \iota)$ is a principal isotopism of a groupoid $(G, \cdot)$ onto a groupoid $(G, \circ)$ means that $(\alpha, \beta, \iota)$ is an isotopism of $(G, \cdot)$ onto $(G, \circ)$.

A binary quasigroup $(A, f)$ is linear over a group if $f(x, y) = \varphi x + a + \psi y$, where $(A, +)$ is a group, $\varphi$ and $\psi$ are automorphisms of the group $(A, +)$ and $a \in A$ is a fixed element.

A quasigroup linear over an Abelian group is also called a T-quasigroup.

Functional equation $s = t$ is quadratic if every variable appears exactly twice in $s = t$. Quadratic equation is balanced (or linear) if every variable appears exactly once in $s$ and once in $t$.

The following are various functional equations.

$$f(f(x, y), z) = f(x, f(y, z)), \tag{1}$$
$$f(f(x, y), f(u, v)) = f(f(x, u), f(y, v)), \tag{2}$$
$$f(f(x, y), f(u, v)) = f(f(f(x, u), y), v), \tag{3}$$
$$f(x, f(y, z)) = f(f(x, y), f(x, z)), \tag{4}$$
$$f(f(x, y), f(y, z)) = f(x, z), \tag{5}$$
$$f(x, x) = x. \tag{6}$$

Associativity (1), mediality (2) and pseudomediality (3) are balanced, transitivity (5) is quadratic but not balanced while left distributivity (4) and idempotency (6) are not even quadratic.

A classification of quadratic level quasigroup functional equations with four variables is given in [6] and [7].

Let us consider the following balanced identities:

$$f(f(x, y), f(u, v)) = f(f(x, u), f(y, v)) \tag{7}$$
$$f(f(x, y), f(u, v)) = f(f(x, u), f(v, y)) \tag{8}$$
$$f(f(x, y), f(u, v)) = f(f(x, v), f(y, u)) \tag{9}$$
$$f(f(x, y), f(u, v)) = f(f(x, v), f(u, y)) \tag{10}$$
$$f(f(x, y), f(u, v)) = f(f(y, u), f(x, v)) \tag{11}$$
$$f(f(x, y), f(u, v)) = f(f(y, u), f(v, x)) \tag{12}$$
$$f(f(x, y), f(u, v)) = f(f(y, v), f(x, u)) \tag{13}$$
$$f(f(x, y), f(u, v)) = f(f(y, v), f(u, x)) \tag{14}$$
$$f(f(x, y), f(u, v)) = f(f(u, x), f(y, v)) \tag{15}$$
$$f(f(x, y), f(u, v)) = f(f(u, x), f(v, y)) \tag{16}$$

$$f(f(x,y),f(u,v)) = f(f(u,y),f(x,v)) \tag{17}$$
$$f(f(x,y),f(u,v)) = f(f(u,y),f(v,x)) \tag{18}$$
$$f(f(x,y),f(u,v)) = f(f(v,x),f(y,u)) \tag{19}$$
$$f(f(x,y),f(u,v)) = f(f(v,x),f(u,y)) \tag{20}$$
$$f(f(x,y),f(u,v)) = f(f(v,y),f(x,u)) \tag{21}$$
$$f(f(x,y),f(u,v)) = f(f(v,y),f(u,x)) \tag{22}$$

A quasigroup is called medial (or paramedial) if it satisfies (7) (or (22)). In the following theorems it is show that every medial (paramedial) quasigroup is a T-quasigroup.

**Theorem 1.1.** *[8] If $(Q,\cdot)$ is a medial quasigroup, then there exists an abelian group, $(Q,+)$, such that*

$$x\cdot y = \varphi(x) + c + \psi(y),$$

*where $\varphi,\psi \in Aut(Q,+)$, $\varphi\psi = \psi\varphi$ and $c \in Q$.*

**Theorem 1.2.** *[9] If $(Q,\cdot)$ is a paramedial quasigroup, then there exists an abelian group, $(Q,+)$, such that*

$$x\cdot y = \varphi(x) + c + \psi(y),$$

*where $\varphi,\psi \in Aut(Q,+)$, $\varphi\varphi = \psi\psi$ and $c \in Q$.*

Equations (7)-(22) are also called non-Belousov level equations.

**Theorem 1.3.** *[10] Let a quasigroup satisfy a non-Belousov level equation. Then it is a T-quasigroup.*

Certain generalizations of medail and paramdial algebras with non-associative operations (quasigroup operations) and their linear representation were obtained in [11, 12].

## 2. MAIN RESULTS

As a generalization of equations (7)-(22), we consider the following functional equations:

$$f(g(x,y),g(u,v)) = g(f(x,u),f(y,v)) \tag{23}$$
$$f(g(x,y),g(u,v)) = g(f(x,u),f(v,y)) \tag{24}$$
$$f(g(x,y),g(u,v)) = g(f(x,v),f(y,u)) \tag{25}$$
$$f(g(x,y),g(u,v)) = g(f(x,v),f(u,y)) \tag{26}$$
$$f(g(x,y),g(u,v)) = g(f(y,u),f(x,v)) \tag{27}$$
$$f(g(x,y),g(u,v)) = g(f(y,u),f(v,x)) \tag{28}$$
$$f(g(x,y),g(u,v)) = g(f(y,v),f(x,u)) \tag{29}$$
$$f(g(x,y),g(u,v)) = g(f(y,v),f(u,x)) \tag{30}$$
$$f(g(x,y),g(u,v)) = g(f(u,x),f(y,v)) \tag{31}$$
$$f(g(x,y),g(u,v)) = g(f(u,x),f(v,y)) \tag{32}$$
$$f(g(x,y),g(u,v)) = g(f(u,y),f(x,v)) \tag{33}$$

Journal of Physics: Conference Series **622** (2015) 012037 doi:10.1088/1742-6596/622/1/012037

$$f(g(x,y),g(u,v)) = g(f(u,y),f(v,x)) \tag{34}$$
$$f(g(x,y),g(u,v)) = g(f(v,x),f(y,u)) \tag{35}$$
$$f(g(x,y),g(u,v)) = g(f(v,x),f(u,y)) \tag{36}$$
$$f(g(x,y),g(u,v)) = g(f(v,y),f(x,u)) \tag{37}$$
$$f(g(x,y),g(u,v)) = g(f(v,y),f(u,x)) \tag{38}$$

**Definition 2.1.** *A pair $(f,g)$ of binary operations is called:*

- *medial (or entropic, in the sense of [13]), if the algebra $(A,f,g)$ satisfies the equation* (23),
- *paramedial, if the algebra $(A,f,g)$ satisfies the equation* (38).

*A binary algebra $\mathbf{A} = (A,F)$ is called:*

- *medial, if every pair of operations of the algebra $\mathbf{A}$ is medial,*
- *paramedial, if every pair of operations of the algebra $\mathbf{A}$ is paramedial.*

A linear representation of a binary medial algebra with non-associative operations (quasigroup operations) and a linear representation of a binary paramedial algebra with non-associative operations (quasigroup operations) are obtained in [14] and [15], respectively.

The aim of this paper is to show that every binary algebra with quasigroup operations, which satisfies one of the (23)-(38), has linear representation over an abelian group.

**Definition 2.2.** *If $(Q,\cdot)$ is a group, then the bijection, $\alpha : Q \to Q$, is called a holomorphism of $(Q,\cdot)$ if*

$$\alpha(x \cdot y^{-1} \cdot z) = \alpha x \cdot (\alpha y)^{-1} \cdot \alpha z,$$

*for every $x,y,z \in Q$.*

Note that this concept is equivalent to the concept of quasiautomorphism of groups [16]. The set of all holomorphisms of $(Q,\cdot)$ is denoted by $Hol(Q,\cdot)$ and it is a group under the superposition of mappings: $(\alpha \cdot \beta)x = \beta(\alpha x)$, for every $x \in Q$.

The following properties of holomorphisms were proved for Muofang loops in [17].

**Lemma 2.3.** *Let for bijections $\alpha_1, \alpha_2, \alpha_3$ on the group $(Q,\cdot)$, the following identity is satisfied:*

$$\alpha_1(x \cdot y) = \alpha_2(x) \cdot \alpha_3(y).$$

*Then $\alpha_1, \alpha_2, \alpha_3 \in Hol(Q,\cdot)$.*

**Lemma 2.4.** *Every holomorphism $\alpha$ of the group $(Q,\cdot)$ has the following form:*

$$\alpha x = \varphi x \cdot k$$

*where, $\varphi \in Aut(Q,\cdot)$ and $k \in Q$.*

**Lemma 2.5.** *Let for bijections $\alpha_1, \alpha_2, \alpha_3, \alpha_4, \alpha_5, \alpha_6$ on the group $(Q,\cdot)$ the following identity be satisfied:*

$$\alpha_1(\alpha_2(x \cdot y) \cdot z) = \alpha_3 x \cdot \alpha_4(\alpha_5 y \cdot \alpha_6 z).$$

*Then $\alpha_1, \alpha_2, \alpha_3, \alpha_4, \alpha_5, \alpha_6 \in Hol(Q,\cdot)$.*

**Lemma 2.6.** *Let $\alpha_0 \in Hol(Q, \cdot)$ and $k \in Q$, then the mapping*

$$\alpha x = \alpha_0 x \cdot k$$

*$x \in Q$, is also a holomorphism of the group $(Q, \cdot)$.*

**Theorem 2.7.** *Let the set $Q$ forms a quasigroup under the binary operations $f$ and $g$. If the pair $(f, g)$ of binary operations satisfies one of the (23)-(38), then there exists a binary operation '+' under which $Q$ forms an abelian group and for arbitrary elements $x, y \in Q$ we have:*

$$f(x, y) = \varphi_1(x) + \psi_1(y) + c_1,$$
$$g(x, y) = \varphi_2(x) + \psi_2(y) + c_2,$$

*where $c_1, c_2$ are fixed elements of $Q$, and $\varphi_i, \psi_i \in Aut(Q, +)$ for $i = 1, 2$. The group $(Q, +)$ is unique up to isomorphisms.*

*Proof.* We'll start with the proof of one case. Other cases can be proved with minor changes.

Let the set $Q$ forms a quasigroup under the binary operations $f$ and $g$ and the pair of non-associative operations (quasigroup operations) $(f, g)$ satisfies the equation (23). Then there exists an operation, '+', under which $Q$ forms an abelian group on which all the following 6 quasigroups $(A_1, \ldots, A_6)$ are isotopic. And there exists 8 one-to-one mappings, $\alpha, \beta, \gamma, \delta, \epsilon, \psi, \varphi, \chi$ of $Q$ onto itself such that:

$$A_1(x, y) = \delta x + \varphi y, \qquad A_2(x, y) = \delta^{-1}(\alpha x + \beta y),$$
$$A_3(x, y) = \varphi^{-1}(\chi x + \gamma y), \quad A_4(x, y) = \psi x + \epsilon y,$$
$$A_5(x, y) = \psi^{-1}(\alpha x + \chi y), \quad A_6(x, y) = \epsilon^{-1}(\beta x + \gamma y),$$

where, $f = A_1 = A_5 = A_6$ and $g = A_2 = A_3 = A_4$. Since $A_1 = A_5$ and $A_1 = A_5$, we have: $\psi, \epsilon \in Hol(Q, +)$.

Therefor, by using the previous lemma: $f(x, y) = \varphi_1(x) + \psi_1(y) + c_1$.

By the same manner the linearity of operation $g$ will be obtain. □

**Corollary 2.8.** *Let $(Q, F)$ be a binary algebra with quasigroup operations which satisfies one of the (23)-(38), then there exists an abelian group $(Q, +)$ such that every operation $f_i \in F$ is represented by the following rule:*

$$f_i(x, y) = \varphi_i(x) + c_i + \psi_i(y),$$

*where $c_i \in Q$ and $\varphi_i, \psi_i \in Aut(Q, +)$. The group $(Q, +)$ is unique up to isomorphisms.*

*Proof.* We'll proof one of th cases and the other cases can be proved with minor changes.

Let $(Q, F)$ be a binary algebra with quasigroup operations which satisfies the equation (23). If $f_0 \in F$ is a fixed operation, then by the theorem 1.1, $f$ is principally isotopic to the abelian group operation '+' on $Q$. Now, if $f_i \in F$ be any operation, then the pair of operations $(f_0, f_i)$ is medial. Hence $f_0$ and $f_i$ are pricipally isotopic to the another abelian group operation '$*$' on

Journal of Physics: Conference Series **622** (2015) 012037 doi:10.1088/1742-6596/622/1/012037

$Q$. Thus, by transitivity of isotopy, any operation $f_i$ is principally isotopic to the same abelian group operation '+'.

Hence, according to the proof of previous theorem, we have:

$$f_i(x, y) = \varphi_i(x) + \psi_i(y) + c_i,$$

where $c_i \in Q$ and $\varphi_i, \psi_i \in Aut(Q, +)$.

□

## 3. Conclusion

Algebras with non-associative operations (specially, algebras with quasigroup operations) which are satisfy in balanced and non-balanced functional equations are important algebraic (combinatorial, geometric, logic) structures which arise in various areas of mathematics and other disciplines. For example, they have applications in combinatorics (as latin squares), in geometry (as nets/webs), in statistics, in special theory of relativity, in coding theory and cryptography, in logic (as second order formulas, hyperidentities and hypervarieties).

We consider 16 balanced quadratic identities (7)-(22) and we generalize them to 16 balanced quadratic functional equations (23)-(38) with two functional variables and four objective variables. If the set $Q$ forms a quasigroup under the binary operations $f$ and $g$ and the pair $(f, g)$ of binary operations satisfies one of the (23)-(38), then there exists a binary operation '+' under which $Q$ forms an abelian group and for arbitrary elements $x, y \in Q$ we have:

$$f(x, y) = \varphi_1(x) + \psi_1(y) + c_1,$$
$$g(x, y) = \varphi_2(x) + \psi_2(y) + c_2,$$

where $c_1, c_2$ are fixed elements of $Q$, and $\varphi_i, \psi_i \in Aut(Q, +)$ for $i = 1, 2$. The group $(Q, +)$ is unique up to isomorphisms.

As a consequence of this result, by putting $f = g$, a straightforward proof of theorems 1.1, 1.2 and 1.3 could be introduced.

Another corollary is the generalization of above result to an algebra $(Q, F)$ with non-associative operations (quasigroup operations). Let $(Q, F)$ be a binary algebra with quasigroup operations which satisfies one of the (23)-(38), then there exists an abelian group $(Q, +)$ such that every operation $f_i \in F$ is represented by the following rule:

$$f_i(x, y) = \varphi_i(x) + c_i + \psi_i(y),$$

where $c_i \in Q$ and $\varphi_i, \psi_i \in Aut(Q, +)$. The group $(Q, +)$ is unique up to isomorphisms.

There exist various balanced and non-balanced quadratic functional equations with two functional variables and four objective variables (approximately, 150 equations). The presence or absence of a linear representation for a pair of quasigroup operations (and as a generalization, for an algebra with quasigroup operations), which are satisfy in these 150 quadratic functional equations, on an abelian group (or on a group) remains unproven.

## References

[1] Dénes J and Keedwell A D 1974 Latin squares and their applications, Acadmiai Kiadó, Budapest.
[2] Belousov V D 1979 Configurations in algebraic nets (Russian), Shtiinca, Kishinev.
[3] Fisher R A 1966 The design of experiments (8th edition), Oliver & Boyd, Edinburgh.
[4] Ungar A 2001 Beyond the Einstein Addition Law and its Gyroscopic Thomas Precession - The Theory of Gyrogroups and Gyrovector Spaces, Kluwer Academic Publishers, Dordrecht, Boston, London.
[5] Shcherbacov V 2003 On some known possible applications of quasigroups in cryptology, manuscript. http://www.karlin.mff.cuni.cz/drapal/krypto.pdf.
[6] Krapež A 2007 Quadratic level quasigroup equations with four variables I Publ. Inst. Math. (Beograd)(N.S.) **81(95)** 53-67.
[7] Krapež A 2013 Quadratic level quasigroup equations with four variables II Publ. Inst. Math. (Beograd)(N.S.) **93(107)** 29-47.
[8] Toyoda K 1941 On Axioms of linear functions Proc. 1941 Imp. Acad. Tokyo Conf. **17** 211-237.
[9] Němec P and Kepka T 1971 T-quasigroups I Acta Univ. Carolin. Math. Phys. **12** 1 39-49.
[10] Förg-Rob W and Krapež A 2005 Equations which preserve the height of variables Aequat. Math. **70** 63-76.
[11] Ehsani A 2014 Characterization of regular medial algebras ScienceAsia **40** 175-181.
[12] Ehsani A and Movsisyan Yu M 2014 A representation of paramedial n-ary groupoids Asian-European J. Math. **7** 1 (17 pages).
[13] Ehsani A 2011 The generalized entropic property for a pair of operations Izv. Nats. Akad. Nauk Armenii Math. **46** 1 29-34 (Russian). Translation in 2011 J. Contemp. Math. Anal. **46** 1 56-60.
[14] Nazari E and Movsisyan Yu M 2011 Transitive modes Demonstratio Math. **44** 3 511-522.
[15] Ehsani A and Movsisyan Yu M 2013 Linear representation of medial-like algebras Communications in Algebra **41** 9 3429-3444.
[16] Belousov V D 1967 Foundations of the theory of quasigroups and loops Nauka. Moscow.
[17] Movsisyan Yu M 1986 Introduction to the theory of algebras with hyperidentities Yerevan State University Press Yerevan (Russian).

Journal of Physics: Conference Series **622** (2015) 012038 doi:10.1088/1742-6596/622/1/012038

# Interval estimates and their precision

**Luboš Marek, Michal Vrabec**

University of Economics, Prague, Faculty of Informatics and Statistics, W. Churchill Sq. 4, CZ – 130 67 Prague 3

marek@vse.cz, vrabec@vse.cz

**Abstract**. A task very often met in in practice is computation of confidence interval bounds for the relative frequency within sampling without replacement. A typical situation includes pre-election estimates and similar tasks. In other words, we build the confidence interval for the parameter value *M* in the parent population of size *N* on the basis of a random sample of size *n*. There are many ways to build this interval. We can use a normal or binomial approximation. More accurate values can be looked up in tables. We consider one more method, based on MS Excel calculations. In our paper we compare these different methods for specific values of *M* and we discuss when the considered methods are suitable. The aim of the article is not a publication of new theoretical methods. This article aims to show that there is a very simple way how to compute the confidence interval bounds without approximations, without tables and without other software costs.

## 1. Introduction

A frequent task undertaken in public opinion polls is that of estimating election preferences of political parties, that is, estimating a proportion of voters who are going to vote for party "xy" (all of our text will strictly avoid references to any particular parties). The tasks of this type, we meet very often and need not be only about elections. From the viewpoint of statistics it means finding a point or interval estimate of an unknown parameter value for hypergeometric distribution – in principle estimating the relative frequency within sampling without replacement. In such types of problems we usually work with approximations, most often those using the normal distribution. In this paper we are, however, going to show that this approach may often lead to inaccurate results. Moreover, agencies always publish only point estimates, completely ignoring interval estimates, which are more informative, especially in this case, even though the interval estimates are easy to find.

Our task therefore is to estimate an unknown parameter value $M$ or $P$ in a parent population of size $N$ on the basis of a random sample of size $n$. There are several approaches to such estimation. In this paper we are going to mutually compare several methods and suggest our own very simple, accurate and efficient method for finding the confidence interval bounds of the unknown parameter value $M$, even without using any specialized statistical software. This means that regarding potential errors we only focus on statistical properties of the estimates (if a correct theory is not used, it would be useless to consider other kinds of errors). In other words, it is the only component of the total error – a discrepancy between the normal distribution (routinely used for the construction of interval estimates) and the exact construction based on the actual distribution of the sample relative frequency (binomial and hypergeometric).

In the following text, we describe the commonly used methods and we compare the results achieved by simulations. We discuss when the considered methods are suitable and we suggest very easy method how to obtain the best results with using MS Excel software.

## 2. Methodology

Let us first focus on the construction of confidence intervals for relative (and possibly absolute) frequencies within the sampling without replacement. We derive a $100(1-\alpha)\%$ confidence interval for parameter $P$ (relative frequency) as

$$P' < P < P'' \tag{1}$$

where $P$ is the relative frequency, $P'$ is the lower bound of the confidence interval, and $P''$ is its upper bound. The confidence interval for absolute frequency $M$ is given by the formula

$$M' < M < M'' \tag{2}$$

where $M$ is the absolute frequency, $M'$ is the lower bound of the confidence interval, and $M''$ is its upper bound. Values $P'$, $P$, $P''$ and $M'$, $M$, $M''$ are governed by relations

$$P = \frac{M}{N}, \quad P' = \frac{M'}{N} \text{ and } P'' = \frac{M''}{N}. \tag{3}$$

Value $1-\alpha$ stands here for the confidence level (most often it is $1-\alpha = 0.95$, that is, $\alpha = 0.05$, hence we speak about the 95% confidence interval). Let us denote by $\alpha_1$ probability $\mathbf{P}(P < P') = \alpha_1$ and by $\alpha_2$ probability $\mathbf{P}(P > P'') = \alpha_2$, where $\alpha_1 + \alpha_2 = \alpha$. Usually it is $\alpha_1 = \alpha_2 = \frac{\alpha}{2}$.

The lower bound of the confidence interval $M'$ (for the absolute frequency), or the lower bound $P'$ (for relative frequency) is given by solving the following equation (for given $m$ and $\alpha_1$; here $m$ is the "success" count in a sample of size $n$)

$$\sum_{x=m}^{n} \frac{\binom{M'}{x}\binom{N-M'}{n-x}}{\binom{N}{n}} \doteq \alpha_1 \quad \text{or} \quad \sum_{x=m}^{n} \frac{\binom{N\cdot P'}{x}\binom{N\cdot(1-P')}{n-x}}{\binom{N}{n}} \doteq \alpha_1 \tag{4}$$

This expression equals $1-F(m-1)$, where $F$ is the distribution function of the hypergeometric distribution with parameters $M'$, $N$, and $n$; or parameters $NP'$, $N$, and $n$.

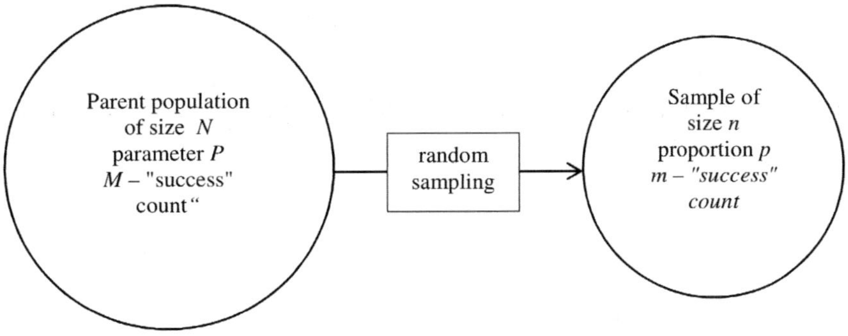


**Figure 1.** Diagram of random sampling without replacement.

The upper bound $M''$ or $P''$ (for the relative frequency) of the confidence is obtained by solving the following equation (for given $m$ and $\alpha_2$; here $m$ is the "success" count in a sample of size $n$)

$$\sum_{x=0}^{m} \frac{\binom{M''}{x}\binom{N-M''}{n-x}}{\binom{N}{n}} \doteq \alpha_2, \quad \text{or} \quad \sum_{x=0}^{m} \frac{\binom{N\cdot P''}{x}\binom{N(1-P'')}{n-x}}{\binom{N}{n}} \doteq \alpha_2. \tag{5}$$

This expression equals $F(m)$, where $F$ is the distribution function of the hypergeometric distribution with parameters $M''$, $N$, and $n$; or parameters $NP''$, $N$, and $n$.

Setting $\alpha_1 = 0$ (i.e., $\alpha_2 = \alpha$), we obtain the *right-hand confidence interval* $0 \le P \le P''$. On the other hand, setting $\alpha_2 = 0$ (i.e., $\alpha_1 = \alpha$), we obtain the *left-hand confidence interval* $P' \le P \le 1$. If $\alpha_1 = \alpha_2 = \frac{\alpha}{2}$ we get the *bilateral confidence interval* $P' < P < P''$. Solving the presented equations is rather complicated, which explains why no "fully adequate" tables of values $P'$ and $P''$ have been published in the literature yet. Moreover, sufficiently detailed tables (for values of *N, n,* and *m*) would be too large. Hence we use in our examples monographs published in [5].

Methods we briefly describe below can be used for determining the lower and upper bounds of the confidence interval:

- tables,
- normal approximation,
- normal approximation with correction to discontinuity,
- binomial approximation,
- Goal Seek procedure in MS Excel spreadsheet.

## 2.1. Tables

The tabled values were calculated numerically (without any approximations) and these values therefore appear to be most accurate among all methods considered in this paper. The values can be found in [4], [5] and [6].

### 2.2. Approximation 1 – normal distribution

If the sample size is sufficiently large and the relative frequency $M / N$ to be estimated is sufficiently close to *1/2*, say, $n \cdot M / N > 30$, cf. [3], the hypergeometric distribution converges, for fixed $M / N$ and $n \to \infty$, to the normal distribution. If this is the case, the bounds of the confidence interval can be calculated with the aid of this approximation. In practice we of course work with a finite but sufficiently large sample of size $n$. The limit distribution of the statistic $p = m / n$ is then normal with parameters:

$$\mu = P = \frac{M}{N} \quad \text{and} \quad \sigma^2 = \frac{1}{n} P(1-P) \frac{N-n}{N-1}.$$

The above-cited equations then enable us to derive that the $100(1-\alpha)\%$ bilateral confidence interval for the unknown parameter $P$ is given as

$$p - u_{1-\alpha/2} \sqrt{\frac{p(1-p)}{n-1} \frac{N-n}{N}} < P < p + u_{1-\alpha/2} \sqrt{\frac{p(1-p)}{n-1} \frac{N-n}{N}}. \tag{6}$$

### 2.3. Approximation 2 – normal distribution with correction to discontinuity

The approximation of the hypergeometric distribution by normal may be made more accurate by taking into account the discontinuity (cause by our approximating a discrete distribution by a continuous one), that is, adding a corrective term $1/(2n)$. More detailed (statistical) reasoning about this term can be found, e.g., in [4]. This implies

$$p - \frac{1}{2n} - u_{1-\alpha/2} \sqrt{\frac{p(1-p)}{n-1} \frac{N-n}{N}} < P < p + \frac{1}{2n} + u_{1-\alpha/2} \sqrt{\frac{p(1-p)}{n-1} \frac{N-n}{N}}. \tag{7}$$

### 2.4. Approximation 3 – binomial distribution

Here an approximation will again be used; this time based on a relationship between the probability functions of the binomial and hypergeometric distributions. It can be proven that, for $N \to \infty$ and fixed $n$, the hypergeometric distribution converges to the binomial distribution "faster" than to normal. The desired confidence interval is then given as

$$p - \frac{1}{2n} - (p - \frac{1}{2n} - p') \sqrt{\frac{N-n}{N}} < P < p + \frac{1}{2n} + (p'' - p - \frac{1}{2n}) \sqrt{\frac{N-n}{N}}, \tag{8}$$

where $p'$ is the lower bound and $p''$ is the upper bound of the confidence interval for the binomial distribution's parameter. The values can be calculated with the aid of known formulas or looked up in statistical tables. Details can be found in [1] and [2].

### 2.5. Goal Seek procedure in v MS Excel

The following formulas are easily derived from (4) and (5). In fact we obtain

$$\alpha_1 = 1 - F(m-1), \tag{9}$$

equivalently

$$1 - \alpha_1 = F(m-1), \tag{10}$$

implying

$$\alpha_1 = 1 - F(m-1), \tag{11}$$

where $F$ is the distribution function of hypergeometric distribution. We utilize these relationships in the Goal Seek procedure. We look for the 95% confidence interval for parameter $M$. The entire situation is illustrated in Figure 2.

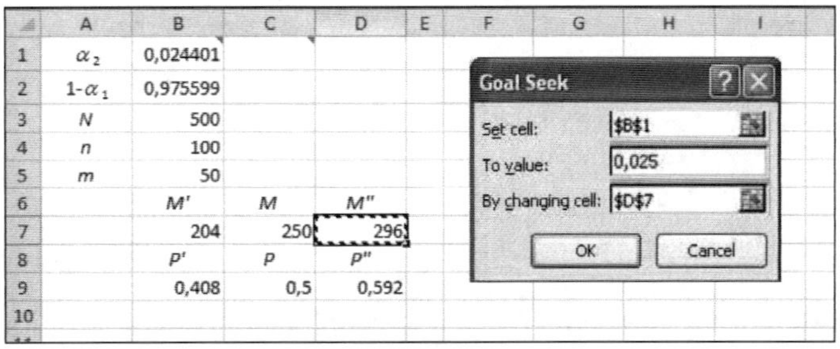


**Figure 2.** Goal Seek procedure in MS Excel.

The theory described above can be studied, e.g., in [1], [2] and [3].

## 3. Simulations

The above theory will be applied to simulated data to mutually compare the results of the considered methods. Such comparison will indicate situations suitable for the use of this or that method.

### 3.1. Simulation 1

Let us assume a random sample of size 100 (sampling without replacement from the parent population of size $N = 500$) in which 50 units pertain to the observed attribute, that is, $n = 100$, $m = 50$, and consequently $p = 0.5$. The unknown value $P$ of the underlying relative frequency lies between the bounds $(P', P'')$ of the confidence interval with the confidence value of 0.95 (alternatively we could consider the absolute frequency value $M$). Most exact values for the bounds of the confidence interval can be looked up in tables. All other methods (based on approximations) provide less accurate results. The obtained results are shown in the Table 1.

**Table 1.** Absolute frequency – Simulation 1.

| Absolute frequency method | lower $M'$ | point $M$ | upper $M''$ | difference $M''-M'$ |
|---|---|---|---|---|
| Tables | 205 | **250** | 296 | 91 |
| Approximation 1 – normal | 206 | **250** | 294 | 88 |
| Approximation 2 – normal with correction | 203 | **250** | 297 | 94 |
| Approximation 3 – binomial | 209 | **250** | 296 | 87 |
| MS Excel | 204 | **250** | 296 | 92 |

**Table 2.** Relative frequency – Simulation 1.

| Absolute frequency method | lower $P'$ | point $P$ | upper $P''$ | difference $P''-P'$ |
|---|---|---|---|---|
| Tables | 0.409 | 0.5 | 0.591 | 0.182 |

| | | | | |
|---|---|---|---|---|
| Approximation 1 – normal | 0.412 | 0.5 | 0.588 | 0.176 |
| Approximation 2 – normal with correction | 0.407 | 0.5 | 0.593 | 0.186 |
| Approximation 3 – binomial | 0.418 | 0.5 | 0.591 | 0.173 |
| MS Excel | 0.408 | 0.5 | 0.592 | 0.184 |

The 95% both sided confidence interval solutions for the relative frequency $P$ can, for all the considered methods, be graphically displayed as line segments representing the interval lengths:

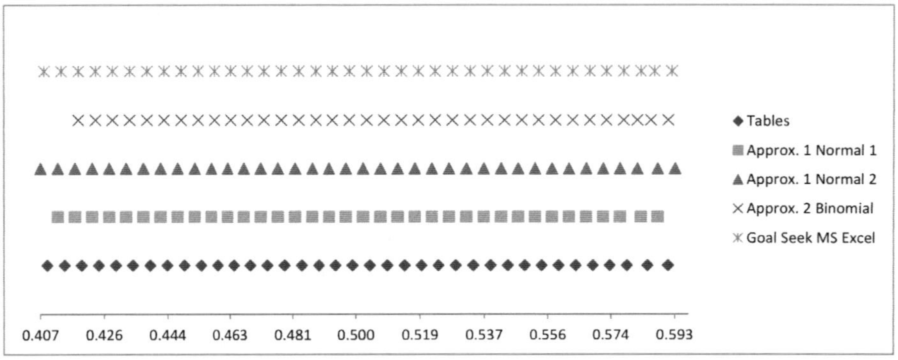


**Figure 3.** Method comparison in Simulation 1.

Intervals are ordered from the bottom upwards. That is, values from the tables are at the bottom, Excel results are at the top. At first sight we can see, from both the Table and the Figure, that at a relative frequency value of 0.5 in the parent population, all intervals are about the same and the results are comparable with minor variations; and the tables and MS Excel provide identical results.

### 3.2. Simulation 2

Let us again assume a random sample of size 100 (sampling without replacement from the parent population of size $N = 10000$), in which 50 units pertain to the observed attribute, that is, $n = 100$, $m = 5$, and consequently $p = 0.05$. The unknown value $P$ of the underlying relative frequency lies between the bounds $(P', P'')$ of the confidence interval with the confidence value of 0.95 (alternatively we could consider the absolute frequency value $M$). Most exact values for the bounds of the confidence interval can again be looked up in tables. All other methods (based on approximations) provide less accurate results. The obtained results are shown in the Table.

**Table 3.** Absolute frequency – Simulation 2.

| Absolute frequency method | lower $M'$ | point $M$ | upper $M''$ | difference $M''$-$M'$ |
|---|---|---|---|---|
| Tables | 170 | 500 | 1130 | 960 |
| Approximation 1 – normal | 100 | 500 | 927 | 827 |
| Approximation 2 – normal with correction | 23 | 500 | 977 | 954 |
| Approximation 3 – binomial | 248 | 500 | 1124 | 876 |

| | | | | |
|---|---|---|---|---|
| MS Excel | 166 | 500 | 1125 | 959 |

**Table 4.** Relative frequency – Simulation 2.

| Absolute frequency method | lower ***P'*** | point ***P*** | upper ***P"*** | difference ***P"-P'*** |
|---|---|---|---|---|
| Tables | 0.017 | 0.05 | 0.113 | 0.096 |
| Approximation 1 – normal | 0.010 | 0.05 | 0.093 | 0.083 |
| Approximation 2 – normal with correction | 0.002 | 0.05 | 0.098 | 0.095 |
| Approximation 3 – binomial | 0.025 | 0.05 | 0.112 | 0.088 |
| MS Excel | 0.017 | 0.05 | 0.113 | 0.096 |

We can again see the 95% both-sided confidence intervals for the relative frequency as line segments representing the interval lengths:

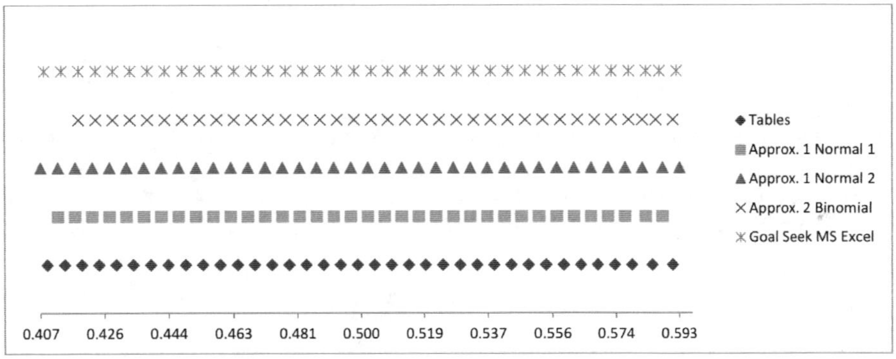


**Figure 4.** Method comparison in Simulation 2.

At first sight we can now see, both from the Table and the Figure, that the achieved results are sharply different and no general comparability takes place. Only two methods, namely, tables and MS Excel, provide identical results.

## 4. Conclusions based on simulations

- If the $P$ parameter value is close to 0.5 (it is even sufficient for it to be between 0.1 and 0.9 (with respect to the sizes of the parent population and the sample), the achieved results are very similar to each other and the methods may be deemed comparable. It is nearly irrelevant which of them is used. No substantial difference exists between the values from tables and those from MS Excel. If we apply this principle to the data from opinion polls, this case covers the instances of parties with large preference proportions, but not those whose preferences are smaller than 10%.

- If parameter $P$ fulfills the condition $P < 0.1$ or $P > 0.9$ (both meaning essentially the same situation), the tabled values should be used, or alternatively the Goal Seek procedure in MS Excel.
- The Goal Seek procedure in MS Excel provides good results in all cases, regardless of the value of $P$. This procedure has another advantage – cost savings on software, because no specialized statistical software is necessary for it. MS Excel is more or less standard installation on any PC, so costs due to its use are practically none. Another argument in favor of the MS Excel procedure is its simplicity and availability.
- the correct procedure (estimation of parameter $P$) has political and economic impacts - estimating election preferences of political parties, market research, public opinion polls and others.

## References

[1] Buonacorsi J P 1987 *A note on confidence intervals for proportions in finite populations* (American Statistician, 41, p. 215 – 218)
This reference has two entries but the second one is not numbered (it uses the 'Reference (no number)' style.
[2] Burstein H 1975 *Finite population correction for binomial confidence limits* (JASA, 70, p. 67 – 69)
[3] Cochran W G 1977 *Sampling Techniques*, 3rd edition (New York, Wiley)
[4] Čermák V and Vrabec M, 1998 *Sampling Survey Theory* (in Czech: Teorie výběrových šetření). University of Economics, Prague. Faculty of Informatics and Statistics)
[5] Chung J H and Delury D B, 1950 *Confidence Limits for the Hypergeometric Distribution* (Toronto, University of Toronto Press)
[6] Quesenberry C P and Hurst D C 1964 Large sample simultaneous confidence intervals for multinomial proportions (Technometrics, 6, p. 191 – 195)

We would like to aknowledge financial support from P402/12/G097.

# The Existence and Stability Analysis of the Equilibria in Dengue Disease Infection Model

**N. Anggriani, A.K. Supriatna, E. Soewono***

Department of Mathematics
Faculty of Mathematics and Natural Sciences, Universitas Padjadjaran
Jl. Raya Bandung-Sumedang km 21 Jatinangor 45363, Indonesia
*Industrial and Financial Mathematics Group
Faculty of Mathematics and Natural Sciences, Institut Teknologi Bandung
Jl. Ganesha 10, Bandung 40132, Indonesia

E-mail: n_anggriani@unpad.ac.id and nsanti_a@yahoo.co.id

**Abstract.**
In this paper we formulate an SIR (Susceptible - Infective - Recovered) model of Dengue fever transmission with constant recruitment. We found a threshold parameter $\Re_0$, known as the Basic Reproduction Number (BRN). This model has two equilibria, disease-free equilibrium and endemic equilibrium. By constructing suitable Lyapunov function, we show that the disease-free equilibrium is globally asymptotic stable whenever BRN is less than one and when it is greater than one, the endemic equilibrium is globally asymptotic stable. Numerical result shows the dynamic of each compartment together with effect of multiple bio-agent intervention as a control to the dengue transmission.

## 1. Introduction

Dengue Haemorrhagic Fever (DHF) is caused by four serotype of viruses. It may result in various clinical manifestations, from an asymptomatic to a fatal one. TDengue virus infection can be devided into two groups : the asymptomatic manifestation (without symptoms) and the symptomatic manifestation (with symptoms). The latter usually appears in common fever (virus syndrome), dengue fever, or dengue hemorrhagic fever (DHF) which includes the dengue shock syndrome (DSS). Common fever occurs in infants, young children, and adults that have been infected by the dengue virus for the first time (primary dengue infection). Common fever cannot be distinguished from the fever caused by infection from other viruses. Basically dengue fever is an acute biphasic fever with symptoms such as headaches and cutaneous rashes. Although dengue fever is not fatal, it may reduce bodily functions. This includes, for instance, aches in joint muscles. DHF and DSS symptoms include fever and signs of blood circulation failure [16]. Preliminary researches on the mathematical model for one-strain DHF epidemiology have been conducted by using DHF models and constant population [6], inconstant population [7], and deterministic models that include vaccination [5],[8],[9]. When ones recover from dengue infection, sometimes later they can be re-infected by the Dengue virus. The reason is due to the reduction of immunity or wanning immunity, as it affects the immunity previously acquired. These patients are categorized under healthy humans who have been infected. Infected humans are divided into two compartments: those who are infected without symptoms and

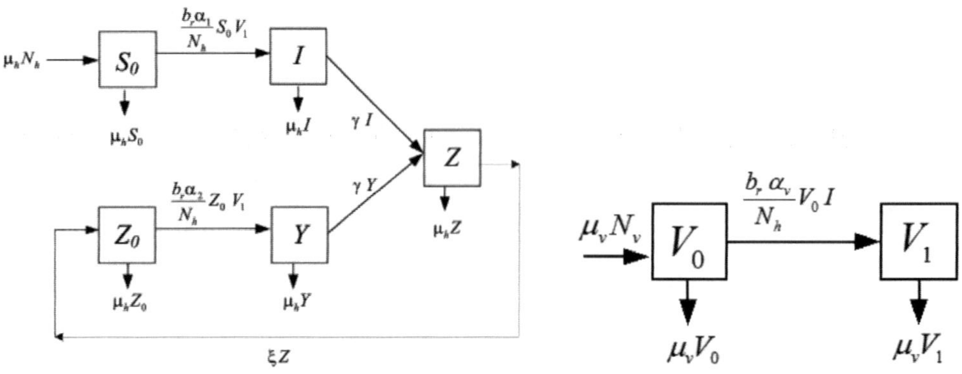

**Figure 1.** Host Transmission and Vector Transmission Diagram.

those who are infected with symptoms. This phenomenon has been modeled in [1] but there is no thorough analysis theorem. The global dynamics of the seventh-dimensional model of dengue disease transmission model in [1] is analyzed using Lyapunov function. In particular, we follow closely the ideas used recently in [13], [14], [15] to establish the global stability of the endemic equilibrium. The organization of this paper is as the following : in section 2, we present the mathematical model describing the transmission of dengue fever with asymptomatic and symptomatic classes [1]. A Mathematical analysis of this model is done in section 3 and section 4. Section 3 discusses the BRN of the model. It also discusses the Local Stability analysis for the endemic equilibrium point, while its global stability analysis is discussed in section 4. The last section shows the numerical analysis of dynamics of each compartment, including the effects of multi agent bio-control. Finally, some conclusions are presented in Section 6.

## 2. Model formulation

In order to derive the equations of the mathematical model, we assumed that host and vector populations have a constant size with birth and death rates for human and vector are constant equal to $\mu_h$ and $\mu_v$ respectively. In the Esteva - Vargas model [6], the majority of the host population eventually becomes permanently immune and the host have long - term immunity. In reality, human usually have one or more than one infection. In our model, individuals in host population became immune or loose immunity.

We assume that the susceptible host is divided into two compartments, $S_0$ (susceptible virgin host) and $Z_0$ (susceptible previously infected host). We consider $S_0$ as a susceptible class with virgin population comes from newborn, while $Z_0$ is a class coming from recovered human with no immunity. The infected class is divided into two classes namely $I$, the asymptomatic of infected human class, and $Y$, the symptomatic of infected human class (DF, DHF, DSS). There is no transmission from $I$ to $Y$ because we assume that the class $Y$ is detectable, hence individuals in $Y$ can be isolated. $Z$ is the class of removed human population. The vector population, due to short life period, is divided into the susceptible $V_0$ and the infective $V_1$. The model is represented by the diagram in Figure 1, with the description of the parameters shown in Table 1. The dynamics equation for host and vector are[1] :

$$
\begin{aligned}
\frac{dS_0}{dt} &= \mu_h N_h - \frac{b_r \alpha_1}{N_h} S_0 V_1 - \mu_h S_0, \\
\frac{dZ_0}{dt} &= \xi Z - \frac{b_r \alpha_2}{N_h} Z_0 V_1 - \mu_h Z_0,
\end{aligned}
$$

$$\begin{aligned}
\frac{dI}{dt} &= \frac{b_r\,\alpha_1}{N_h}\,S_0\,V_1 - (\gamma + \mu_h)\,I,\\
\frac{dY}{dt} &= \frac{b_r\,\alpha_2}{N_h}\,Z_0\,V_1 - (\gamma + \mu_h)\,Y,\\
\frac{dZ}{dt} &= \gamma\,(Y + I) - (\xi + \mu_h)\,I,\\
\frac{dV_0}{dt} &= \mu_v\,N_v - \frac{b_r\,\alpha_v}{N_h}\,V_0\,I - \mu_v\,V_0,\\
\frac{dV_1}{dt} &= \frac{b_r\,\alpha_v}{N_h}\,V_0\,I - \mu_v\,V_1,
\end{aligned} \tag{1}$$

where : $N_h = S_0 + Z_0 + I + Y + Z$ is the total population of host, and $N_v = V_0 + V_1$ is the total population of vectors.

Note that all trajectories in the first quadrant enter or stay inside the region

$$\Omega \;=\; \{(S_0, Z_0, I, Z, V_0, V_1) \in [0,1]^6 \mid 0 \le V_0 + V_1 \le N_v, 0 \le S_0 + Z_0 + I + Z \le N_h\}. \tag{2}$$

| Parameter | Description |
|---|---|
| $\mu_h$ | Host death rate |
| $\mu_h$ | Vector death rate |
| $\alpha_i$ | Probability of successful transmission from host to vector |
| $\alpha_v$ | Probability of successful transmission from vector to host |
| $\xi$ | Wanning immunity |
| $b_r$ | bitting rate per population human |
| $\gamma$ | Recovery rate |
| $N_h$ | Total population of host |
| $N_v$ | Total population of host |

**Table 1.** Definition of parameter

## 3. ANALYSIS OF THE MODEL

### *3.1. Basic Reproduction Number*

As shown in [1] the BRN of the model without protection is:

$$\Re_0 = \sqrt{\frac{b_r^2\,\alpha_1\,\alpha_v\,N_v}{\mu_v\,(\mu_h + \gamma) N_h}}. \tag{3}$$

The value of the basic reproduction number depends on the value of the biting rate per person. This is the reason that if we want to control dengue disease, we can reduce the biting rate per person, for example by applying insect repellent. We can also control the disease by reducing the life time of mosquitoes.

### *3.2. Stability of the endemic equilibrium*

Before proving the local stability of the disease equilibrium, we present some result concerning the existence of equilibrium points of system (1).

PROPOSITION 3.1 *The model 1 has two equilibrium points, the disease free equilibrium* $E_0 = (S_0^0, Z_0^0, I^0, Y^0, Z^0, V_0^0, V_1^0) = (N_h, 0, 0, 0, 0, N_v, 0)$, *and the endemic equilibrium* $E_e = (S_0^*, Z_0^*, I^*, Y^*, Z^*, V_0^*, V_1^*)$, *where:*

$$S_0^* = \frac{N_h(A_1 + \Re_0^2 K)}{\Re_0^2(A_1 + K)}, \tag{4}$$

$$Z_0^* = \frac{A_1 C(\Re_0^2 - 1)(A_1 + K\Re_0^2)\xi}{M(A_1 + K)\Re_0^2}, \tag{5}$$

$$I^* = \frac{A_1(\Re_0^2 - 1)K}{\Re_0^2(A_1 + K)C}, \tag{6}$$

$$Y^* = \frac{A_1 A_2 K(\Re_0^2 - 1)^2 \xi\gamma}{\Re_0^2(A_1 + K)CM}, \tag{7}$$

$$Z^* = \frac{A_1\gamma(\Re_0^2 - 1)(A_1 + A_1(\Re_0^2 - 1) + K\Re_0^2}{M(A_1 + K)\Re_0^2}, \tag{8}$$

$$V_0^* = \frac{N_v(A_1 + K)}{A_1 + K\Re_0^2}, \tag{9}$$

$$V_1^* = \frac{K(\Re_0^2 - 1)N_v}{A_1 + K\Re_0^2}, \tag{10}$$

$$A_i = b_r\,\alpha_i\,N_v;\ i = 1, 2, v,\ K = N_h\,N_v,$$

$$C = \gamma + \mu_h,\ D = \xi + \mu_h \tag{11}$$

$$M = K\,\Re_0^2\,D\,C + \mu_h\,[A_1 + A_2\,(\Re_0^2 - 1)]\,D + A_2\,(\Re_0^2 - 1)\,\gamma\,\mu_h + A_1\,D \tag{12}$$

It can be verified that if $\Re_0 > 1$, then there exists an endemic equilibrium point, and if $\Re_0 < 1$ then the endemic equilibrium disappears. It indicates that to reduce the outbreak of DHF, we should make the value of $\Re_0 < 1$ lower.

The next proposition gives a condition for the local stability of the endemic equilibrium state $E_0$, the local stability of the disease free equilibrium proven in [1].

PROPOSITION 3.2 *If* $\Re_0 > 1$, *then the endemic equilibrium* $E_e$ *is locally asymptotically stable if the inequality*

$$C\,(A_1 + K)N_h(A_1 + K\Re_0^2) < A_1\,K\,(\Re_0^2 - 1)^2 \tag{13}$$

*is also satisfied, otherwise* $E_e$ *is unstable.*

**Proof.** Let the equilibrium $E_e = (S_0^*, Z_0^*, I^*, Y^*, Z^*, V_0^*, V_1^*)$, exists, that is when $\Re_0 > 1$. By using the coordinate system $E_e$, of model (1) at point $E_e$ gives the Jacobian matrix $J(E_e)$:

$$J(E_e) = \begin{bmatrix} J_1 & J_2 \\ J_3 & J_4 \end{bmatrix}$$

where

$$J_1 = \begin{bmatrix} -\frac{(b_r V_1^* \alpha_1 + K)}{N_h} & 0 & 0 \\ 0 & -\frac{(b_r V_1^* \alpha_2 + K)}{N_h} & 0 \\ \frac{(b_r V_1^* \alpha_1)}{N_h} & 0 & -\gamma - \mu_h \end{bmatrix};\ J_2 = \begin{bmatrix} 0 & 0 & 0 & -\frac{(b_r S_0^* \alpha_1)}{N_h} \\ 0 & \xi & 0 & -\frac{(b_r Z_0^* \alpha_2)}{N_h} \\ 0 & 0 & 0 & \frac{(b_r S_0^* \alpha_1)}{N_h} \end{bmatrix}$$

$$J_3 = \begin{bmatrix} 0 & \frac{(b_r V_1^* \alpha_2)}{N_h} & 0 \\ 0 & 0 & \gamma \\ 0 & 0 & -\frac{(b_r V_0^* \alpha_v)}{N_h} \\ 0 & 0 & \frac{(b_r V_0^* \alpha_v)}{N_h} \end{bmatrix}, J_4 = \begin{bmatrix} -\gamma - \mu_h & 0 & 0 & \frac{(b_r Z_0^* \alpha_2)}{N_h} \\ \gamma & -\gamma - \xi & 0 & 0 \\ 0 & 0 & -\frac{(b_r I^* \alpha_v + K)}{N_h} & 0 \\ 0 & 0 & \frac{(b_r I^* \alpha_v)}{N_h} & -\mu_v \end{bmatrix}$$

and $E_e$=$(S_0^*, Z_0^*, I^*, Y^*, Z^*, V_0^*, V_1^*)$ is derived from (4) - (10)

The eigenvalues of $J(E_e)$ are $-\mu_h$, $-\mu_v$, the roots of polynomial $p_2(x) = a_2 x^2 + b_2 x + c_2$, and the roots of polynomial $p_3(x) = a_3 x^3 + b_3 x^2 + c_3 x + d_3$ where

$$\begin{aligned}
a_2 &= N_h(A_1 + \Re_0^2 K)(\gamma + \mu_h)\mu_h + \xi) + b_r K N_v (\Re_0^2 - 1)\alpha_2(\gamma + \mu_h + \xi,) \\
b_2 &= N_h A_1 (\gamma + 2\mu_h + \xi) + K[b_r N_v (\Re_0^2 - 1)\alpha_2 + N_h \Re_0^2 \gamma + 2\mu_h + \xi)], \\
c_2 &= N_h (A_1 + K \Re_0^2) \\
a_3 &= C(A_1 + K) N_h^2 \Re_0^2 (A + K \Re_0^2), \\
b_3 &= b_{3A} + b_{3B} + b_{3C} + a_3 \mu_v, \\
c_3 &= b_{3A}(C + \mu_v) + b_{3B} C + b_{3C} [\gamma + \frac{1}{2}(\mu_h + \mu_v)] + G_{cd}, \\
d_3 &= d_{3A} + b_{3A}(C \mu_v) + b_{3B}(\gamma \mu_v + \mu_v^2) + b_{3C}(\frac{1}{2}\mu_v \mu_h) + G_{cd}. \\
b_{3A} &= b_r C K (A_1 + K) N_h N_v (\Re_0^2 - 1)\Re_0^2 \alpha_1, \\
b_{3B} &= A_1 b_r N_h K (A_1 + K \Re_0^2)\alpha_v (\Re_0^2 - 1), \\
b_{3C} &= 2C(A_1 + K) N_h^2 \Re_0^2 (A_1 + K \Re_0^2)\mu_h, \\
d_{3A} &= A_1 n_r^2 K^2 N_v \gamma (\Re_0^2 - 1)^2 \alpha_1 \alpha_v, \\
G_{cd} &= -b_r^2 N_v \alpha_1 \alpha_v \mu_v Q, \\
Q &= C(A_1^2 + K^2 \Re_0^2) N_h + A_1 K[-K(\Re_0^2 - 1)^2 + C N_h (\Re_0^2 + 1)]
\end{aligned}$$

The values of $C$ and $K$ are in the equation (11). Note that the coefficient $a_2, b_2, c_2 > 0$ if $\Re_0 > 1$, hence all the roots of polynomial $p_2$ have negative real part when $\Re_0 > 1$. If $\Re_0 > 1$, then coefficient $a_3, b_3 > 0$, for the coefficient $c_3, d_3$ will be greater than zero with the condition $G_{cd} > 0$, then $Q < 0$. Then the condition $C(A_1 + K)N_h(A_1 + K\Re_0^2) < A_1 K (\Re_0^2 - 1)^2$ will be satisfied.

Applying Routh - Hurwitz criteria to the polynomial with third degree will ensure that be stability is acquired if the condition $a_3 d_3 < b_3 c_3$ is satisfied. From equation (11), it can be seen that

$$\begin{aligned}
a_3 d_3 &= \frac{1}{2} a_3 (2 b_{3A} C \mu_v + 2 b_{3B} (\mu_h^2 + \gamma \mu_v) + b_{3C} \mu_v \mu_h + 2 G_{cd} + 2 d_{3A} \\
a_3 d_3 &< \frac{1}{2}(b_{3A} + b_{3B} + b_{3C} + a_3 \mu_v)(2 b_{3A}(C + \mu_v) + 2 b_{3B} C + b_{3C}(2\gamma + \mu_h + \mu_v) + 2 G_{cd}) \\
&< b_3 c_3
\end{aligned}$$

this means that $E_e$ is a locally asymptotically stable when $\Re_0 > 1$, and a saddle point where $\Re_0 < 1$. This proves the proposition. ■

## 4. GLOBAL STABILITY ANALYSIS

### *4.1. Disease-free equilibrium*

We now study the global behavior of disease - free equilibrium for system (1).

Journal of Physics: Conference Series **622** (2015) 012039 doi:10.1088/1742-6596/622/1/012039

PROPOSITION 4.1 *If $\Re_0 \leq 1$, then the disease - free equilibrium $E_0$ model (1) is globally asymptotically stable.*

**Proof.**
We follow closely the ideas used recently in [13, 14, 15] to establish the global stability of the non endemic equilibrium. In proving the global asymptotical stability of $E_0$ in $\Omega$ for $\Re_0 \leq 1$, we construct a Lyapunov function $V : \Omega \to R$, where

$$V(t) = S_0^0\left(\frac{S_0}{S_0^0} - S_0^0 \ln\frac{S_0}{S_0^0}\right) + \frac{N_h(\gamma+\mu_h)}{N_v b_r \alpha_v}V_1 + I + \frac{N_h(\gamma+\mu_h)}{N_v b_r \alpha_v}V_0^0\left(\frac{V_0}{V_0^0} - V_0^0 \ln\frac{V_0}{V_0^0}\right). \quad (14)$$

Differentiating with respect to time yields and from model (1) we have $S_0^0 = N_h$ and $V_0^0 = N_v$, then equation(14) becomes :

$$\begin{aligned}\dot{V}(t) &= \mu_h N_h\left(2 - \frac{S_0^0}{S_0} - \frac{S_0}{S_0^0}\right) + \left(\frac{N_h(\gamma+\mu_h)}{N_v b_r \alpha_v}\right)\mu_v N_v\left(2 - \frac{V_0^0}{V_0} - \frac{V_0}{V_0^0}\right) + V_1(\Re_0^2 - 1)\left(\frac{N_h(\gamma+\mu_h)}{N_v b_r \alpha_v}\right)\\ &= -\mu_h N_h\frac{(S_0 - S_0^0)^2}{S_0 S_0^2} - \left(\frac{N_h(\gamma+\mu_h)}{N_v b_r \alpha_v}\right)\mu_v N_v\frac{(V_0 - V_0^0)^2}{V_0 V_0^2} + V_1(\Re_0^2 - 1)\left(\frac{N_h(\gamma+\mu_h)}{N_v b_r \alpha_v}\right) \quad (15)\\ &\leq 0\end{aligned}$$

From the inspection of model (1) , it can be seen that $E_0$ is the largest invariant set under the flow generated by the vector field of model (1) for which $S_0 = S_0^0, Y = Z = 0, V_0 = V_0^0$. Therefore, by [12] the equilibrium $E_0$ is globally asymptotically stable in $\Omega$ for $\Re_0 \leq 1$. This concludes the proof. ■

### 4.2. Endemic equilibrium

PROPOSITION 4.2 *Assume that*

$$N_h = S_0^*, N_v = V_0^*, b_r = \frac{\mu_v\,(\gamma+\mu_h)}{\alpha_1\, p_i\, V_0^*} \quad (16)$$

*If $\Re_0 > 1$, then the disease equilibrium $E_e^*$ is globally asymptotically stable.*

**Proof.** The same approach to proposition 4.1 is applied, In proving the global asymptotical stability of $E_e^*$ in $\Omega$ for $\Re_0 > 1$, we construct a Lyapunov function $L : \Omega \to R$, where

$$L(t) = (S_0 - S_0^* \ln S_0) + \frac{(\gamma+\mu_h)}{p_v V_0^*}(V_0 - V_0^* \ln V_0) + I + \frac{(\gamma+\mu_h)}{p_i V_0^*}V_1. \quad (17)$$

with : $p_i = \frac{b_r\,\alpha_i}{N_h}$, $i = 1, v$. Differentiating with respect to time yields

$$\begin{aligned}\dot{L}(t) &= \mu_h N_h\left(1 - \frac{S_0^*}{S_0}\right) + \mu_h S_0^*\left(1 - \frac{S_0^*}{S_0}\right) + \left(\frac{(\gamma+\mu_h)}{p_v V_0^*}\right)(p_v\, V_0^*\, I) - (\gamma+\mu_h)I\\ &+ \frac{(\gamma+\mu_h)}{p_v V_0^*}\left[\mu_v N_v\left(1 - \frac{V_0^*}{V_0}\right) + \mu_v V_0^*\left(1 - \frac{V_0^*}{V_0}\right)\right]\\ &+ \left(\frac{\mu_v(\gamma+\mu_h)}{\alpha_1\, p_v\, V_0^*}\right)\frac{\alpha_1}{N_h}N_h\, V_1 - \left(\frac{(\gamma+\mu_h)}{p_v V_0^*}\right)\mu_v\, V_1 \quad (18)\end{aligned}$$

From model (1) and condition (16) we have $S_0^* = N_h$ and $V_0^* = N_v$, then equation(19) becomes :

$$\begin{aligned} \dot{L}(t) &= \mu_h N_h (2 - \frac{S_0^*}{S_0} - \frac{S_0}{S_0^*}) + \mu_v N_v (2 - \frac{V_0^*}{V_0} - \frac{V_0}{V_0^*}) \\ &= -\mu_h N_h \frac{(S_0 - S_0^*)^2}{S_0 \, S_0^*} - \mu_v N_v \frac{(V_0 - V_0^*)^2}{V_0 \, V_0^*} \\ &\leq 0 \end{aligned} \tag{19}$$

From inspection of system (1), it can be proved readily, that $E_e$ is the only positively invariant subset of the set where $\dot{L}(t) = 0$. Therefore, by [12] the equilibrium $E_e$ is globally asymptotically stable in $\Omega$. This proves the theorem. ■

## 5. Numerical Examples and Remarks

Numerical examples are shown in this section. The parameters are determined by real life observation as follows. The life expectancy of human is 60 years. The life time of vector is 30 days. The bitting rate of vector is 1/3 or 1/5 per day. The time for each person can recover is 7 days. The total human population is 100,000 person. Finally, the total vector population is 10,000. The transmission probability are arbitrarily chosen. Figure (2, left) is a dynamic projection model (1) in $IY$ field with fixed values of $b_r$. There is indication of oscillations at $b_r$ greater than 3. Figure (2, right) shows the dynamics of a subpopulation of healthy humans but has the possibility of infection. The figure includes the projected dynamics model (1) into the phase plane of $Z_0 - S_0$. For this numerical example we used a set of values of the parameters with the same initial conditions and the average value of unity mosquitoes at different times. The integration time used was 100 days. In this case both population approaching certain positive numbers, or in other word, $S_0$ and $Z_0$ are co - exist. The increased of $b_r$ causes the increase of $Z_0$ and decrease $S_0$. We can explore the effect of bio-control in a multi-agents intervention by substituting $N_v$ with $u_1, N_v$ and $\mu_v$ by $u_2\mu_v$. Here $u_1 \leq 1$ indicates the effect of bio control, such as predatory in larvae stage of the mosquitoes. For $u_2 \leq 1$ indicates the effect of wolbachia infection in reducing life span of mosquitoes.

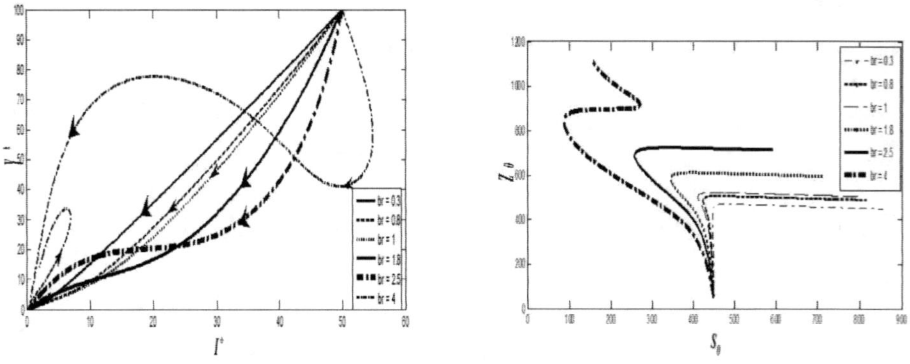

**Figure 2.** Left figure represent the numerical simulation of system (1) on the phase of $I - Y$ and right figure the phase of $S_0 - Z_0$ with parameter values $\gamma = 0.071, \beta_1 = 0.1, \beta_2 = 0.1, \alpha_1 = 0.2, \alpha_2 = 0.2, b = 3$, and different values of $b_r$.

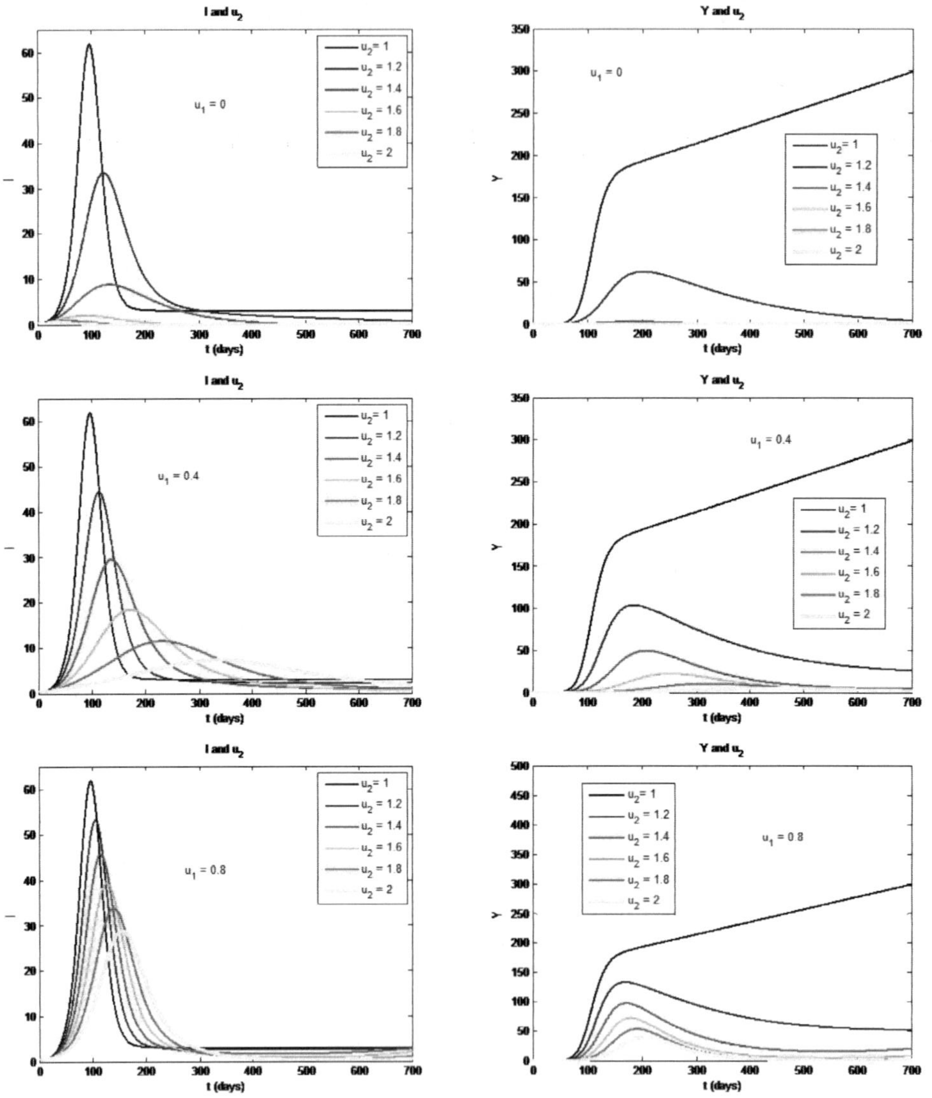


**Figure 3.** The dynamics of Compartment I dan Y with wolbachia control for fixed value of $u_1$

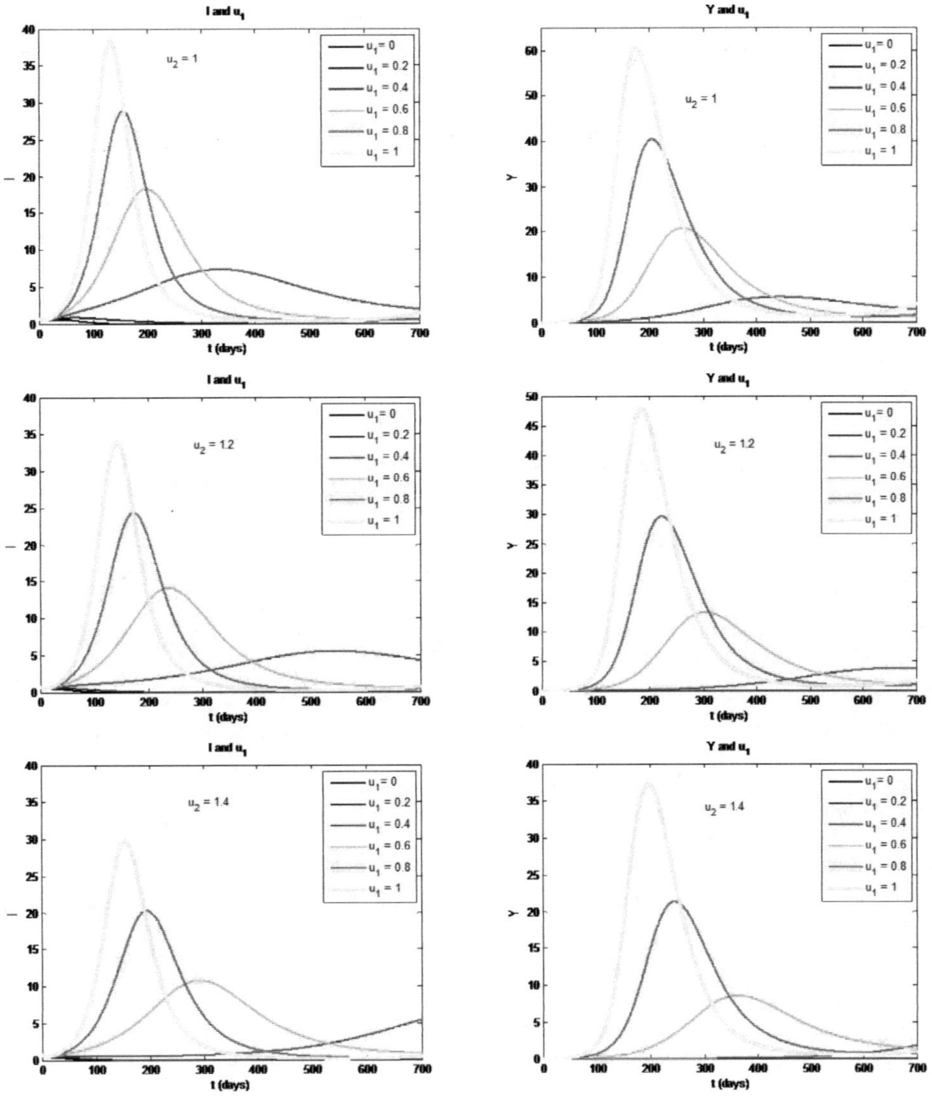


**Figure 4.** Dynamical Compartment I and Y with wolbachia control for fixed value of $u_2$

## 6. Conclusion

In this paper we construct and analyze a model of the spread of Dengue Hemorrhagic Fever, with and without symptoms, regardless the infecting virus serotype. We found a threshold for the model, $R_0$. The threshold associated with the BRN and it determined the existence and the stability of the equilibrium point. The model has two equilibrium points. The first, the disease-free equilibrium that always exists and globally asymptotically stable, if $R_0 < 1$. The second, the endemic equilibrium which its existence is guaranteed when the threshold $R_0 > 1$. In other words, if the threshold value or $R_0$ is less than or equal to one, then the dengue disease will disappear from the population. This shows the relationship between the existence or stability of the endemic point and the value of Basic Reproduction Number. Numerical examples show that the implementation of multiple bio-agent is prospective in reducing dengue transmission. Further analysis is being undertaken.

### Acknowledgment
Part of the work was funded by the Indonesian Government through Penelitian Hibah Kompetensi 2015.

### References
[1] N. Anggriani, A.K. Supriatna, & E. Soewono,*A Critical Protection Level Derived from Dengue Infection Mathematical Model Considering Asymptomatic and Symptomatic Classes*, Journal of Physics: Conference Series 423(2013)012056.
[2] C. Chavez, & Z. Feng *On The Computation of $\Re_0$ and its Role On Global Stability*, In Mathematical Approaches for Emerging and Reemerging Infectious Disease: An Introduction., IMA (2002) 125 : 229 - 250, Spriger-Verlag
[3] R.M. Anderson, & R.M. May *Infectious Disease of Humans: Dynamics and Control*, Oxford
[4] D. Gubler, *Dengue and Dengue Hemorrhagic Fever*, J.Clinic Microbiol (1998) 11: 480-496
[5] E. Soewono, & A. Supriatna, *A two - dimensional Model for The Transmission of Dengue Fever Disease*, Bull. Malay. Math Sci. Soc (2001) 24 : 49 - 57
[6] L. Esteva, & C. Vargas, *Analysis of a dengue disease transmission model*, J. Math. Biosci (1998) 150:131-151
[7] L. Esteva, & C. Vargas, *A Model for Dengue Disease with variable human population*, J. Math. Biol (1999) 38:220-240.
[8] M. Derouich, & A. Boutayeb, *A Model of Dengue Fever*, BioMed. Engineer online (2003) 2: 4 (doi:10.1186/1475-925x-2-4)
[9] M. Derouich, & A. Boutayeb, *Dengue Fever : Mathematical Modelling and Computer Simulation*, Applied Mathematics And Computation (2006), 177, 528-544.
[10] O. Diekmann, & J.A.P. Heesterbeek, *Mathematical Epidemiology of Infectious Diseases*, John Wiley & Son., New York, 2000
[11] Feng, & K.X. Velasco - Hernandez, *Competitive exclusion in a vector-host model for the dengue fever*, J. math. Biol. (1997) 35 : 523 - 544
[12] LaSalle JP, *The stability of dynamical systems*, Philadelphia:SIAM (1976)
[13] J. Tewa, & J. Luc Dimi, & S. Bowong, *Lyapunov function for a dengue disease transmission model*, Chaos Solitons & Fractals (2009) 39:936 - 941
[14] A. Korobeinikov, *Lyapunov function and global properties for SEIR and SEIS epidemic models*, Math. Med Biol (2004) 21:75 - 83
[15] A. Korobeinikov, & P.K. Maini, *A Lyapunov function and global properties for SIR and SEIR epidemic models*, Math. Biosci Eng (2004) 1:57 - 60
[16] WHO, *Dengue Haemorrhagic Fever, Diagnosis, Treatment and Control*,France(2009). http://whqlibdoc.who.int/publications/2009/9789241547871$_eng$.pdf

# Preparation of Biodiesel from Microalgae and Palm Oil by Direct Transesterification in a Batch Microwave Reactor

**Marwan[1], Suhendrayatna[1], and E Indarti[2]**

[1]Catalysis and Reaction Engineering Laboratory, Chemical Engineering Department, Faculty of Engineering, Syiah Kuala University, Darussalam, Banda Aceh-23111, Indonesia.
[2]Agricultural Product Technology Department, Faculty of Agriculture, Syiah Kuala University, Darussalam, Banda Aceh-23111, Indonesia.

Corresponding author: marwan_husin@yahoo.com

**Abstract.** The present work was aimed to study the so-called direct transesterification of microalgae lipids to biodiesel in a batch microwave reactor. As a comparison, preparation of palm oil to biodiesel by alkaline catalyzed ethanolysis was also carried out. Palm oil biodiesel was recovered close to an equilibrium conversion (94-96% yield) under microwave heating for at least 6 min, while the conventional method required more than 45 minutes reaching the same yield. A very short reaction time suggests the benefit of microwave effect over conventional heating method in making biodiesel. FTIR analysis revealed the presence of fatty acid ethyl esters with no undesired chemical groups or compounds formed due to local heat generated by microwave effect, thus the conversion only followed transesterification route. Oil containing microalgae of *Chlorella sp.* isolated from the local brackish water pond was used as a potential source of biodiesel. High yield of biodiesel (above 0.6 g/g of dried algae) was also attainable for the direct transesterification of microalgae in the microwave reactor. Effect of water content of the algae biomass became insignificant at 11.9 %(w/w) or less, related to the algae biomass dried for longer than 6 h. Fast transesterification of the algal oil towards equilibrium conversion was obtained at reaction time of 6 min, and at longer times the biodiesel yield remains unchanged. FAME profile indicates unsaturated fatty acids as major constituents. It was shown that microwave irradiation contributes not only to enhance the transeseterification, but also to assist effective release of fatty acid containing molecules (e.g. triacylglycerol, free fatty acids and phospholipids) from algal cells. **Keywords**: biodiesel, transesterification, ethanolysis, microwave, palm oil, microalgae, *Chlorella sp.*

## 1. Introduction

Biodiesel is a non-petroleum-derived fuel that comprises alkyl esters produced by the transesterification of triacylglycerol. Biodiesel is renewable and environmental-friendly, with calorific value equivalent to petroleum-based fuel. Transesterification has been recognized as the simplest method to prepare biodiesel from edible and non-edible lipids. However, energy consumption inefficiencies in the process which result in higher cost of biodiesel is still a great concern. This is generally related to the heating method employed in the process. Recently, many investigations have applied microwave irradiation in biodiesel production due to due to inherent advantages of shorter reaction time and lower energy requirements found to complete numerous inorganic and organic chemical reactions [1]. Microwave assisted reactions not only reduce the reaction time and increase the biodiesel yield but also could reduce use of alcohol and the product separation time significantly [2,3].

The present study deals with the use of microwave reactor operated non-isothermally to convert microalgae lipid to biodiesel by in-situ or direct transesterification, an approach that has been widely used as an analytical procedure to prepare FAME for the determination of the fatty acid composition of lipid containing tissues [4,5]. The direct transesterification method can produce FAME in higher quantity than can be achieved in the two-step method of traditional solvent extraction and transesterification [5,6,7]. The method was also successfully applied for pure and mixed cultures of microorganism. Therefore, a combination of microwave heating and direct transesterification is expected to enhance biodiesel yield from microalgae at a very short time. As a comparison, palm oil biodiesel was also prepared by alkaline catalyzed transesterification under conventional and microwave heating.

## 2. Experimental Method

### *2.1 Materials*

Palm frying oil (Bimoli) was purchased from a local grocery store. It is the only branded frying oil sold bulky in the market. The green microalgae strain used in this experiment was *Chlorella sp.* which was collected from 5 L cultivation tanks. The alga was grown in domestic wastewater medium. Reagent grade methanol, chloroform, sulfuric acid, technical grade alcohol (96% ethanol), and sodium hydroxide were used as purchased without any further purification. Distilled water was used to wash biodiesel. The reaction was carried out in a 250 ml screw-capped bottle, placed inside a household microwave oven (Model NN-ST 342M, Panasonic) operating at frequency 2.45 GHz with maximum output 800 W.

### *2.2 Transesterification of Palm Oil*

A ratio of oil and ethanol of 100:50 (ml/ml) and catalyst loadings of 0.25-1.0 g NaOH/100 ml of oil was used throughout the experiments. The oil, ethanol, and NaOH were charged into a 250 ml bottle with screw-cap. Vigorous mixing was applied for 2 minutes to obtain a good dispersion of oil-ethanol. The reaction bottle was placed inside the microwave oven with heating set at low–micropower. The reaction time was adjusted for 1 to 10 minutes without temperature control, thus the temperature rise was observed (referred to the temperatures measured at the end of each reaction time). Upon cooling to room temperature the reaction bottle was removed from the microwave and the reaction mixture was settled in a separatory funnel. The final biodiesel was obtained after the upper layer separated and washed with water three times. As a comparison, palm biodiesel was also prepared by conventional method; the transesterification in a stirred glass reactor by electric heating at 60°C for 15, 30 and 60 min. The stirring speed was maintained at 200 rpm. Biodiesel produced by microwave heating was characterized to determine density and viscosity. The chemical changes of the PO biodiesel were analyzed by Transform Infrared (FTIR) Spectrophotometer (Model 8400S, Shimadzu) equipped with Interferometer to eliminate the effect of $H_2O$ and $CO_2$ in the surrounding atmosphere.

### *2.3 Direct Transesterification of Microalgae*

Wet microalgae biomasses were oven-dried at varying times to obtain certain water content. Initial work on algae suggests that drying time of 6 h resulted in water content of 0.113 g/g biomass which did not significantly affect the biodiesel yield. Therefore, the rest experiments on microalgae used 6 h dried biomass. The microalgae was poured into a 250 ml reaction bottle containing a mixture of methanol, $H_2SO_4$, and chloroform (1.7:0.3:2.0 v/v/v) as prescribed by Indarti *et al.* [5]. The bottle was tightly closed with screw-cap and placed inside the microwave chamber. The mixture was heated by microwave irradiation set at low-micropower for a given time. After the heating treatment, the mixture was allowed to cool for 30 min and transferred to a separatory funnel. About 10 ml distilled water was added to the mixture, which was then shaken and allowed to separate. The yield of biodiesel was evaluated by gravimetric method. The bottom layer containing FAME was poured into a pre-weighed petri dish, which was evaporated in an oven at 60°C overnight. The remaining liquid was weighed as biodiesel mass.

Fatty acid methyl ester (FAME) content of the algal biodiesel sample was determined with a gas chromatography-mass spectrometer (GC-MS) (Model QP2010 Plus, Shimadzu) equipped with flame ionization detector (FID) and an Rtx-5MS column (0.25 mm ID and 30 m long, Restek). One microliter of the bottom layer was injected into the GC unit. The fatty acids were identified by comparing the retention times with those of standard fatty acids. Their composition was calculated based on the corresponding peak areas in the chromatogram.

## 3. Results and Discussion

### *3.1 Preparation of Palm Oil Biodiesel under Microwave Heating*

The ability of a molecule to couple with the microwave radiation is a function of its molecular polarisability (i.e. a function of its dipole moment), thus only polar molecules interact with microwave energy [1]. Palm oil and ethanol as reacting fluids involved in the transesterification are polar compounds and their interaction with the microwave radiation has resulted in temperature increase to more than 80°C after 10 min. The reaction time was then limited to 10 min to avoid excessive boiling of compounds in the reaction mixture.

Figure 1 indicates that the yield of biodiesel reached a maximum value in less than 10 min depending on the catalyst amount added. The maximum yield of 94-96% suggests the reaction was in the equilibrium. Beyond the maxima, a slight decrease in biodiesel yield observed probably due to the favor of the equilibrium in the reverse reaction. It is in good agreement with the findings of Hernando *et al.* [8] and Suppalakpanya *et al.* [9] in producing biodiesel from rapeseed oil and crude palm oil, respectively. Compared with the conventional heating method, the yields of biodiesel reached close to maximum after 45 min as shown in insert of Figure 1. A very short reaction time with microwave heating to obtain the optimum yield suggests the benefit of microwave over conventional heating method in making biodiesel.

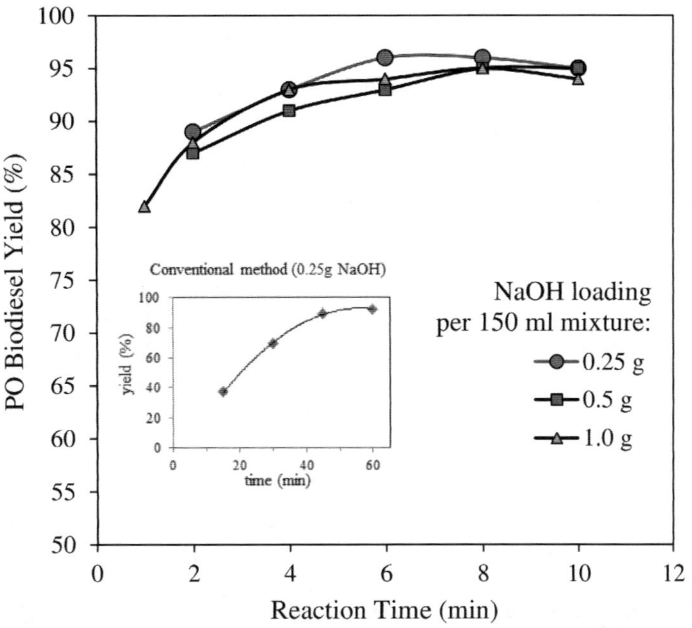


**Figure 1.** Change of PO biodiesel yield during microwave heating.

After the transesterification, the reaction mixture was allowed to separate into the upper layer containing biodiesel and the bottom layer. The upper layer was then washed with warm water three times. Further separation of dispersed phase was carried out by centrifugation. Later, the remaining biodiesel layer was washed again three times. The quality of palm biodiesel produced in this work was observed for its physical and chemical parameters.

The density of PO biodiesel at 40°C was 0.880-0.888 g/ml which is equivalent to petroleum-based diesel (0.834 g/ml) and within ASTM D6751 and SNI 04-7182-2006 limits for biodiesel. Viscosity has been the most crucial property of biodiesel since its effect on the fluidity of the fuel determines the fuel injection system, particularly at low temperature. The viscosity of PO biodiesel 40°C was 3.825-4.655 cps which is slightly higher than ASTM D6751 limit for biodiesel, but within the SNI 04-7182-2006 value for Indonesian biodiesel.

FTIR analysis has been shown as a versatile method to determine the quality of biodiesel [10,11]. Figure 2 shows that the FTIR spectra of origin oil and the produced biodiesel are very similar since the reaction did not cause significant change of chemical or functional bonds. The hot spot effect of microwave heating did not lead to undesired reaction routes. The triacylglycerol was broken down into fatty acid ethyl esters and glycerol. The FTIR spectra of the origin oil and biodiesel showed peaks from carbonyl (C=O) at 1744 $cm^{-1}$ and that of C-O at 1163 $cm^{-1}$ similar to that reported in the literature [12], indicating that the carbonyl and C-O in the oil is retained in the biodiesel. The stretching vibrations of CH, $CH_2$, and $CH_3$ appear at 3003, 2854, and 2922 $cm^{-1}$, whereas the bending vibrations ($\rho CH_2$) of these groups are found at 1375, 1163 and 723 $cm^{-1}$, respectively.

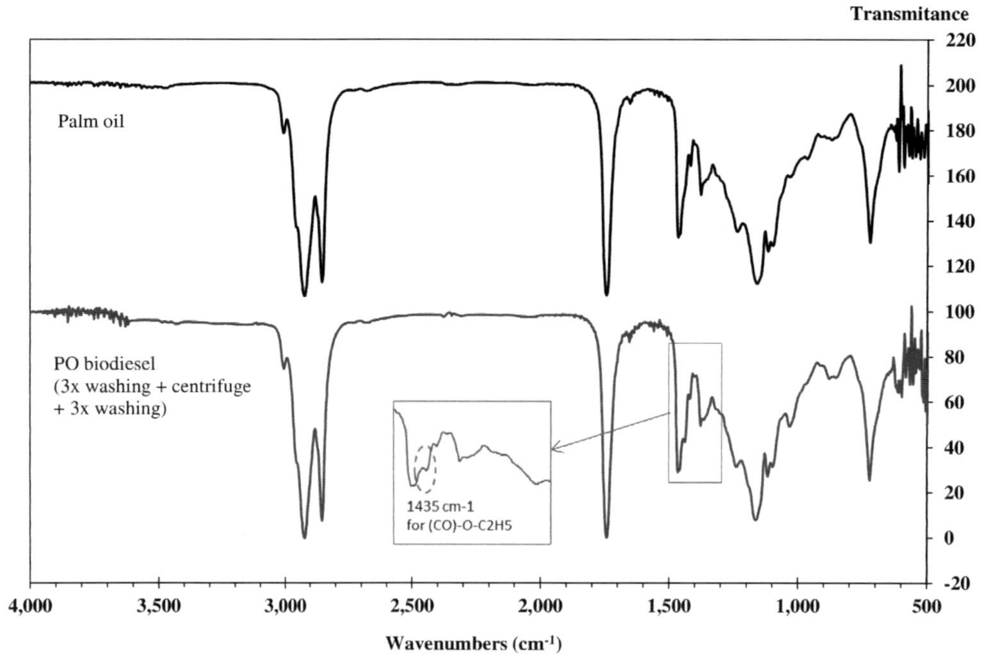


**Figure 2.** FTIR spectra of palm oil and PO biodiesel

Some specific peaks of biodiesel are possibly to be identified with close spectra observation as given in Figure 2. The most influence as a result from transesterification was to find the new signal at 1435 cm-1 which is definitely the ethyl ester group with its deformation vibration (Shimadzu, SCA_110_035). Tahvildari and Mohammadi [13] in preparing ethyl esters from linseed oil reported the shifting of absorption peaks of the oil sample at 1744 and 1162 cm−1 to 1739 and 1178 cm−1 in the biodiesel, respectively, but it was not clearly observed in this work.

### *3.2 Direct Transesterification of Microalgal Oil*

*Chlorella sp.* used as a source of oil was isolated from brackish water pond in Tanjong Selamat (Aceh Besar) by serial dilution in the modified Detmer medium. The stock culture of *Chlorella sp.* was then maintained in BG-11 medium with light irradiation and $CO_2$-enriched air flow. Previous study under optimized condition, the microalgae could accumulate crude lipid as high as 0.821 g/g dried biomass. For the purpose of this experiment, the algal biomass were then cultured in 5 L glass tanks filled with 50% diluted municipal domestic wastewater, and harvested in their stationary growth phase. The collected algal biomass was drained for 2 h before subject to drying.

Water present in the biomass has a damaging effect on the yield of biodiesel [14,15]. In the present work, different water content of the algal biomass was used to examine the water effect on the biodiesel yield from *Chlorella sp.* using the direct transesterification method. Freshly harvested biomass (after 2 h draining) contained water of 48.0% (wet basis). Figure 3 shows the change of water level of the algal biomass upon drying at 60°C for various times. The biomass was found completely dried after 12 h, assuming that the biomass was free of unbounded water. Water significantly inhibits the algal oil to biodiesel conversion with water content greater than 11.9%. When the water content less than 11.9%, biodiesel yield was relatively constant (0.627-0.629 g/g dried biomass). It is probably due to excess methanol of the reacting mixture minimizing water effect. Similar finding was also found in the direct transesterification of *Chaetoceros gracilis* since a higher water content of the algal biomass was compensated by adding more methanol [14].

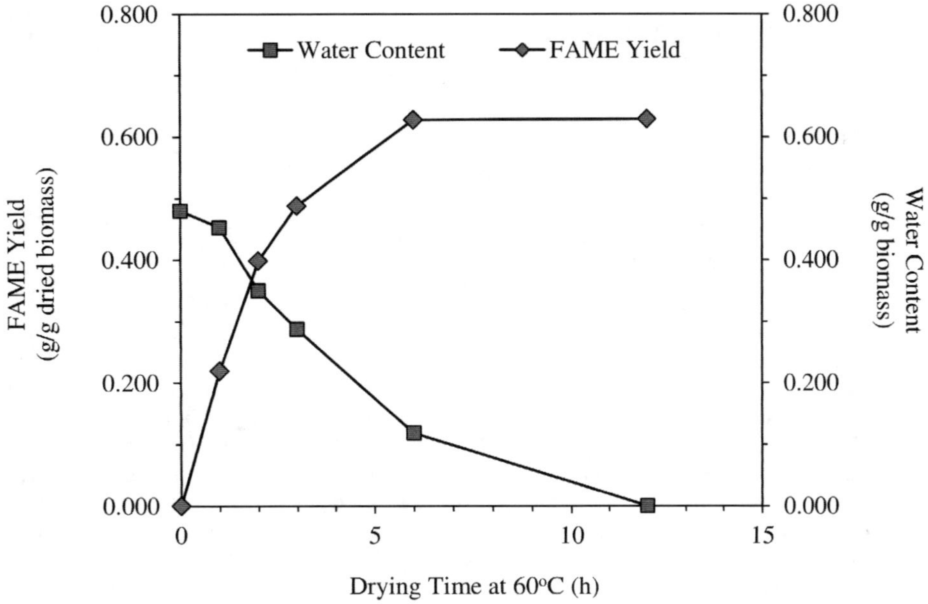


**Figure 3**. Effect of drying time on water content and the biodiesel yield (10 min reaction)

To investigate the influence of heating time, the algal biomass with a water content of 0.113 g/g of wet algae was used since it was found that no considerable differences in the biodiesel yield were observed with the use of lower water content. Figure 4 shows an increasing reaction time allows the transesterification reaction to proceed to completion and results in a higher yield of biodiesel from the microalgae. A very short conversion time was also reached to produce the algal biodiesel as high yield was attained at least after the microwave heating of 6 min, which corresponding to the yield of 0.591 g of biodiesel per g of dried biomass. Higher amount of biodiesel yield (>0.620 g/g of dried biomass) was obtained at slightly longer reaction times. Patil *et al.* [16] found the optimum reaction time in preparing biodiesel from *Nannochloropsis* by microwave assisted direct transesterification was 4-5 min, since microwave irradiation contributes to more effective in the destruction of the cells and accelerating better the transesterification reaction in a shorter reaction time. With sulfuric acid and extraction agents (methanol and chloroform), 2 min was sufficient for cell disruption under microwave irradiation at 60°C [17]. However, longer reaction time to reach maximum yield was reported in transesterifying *Chlorella pyrenoidosa* (30 min) and *Chaetoceros gracilis* (more than 125 min) [14,18].

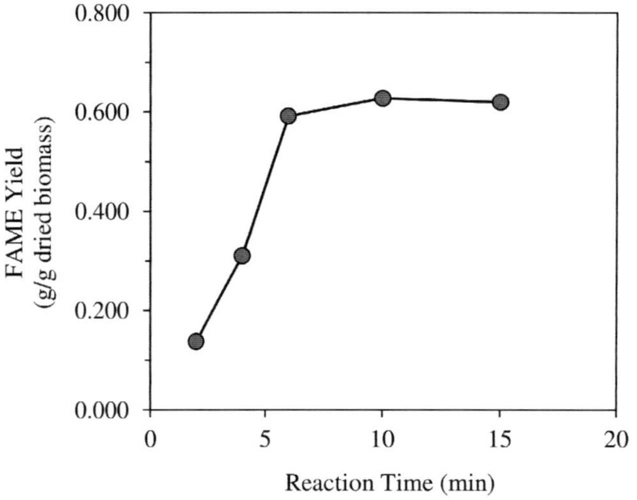


**Figure 4**. Effect of reaction time on the algal biodiesel yield

Fatty acids of the algal biodiesel were determined by GC analysis and given in Table 1. In general, FAME composing the algal biodiesel was not only from the triacylglycerol and the free fatty acids, but also from the other forms of lipid due to the extractable effect of methanol and chloroform as co-solvent in the direct transesterification. Both simultaneously dissolve phospholipids, glycolipids, sphingolipids, and sterols from cell membranes. Table 1 shows that the algal biodiesel comprises 28% saturated fatty acids and 72% unsaturated ones. The relatively high content of unsaturated fatty acids was also reported for other algal biodiesel [15,16,18]. Although fatty acid profile and degree of saturation does not have much of an impact on the transesterification reaction, they do affect the biodiesel properties and then determine its uses. Due to inherent properties of unsaturated fatty acids, biodiesel dominated by such fatty acids has good cold-flow properties. Unfortunately, these fatty acids are mostly vulnerable to oxidation. Biodiesel produced from *Chlorella sp.* cultivated for this work, therefore, may have an instability problem during storage for a long period.

**Table 1**. Fatty acid profile of the biodiesel derived from *Chlorella sp.*

| No. | Fatty acids | Area (%) |
|---|---|---|
| 1. | Myristic acid C14:0 | 3.7 |
| 2. | Palmitic acid C16:0 | 19.4 |
| 3. | Palmitoleic acid C16:1 | 2.4 |
| 4. | Oleic acid C18:1n9 | 18.0 |
| 5. | Linoleic acid C18:2n6 | 26.8 |
| 6. | Linolenic acid C18:3n3 | 24.7 |
| 7. | Arachidic acid C:20 | 4.9 |

## 4. Conclusion

A combination of microwave irradiation and direct transesterification was successfully implemented in converting palm oil and microalgal oil to biodiesel in high yield. Microalgae used were *Chlorella sp.* cultured in diluted municipal wastewater. Using pure oil such as palm oil, it was revealed that no undesired species was formed during microwave irradiation, thus the transesterification was the only reaction mechanism taking place. Transesterification of palm oil and microalgae took place at fast rate, reaching an equilibrium conversion within 6-10 min. When using microalgae biomass, water content should be taken into account. Water content of 11.9% (wet basis) or less was required to attain high biodiesel yield. High yield for the biodiesel from microalgae was achieved due to simultaneous contribution of fast transesterification, enhanced releases of lipids and role of chloroform as co-solvent during microwave irradiation. This study suggests that use of microwave in the direct transesterification of lipids contained in microalgae is an effective approach to reduce reaction time and costs as some operation steps are eliminated. Further cost reduction is also feasible by reducing feedstock cost as the microalgae can be cultured in wastewater streams or lagoon systems.

## Acknowledgements

The authors would like to acknowledge The Ministry of National Education and Culture, Indonesia for funding assistance (National Strategic Research Scheme, Agreement No. 174/UN11.2/LT/SP3/2014) and Syiah Kuala University for providing access to research facilities. The authors wish also to thank Vera Viena and Ichwanul Ikhsan for kind assistance during the project.

## References

[1] Surati M A, Jauhari S and Desai K R 2012 *Arc. Appl. Sci. Res.* **4** 645.
[2] Vyas A P, Verma J L and Subrahmanyam N 2010 *Fuel* **89** 1.
[3] Motasemi F and Ani F N 2012 *Renew. Sustain. Energy Rev.* **16** 4719.
[4] Rodríguez-Ruiz J, Belarbi E, Sánchez J L G, Alonso D L 1998 *Biotechnol. Tech.* **12** 689.
[5] Indarti E, Majid M I A, Hashim R and Chong A 2005 *J. Food Comp. Analysis* **18** 161.
[6] Lewis T, Nichols P D and McMeekin T A 2000 *J. Microbiol. Method* **43** 107.
[7] Vicente G, Bautista L F, Rodríguez R, Gutiérrez F J, Sádaba I, Ruiz-Vázquez R M, Torres-Martínez S and Garre V 2009 *Biochem. Eng. J.* **48** 22.
[8] Hernando J, Leton P, Matia M P, Novella J L and Alvarez-Builla J 2007 *Fuel* **86** 1641.
[9] Suppalakpanya K, Ratanawilar S, Nikhom R and Tongurai C 2011 *Songklanakarin J. Sci. Technol.* **33** 79.
[10] Mirghani M E S, Kabbshi N A, Alam M Z, Duqsieh I Y and Alkatib M F R 2011 *J. Am. Oil. Chem. Soc.* **88** 1897.
[11] Yuan T, Akochi-Koble E, Pinchuk D and van de Voort V R 2014 *Intern. J. Renew. Energy Biofuels* Article ID 178474.

Journal of Physics: Conference Series **622** (2015) 012040 doi:10.1088/1742-6596/622/1/012040

[12] Tariq M, Ali S, Ahmad F, Ahmad M, Zafar M, Khalid N and Khan M A 2011 *Fuel Proces. Techn.* **92** 336.
[13] Tahvildari K and Mohammadi S 2014 *Acad. Res. Intern.* **5** 27.
[14] Wahlen B D, Willis R M and Seefeldt L C 2011 *Biores. Techn.* **102** 2724.
[15] Ehimem E A, Sun Z F and Carrington, C G 2010 *Fuel* **89** 677.
[16] Patil P D, Gude V G, Mannarswamy A, Cooke P, Nimalakhandan N, Lammers P and Deng S G 2012 *Fuel* **97** 822.
[17] Yuan H, Yang B L and Zhu G L 2009 *Energy Fuel* **23** 548.
[18] Cheng J, Yu T, Li T, Zhou J and Cen K 2013 *Biores. Techn.* **131** 531.

# An application of input-output analysis in analyzing the impacts of final demands changes on the total outputs of Japanese energy sectors: A further study

**Ubaidillah Zuhdi**

Department of Industrial Administration, Tokyo University of Science, Noda 278-8510, Japan
School of Business and Management, Institut Teknologi Bandung, Bandung 40132, Indonesia

ubaidillah.zuhdi@gmail.com

**Abstract**. The purpose of this study is to continue the previous study which discussed the impacts of final demands changes on the total outputs of Japanese energy sectors. More specifically, this study aims to conduct a deeper analysis regarding these impacts. This study employs a demand-pull Input-Output (IO) quantity model, one of the calculation instruments in IO analysis, as a tool of analysis. This study focuses on two sectors, namely (1) petroleum refinery products, and (2) non-ferrous metals. Two conditions are considered in the analysis part, namely (1) "whole sector change", and (2) "pure change". The results show that in both conditions, both discussed sectors have similar patterns. The results also explain that, in both conditions, the biggest positive impact for the sector of petroleum refinery products is given by scenario 4, the modification of consumption expenditures of private.

## 1. Introduction

In this day and age, human dependence on energy can be easily observed. The simplest evidence is human needs energy when he or she does the activities. This dependency can also be seen in broader aspects. For example, the government usually considers the availability of energy sources when new economic policies must be made. Besides, countries also consider this availability before executing the economic activities such as an international trade. Therefore, based on these facts, one can argue that energy is also economically important.

Many previous studies discussed the energy topic. For example, [1] described the connections between the prices of oil and Euro area's inflation. [2] investigated the effect of oil price shocks on the economic growth of some Middle East and North Africa (MENA) countries. [3] analyzed the relationship between the changes of price of oil and real Gross Domestic Product (GDP) which the focus was the economy of US. Besides, [4] analyzed the ways to increase the total outputs of Japanese energy sectors in the future. He used Input-Output (IO) analysis and some scenarios of final demands change in his study.

The study analyzes the impacts of final demands changes on the total outputs of energy sectors of specific country, viewed from above previous studies, however, is still thin. This analysis is important because it will describe the characteristics of these sectors when the changes of national economy condition happen. Further, it can give the understanding regarding the improvements needed for enhancing the sectors. This study is conducted in order to fulfill the gap.

The purpose of this study is to continue the previous study which discussed the impacts of final demands changes on the total outputs of Japanese energy sectors. More specifically, this study aims to conduct a deeper analysis regarding these impacts. This study employs IO approach as a tool of analysis.

Journal of Physics: Conference Series **622** (2015) 012041 doi:10.1088/1742-6596/622/1/012041

## 2. Methodology

The methodology of this study is similar with the one used in a previous study, the study conducted by [4], and described as follows. The first step is to explain the data used. This study uses an aggregated IO table of Japan for 2005. This table consists of 89 industrial sectors. These industries are described in Appendix.

The second step is to describe Japanese energy sectors used. Table 1 shows these sectors. As with a previous study, this study also focuses on two sectors. These sectors are (1) petroleum refinery products, and (2) non-ferrous metals.

The third step is to conduct the calculation in order to know the impacts of final demand changes on the total outputs of analyzed sectors. A demand-pull IO quantity model, one of the calculation instruments in IO analysis, is used in this calculation. [5] explained that following equation is a representation of this model:

$$\mathbf{x}^1 = \mathbf{L}^0\mathbf{f}^1 \qquad (1)$$

where x, L, and f are matrices of total outputs of sectors, Leontief inverse, and final demands of sectors, respectively. 0 and 1 describe initial and future periods, respectively. An initial period in this study is 2005. Table 2 explains the scenarios of final demands modification used. These scenarios, compared with previous ones, are slightly different. The difference can be seen on the existence of scenario 4, the change of consumption expenditures of private. Simultaneously, the deeper understanding regarding the impacts is obtained through this extension.

The conditions of "whole sector change" and "pure change" are considered in above calculation. The former situation explains the condition which the changes of final demands are addressed to all Japanese industrial sectors while the latter one only focuses on the discussed sectors. In this study, the former one will be called "condition A" while the term of "condition B" is used to explain the latter condition. The analysis regarding above impacts is discussed on the next step. Conclusions of this study and suggestions for further researches are described on the final step.

**Table 1.** Japanese energy sectors used in this study.

| No. | Sector Number | Sector Name |
|---|---|---|
| 1 | 6 | Metallic ores |
| 2 | 7 | Non-metallic ores |
| 3 | 8 | Coal mining, crude petroleum and natural gas |
| 4 | 26 | Final chemical products, n.e.c.[a] |
| 5 | 27 | Petroleum refinery products |
| 6 | 28 | Coal products |
| 7 | 36 | Pig iron and crude steel |
| 8 | 37 | Steel products |
| 9 | 38 | Steel castings and forgings, and other steel products |
| 10 | 39 | Non-ferrous metals |
| 11 | 40 | Non-ferrous metal products |
| 12 | 41 | Metal products for construction and architecture |
| 13 | 42 | Other metal products |

[a] Not elsewhere classified.

(Source: [6] with slight modifications)

**Table 2.** The scenarios of final demands modification used in this study.

| Component of Final Demand | Scenario | | | |
|---|---|---|---|---|
| | 1 | 2 | 3 | 4 |
| | Exports Modification | Imports Modification | The Modification of Consumption Expenditures of Outside Households | The Modification of Consumption Expenditures of Private |
| Exports | Increase 30% | Constant | Constant | Constant |
| Imports | Constant | Increase 30% | Constant | Constant |
| Consumption expenditures of outside households | Constant | Constant | Increase 30% | Constant |
| Consumption expenditures of private | Constant | Constant | Constant | Increase 30% |

## 3. Results and analysis

Table 3 describes the total outputs of discussed sectors for each scenario on condition A. Figures 1 and 2 explain in more details the dynamics happen on these total outputs on this condition. Based on these results, one can argue that, on this condition, the biggest positive impact on the total output of petroleum refinery products sector is given by scenario 4, the modification of consumption expenditures of private. Meanwhile, scenario 1, the exports modification, has the biggest positive effect on the total output of non-ferrous metals sector. On the contrary, the negative impact is given by scenario 2, the change of imports.

On the other hand, Table 4 describes the total outputs of analyzed sectors for each scenario on condition B. Figures 3 and 4 explain in more details the dynamics happen on these total outputs on this condition. Based on these results, one can say that, on this condition, the biggest positive impact on the total output of petroleum refinery products sector is given by scenario 4, the modification of consumption expenditures of private. Meanwhile, scenario 1, the change of exports, has the biggest positive effect on the total output of non-ferrous metals sector. On the contrary, the negative impact is given by scenario 2, the imports change.

Above phenomena show that, in both conditions, both discussed sectors have similar patterns, namely these industries receive the positive impacts from scenarios 1, 3, and 4 while the opposite impact is obtained from scenario 2. This negative impact also appeared in the previous study. Above phenomena also explain that, in both conditions, the biggest positive impact for the sector of petroleum refinery products is given by scenario 4, the modification of consumption expenditures of private. This is a new finding compared with a previous study. Based on these results, one can argue that the effective ways to increase the total output of this sector in the future are to open greater opportunities for private sectors in processing products of this industry, and to restrict import activities regarding these products.

**Table 3.** Total outputs of discussed sectors for each scenario on condition A (100 million Yen).

| Sector Number | Sector Name | $X_t$ | $X_{t+1}$, Scenario 1 | $X_{t+1}$, Scenario 2 | $X_{t+1}$, Scenario 3 | $X_{t+1}$, Scenario 4 |
|---|---|---|---|---|---|---|
| 27 | Petroleum refinery products | 156,740.17 | 170,203.33 | 138,594.26 | 158,126.05 | 193,243.18 |
| 39 | Non-ferrous metals | 21,527.12 | 29,139.76 | 11,014.71 | 21,820.99 | 24,959.84 |

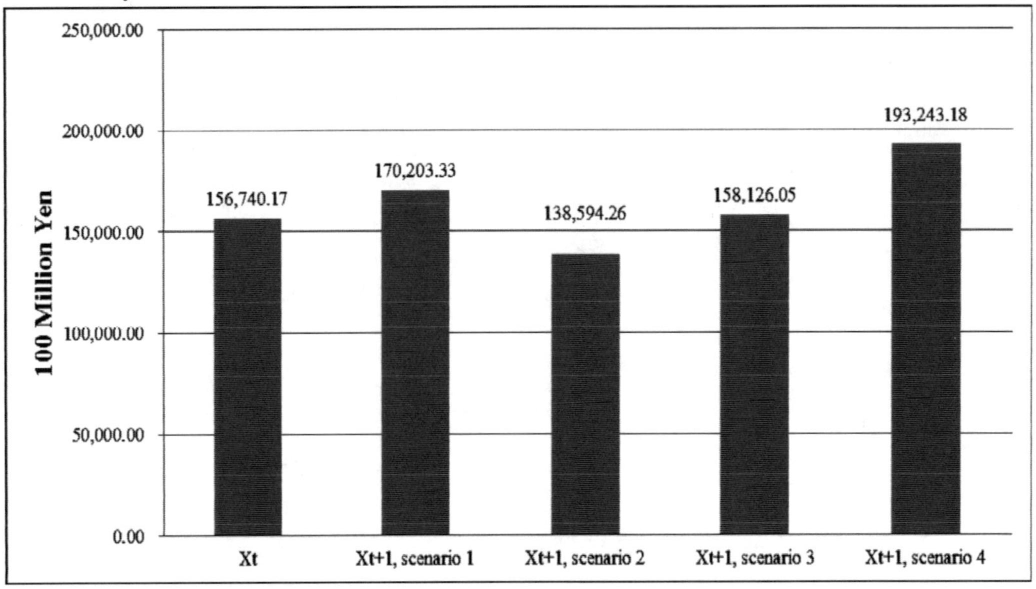


**Figure 1.** The dynamics happen on the total output of petroleum refinery products sector (condition A).

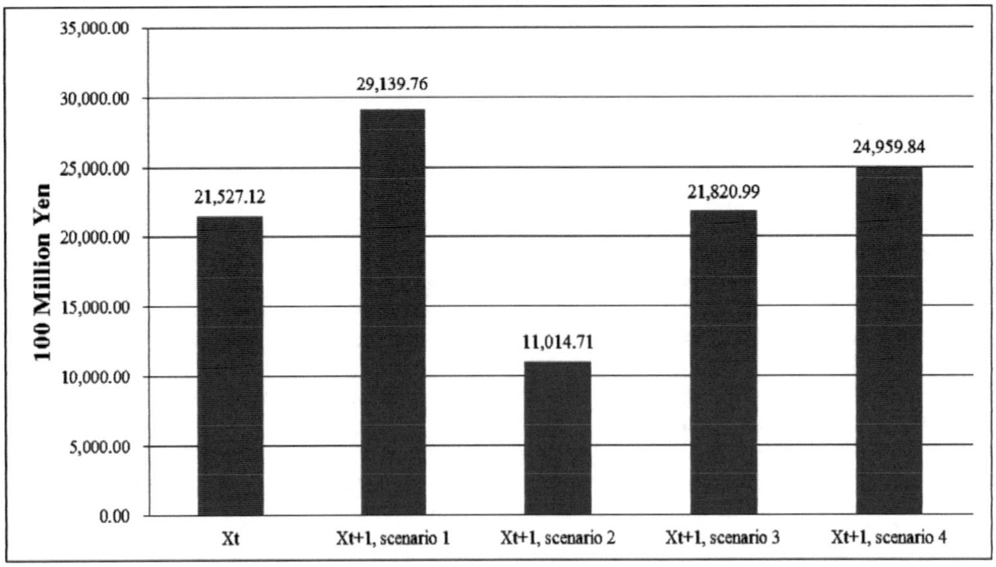


**Figure 2.** The dynamics happen on the total output of non-ferrous metals sector (condition A).

Journal of Physics: Conference Series **622** (2015) 012041 doi:10.1088/1742-6596/622/1/012041

**Table 4.** Total outputs of discussed sectors for each scenario on condition B (100 million Yen).

| Sector Number | Sector Name | $X_t$ | $X_{t+1}$, Scenario 1 | $X_{t+1}$, Scenario 2 | $X_{t+1}$, Scenario 3 | $X_{t+1}$, Scenario 4 |
|---|---|---|---|---|---|---|
| 27 | Petroleum refinery products | 156,740.17 | 159,467.79 | 148,055.30 | 156,843.72 | 175,229.98 |
| 39 | Non-ferrous metals | 21,527.12 | 22,943.73 | 14,992.47 | 21,527.36 | 21,896.06 |

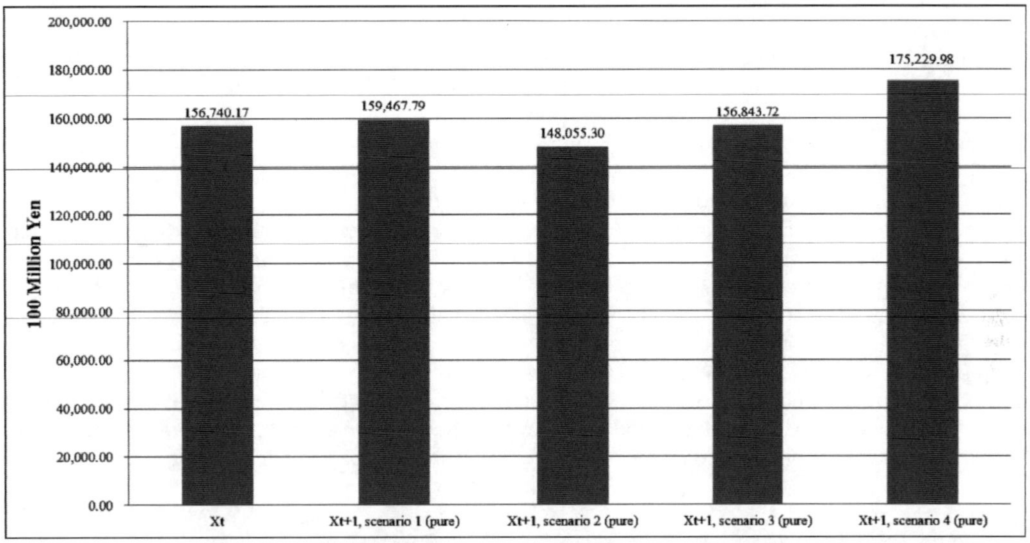


**Figure 3.** The dynamics happen on the total output of petroleum refinery products sector (condition B).

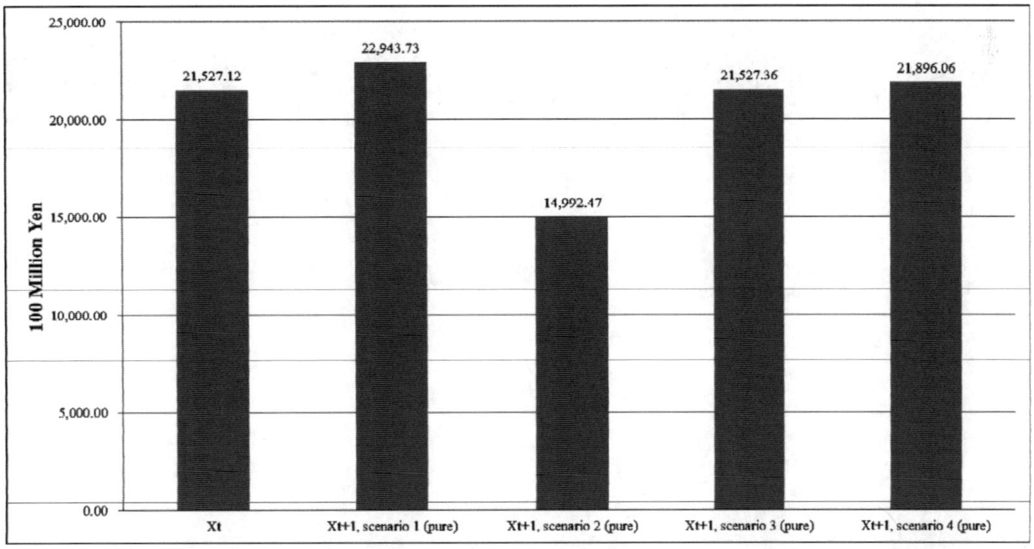


**Figure 4.** The dynamics happen on the total output of non-ferrous metals sector (condition B).

## 4. Conclusions and further researches

This study, as a continuation of a previous study, conducted the deeper analysis regarding the impacts of final demands changes on the total outputs of Japanese energy sectors. This study employed a demand-pull IO quantity model, one of the calculation instruments in IO analysis, as a tool of analysis. This study focused on two sectors, namely (1) petroleum refinery products, and (2) non-ferrous metals. Two conditions were considered in the analysis part, namely (1) "whole sector change", and (2) "pure change". An initial period in this study was 2005. The difference between current and previous studies could be seen on the existence of scenario 4, the change of consumption expenditures of private.

The results showed that in both conditions, both discussed sectors had similar patterns, namely these industries obtained the positive impacts from scenarios 1, 3, and 4 while the opposite impact was received from scenario 2. This negative impact also appeared in the previous study. The results also explained that, in both conditions, the biggest positive impact for the sector of petroleum refinery products was given by scenario 4, the modification of consumption expenditures of private. This was a new finding in this topic. Based on these results, the suggestions from this study regarding the effective ways to increase the total output of this sector in the future were to open greater chances for private sectors in processing products of this industry, and to limit import activities regarding these products.

The deeper understanding regarding the impacts of final demands changes on the total outputs of Japanese energy sectors was obtained from this study. However, this study only analyzed specific energy sectors of Japan. In other words, the comprehensive view regarding the impacts on the national economy of Japan did not appear in this study. This view is needed in order to make comprehensive policies for enhancing the economic condition of Japan in the future. Therefore, as a further research, this study proposes the same analysis for other Japanese industrial sectors.

The other suggested further research from this study is to conduct the international comparison on the current topic. This comparison will describe the characteristics of industries of analyzed countries when the final demands changes happen. A good example is to compare developed and developing countries.

**References**

[1] Jacquinot P, Kuismanen M, Mestre R and Spitzer M 2009 An assessment of the inflationary impact of oil shocks in the euro area *The Energy Journal* **30** 49–84

[2] Berument H, Ceylan N B and Dogan N 2010 The impact of oil price shocks on the economic growth of selected MENA countries *The Energy Journal* **31** 149–176

[3] Jimenez-Rodriguez R 2009 Oil price shocks and real GDP growth: testing for non-linearity *The Energy Journal* **30** 1–24

[4] Zuhdi U 2014 An input-output approach to analyze the ways to increase total output of energy sectors: The case of Japan *IOP Conference Series: Earth and Environmental Science* **19** 012015

[5] Miller R E and Blair P D 2009 *Input-Output Analysis: Foundations and Extensions* (Cambridge: University Press)

[6] Zuhdi U and Mori S 2012 Analyzing the gross output change of energy sector industry in Indonesia between 1990 and 2005 by decomposition analysis *Proc. The 3rd Int. Associaton for Energy Economics Asian Conf. (Kyoto, Japan, 20 – 22 February 2012)* p 110

[7] Zuhdi U 2014 Analyzing the role of creative industries in national economy of Japan: 1995-2005 *Open Journal of Applied Sciences* **4** 197–211

Journal of Physics: Conference Series **622** (2015) 012041 doi:10.1088/1742-6596/622/1/012041

**Appendix. Japanese Industrial Sectors (89 Sectors)**

| No. | Sector Name |
|---|---|
| 1 | Crop cultivation |
| 2 | Livestock |
| 3 | Agricultural services |
| 4 | Forestry |
| 5 | Fisheries |
| 6 | Metallic ores |
| 7 | Non-metallic ores |
| 8 | Coal mining, crude petroleum and natural gas |
| 9 | Foods |
| 10 | Beverage |
| 11 | Feeds and organic fertilizer, n.e.c. |
| 12 | Tobacco |
| 13 | Textile products |
| 14 | Wearing apparel and other textile products |
| 15 | Timber and wooden products |
| 16 | Furniture and fixtures |
| 17 | Pulp, paper, paperboard, building paper |
| 18 | Paper products |
| 19 | Printing, plate making and book binding |
| 20 | Chemical fertilizer |
| 21 | Industrial inorganic chemicals |
| 22 | Petrochemical basic products and intermediate chemical products |
| 23 | Synthetic resins |
| 24 | Synthetic fibers |
| 25 | Medicaments |
| 26 | Final chemical products, n.e.c. |
| 27 | Petroleum refinery products |
| 28 | Coal products |
| 29 | Plastic products |
| 30 | Rubber products |
| 31 | Leather, fur skins and miscellaneous leather products |
| 32 | Glass and glass products |
| 33 | Cement and cement products |
| 34 | Pottery, china and earthenware |
| 35 | Other ceramic, stone and clay products |
| 36 | Pig iron and crude steel |
| 37 | Steel products |
| 38 | Steel castings and forgings, and other steel products |
| 39 | Non-ferrous metals |
| 40 | Non-ferrous metal products |

Journal of Physics: Conference Series **622** (2015) 012041 doi:10.1088/1742-6596/622/1/012041

| | |
|---|---|
| 41 | Metal products for construction and architecture |
| 42 | Other metal products |
| 43 | General industrial machinery |
| 44 | Special industrial machinery |
| 45 | Other general machines |
| 46 | Machinery for office and service industry |
| 47 | Electrical appliance |
| 48 | Motor vehicles |
| 49 | Ships and repair of ships |
| 50 | Other transportation equipment and repair of transportation equipment |
| 51 | Precision instruments |
| 52 | Miscellaneous manufacturing products |
| 53 | Building construction |
| 54 | Repair of construction |
| 55 | Civil engineering |
| 56 | Electricity |
| 57 | Gas and heat supply |
| 58 | Water supply |
| 59 | Waste management service |
| 60 | Commerce |
| 61 | Finance and insurance |
| 62 | Real estate agencies and rental services |
| 63 | House rent |
| 64 | Railway transport |
| 65 | Road transport (except transport by private cars) |
| 66 | Self-transport by private cars |
| 67 | Water transport |
| 68 | Air transport |
| 69 | Freight forwarding |
| 70 | Storage facility service |
| 71 | Services relating to transport |
| 72 | Communication |
| 73 | Broadcasting and information services |
| 74 | Public administration |
| 75 | Education |
| 76 | Research |
| 77 | Medical service and health |
| 78 | Social security |
| 79 | Other public services |
| 80 | Advertising, survey and information services |
| 81 | Goods rental and leasing services |
| 82 | Repair of motor vehicles and machine |
| 83 | Other business services |
| 84 | Amusement and recreational services |
| 85 | Eating and drinking places |

| | |
|---|---|
| 86 | Accommodations |
| 87 | Other personal services |
| 88 | Office supplies |
| 89 | Activities not elsewhere classified |

(Source: [7] with slight modifications)

ScieTech 2015 IOP Publishing
Journal of Physics: Conference Series **622** (2015) 012042 doi:10.1088/1742-6596/622/1/012042

# Wavelet Types Comparison for Extracting Iris Feature Based on Energy Compaction

**R. Rizal Isnanto**[1]

[1] Department of Computer Engineering Faculty of Engineering, Diponegoro University
Jalan Prof. Soedarto, S.H., Tembalang, Semarang, 50278 INDONESIA
e-mail: rizal_isnanto@undip.ac.id; rizal_isnanto@yahoo.com

**Abstract**. Human iris has a very unique pattern which is possible to be used as a biometric recognition. To identify texture in an image, texture analysis method can be used. One of method is wavelet that extract the image feature based on energy. Wavelet transforms used are Haar, Daubechies, Coiflets, Symlets, and Biorthogonal. In the research, iris recognition based on five mentioned wavelets was done and then comparison analysis was conducted for which some conclusions taken. Some steps have to be done in the research. First, the iris image is segmented from eye image then enhanced with histogram equalization. The features obtained is energy value. The next step is recognition using normalized Euclidean distance. Comparison analysis is done based on recognition rate percentage with two samples stored in database for reference images. After finding the recognition rate, some tests are conducted using Energy Compaction for all five types of wavelets above. As the result, the highest recognition rate is achieved using Haar, whereas for coefficients cutting for $C(i) < 0.1$, Haar wavelet has a highest percentage, therefore the retention rate or significan coefficient retained for Haar is lower than other wavelet types (db5, coif3, sym4, and bior2.4)

**Keywords**: iris feature, energy compaction, recognition rate, wavelet coefficient.

## 1. Introduction

Humans as individuals, have unique characteristics and distinctive. These characteristics can be used to recognize or identify persons. This is known as biometric recognition. Iris is the part of the circle around eye pupil. Although iris has a relatively narrow region compared with entire area of the human body, iris has a very unique pattern, different in each individual and the pattern will remain stable. For those reasons, iris can be used as the basis for the recognition in biometrics.

Many algorithms have been applied as a method of iris recognition, such as PCA (Principal Component Analysis), ICA (Independent Component Analysis), Gabor-Wavelet algorithm [6], characterizing Key Local Variation, Laplace Pyramid, Gray Level Co-ocurrence Matrix (GLCM) [7] and others. Haar wavelet transform as a method to analyze the texture is still rarely used as a feature extractor on iris pattern. In this research, texture-based recognition methods were analyzed using the method of characterizing the transforms of Haar, Daubechies, Coiflets, Symlets, and Biorthogonal wavelets, which then the comparison of these five types of wavelets based on energy compaction will be conducted. The kind of research has not been done by other researchers, for which this research can show its novelty aspect. The previous research was done by the author and published in [7].

In order not to deviate far from the problems exist, the research problems are limited as follows:
1. Iris image used is the image that has been available, which is using the database of CASIA V1.0.
2. Research is focused on the compaction energy of Haar Daubechies (db5), Coifles (coif3), Symlets (sym4), and Biorthogonal (bior2.4) wavelets transform as feature extractor and feature used is energy
3. Recognition method used is the normalized Euclidean distance method.

The purpose of doing this research is to analyze some types of wavelets that can perform a comparative analysis on iris recognition rate using the Haar, Daubechies (db5), Coifles (coif3), Symlets (sym4), and Biorthogonal (bior2.4) wavelets with Energy Compaction calculations. Therefore those results then are compared and analyzed against their recognition rates. While the motivation of the research is that the iris can be used as an organ in biometric system with higher accuracy level and the fact that, up until now, there is no adequate research done for comparing these five wavelets mentioned above to extract iris features based on iris compaction.

Iris can serve as the basis for biometric systems. Each iris has a texture that is very detailed and unique to each person and remain stable for decades. The eye cannot be altered through surgery without causing any damage to eyesight. Fig.1 shows the anatomy of the eye, and examples of human iris [2].

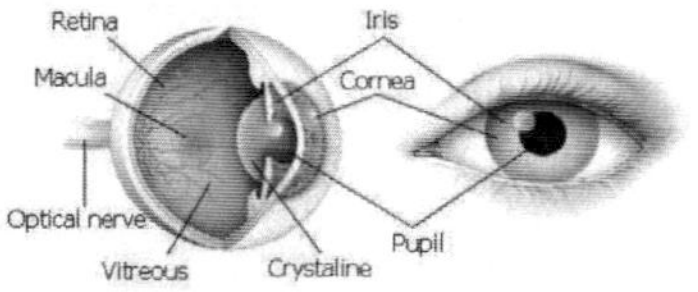


**Figure 1.** Anatomy of eyes and example of iris region.

The advantages of using the iris for reliable identification system are [3] as follows.
1. Iris is insulated and shielded from the outside environment.
2. In the iris, it is not possible to do some surgeries without causing defects in the eye.
3. Iris has a physiological response to light, which allows testing of the natural use of the possibility of fraud and faked eye lenses and so forth.

Wavelets are mathematical functions that satisfy certain requirements are able to perform the decomposition of a function [5]. Hierarchically, wavelet is used to represent data or other functions. Wavelet can be used to describe a model or the original image into a mathematical function regardless of the shape of the model in the form of image, a curve or a plane. Wavelet transform is a function that converts the signals from region to region the frequency or time scale. The most appropriate wavelet transform used in image processing because there is not much information is lost during the reconstruction.

Wavelet is a base derived from a wavelet basis function which is also said to be a scaling or scaling function. Scaling function has properties which can be assembled from a number of copies that have been dilated, translated and scaled. This function is derived from the dilation equation, which is considered as the basis of wavelet theory.

The wavelet decomposition can be continued using lowpass-lowpass (LL) as its input for getting next decomposition stages. Figure 4 shows a decomposition image from level 1 until level 3. Maximum level of decomposition in image processing using wavelet (except biorthogonal wavelets) can be formulated as indicated in Eq. (1) [4]. This equation is derived from how many times we can divide the data length (i.e. the minimum size of either horizontal or vertical pixels of an image) by number of 2 (two).

Journal of Physics: Conference Series **622** (2015) 012042 doi:10.1088/1742-6596/622/1/012042

$$\text{Level}_{max} = \frac{\log(\text{datalength}/(\text{filterlength}-1))}{\log(2)} \quad (1)$$

Energy is defined as the sum of the squares of the values. So the energy of an image is the sum of the squares of the pixel values, the energy in the wavelet transform of an image is the sum of the squares of the transform coefficients. During wavelet analysis the energy of a signal is divided between approximation and details signals but the total energy does not change. During compression, however, energy is lost because thresholding changes the coefficient values and hence the compressed version contains less energy.

The compaction of energy describes how much energy has been compacted into the approximation signal during wavelet analysis. Compaction will occur wherever the magnitudes of the detail coefficients are significantly smaller than those of the approximation coefficients. Compaction is important when compressing signals because the more energy that has been compacted into the approximation signal the less energy can be lost during compression.

Energy compaction can be obtained from the number of least significant coefficients, for which those coefficients can be cut or removed. In this case, it can be said that insignificant coefficients mean that there are zero valued coefficients and which can be seemed less enough to be cut or removed. In energy compaction, the term of 'retention' is also used. Retention indicates coefficient values which can be retained in calculation.

## 2. Research Method

Method of iris recognition system can be depicted in a flowchart which is shown in Figure 2, as an enhancement procedure introduced by Daugman [2], which is improved by Isnanto, *et al.* [7]. Broadly explained, the processes can be classified into some steps which can be depicted in pseudocodes below.

```
PROCEDURE Iris_Recognition

CASE wavelet_types
   1: Haar
   2: db5
   3: coif3
   4: sym4
   5: bior2.4

INPUT OriginalEyeImage
   DO segmenting OriginalEyeImage to separate IrisImage from its OriginalEyeImage
   DO normalizing the IrisImage for getting NormalizedIrisImage
   MinEuclideanDistance := 100 % impossible distance in the research
   FOR i = 1 TO NumberOfStoredImage
        DO extracting the features using CASE 1,2,3,4,5; OUTPUT feature(CASE)
        CALCULATE EuclideanDistance(feature(CASE) , feature(i))
        IF EuclideanDistance(feature(CASE) , feature(i)) < MinEuclideanDistance ...
        ... THEN MinEuclideanDistance = EuclideanDistance(feature(CASE),feature(i))
        DO plotting graphics of correlation between the number of coefficient...
        ...(x-axis) against coefficients values (y-axis) for CASE 1,2,3,4,5
        CALCULATE energy compaction with some coefficient values of CASE...
        ...1,2,3,4,5 which results of retention for CASE 1,2,3,4,5

   NEXT i
OUTPUT recognized as feature which has MinEuclideanDistance
ANALYZE the results of PROCEDURE executes
```

Iris image used is the image from database of CASIA V1.0 – Portions of the research in this paper use the CASIA-IrisV1 collected by the Chinese Academy of Sciences' Institute of Automation (CASIA) – and iris images captured using Irdosoft 4.0 camera [1].

In database, pupil circle and iris circle are not always perfect circle for which for getting only iris needs a complex computation. For simplifying the computation, it is assumed that pupil as well as iris has a perfect-circle form. First step is finding pupil circle, centre point and its radius. Some subprocesses have to be done are thresholding, smoothing, and obtaining the centre point of pupil and its radius using Circular Hough Transform.

Pupil and iris images which are circle-form with their diameters are varied, are then converted into a fix-sized rectangular form. The size is 60×512 pixels. This conversion also for simplifying both coding as well as computation.

The presence of eyelids and eyelashes also can can disrupt the process of iris recognition and reduce the level of accuracy. Therefore, not all parts of the iris is taken. For the iris image of CASIA, the top part of the image is cut so that the bottom of the iris of the eye are taken. Image size is 60 × 384 pixels. Iris image which has been converted into a rectangular shape has a low contrast level so that the resulting level of accuracy is less good. Therefore, the contrast then enhanced using adaptive histogram equalization. First, image is is split into 3 × 12 parts of each section measuring 20 × 32 pixels. The purpose is to obtain images with good contrast but will not damage the overall image quality.

Iris image is split into 6 slices. Feature extraction is per-formed to all images which have been split before using 5 (five) wavelet types. Features obtained are energies of $E_a$, $E_h$, $E_v$, dan $E_d$.

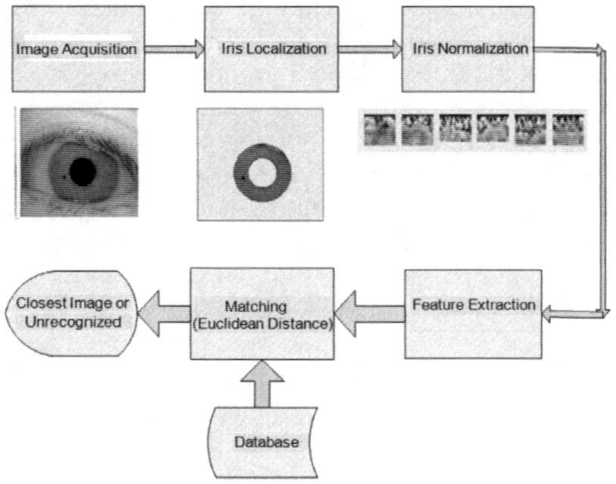


**Figure 2**. Flowchart of iris recognition system using feature extraction from different wavelets.

## 3. Results and Discussion

### *3.1 Recognition Testing*

The purpose of recognition test is to obtain what level of recognition rate of the system, when different decomposition levels are used from 5 types of wavelet filter, i.e. Haar, db5, coif3, sym4, and bior2.4.

Some tests were conducted for 128 images, originated from 64 persons. While the stored images in database are 2 samples for each person. Maximum decomposition levels for Haar, db5, coif3, sym4, and bior2.4 are 5, 2, 1, 3, and 2, respectively. From all allowed decomposition level tests, it can be inferred that: (a) The best recognition level is 84.375% which is achieved by Haar with decomposition level 3. (b) The decreasing order of average of recognition rate is: Haar, sym4, db5, and bior2.4 for the least. Table 1 shows all wavelet tests up to the maximum level achieved. From this table, it can be shown that the higher level of decomposition, the recognition level tends to be better.

**Table 1.** Recognition rate from different decomposition level.

| Decomposition Level | Recognition Rate (%) | | | | |
|---|---|---|---|---|---|
| | **Haar** | **db5** | **coif3** | **sym4** | **bior2,4** |
| 1 | 68,750 | 38,231 | 25,000 | 42,188 | 42,969 |
| 2 | 76,563 | 68,750 | - | 60,938 | 60,156 |
| 3 | 84,375 | - | - | 65,625 | - |
| 4 | 83,594 | - | - | - | - |
| 5 | 83,594 | - | - | - | - |
| **Average of recognition rate (%)** | **79,375** | **53,491** | **25,000** | **56,250** | **51,563** |

*3.2 Energy Compaction Testing*

In this test, wavelet coefficient threshold $C(i)$ of 0.001 and 0.1 respectively are applied. We can use 2 (two) scenarios to analyze the comparison of energy compaction for 5 types of wavelet tested. *First*, we can visually analyze from graphics which depicting the relationship between $i$-th coefficient and its corresponding wavelet coefficient value $C(i)$. Figures 3-7 show these relationships and we therefore can observe the existing trends of the graphics. It can be shown that wavelet coefficient distribution of Haar wavelet gives the best retention because of most coefficients have a small value, and the first values are very high. It means that the approximation values are high enough if compared with detail values. *Second*, we can use the retention concept which means coefficient values which can be retained in calculation. The more coefficients can be cut or removed, the less retention can be produced. The retention is better when its value is lower. Table 2 and Table 3 depict the retention level for threshold value 0.001 and 0.1, respectively.

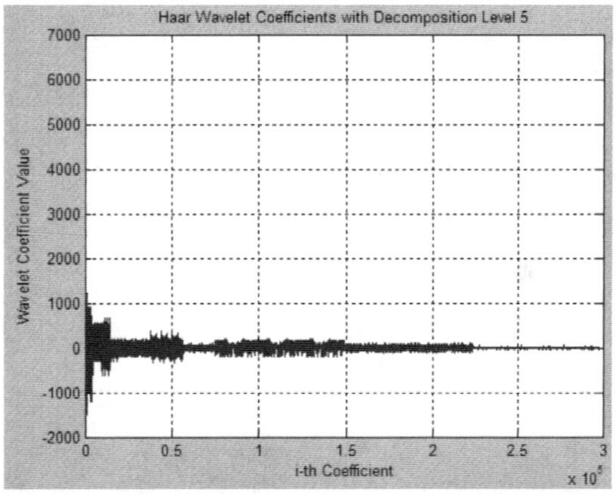


**Figure 3.** Wavelet coefficient values for Haar, decomposition level 5.

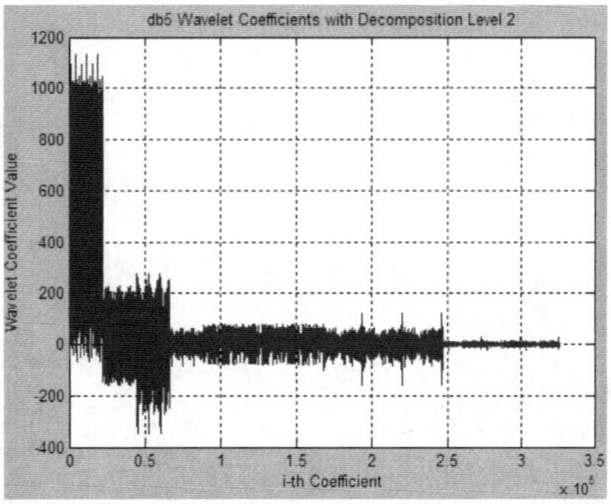


**Figure 4.** Wavelet coefficient values for db5, decomposition level 2.

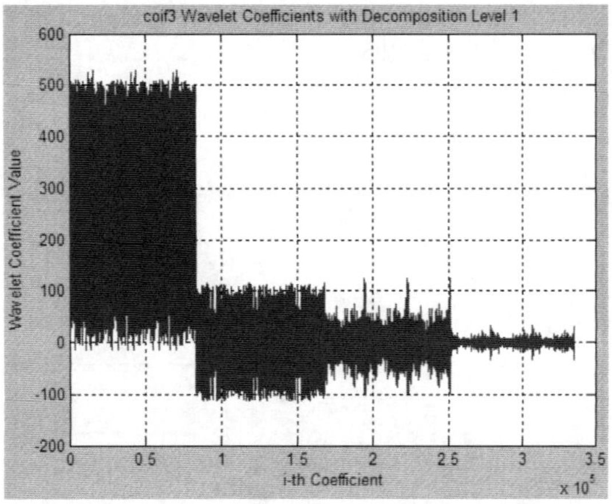


**Figure 5.** Wavelet coefficient values for coif3, decomposition level 1.

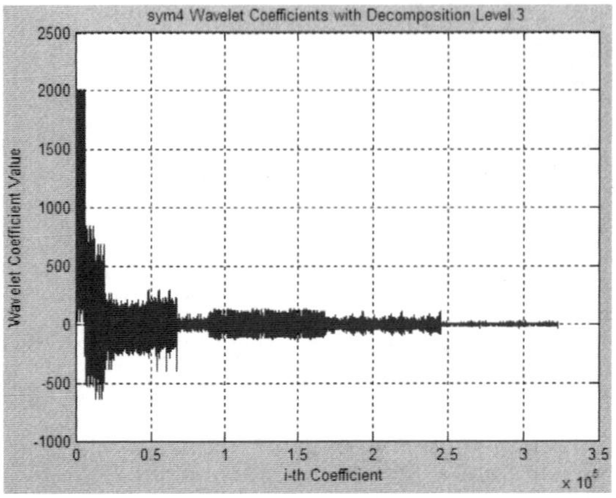


**Figure 6.** Wavelet coefficient values for sym4, decomposition level 3.

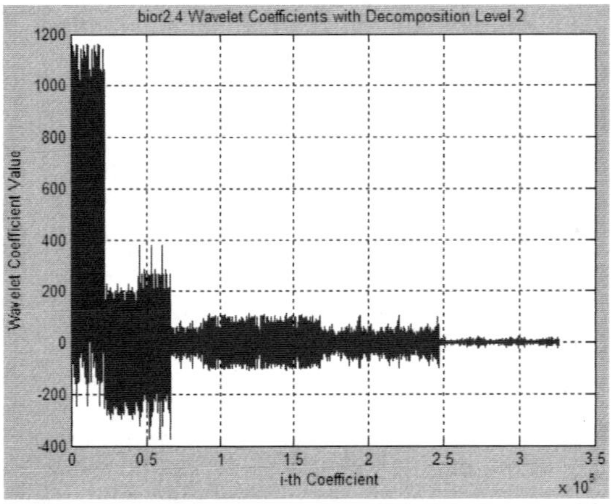


**Figure 7.** Wavelet coefficient values for bior2.4, decomposition level 2.

**Table 2.** Energy Compaction with Threshold C(i) > = 0.001

| Wavelet | Energy compaction | Decomposition Level | | | | | Average | |
|---|---|---|---|---|---|---|---|---|
| | | 1 | 2 | 3 | 4 | 5 | | |
| **Haar** | Retention (%) | 58.7 | 49.3 | 46.8 | 46.3 | 46.1 | 49.44 | |
| | Removed (%) | 41.3 | 50.7 | 53.2 | 53.7 | 53.1 | *50.40* | Highest |
| **db5** | Retention (%) | 63.1 | 54.2 | - | - | - | 58.65 | |
| | Removed (%) | 36.9 | 45.8 | - | - | - | 41.35 | |
| **coif3** | Retention (%) | 62.3 | - | - | - | - | 62.30 | |
| | Removed (%) | 37.7 | - | - | - | - | 37.70 | |
| **sym4** | Retention (%) | 62.5 | 53.3 | 51.0 | - | - | 55.60 | |
| | Removed (%) | 37.5 | 46.7 | 49.0 | - | - | 44.40 | |
| **bior2.4** | Retention (%) | 60.4 | 51.3 | - | - | - | 55.85 | |
| | Removed (%) | 39.6 | 48.7 | - | - | - | 44.15 | |

**Table 3.** Energy Compaction with Threshold C(i) > = 0.1

| Wavelet | Energy compaction | Decomposition Level | | | | | Average | |
|---|---|---|---|---|---|---|---|---|
| | | 1 | 2 | 3 | 4 | 5 | | |
| **Haar** | Retention (%) | 58.8 | 49.3 | 46.9 | 46.3 | 46.1 | 49.48 | |
| | Removed (%) | 41.2 | 50.7 | 53.1 | 53.7 | 53.9 | *50.52* | Highest |
| **db5** | Retention (%) | 60.9 | 50.9 | - | - | - | 55.90 | |
| | Removed (%) | 39.1 | 49.1 | - | - | - | 44.10 | |
| **coif3** | Retention (%) | 59.8 | - | - | - | - | 59.80 | |
| | Removed (%) | 40.2 | - | - | - | - | 40.20 | |
| **sym4** | Retention (%) | 60.0 | 50.7 | 48.4 | - | - | 53.03 | |
| | Removed (%) | 40.0 | 49.3 | 51.6 | - | - | 46.97 | |
| **bior2.4** | Retention (%) | 59.1 | 49.8 | - | - | - | 54.45 | |
| | Removed (%) | 40.9 | 50.2 | - | - | - | 45.55 | |

For removing the coefficients $C(i) < 0.1$ as well as for $C(i) < 0.001$, Haar wavelet has the highest percentage of coefficients removed, there are 50.54% and 50.52% for which significant coefficients (retention) are less than other wavelets (db5, coif3, sym4, dan bior2.4).

According the previous mentioned reason, based on recognition level as well as its energy compaction, Haar wavelet has good performance for extracting the iris feature when compared with other wavelets. Based on recognition level, Haar indicates the highest level of recognition, i.e. 79.38%. Whereas, according to energy compaction, Haar also indicates the highest percentage of coefficients removed as mentioned in previous paragraph.

## 4. Conclusion

From results of tests and analysis, it can be concluded that: The higher level of decomposition, the recognition level tends to be better. Applying Haar wavelet produces the highest recognition rate in all allowed decomposition level (from level 1 up to 5). On the other hand, for removing the coefficients $C(i) < 0.1$ as well as for $C(i) < 0.001$, Haar wavelet has the highest percentage of coefficients removed, for which significant coefficients (retention) are less than other wavelets (db5, coif3, sym4, and bior2.4). Therefore, from the previous mentioned, based on recognition level as well as its energy compaction, Haar wavelet has good performance for extracting the iris feature when compared to other wavelets.

## Acknowledgments

The author would like to express their sincere gratitude to the Directorate General of Higher Education (DIKTI) for the financial support received under the project of Penelitian Hibah Bersaing BOPTN Undip Year 2014 by Contract No. DIPA – 023.04.02.189185/2014.

## References

[1] CASIA, *Iris Image Database*, www.sinobiometrics.com.

[2] Daugman, J. "How Iris Recognition Works", IEEE Transaction on Circuits and system for Video Technology, vol 14, no.1, January 2004.

[3] Daugman, J. "*Iridology*", http://www.cl.cam.ac.uk/ users/jgd1000/, Februari, 2009.

[4] Misiti, M., Y. Misiti, G. Oppenheim, and J.M. Poggi, *Wavelet Toolbox For Use with Matlab*, User's Guide Version 3, The Math Works, Natick, MA, 2004.

[5] Moreno, R.P. and A. Gonzaga, *Features Vector For Personal Identification Based On Iris Texture*. Departamento de Engenharia Elétrica - EESC – USP, Februari, 2009.

[6] Rahmawati, I., "*Pemampatan Citra Digital Dengan Wavelet Paket*", Skripsi S-1, Universitas Diponegoro, 2007.

[7] Isnanto, R.R., Widodo, T.S., and Suhardjo, *Implementation of Gray Level Cooccurrence Matrix (GLCM) for Human Iris Recognition System*, Proceedings of Conference on Information Technology and Electrical Engineering 2010 (CITEE 2010), Yogyakarta, Indonesia, ISSN: 2085-6350, pp. 95-101

Journal of Physics: Conference Series **622** (2015) 012043 doi:10.1088/1742-6596/622/1/012043

# The homotopic mapping solutions for the generalized Schrödinger equation

**Dianhen Lu** [1,a], **Xinyu Cui**[1,b], **Baojian Hong** [1,c]

[1]Nonlinear Scientific Research Center, Faculty of Science, Jiangsu University, Zhenjiang Jiangsu, 212013,P.R.China

E-mail: [a]dclu@ujs.edu.cn, [b]2314200389@qq.com, [c]hbj@njit.edu.cn

**Abstract.** This work is concerned on how to find the approximation solutions of the generalized Schrödinger equation by using the homotopic mapping method. First, we give the basic idea of Homotopic mapping method. In the following, we introduce a homotopic mapping, and make the generalized Schrödinger equation do a homotopic deformation, then we get the approximation solutions in given initial conditions. Finally we give two examples about derivative Schrödinger equation and Hirota equation to illustrate efficiency of this method, and the exact solutions of these equations can be solved in the condition of selecting proper auxiliary function $v$.

## 1. Introduction

The investigation of solutions to the nonlinear partial differential equation(NLPDEs) play an important role in natural science, many important nonlinear phenomenon in various field such as fluid dynamic, mathematic biology and plasma physics can be described NLPDEs. Therefore, more and more scholars devote themselves to obtain the exact solutions or approximate solutions of NLPDEs. In recent years, many searching for exact solutions or approximation solutions of NLPDEs methods have also been developed and improved, including the method of homogeneous balanced [1], the method of trial function [2], the method of Adomian decomposition [3-4], the method of perturbation [5], etc. The homotopic mapping method [6-9] is a new method and well performed in searching for analytical approximate solutions of NLPDEs. For example, the variable coefficient KdV equation[10], the two-dimensional KdV-Burgers equation [11], the epidemic model [12], etc.

The object of this study is the generalized Schrödinger equation:

$$iu_t + \alpha u_{xx} + i\beta u_{xxx} + i\gamma(|u|^2 u)_x + \delta |u|^2 u = 0. \tag{1.1}$$

As $\alpha = -1, \beta = 0, \gamma = 1, \delta = 0$, Eq.(1.1) is the derivative Schrödinger equation:

$$u_t + iu_{xx} + (|u|^2 u)_x = 0.$$

As $\alpha = 1, \beta = 2, \gamma = 1, \delta = 1$, Eq.(1.1) is Hirota equation:

$$iu_t + u_{xx} + 2iu_{xxx} + i(|u|^2 u)_x + |u|^2 u = 0.$$

The generalized Schrödinger equation is a nonlinear equation which is often encountered in modern physics study, it is also a basic equation in researching light pulse transmission problem in nonlinear dispersive optical medium. For this equation, there are already a number of ways to solve the solution, such as the split-step Fourier method [13], trigonometric function expansion method [14], general mapping approach [15] etc. But these methods are just applied to some special forms of this equation. We can get the approximation solution of this equation with initial condition by using the homotopic mapping method. The object of this study is general, the consequence has a significant meaning.

## 2. Basic idea of homotopic mapping method

We consider the following nonlinear differential equation:

$$A(u) - f(r) = 0, r \in \Omega,$$

and boundary conditions

$$B(u, \partial u/\partial n) = 0, r \in \Gamma,$$

where $A$ is a general differential operator, $B$ is a boundary operator, $f(r)$ is a known analytical function, $\Gamma$ is the boundary of the domain $\Omega$, and $\partial/\partial n$ denoted differential along the normal drawn outwards from $\Omega$. The operator $A$ can be divided into two parts $L$ and $N$. Therefore $A(u) - f(r) = 0$ can be rewritten as follow:

$$L(u) + N(u) - f(r) = 0$$

Now We construct a homotopic mapping

$$H(u, p) : \Omega \times [0, 1] \to R,$$

$$H(u, p) = L(u) - L(v) + p[L(v) + N(u) - f(r)], \tag{2.1}$$

where $p$ is a parameter, $v$ is an auxiliary function.

According to Eq.(2.1), we have

$$H(u, 0) = L(u) - L(v),$$

$$H(u, 1) = A(u) - f(r),$$

and the changing process of $p$ from 0 to 1, is just that of $H(u, p)$ from $L(u) - L(v)$ to $A(u) - f(r)$. In topology, that is called deformation $L(u) - L(v)$ and $A(u) - f(r)$ are called homotopic. We can set that the solution of $H(u, p) = 0$ can be expressed as a series in $p$ as follow:

$$\tilde{u}(x, t, p) = \sum_{i=0}^{\infty} u_i(x, t) p^i = u_0 + pu_1 + p^2 u_2 + \cdots$$

When $p = 0$, $\tilde{u}(x, t, 0) = u_0(x, t)$ is the solution of $L(u) - L(v) = 0$. When $p \to 1$, we can obtain the approximation solution of $A(u) - f(r) = 0$ as follows

$$\tilde{u}(x, t, 1) = u_0 + u_1 + u_2 + \cdots$$

## 3. The approximation solution of the generalized Schrödinger equation

The Eq(1.1) can be rewritten as the following:

$$u_t - i\alpha u_{xx} + \beta u_{xxx} + \gamma(|u|^2 u)_x - i\delta |u|^2 u = 0. \tag{3.1}$$

In order to obtain the solution of Eq.(3.1), we constructed a homotopic mapping

$$H(u,p) : \Omega \times [0,1] \to R,$$

$$H(u,p) = L(u) - L(v) + p[L(v) + N(u) - f(r)], \tag{3.2}$$

where $R = (-\infty, +\infty)$, $L(u) = u_t - i\alpha u_{xx} + \beta u_{xxx}$, $v$ is an auxiliary function.

According to Homotopic mapping, $H(u,1) = 0$ is equal to Eq.(3.1). Therefore, as $p \to 1$, the solution of $H(u,p) = 0$ is the solution of Eq.(3.1).

Set

$$\tilde{u}(x,t,p) = \sum_{i=0}^{\infty} u_i(x,t)p^i = u_0 + pu_1 + p^2u_2 + \cdots \tag{3.3}$$

According to [16], when $p \in [0,1]$, the series (3.3) is uniform convergence.

Let $u_0 = v$, substituting Eq.(3.3) in $H(u,p) = 0$, we have

$$L\left(\sum_{i=0}^{\infty} u_i p^i\right) - L(v) + p\left[L(v) + \gamma\sum_{i=0}^{\infty} A_i p^i - i\delta \sum_{i=0}^{\infty} B_i p^i\right] = 0, \tag{3.4}$$

where $|u|^2 u = \sum\limits_{n=0}^{\infty} B_n p^n$, $\left(|u|^2 u\right)_x = \sum\limits_{n=0}^{\infty} A_n p^n$.

Because $|u|^2 u$ and $\left(|u|^2 u\right)_x$ are complex function, the corresponding polynomials are

$$\begin{aligned} |u|^2 u &= u\bar{u}u = u^2\bar{u} \\ &= (u_0 + u_1p + u_2p^2 + \cdots)^2(\bar{u}_0 + \bar{u}_1p + \bar{u}_2p^2 + \cdots) \\ &= (C_0 + C_1p + C_2p^2 + C_3p^3 + \cdots)(\bar{u}_0 + \bar{u}_1p + \bar{u}_2p^2 + \cdots) \\ &= B_0 + B_1p + B_2p^2 + B_3p^3 + \cdots, \end{aligned}$$

where

$$C_n = \begin{cases} 2\sum\limits_{i=0}^{k} u_i u_{n-i}, & n = 2k+1, \\ 2\sum\limits_{i=0}^{k} u_i u_{n-i} - u_k^2, & n = 2k, \end{cases} \qquad k = 0,1,2,3\cdots$$

$$B_j = \sum_{n=0}^{j} C_n \bar{u}_{j-n}, \qquad j = 0,1,2,3\cdots \tag{3.5}$$

According (3.5), we have

$$\begin{aligned} B_0 &= u_0^2\bar{u}_0, \\ B_1 &= u_0^2\bar{u}_1 + 2u_0u_1\bar{u}_0, \\ B_2 &= u_0^2\bar{u}_2 + 2u_0u_1\bar{u}_1 + 2u_0u_2\bar{u}_0 + u_1^2\bar{u}_0 \\ &\vdots \end{aligned} \tag{3.6}$$

Similarity

$$\begin{aligned} (|u|^2 u)_x &= (u\bar{u}u)_x = (u^2\bar{u})_x \\ &= [B_0 + B_1p + B_2p^2 + B_3p^3 + \cdots]_x \\ &= A_0 + A_1p + A_2p^2 + A_3p^3 + \cdots, \end{aligned}$$

where

$$A_j = (B_j)_x = (\sum_{n=0}^{j} C_n \bar{u}_{j-n})_x, \qquad j = 0, 1, 2, 3 \ \cdots \tag{3.7}$$

From (3.7), we have

$$\begin{aligned}
&A_0 = 2u_0u_{0x}\bar{u}_0 + u_0^2\bar{u}_{0x},\\
&A_1 = 2u_0u_{0x}\bar{u}_1 + u_0^2\bar{u}_{1x} + 2u_{0x}u_1\bar{u}_0 + 2u_0u_{1x}\bar{u}_0 + 2u_0u_1\bar{u}_{0x},\\
&A_2 = 2u_0u_{0x}\bar{u}_2 + u_0^2\bar{u}_{2x} + 2u_{0x}u_1\bar{u}_1 + 2u_0u_{1x}\bar{u}_1 + 2u_0u_1\bar{u}_{1x} + 2u_{0x}u_2\bar{u}_0\\
&+2u_0u_{2x}\bar{u}_0 + 2u_0u_2\bar{u}_{0x} + 2u_1u_{1x}\bar{u}_0 + u_1^2\bar{u}_{0x},\\
&\vdots
\end{aligned} \tag{3.8}$$

Arranging the coefficients of $p$ power in (3.4), we have

$$\begin{aligned}
&p^0 : L(u_0) = L(v),\\
&p^1 : L(u_1) = -L(v) - \gamma A_0 + i\delta B_0,\\
&P^2 : L(u_2) = -\gamma A_1 + i\delta B_1,\\
&\vdots\\
&p^j : L(u_j) = -\gamma A_{j-1} + i\delta B_{j-1},\\
&\vdots
\end{aligned} \tag{3.9}$$

We let $u_0 = v$ and substitute it in (3.9), $u_j$ can be concluded step by step. So the approximation solution of (1.1) can be expressed as follows:

$$\tilde{u} = u_0 + u_1 + u_2 + \cdots$$

## 4. The approximation solution of derivative Schrödinger equation

Consider the following derivative Schrödinger equation with initial condition:

$$\begin{cases} u_t + iu_{xx} + (|u|^2 u)_x = 0,\\ u(x,0) = e^{\frac{1}{2}ix}. \end{cases} \tag{4.1}$$

The homotopic deformation of Eq.(4.1) is

$$L(u) - L(v) + p\left[L(v) + \left(|u|^2 u\right)_x\right] = 0, \tag{4.2}$$

where $L(u) = u_t + iu_{xx}$.

Substituting $u = u_0 + u_1p + u_2p^2 + \cdots$ in Eq.(4.2), we can obtain

$$L\left(\sum_{i=0}^{\infty} u_i p^i\right) - L(v) + p\left[L(v) + \sum_{i=0}^{\infty} A_i p^i\right] = 0, \tag{4.3}$$

where

$$\begin{aligned}
&A_0 = 2u_0u_{0x}\bar{u}_0 + u_0^2\bar{u}_{0x},\\
&A_1 = 2u_0u_{0x}\bar{u}_1 + u_0^2\bar{u}_{1x} + 2u_1u_{0x}\bar{u}_0 + 2u_0u_{1x}\bar{u}_0 + 2u_0u_1\bar{u}_{0x},\\
&A_2 = 2u_0u_{0x}\bar{u}_2 + u_0^2\bar{u}_{2x} + 2u_1u_{0x}\bar{u}_1 + 2u_0u_{1x}\bar{u}_1 + 2u_0u_1\bar{u}_{1x} + 2u_2u_{0x}\bar{u}_0\\
&+2u_0u_{2x}\bar{u}_0 + 2u_0u_2\bar{u}_{0x} + 2u_1u_{1x}\bar{u}_0 + u_1^2\bar{u}_{0x},\\
&\vdots
\end{aligned} \tag{4.4}$$

Arranging the coefficients of $p$ power in (4.3), we have

$$\begin{aligned}
&p^0 : L(u_0) = L(v),\\
&p^1 : L(u_1) = -L(v) - A_0,\\
&p^2 : L(u_2) = -A_1,\\
&\vdots\\
&p^j : L(u_j) = -A_{j-1}\\
&\vdots
\end{aligned}$$

Let $u_0 = v = e^{\frac{1}{2}i(x-ct)}$, according (4.4) we have $A_0 = \frac{i}{2}e^{\frac{1}{2}i(x-ct)}$, so

$$u_{1t} + iu_{1xx} = -\left(-\frac{1}{2}ci + \frac{1}{4}i\right)e^{\frac{1}{2}i(x-ct)}.$$

If we set $c = \frac{1}{2}$, then $u_1 = 0$, and $A_1 = 0, u_2 = 0, A_2 = 0, \cdots$, we can obtain

$$\tilde{u} = u_0 = e^{\frac{1}{2}i(x-\frac{1}{2}t)}.$$

**Remark 1**: It is easy to prove that the above $\tilde{u}$ is the exact solution of Eq.(4.1).

## 5. The approximation solution of Hirota equation

Considering the following Hirota equation with initial condition:

$$\begin{cases} iu_t + u_{xx} + 2iu_{xxx} + i(|u|^2 u)_x + |u|^2 u = 0, \\ u(x,0) = e^{ix}. \end{cases} \tag{5.1}$$

The homotopic deformation of Eq.(5.1) is

$$L(u) - L(v) + p\left[L(v) + \left(|u|^2 u\right)_x - i|u|^2 u\right] = 0, \tag{5.2}$$

where $L(u) = u_t - iu_{xx} + 2u_{xxx}$.

Substituting $u = u_0 + u_1 p + u_2 p^2 + \cdots$ in Eq.(5.2), and arranging the coefficients of $p$ power, we have

$$\begin{aligned}
&p^0 : L(u_0) = L(v),\\
&p^1 : L(u_1) = -L(v) - A_0 + iB_0,\\
&p^2 : L(u_2) = -A_1 + iB_1,\\
&\vdots\\
&p^j : L(u_j) = -A_{j-1} + iB_{j-1},\\
&\vdots
\end{aligned}$$

where

$$\begin{aligned}
&A_0 = 2u_0u_{0x}\bar{u}_0 + u_0^2\bar{u}_{0x},\\
&A_1 = 2u_0u_{0x}\bar{u}_1 + u_0^2\bar{u}_{1x} + 2u_1u_{0x}\bar{u}_0 + 2u_0u_{1x}\bar{u}_0 + 2u_0u_1\bar{u}_{0x},\\
&A_2 = 2u_0u_{0x}\bar{u}_2 + u_0^2\bar{u}_{2x} + 2u_1u_{0x}\bar{u}_1 + 2u_0u_{1x}\bar{u}_1 + 2u_0u_1\bar{u}_{1x} + 2u_2u_{0x}\bar{u}_0\\
&+2u_0u_{2x}\bar{u}_0 + 2u_0u_2\bar{u}_{0x} + 2u_1u_{1x}\bar{u}_0 + u_1^2\bar{u}_{0x},\\
&\vdots
\end{aligned} \tag{5.3}$$

$$\begin{aligned}
&B_0 = u_0^2\bar{u}_0,\\
&B_1 = u_0^2\bar{u}_1 + 2u_0u_1\bar{u}_0,\\
&B_2 = u_0^2\bar{u}_2 + 2u_0u_1\bar{u}_1 + 2u_0u_2\bar{u}_0 + u_1^2\bar{u}_0,\\
&\vdots
\end{aligned} \tag{5.4}$$

Let $u_0 = v = e^{i(x-ct)}$, according (5.3) and (5.4), we know $A_0 = ie^{i(x-ct)}, B_0 = e^{i(x-ct,)}$, So

$$u_{1t} - iu_{1xx} + 2u_{1xxx} = (ci + i)\, e^{i(x-ct)}.$$

If we set $c = -1$, then $u_1 = 0$, and $A_1 = 0, B_1 = 0, u_2 = 0, A_2 = 0, B_2 = 0, \cdots$, we can obtain

$$\tilde{u} = u_0 = e^{i(x+t)}.$$

**Remark 2**: It is easy to prove that above $\tilde{u}$ is the exact solution of Eq.(5.1).

## 6. Conclusion

From section 4 and section 5, if we choose proper auxiliary function $v$, a great quantity of compute work can be reduced, we even conclude the exact solution of equations. In this paper, we get the approximation solution of the generalized Schrödinger equation by using the homotopic mapping method. Then, two examples of the derivative Schrödinger equation and Hirota equation are given. Through these examples, the homotopic mapping method has advantage of brief and efficient.

## Acknowledgements

The work was supported by the National Nature Science Foundation of China (No. 61070231), and the Out standing Personnel Program in Six Fields of Jiangsu (No.2009188), and the Natural Science Foundation of Jiangsu Province (BK20140525).

## References

[1] Fan En-gui, Zhang Hong-qing. Homogeneous balance method of nonlinear soliton equations [J]. ACTA PHYSICA SINICA. 47(3) (1998) 353-10.
[2] Xie Yuan-xi, Tang Jia-shi. Solving a class of nonlinear partial differential equations the exact solution by the simplify trial function method [J]. Journal of Dynamics and Control. 3(1) (2005) 15-18.
[3] Niu Hong-ling, Hao Ling, Yu Zhi-xian, Yin Jian-hua. The solutions for some fractional Nonlinear Partial Differential Equation via the Adomian Decomposition Method[J]. Journal of Liaoning Technical University (Natural Science). 2013 (2013) 010132-04.
[4] Jin Yi, Chen Liang. The exact solution for a class of nonlinear Schrödinger equation via the Adomian decomposition Method [J]. Journal of Hangzhou Normal University(Natural Science Edition). 7(7) (2008) 256-260.
[5] Wang Ai-hua. The Application of Perturbation Method in the Problen about Dynamic Plastic Sheet[J]. Journal of JinZhou Teachers College(Natural Sciences Edition)). 2000 (2000) 020010-04.
[6] Zhang Ben-gong,Li Shao-yong,Liu Zheng-rong. Homotopy perturbation method for modified Camassa-Holm and Degasperis-Procesi equations[J]. Physics Letters A. 372 (2008) 1867-1872.
[7] N.H.Sweilam, M.M.Khader. Exact solutions of some coupled nonlinear partial differential equations using the homotopy perturbation method[J]. Computers and Mathematics with Applications. 58 (2009) 213-2141.
[8] Lin Jin. Application of variational iteration method and homotopy perturbationmethod to the modified Kawahara equation[J]. Mathematical and Computer Modelling 49 (2009) 573-578
[9] J.H.He, Application of Homotopy perturbation method to nonlinear equation, Chaos, Solitons and Fractals. 26 (2005) 695-700.
[10] Lu Dian-chen, Chen Ting-ting, Hong Bao-jian. Homotopic Approximation Solutions for the Perturbed CKdV Equation with Variable Coefficients[J]. The Scientific World Journal. 2014 (2014), Article ID 593642, 5 pages.
[11] A.Molabahrami, F.Khani, S.Hamedi-Nezhad. Soliton solutions of the two-dimensional KdV-Burgers equation by homotopy perturbation method[J]. Physics Letters A. 370 (2007) 433-436.
[12] M.Rafei, D.D.Ganji, H.Daniali. Solution of the epidemic model by homotopy perturbation method[J]. Applied Mathematics and Computation. 187 (2007) 1056-1062.
[13] Zhao Lei, Sui Zhan, Zhu Qi-hua, Zhang Ying, Zuo Yan-lei. Improvement and precision analysis of the split -step Fourier method in solving the general nonlinear Schrödinger equation[J]. ACTA PHYSICA SINICA. 58(07) (2009) 4731 -4737.

[14] Zhao Ming-zhuo, Liu Xiao-juan. Optical soliton solution to higher order dispersion modified non-linear Schrödinger equation[J]. Journal of Hunan University of Science and Technology(N atural Science Edition). 24(3) (2009) 126-128.
[15] Fang Jian-ping. Study on the nonlinear [J]. Schrödinger system by a general mapping approach[J]. Journal of Lishui University. 27(5) (2005) 26-32.
[16] Liao S J. Beyond Perturbation: Introduction to the Homotopy Analysis Method. New York: CRC Press. 2004.

# Optimization of reactor network design problem using Jumping Gene Adaptation of Differential Evolution

**Ashish M. Gujarathi*[a], S. Purohit[b] and B. Srikanth[c]**

[a]Petroleum and Chemical Engineering Department

Sultan Qaboos University, Muscat, PC 123, Sultanate of Oman

[b]Birla Institute of Technology and Science, Pilani, India.

ashishg@squ.edu.om; ashishgujrathi@gmail.com

**Abstract**. Detailed working principle of jumping gene adaptation of differential evolution (DE-JGa) is presented. The performance of the DE-JGa algorithm is compared with the performance of differential evolution (DE) and modified DE (MDE) by applying these algorithms on industrial problems. In this study Reactor network design (RND) problem is solved using DE, MDE, and DE-JGa algorithms: These industrial processes are highly nonlinear and complex with reference to optimal operating conditions with many equality and inequality constraints. Extensive computational comparisons have been made for all the chemical engineering problems considered. The results obtained in the present study show that DE-JGa algorithm outperforms the other algorithms (DE and MDE). Several comparisons are made among the algorithms with regard to the number of function evaluations (*NFE*)/CPU-time required to find the global optimum. The standard deviation and the variance values obtained using DE-JGa, DE and MDE algorithms also show that the DE-JGa algorithm gives consistent set of results for the majority of the test problems and the industrial real world problems.

## 1. Introduction

Optimization refers to finding the best value of solution among the set of feasible solutions. In case of single objective optimization, there may exist a single global optimum solution and several near global or the local optimal solutions. Thus, the ideal outcome of a single objective optimization problem is a single global solution. The traditional algorithms often start with a single point (guess value) and end up with a single point solution. However, the outcome of gradient based traditional algorithms largely depends on its control parameters such as the step size and the direction of search that are being used. In a complex and non-linear search space, which may involve multiple solutions (such as local optima and a global optimum solution), an inefficient search algorithm may get trapped at local optimal solution. In contrast, evolutionary algorithms which mimic nature's principle of survival of the fittest start with multiple population points [1-4]. Due to the strong genetic operators, evolutionary algorithms are found to achieve the global optimum in majority of industrial applications both for single objective [5] and multi-objective optimization [6]. Thus it is necessary to develop an optimization algorithm which not only captures and global solutions, but also acquires it at a faster rate.

Journal of Physics: Conference Series **622** (2015) 012044 doi:10.1088/1742-6596/622/1/012044

In the recent years, non-traditional search and optimization methods based on natural phenomena, such as Simulated Annealing (SA), Genetic algorithms (GA), Differential evolution (DE), Self Organizing Migrating Algorithms (SOMA), Particle Swarm Optimization (PSO), Tabu Search (TS), Scatter Search (SS), and Ant Colony Optimization (ACO) have been developed to overcome the problems of traditional optimization methods. These algorithms are stochastic in nature and are comparatively new and are gaining popularity due to certain properties which the deterministic algorithms do not have. Thus these are found to have better global perspective than the traditional methods [4, 7].

### *1.1. Differential evolution in brief*
Differential evolution (DE), an evolutionary algorithm is an extension of genetic algorithm (GA) for solving single objective optimization problems. Simple GA [3] uses a binary coding for representing problem parameters whereas DE uses real coding of floating point numbers. DE is an exceptionally simple evolution strategy, which is significantly faster & robust at numerical optimization and is more likely to find a function's true global optimum [8].
It is found that DE is a novel optimization method which is capable of handling nonlinear, multimodal and non-differentiable optimization problems [5]. Original DE dealt with a single strategy [8]. Later on ten different strategies have been suggested by [9]. A set of control parameters that works out to be the best for a given problem may not work well when applied for a different problem. The best value of control parameters to be adopted for each problem is to be determined separately by trial & error. Similarly, the strategy that works out to be the best for a given problem may not be effective when applied to some other problem [5]. Among the DE's advantages are its simple structure, ease of use, speed and robustness. The details of DE algorithm and pseudo code are available in literature [5, 7, 9, 10]. DE is also extended to solve multi-objective optimization problems. Several variants of multi-objective differential evolution (MODE) algorithm (namely, MODE-II, MODE-III, Hybrid-MODE, Elitist-MODE, Trigonometric mutation-MODE and jumping gene variants of MODE) are available in literature. The following sections give the detailed working principles of differential evolution with jumping gene adaptation algorithm [11-17].

### *1.2. Differential evolution with jumping gene adaptation*
Transposons, also called jumping genes, are sequences of DNA that can move around to different positions within the genome of a single cell, a process called transposition. In the process, they can cause mutations and change the amount of DNA in the genome. Thus, they promote inter and intra chromosome movement [18].
A new jumping gene based differential evolution algorithm is developed which is an improvement to the modified differential evolution [10]. Jumping gene phenomenon has already been applied to evolutionary techniques both binary and real coded like genetic algorithm and simulated annealing for multi-objective optimization [19-21]. It was observed that implementation of jumping gene operation to the original evolutionary algorithm helps in achieving the optimum solution at a faster rate. Though the concept of jumping gene is applied on genetic algorithms and simulated annealing algorithm, to the best of our knowledge this is the first attempt to implement the jumping gene operation in differential evolution algorithm.
According to the experimental observation jumping genes can move around the genome in two ways [18]: Cut and paste, and Copy and paste.
Cut and paste means a piece of DNA is cut and pasted somewhere else. The latter means the genes remain at the same location while the message in the DNA is copied into RNA and then copied back into DNA at another place in the genome. Whichever the process of transposition, the jumping genes in one place can affect the genetic behaviour at other places in the genome [18]. Jumping gene may allow searching the entire search space (during the initial generations) and thus it creates more chances of achieving a better convergence and diversity and avoiding convergence to a local maxima/minima.

The jumping gene adaptation is applied immediately after the replacement of target vector by the trial vector (i.e., within the population loop). Following pseudo code gives the working of jumping gene algorithm based on single chromosome. Two kinds of mechanisms are considered, viz. cut and copy as given below.

%% Cut and paste in single chromosome
Step1: Identify ins_pos, gene_pos and gene_length
X_initial is from the current population
Copy X_initial to X_jump
Step2: Check for bounds at positions gene_pos to gene_pos + gene_length-1 and corresponding values at ins_pos to ins_pos+gene_length-1
If the bounds are different then map the elements at position gene_pos to gene_pos+gene_length-1 and copy to X_gene else copy the original X_initial
Step3: Remove elements from gene_pos to gene_pos+gene_length-1 from X_jump
Step4: Copy elements at position 1 to (ins_pos-1) in X_jump to xjg1
Copy Elements at position ins_pos:end to xjg2. Now we have 3 vectors. Con-catenate these vectors to form X_jump_trial
Step5 : Thus, X_jump_trial = xjg1 + X_gene + xjg2
Step6: Compare the cost of X_jump_trial with the cost of target vector and replace the target if better point is achieved
%%Copy and paste in single chromosome
Step1: Identify ins_pos, gene_pos and gene_length
X_initial is from the current population
Step2: Copy elements at position gene_pos to gene_pos+gene_length-1 to X_gene
Step3: Change elements from ins_pos to ins_pos+gene_length -1 with values from X_gene and name this new vector as X_jump_trial
No shifting of elements of an array is required
Step4: Compare the cost of X_jump_trial with the cost of target vector and re-place the target if better point is achieved

More variants of cut and paste and copy and paste in double chromosomes are already designed for both single and multi-objective differential evolution and is available in literature [22]. However in this study results obtained for only single chromosome based JG are reported. In our previous study it was observed that the differential evolution with jumping gene adaptations (DE-JGa algorithm) is able to consistently achieve a value of standard deviation close to zero for majority of the test problems [23]. In terms of best and worst solutions (among the finally obtained solutions) DE-JGa algorithm was able to provide best and worst solutions which are closest to the global solutions. The mean values of solutions obtained using DE-JGa algorithm were consistently close to global optimum solutions (exact global solutions for many case studies) and were achieved at a faster rate compared to those obtained using DE and MDE algorithms. Inspiring from the results obtained on test problems, DE-JGa algorithm is used solve industrial real world problems

## 2. Problem Formulation

In this study reactor network design (RND) problem is considered for optimization study. RND problem is optimized using variants of differential evolution algorithms namely, DE, MDE, and DE-JGa.

This example, taken from Ryoo and Sahinidis [24], is a reactor network design problem, describing the system shown in Figure 1. It involves the design of a sequence of two CSTR reactors where the consecutive reaction A→B→C takes place. The objective function considered in this study is the maximization of the concentration of product B (i.e., $x_4 = C_{B2}$) in the exit stream. This problem is known to have caused difficulties for other global optimization methods. Mathematical, the problem is represented as given through Eqs. 1-2f:

Journal of Physics: Conference Series **622** (2015) 012044 doi:10.1088/1742-6596/622/1/012044

$$Min.\ f = -x_4 \quad (1)$$

Subject to

$$x_1 + k_1 x_2 x_5 = 1 \quad (2a)$$

$$x_2 - x_1 + k_2 x_2 x_6 = 0 \quad (2b)$$

$$x_3 + x_1 + k_3 x_3 x_5 = 1 \quad (2c)$$

$$x_4 - x_3 + x_2 - x_1 + k_4 x_4 x_6 = 0 \quad (2d)$$

$$x_5^{0.5} + x_6^{0.5} \leq 4 \quad (2e)$$

$$(0, 0, 0, 0, 10^{-5}, 10^{-5}) \leq (x_1, x_2, x_3, x_4, x_5, x_6) \leq (1, 1, 1, 1, 16, 16) \quad (2f)$$

Where

$k_1 = 0.09755988$

$k_2 = 0.99k_1$

$k_3 = 0.0391908$

$k_4 = 0.9k_3$

The global optimum is: $(x_1, x_2, x_3, x_4, x_5, x_6; f)$ = (0.771462, 0.516997, 0.204234, 0.388812, 3.036504, 5.096052; -0.388812).

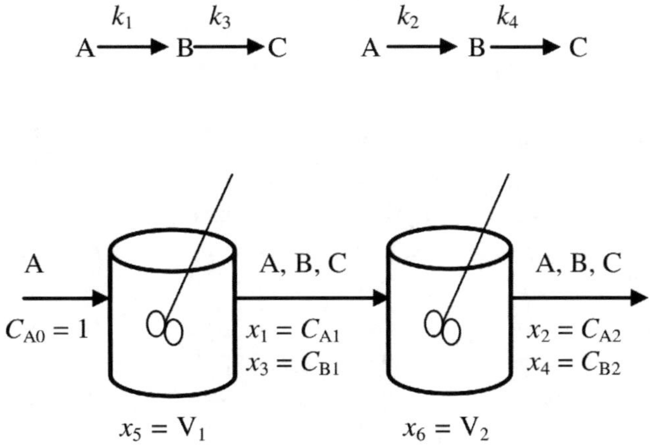


**Figure. 1.** Reactor network design problem

This example constitute a very difficult test problem as it possesses a local minimum with an objective function value that is very close to that of the global solution. The local solutions are with $f$ = -0.375 and $f$ = -0.3881 [5, 10]. Interestingly enough, the two local solutions utilize only one of the two reactors whereas the global solution makes use of both reactors. This problem can be reformulated by eliminating equality constraint as follows (Eqs. 3-5):

$$Max.\ f = \frac{k_2 x_6 (1 + k_3 x_5) + k_1 x_5 (1 + k_2 x_6)}{(1 + k_1 x_5)(1 + k_2 x_6)(1 + k_3 x_5)(1 + k_4 x_6)} \quad (3)$$

Subject to

$$x_5^{0.5} + x_6^{0.5} \leq 4 \quad (4)$$

$$(10^{-5}, 10^{-5}) \leq (x_5, x_6) \leq (16, 16) \quad (5)$$

Journal of Physics: Conference Series **622** (2015) 012044 doi:10.1088/1742-6596/622/1/012044

Global optimum is same after reformulation. In the present study, the reformulated problem is solved using DE, MDE and Jumping gene adaptation of DE.

## 3. Results and Discussion

Table 1 show the results obtained using DE, MDE and DE-JGa for the optimization of reactor network design problem. The results are shown considering average best, worst, mean, median, and standard deviation of the optimum values of 25 runs. The term 'Error' considered in this study is the difference between average cost of population and cost or objective function value corresponding to known global optimum in a generation. Though the experimental runs are taken for 500 numbers of generations, the comparison of results is carried out at intermediate numbers of functions evaluations (FES) i.e., FES value of 10000,15000,20000,50000 and 100000 (Corresponding to 50, 75, 100, 250 and 500 generations respectively).

**Table 1.** Results obtained for RND problem using various algorithms

| Algorithm | **FES** | **Best** | **Worst** | **Mean** | **Median** | **Std-Dev** |
|---|---|---|---|---|---|---|
| DE | 10000 | 0.388811 | 0.388453 | 0.388789 | 0.388793 | 1.81E-05 |
| | 15000 | 0.388811 | 0.388804 | 0.388811 | 0.388811 | 2.46E-07 |
| | 20000 | 0.388811 | 0.388811 | 0.388811 | 0.388811 | 2.88E-09 |
| | 50000 | 0.388811 | 0.388811 | 0.388811 | 0.388811 | 0 |
| | 100000 | 0.388811 | 0.388811 | 0.388811 | 0.388811 | 0 |
| MDE | 10000 | 0.388811 | 0.388723 | 0.388807 | 0.388806 | 4.62E-06 |
| | 15000 | 0.388811 | 0.388811 | 0.388811 | 0.388811 | 2.92E-08 |
| | 20000 | 0.388811 | 0.388811 | 0.388811 | 0.388811 | 2E-10 |
| | 50000 | 0.388811 | 0.388811 | 0.388811 | 0.388811 | 0 |
| | 100000 | 0.388811 | 0.388811 | 0.388811 | 0.388811 | 0 |
| DE-JGa | 10000 | 0.38881 | 0.37583 | 0.387298 | 0.387632 | 0.001278 |
| | 15000 | 0.38881 | 0.38647 | 0.38863 | 0.388671 | 0.000155 |
| | 20000 | 0.38881 | 0.38857 | 0.388793 | 0.388798 | 1.63E-05 |
| | 50000 | 0.38881 | 0.38881 | 0.38881 | 0.38881 | 0 |
| | 100000 | 0.38881 | 0.38881 | 0.38881 | 0.38881 | 0 |

Taking standard deviation to be one of the measures of the effectiveness of any algorithm, it is observed that MDE produces better results than those obtained using DE and DE with jumping gene adaptation. The consistency with which DE and MDE has performed is evident from the fact that standard deviation achieved a value of 0 (zero) or close to zero compared to that obtained using DE combined with jumping gene (DE-JGa) algorithm. For most of the test problems and industrial case studies, the jumping gene strategy of DE algorithm is able to give consistent results at a faster rate as compared to those obtained using differential evolution and modified differential evolution algorithms. Table 1 show the results obtained using DE, MDE and DE-JGa for the optimization of reactor network design problem. The standard deviation values at FES value of 20000 obtained using DE, MDE and DE-JGa algorithms are $2.88 \times 10^{-09}$, $2 \times 10^{-10}$, and $1.63 \times 10^{-05}$ respectively. For reactor network design problem MDE algorithm is able to produce better results that those obtained using DE and MDE and DE-JGa algorithms. The consistency with which all the strategies of DE have performed is evident from the fact that standard deviation achieved a value of 0 (zero).

2 show that error reduces faster in MDE compared to DE and DE-JGa in the first few generations. But after 36 generations, DE-JGa converges to zero faster than MDE. However when given enough opportunity, we see that jumping gene algorithm performs exceptionally well among all the three. Using both DE and MDE algorithms the final value of error converges to $5.67 \times 10^{-7}$ and it converge to the value of zero using DE-JGa algorithm. Thus the global optimum value is obtained using DE-JGa algorithm and a value close to global optimum is obtained using by DE and MDE algorithms in this case study (RND problem).

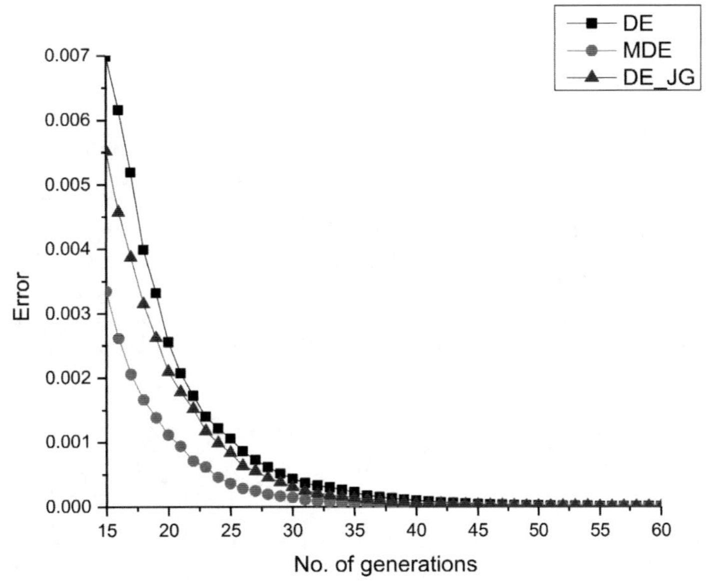


**Figure 2.** Error against generation number for RND problem

### 3.1. Algorithm Specific Study

In this section each of the algorithm is tested for its robustness using various values of control parameters such as (crossover constant) CR and F (scaling factor). The important evaluation parameters such as FES, CPU time and predicted function value are also obtained for each of the problems using strategies of differential evolution algorithm.

#### *3.1.1. Effect of control parameters (CR and F)*

It is possible that the output of stochastic algorithms such as DE and its variants may depend on its control parameters (CR and F). Thus an attempt is made to check the effects of those control parameters on the performance of algorithms on the selected test problems. During the study of effect of CR, the value of scaling factor was generated randomly whereas during the study of effect of scaling factor, the value of CR was set to 0.8 (based on our previous experience). Figures 3a-3b shows the effect of *CR* and *F* on the objective functions cost (function value). Each point on the figures (Figure 3a-3b) represents the average of 25 experiments.

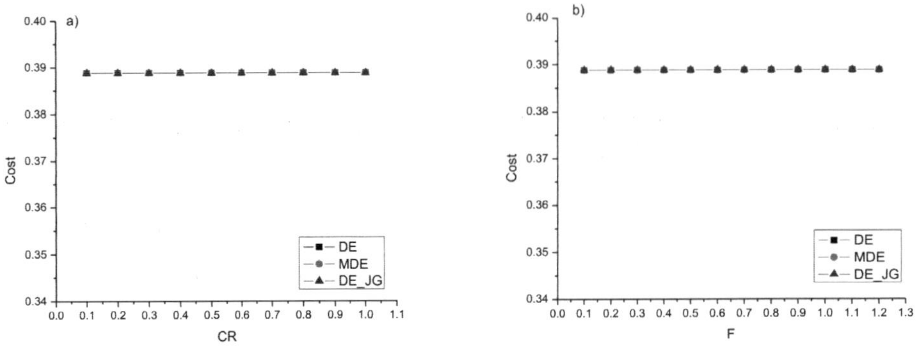


**Figure. 3.** Effect of (a) CR and (b) F on cost of reactor network design problem

Table 3 shows FES, CPU time and function value for various problems using strategies of DE. In all these problems, MDE is able to locate the global optimum in lesser number of FES than DE and MDE. DE is able to locate the global optimum in lesser number of FES than DE-JGa. The CPU time taken by MDE is much lesser compared to DE and DE-JGa. Both DE and MDE are is able to converge to optimum value faster than DE-JGa. The DE-JGa algorithm converges to optimum for various other industrial problems (which are not reported in this study).

**Table 3.** Number of function evaluations, CPU time and function value

| Problem | | **DE** | **MDE** | **DE-JGa** |
|---|---|---|---|---|
| RND | FES | 14283 | 12699 | 23488 |
| | CPU Time (s) | 3.634823 | 2.979619 | 4.58643 |
| | Function Value (Cost) | 0.3888119 | 0.3888119 | 0.3888119 |

## 4. Conclusions:

A new algorithm, differential evolution with jumping gene adaptations (DE-JGa) is discussed. The performance of the newly developed algorithm is compared with the performance of existing algorithms (DE and MDE) by applying these algorithms to some selected industrial problems. Detailed algorithm specific study is carried out to check the robustness of individual algorithms with respect to its control parameters. Following conclusions are drawn based on the results obtained in the present study.

1. In case of problems which are having less number of decision variables the output of all the algorithms (considered in this is study) is independent of the value of the CR.
2. MDE is able to consistently achieve value of standard deviation close to zero for majority of the test problems considered in this study.
3. DE-JGa algorithm is able to give the objective function value close to optimum over a wide range of CR and F compared to DE and MDE algorithms.
4. In terms of FES and CPU time, MDE algorithm outperforms DE and DE-JGa algorithms

## References

1. Goldberg DE 1989 *Genetic Algorithms for Search, Optimization, and Machine Learning.* (Reading, M.A.: Addison-Wesley)

2. Deb K 2001 *Multi-Objective Optimization using Evolutionary Algorithms* (New York: John Wiley & Sons Limited).
3. Deb K 2005 *Optimization for engineering design algorithms and examples* (Prentice-Hall of India Pvt. Ltd)
4. Babu BV 2004 *Process plant simulation*. (New York: Oxford University Press)
5. Angira R 2006 *Evolutionary Computation for Optimization of Selected Non-linear Chemical Processes. Ph.D. Thesis* (BITS-Pilani, India)
6. Gujarathi A M 2010 *Pareto Optimal Solutions in Process Design Decisions using Evolutionary Multi-objective Optimization. Ph.D. Thesis* (BITS-Pilani, India)
7. Corne D, Dorigo M and Glover F 1999 *New Ideas in Optimization* (Mc-graw-Hill Publications, London)
8. Price K V and Storn R 1997 *Dr. Dobb's J.* **22** 18
9. Price K. and Storn R. 2014.. URL: http://www.ICSI.Berkeley.edu/~storn/code.html.
10. Babu B V and Angira R 2006 *Compu. Che. Engng..* **30** 989
11. Gujarathi AM and Babu BV 2009 *Chem. Eng. Sci.* **65** 2009
12. Gujarathi AM and Babu BV 2009 *Ind. Eng. Chem. Res.* **48** 11115
13. Gujarathi AM and Babu BV 2010 *Int. J Bio-insp. Compu.* **2** 213
14. Gujarathi AM and Babu BV 2009 *Mater. Manuf. Processes.* **24** 303
15. Gujarathi AM and Babu BV 2011) *Mater. Manuf. Processes.* **26** 455
16. Gujarathi AM and Babu BV 2011 App. Comp. Intel. Soft. Comp. in press.
17. Gujarathi AM, Babu BV (2011) *Compu. Meth. Mater. Sci.* 11 463
18. Ripon KSN, Kwong S and Man KF 2007 *Info. Sci.* **177** 632
19. Kasat RB and Gupta SK 2003 *Compu. Che. Engng..* **27** 1785
20. Sankarrao and Gupta SK 2007 *Compu. Che. Engng.***31** 1496
21. Agrawal N, Rangaiah GP, Ray AK and Gupta SK 2006 *Ind. Eng. Chem. Res.* **45** 3182
22. Gujarathi AM and Purohit S 2014. Under preparation.
23. Gujarathi AM, Porohit S, Manachanda R and Ghosh D 2011 Differential Evolution using Jumping Genes Adaptation. *Proc. of Indian Int. Conf. on Artificial Intel. (IICAI 2011)*, Bangalore, India.
24. Ryoo HS and Sahinidis NV 1995 *Compu. Che. Engng.* 19: 551–566.

# Determined position dependent Mass of the Rosen-Morse potential and its Bound state

**Ali R. Amani**

Department of Physics, Ayatollah Amoli Branch, Islamic Azad University, Amol, Iran
P.O. Box 678, Amol, Mazandaran, Iran

E-mail: a.r.amani@iauamol.ac.ir

**Abstract.** In this paper, Schrödinger equation has been solved analytically with position dependent mass by the Rosen-Morse potential. The position dependent mass defined as the function $1/(1-\tanh(\eta x))$. Then corresponding position dependent mass substituted into Schrödinger equation. After that the obtained equation compared with associated Jacobi differential equation. Therefore, the eigenvalue and eigenfunction have been calculated, and from there the bound state has been found in terms of quantum numbers and the coefficients of potential. Also, the wave function has been obtained in terms of the associated Jacobi equation by $\mu = -3$ and $\nu = 2$.

## 1. Introduction

One of the fundamental wave equations is the Schrödinger equation that one used in physics and chemistry. Solutions of Schrödinger equation for some physical potential have important applications in molecular physics, quantum chemistry, nuclear, condensed matter physics, high energy physics and particle physics. These potentials are such as, Hulthén [1], Morse [2], Rosen-Morse [3], Pseudo-harmonic [4], Mie [5], Poschl-Teller [6], Kratzer-Fues [7] and Woods-Saxon [8]. We usually solve these potentials as analytically and we can calculate energy eigenvalues and the corresponding wave functions exactly. Application of the mentioned potentials and Woods-Saxon potential are very useful to describe molecular structures and the interaction of a neutron with a heavy nucleus, respectively.

Exact solutions of Schrödinger equation for the mentioned potentials are interesting in the fields of material science and condense matter physics. We note that, there are various methods to obtain exact solutions energy eigenvalues and corresponding wave function. One of the methods is a factorization method that can factorize the Hamiltonian of corresponding system in terms of production of two first order differential operators, in which so-called lowering and raising operators. The two obtained Hamiltonian are partner of each other. This method allowed to classify the problems according to the characteristics involved in the potentials and it is closely related to the supersymmetric quantum mechanics [9, 10, 11].

In quantum mechanics, all of the analytically solvable potentials have a feature that be called as shape invariance. In fact, the shape invariance is an integrability condition and an interesting feature of supersymmetric quantum mechanics. This method is an exact and elegant technique to determine the eigenvalues and eigenfunctions of quantum mechanical problems [9].

In this paper, we are going to solve the Schrödinger equation with position dependent mass by the Rosen-Morse potential.

We have organized the present work as follows:

In section 2 we will present the general form of Schrödinger equation with position dependent mass by the Rosen-Morse potential. In section 3 we consider the position dependent mass as a function of position, and from there by using associated Jacobi polynomial we can obtain the eigenvalue and wave function for the corresponding system. In section 4 we write the corresponding Schrödinger equation as raising and lowering operators in terms of the quantum numbers of radial and orbital angular momentum, and as well as we investigate the shape invariance condition. Finally, we will express a brief report of the present work, in section 5.

## 2. Effective Schrödinger equation background

In order to obtain the eigenvalues and wave function with position dependent mass by the Rosen-Morse potential, the effective Hamiltonian which was first introduced by von Roos [12] is written by Weyl ordering in terms of the kinetic energy operator as follows [13],

$$H = \tfrac{1}{4(\alpha+1)}\Big\{a\Big(m^{-1}(\vec{r})\,\hat{p}^2 + \hat{p}^2\,m^{-1}(\vec{r})\Big) + m^{\alpha}(\vec{r})\,\hat{p}\,m^{\beta}(\vec{r})\,\hat{p}\,m^{\gamma}(\vec{r}) + m^{\gamma}(\vec{r})\,\hat{p}\,m^{\beta}(\vec{r})\,\hat{p}\,m^{\alpha}(\vec{r})\Big\}, \tag{1}$$

We should point out that several authors considered a constraint as $\alpha + \beta + \gamma = -1$ for the effective Hamiltonian [14, 15, 16, 17]. In that case, the one-dimensional Schrödinger equation be written by the corresponding constraint as the following expression,

$$-\frac{\hbar^2}{2m(x)}\frac{d^2\psi(x)}{dx^2} + \frac{\hbar^2}{2}\left(\frac{1}{m^2(x)}\frac{dm(x)}{dx}\right)\frac{d\psi(x)}{dx} + \left[V(x) + U_{\alpha\beta\gamma a}(x) - E\right]\psi(x) = 0, \tag{2}$$

where $U_{\alpha\beta\gamma a}$ is equal to,

$$U_{\alpha\beta\gamma a}(x) = -\tfrac{\hbar^2}{4m^3(x)(a+1)}\left[(\alpha+\gamma-a)m(x)\tfrac{d^2m(x)}{dx^2} + 2(a-\alpha\gamma-\alpha-\gamma)\left(\tfrac{dm(x)}{dx}\right)^2\right]. \tag{3}$$

It should be noted that several authors solved the Schrödinger equation with position dependent mass for different cases such as: ($a = \beta = \gamma = 0$ and $\alpha = -1$ [18]), ($a = \alpha = \gamma = 0$ and $\beta = -1$ [19]), ($a = 0$, $\alpha = \gamma = -\frac{1}{2}$ and $\beta = 0$ [20]) and ($a = \alpha = 0$ and $\beta = \gamma = -\frac{1}{2}$ [22]).

In this job, we are going to solve the corresponding system by the Rosen-Morse potential. We note that this potential describes the vibrations of diatomic molecule, and one be written as trigonometric functions in the following form,

$$V(x) = V_1\tanh^2(\eta x) + V_2\tanh(\eta x) - V_1, \tag{4}$$

where $V_1$ and $V_2$ are the depth of the potential and $\eta$ is the range of the potential. The corresponding potential has a minimum in equilibrium distance or bond length i.e., $r_e = -\frac{1}{\eta}\mathrm{arctanh}(\frac{V_2}{2V_1})$.

By substituting (3) and (4) into (2), and by making a variable change as $y = \tanh(\eta x)$, the Schrödinger equation (2) be rewritten in terms of variable $y$ as,

$$\begin{aligned}(1-y^2)\psi''(y) &+ \Big[-2y - (1-y^2)\tfrac{m'(y)}{m(y)}\Big]\psi'(y) + \Big[-\tfrac{2V_1}{\hbar^2\eta^2}\tfrac{y^2}{1-y^2}m(y) - \tfrac{2V_2}{\hbar^2\eta^2}\tfrac{y}{1-y^2}m(y) \\ &+ \tfrac{2(V_1+E)}{\hbar^2\eta^2}\tfrac{1}{1-y^2}m(y) + \tfrac{\alpha+\gamma-a}{2(a+1)}(1-y^2)\tfrac{m''(y)}{m(y)} - \tfrac{\alpha+\gamma-a}{(a+1)}y\tfrac{m'(y)}{m(y)} \\ &+ \tfrac{a-\alpha\gamma-\alpha-\gamma}{(a+1)}(1-y^2)\tfrac{m'^2(y)}{m^2(y)}\Big]\psi(y) = 0,\end{aligned} \tag{5}$$

where the prime displays the derivative with respect to variable $y$. We note that the coefficients $\alpha$, $\beta$ and $\gamma$ obey the constraint $\alpha+\beta+\gamma=-1$.

In next section, we will solve the above equation, and we will obtain the eigenvalues and wave function for $l$-wave cases.

## 3. The Eigenvalues and the wave function

In this section, we are going to solve Eq. (5), the motivation of the job is based on analytically solution. In that case, we refer to use position dependent mass in the following form,

$$m(y)=\frac{1}{1-y}, \tag{6}$$

This choice will lead to an analytical solution. So that $m(y)$ is a function without singularity, because there is a definite value for $m(y)$ in all values $y$.

Now we are inserting Eq. (6) into (5), then the Schrödinger equation becomes,

$$\begin{aligned}(1-y^2)\psi''(y)-(3y+1)\,\psi'(y)+\Big[-\tfrac{2V_1}{\hbar^2\eta^2}\tfrac{y^2}{(1-y)^2(1+y)}-\tfrac{2V_2}{\hbar^2\eta^2}\tfrac{y}{(1-y)^2(1+y)}\\+\tfrac{2(V_1+E)}{\hbar^2\eta^2}\tfrac{1}{(1-y)^2(1+y)}+\tfrac{\alpha+\gamma-a}{(a+1)}\tfrac{(1+y)}{(1-y)}-\tfrac{\alpha+\gamma-a}{(a+1)}\tfrac{y}{(1-y)}+\tfrac{a-\alpha\gamma-\alpha-\gamma}{(a+1)}\tfrac{2(1+y)}{(1-y)}\Big]\,\psi(y)=0,\end{aligned} \tag{7}$$

We can get wave function $\psi(y)$ by making separation of variables method as multiplication of two functions $u(y)$ and $P(y)$. Here we note that function $P(y)$ is as a special function [21]. This method helps us that we have an analytical solution, so we have,

$$\psi(y)=u(y)P(y), \tag{8}$$

by substituting Eq. (8) into Eq. (7) leads to,

$$\begin{aligned}(1-y^2)P''(y)+\Big[-2\tfrac{u'}{u}(1-y^2)-(3y+1)\Big]P'(y)\\+\Big[(1-y^2)\tfrac{u''}{u}-(3y+1)\tfrac{u'}{u}-\tfrac{2V_1}{\hbar^2\eta^2}\tfrac{y^2}{(1-y)^2(1+y)}-\tfrac{2V_2}{\hbar^2\eta^2}\tfrac{y}{(1-y)^2(1+y)}\\+\tfrac{2(V_1+E)}{\hbar^2\eta^2}\tfrac{1}{(1-y)^2(1+y)}+\tfrac{\alpha+\gamma-a}{(a+1)}\tfrac{(1+y)}{(1-y)}-\tfrac{\alpha+\gamma-a}{(a+1)}\tfrac{y}{(1-y)}+\tfrac{a-\alpha\gamma-\alpha-\gamma}{(a+1)}\tfrac{2(1+y)}{(1-y)}\Big]\,P(y)=0.\end{aligned} \tag{9}$$

On the other hand, the associated Jacobi polynomial is an orthogonal polynomial in the interval $[-1,\,1]$, so one to be written as,

$$\begin{aligned}(1-y^2)P^{\mu,\nu\prime\prime}_{n,l}(y)-[\mu-\nu+(\mu+\nu+2)y]\,P^{\mu,\nu\prime}_{n,l}(y)\\+\Big[n(\mu+\nu+n+1)-\tfrac{l(\mu+\nu+l+(\mu-\nu)y)}{1-y^2}\Big]\,P^{\mu,\nu}_{n,l}(y)=0,\end{aligned} \tag{10}$$

where $n\geq l\geq 0$, so $n$ and $l$ are positive integers. The polynomial $P^{\mu,\nu}_{n,l}(y)$ is the eigenfunction of Eq. (10) and its Rodrigues formula is written in the following form,

$$P^{\mu,\nu}_{n,l}(y)=\sqrt{\frac{\Gamma(\mu+\nu+n+l+1)}{\Gamma(n-l+1)\Gamma(\mu+n+1)\Gamma(\nu+n+1)}}\,\frac{C\ (-1)^l\ 2^{-n}}{(1-y)^{\mu+\frac{l}{2}}(1+y)^{\nu+\frac{l}{2}}}\frac{d^{n-l}}{dy^{n-l}}\left((1-y)^{\mu+n}(1+y)^{\nu+n}\right), \tag{11}$$

where $C$ is an arbitrary function in terms of $n$ and $l$.

By considering function $P(y)$ in Eq. (9) as the associated Jacobi polinomial, and to compare Eqs. (9) and (10) together we can obtain the parameters of eigenvalue and eigenfunction for the Rosen-Morse potential. For this purpose, we will achieve the associated Jcobi parameters as $\mu = -3$ and $\nu = 2$. Therefore, we can obtain function $u(y)$ in the following form,

$$u(y) = \frac{(1+y)}{(1-y)^2}. \tag{12}$$

Comparing the third terms (9) and (10), we get the following relations,

$$\alpha = \frac{n^2(a+1)}{\gamma+1} - 1, \tag{13}$$

$$V_1 = \frac{\eta^2\hbar^2}{2}\left(n^2 + 5l - 3 + \frac{\gamma\alpha}{a+1}\right), \tag{14}$$

$$V_2 = -\frac{\eta^2\hbar^2}{2}\left(-n^2 + l^2 + 4l + \frac{4a + \alpha\gamma + \alpha + \gamma + 5}{a+1}\right), \tag{15}$$

by aforesaid equations, the energy eigenvalue becomes,

$$E_{n,l} = \tfrac{\eta^2\hbar^2}{(1+\gamma)}\left[\tfrac{\gamma(2a-\gamma)}{2(a+1)} + (\gamma + \tfrac{1}{2})(n^2 - l(l-1) - 7) + \gamma(-\tfrac{5}{2}l + 1) + \tfrac{2V_2}{\eta^2\hbar^2}\right] + 2V_1, \tag{16}$$

we can see from Eq. (16) that the obtained energy values depend on the quantum numbers, the potential coefficients and constants $a$ and $\gamma$.

The corresponding eigenfunction will rewrite for Rosen-Morse potential with position dependent mass as the following form,

$$\psi(x) = \frac{(1+\tanh(\eta x))}{(1-\tanh(\eta x))^2} P_{n,l}^{-3,2}(\tanh(\eta x)). \tag{17}$$

## 4. Ladder Operators and shape invariance condition

In this section, we are going to factorize the corresponding wave function in terms of two first-order differential equations. These first-order equations so-called the raising and lowering operators [23, 24, 25]. For this purpose, we use from the associated Jocobi differential equation. Then, we will write corresponding operators with respect to quantum numbers in he following forms,

### *4.1. Ladder Operators in terms of n and l*

The corresponding operators are [26],

$$A^+_{n,l}(y)A^-_{n,l}(y)P^{\mu,\nu}_{n,l}(y) = \frac{4(n-l)(\mu+n)(\nu+n)(\mu+\nu+n+l)}{(\mu+\nu+2n)^2}P^{\mu,\nu}_{n,l}(y), \tag{18}$$

$$A^-_{n,l}(y)A^+_{n,l}(y)P^{\mu,\nu}_{n-1,l}(y) = \frac{4(n-l)(\mu+n)(\nu+n)(\mu+\nu+n+l)}{(\mu+\nu+2n)^2}P^{\mu,\nu}_{n-1,l}(y), \tag{19}$$

where

$$A^+_{n,l}(y) = (1-y^2)\frac{d}{dy} - (\mu+\nu+n)y - \frac{(\mu-\nu)(\mu+\nu+n+l)}{\mu+\nu+2n}, \tag{20}$$

$$A^-_{n,l}(y) = -(1-y^2)\frac{d}{dy} - ny + \frac{(\mu-\nu)(n-l)}{\mu+\nu+2n}, \tag{21}$$

by comparing the raising and lowering operators with the Jacobi differential equation (10) and Schrödinger equation, we can obtain the raising and lowering operators in terms of the parameters $n$ and $l$ against position $x$ in the following form,

$$A^{+}_{n,l}(x) = \frac{1}{\eta}\frac{d}{dx} - (\mu + \nu + n)\tanh(\eta x) - \frac{(\mu - \nu)(\mu + \nu + n + l)}{\mu + \nu + 2n}, \tag{22}$$

$$A^{-}_{n,l}(x) = -\frac{1}{\eta}\frac{d}{dx} - n\tanh(\eta x) + \frac{(\mu - \nu)(n - l)}{\mu + \nu + 2n}. \tag{23}$$

By inserting $\mu = -3$ and $\nu = 2$ into the above equation be yielded as,

$$A^{+}_{n,l}(x) = \frac{1}{\eta}\frac{d}{dx} - (n - 1)\tanh(\eta x) + \frac{5(n + l - 1)}{2n - 1}, \tag{24}$$

$$A^{-}_{n,l}(x) = -\frac{1}{\eta}\frac{d}{dx} - n\tanh(\eta x) - \frac{5(n - l)}{2n - 1}. \tag{25}$$

On the one hand, in order to study the shape invariance condition we have to obtain the partner Hamiltonian as multiplying between the raising and lowering operators i.e., $H^{+}_{n,l} = A^{+}_{n,l}A^{-}_{n,l}$ and $H^{-}_{n,l} = A^{-}_{n,l}A^{+}_{n,l}$. On the other hand, the shape invariance condition satisfies by condition $H^{+}_{n,l} - H^{-}_{n',l'} =$ constant ($n'$ and $l'$ are shifted integers $n$ and $l$). Then we can obtain the corresponding condition as,

$$H^{+}_{n,l} - H^{-}_{n+1,l} = -2n + \frac{50(n^2 + l^2 - l)}{(2n - 1)^2(2n + 1)}, \tag{26}$$

thus we have obtained the shape invariance condition with respect to $n$ and $l$.

### *4.2. Ladder Operators with quantum number l*

We can factorize the associated Jacobi's differential equation by raising and lowering operators with respect to the parameter $l$ as,

$$A^{+}_{l}(y)A^{-}_{l}(y)P^{\mu,\nu}_{n,l}(y) = (n - l + 1)(\mu + \nu + n + l)P^{\mu,\nu}_{n,l}(y), \tag{27}$$

$$A^{-}_{l}(y)A^{+}_{l}(y)P^{\mu,\nu}_{n,l-1}(y) = (n - l + 1)(\mu + \nu + n + l)P^{\mu,\nu}_{n,l-1}(y), \tag{28}$$

where

$$A^{+}_{l}(y) = \sqrt{1 - y^2}\frac{d}{dy} + \frac{(l - 1)y}{\sqrt{1 - y^2}}, \tag{29}$$

$$A^{-}_{l}(y) = -\sqrt{1 - y^2}\frac{d}{dy} + \frac{(\mu - \nu) + (\mu + \nu + l)y}{\sqrt{1 - y^2}}, \tag{30}$$

On the other hand, by comparing the above Jacobi's differential equation and corresponding Schrödinger equation we will achieve to the following expressions,

$$A^{+}_{l}(x) = \frac{\cosh(\eta x)}{\eta}\frac{d}{dx} + (l - 1)\sinh(\eta x), \tag{31}$$

$$A^{-}_{l}(x) = -\frac{\cosh(\eta x)}{\eta}\frac{d}{dx} + (\mu - \nu)\cosh(\eta x) + (\mu + \nu + l)\sinh(\eta x). \tag{32}$$

By substituting $\mu = -3$ and $\nu = 2$ into the corresponding operators, we will have

$$A_l^+(x) = \frac{\cosh(\eta x)}{\eta}\frac{d}{dx} + (l-1)\sinh(\eta x), \tag{33}$$

$$A_l^-(x) = -\frac{\cosh(\eta x)}{\eta}\frac{d}{dx} - 5\cosh(\eta x) + (l-1)\sinh(\eta x). \tag{34}$$

Similar to the above approach, the shape invariance condition obtain with the multiplicity of the raising and lowering operators as the partner Hamiltonian as $H_l^+ = A_l^+ A_l^-$ and $H_l^- = A_l^- A_l^+$. Therefore, we can obtain this condition as,

$$H_l^+ - H_{l+1}^- = 2l - 1, \tag{35}$$

this relation shows us that the shape invariance condition is established with respect to parameters $l$. Therefore, there is the shape invariance condition for issue of position dependent mass.

## 5. Conclusion

In this paper, we have discussed on the solution of the Schrödinger equation with position dependent mass by Rosen-Morse potential. In that case we have chosen a position dependent mass as functional of $\tanh(\eta x)$. This choice leads to an analytically solution for wave function and eigenvalues. By using the factorization method and comparing it with associated Jacobi differential equation, we have obtained energy eigenvalue and wave function. The raising and lowering operators have calculated with respect to parameters $n$ and $l$. By computing the partner Hamiltonian of system, the shape invariance condition has been satisfied with respect to parameters $n$ and $l$.

[1] W. C. Qiang, Y. Gao, and R. S. Zhou, Cent. Eur. J. Phys. 6, no. 2 (2008): 356-362. 1
[2] J. Lu, Physica Scripta 72, no. 5 (2005): 349. 1
[3] N. Rosen and P. M. Morse, Physical Review 42, no. 2 (1932): 210. 1
[4] R. Sever, C. Tezcan and M. Aktaş, Ö. Yeşiltaş, J. Math. Chem., 43, (2007) 845. 1
[5] Y. Weissman and J. Jortner, Phys. Lett. A 70, (1979) 177. 1
[6] G. Pöschl and E. Teller, Z. Phys. 83, (1933) 143. 1
[7] A. Kratzer, Z. Phys. 3, (1920) 289. 1
[8] H. Fakhri and J. Sadeghi, J., Mod. Phys. Lett. A 19, (2004) 615. 1
[9] F. Cooper, A. Khare and U. Sukhatme, Phys. Rep. 251 (1995) 268. 1
[10] A. R. Amani and H. Ghorbanpour, Acta Physica Polonica-Series B 43.9 (2012): 1795. 1
[11] A. R. Amani, J. Sadeghi and M. Rostami, Proc. of International Conference on Science and Technology, Universiti Teknologi MARA, Pulau Pinang, Malaysia (2008) 1620-1626. 1
[12] O. von Roos, 1993 Phys. Rev. B 27 7547. 2
[13] J. Sá Borges, L. N. Epele, H. Fanchiotti, C. A. Garcia Canal and F. R. A. Simão, Phys. Rev. A 38 (1988)3101. 2
[14] O. Rojo and J. S. Levinger, Phys. Rev. 123 (1961) 2177. 2
[15] M. Razavy, G. Field and J. S. Levinger, Phys. Rev. 125 (1962) 269. 2
[16] A. S. Dutra and C. A. S. Almeida, Phys. Lett. A 275, 25 (2000). 2
[17] A. Arda and R. Sever, Commun. Theor. Phys. 56 (2011) 51. arxiv:[math-ph]1012.1977. 2
[18] T. Gora and F. Williams, Phys. Rev. 177 (1969) 1179. 2
[19] D. J. Ben Danial and C. B. Duke, Phys. Rev. 152 (1966) 683. 2
[20] Q. G. Zhu and H. Kroemer, Phys. Rev. B 27 (1983) 3519. 2
[21] J. Sadeghi, A. Amani and A. Pourdarvish, Applied Sciences-APPS, 8(1) (2006) 146-152. 3
[22] T. Li and K. J. Kuhn, Phys. Rev. B 47 (1993) 12760. 2
[23] M. A. Jafarizadeh and H. Fakhri, Phys. Lett. A, 230, (1997) 164. 4
[24] M. A. Jafarizadeh and H. Fakhri, Annals of Physics 266, no. 1 (1998): 178-206. 4
[25] A. R. Amani and H. Ghorbanpour, Acta Phys. Pol. B 43, no. 9 (2012): 1795. 4
[26] A. R. Amani, M. A. Moghrimoazzen, H. Ghorbanpour, and S. Barzegaran, Afr. J. Math. Phys. 10 (2011): 31-37. 4

# Topological defect solutions for a system of three scalar fields

**M. R. Niazian**[1,*] **and Ali R. Amani**[1,†]

[1] Department of Physics, Ayatollah Amoli Branch, Islamic Azad University, Amol, Iran
P.O. Box 678, Amol, Mazandaran, Iran

E-mail: [*] m.reza.niazian@gmail.com

E-mail: [†] a.r.amani@iauamol.ac.ir

**Abstract.** In this paper, we studied on the defect structures as topological by non-linear three scalar fields. By using modified Adomian decomposition method (MADM), and Adomian decomposition method (ADM) we have found the solutions of three scalar fields. Then we compared the obtained results each other by numerical solution. Also, we consider the static case and draw $\phi(x)$, $\chi(x)$, and $\rho(x)$ with the choice of different values for parameter $r$.

## 1. Introduction

Recently, a lot of literatures have been considered nonlinear equations as ordinary differential equation (ODE) and partial differential equation (PDE). These nonlinear equations are used in the different branches of physics, engineering and the other sciences. One of applications is defect structures that plays an important role in cosmology and high energy physics and can be of topological or non-topological nature [1, 2, 3]. Kink-like and lump-like are topological and non-topological structures respectively, which can be described by real scalar fields in $1+1$ space-time dimensions under the action of nonlinear interactions. One of main applications of topological defects is in cosmology especially for formation of structure in the early Universe, because topological defects are as carriers of attractive gravitational force [4, 5].
So far, the defect structures solutions have investigated by orbit method [6, 7]. In the present work, we focus attention on three coupled real scalar fields which are topological or kink-like defects [8]. For example the single real scalar field support just single defect as tanh-like kink, and the double sin-Gordon model may support two different defects, large and small kink [8]. In other words, a system containing two or more real scalar fields give rise to at least two other classes of systems those that support defects that engender internal structure and those that support junctions of defects [6, 7, 9, 10]. Also, we know that the regular hexagonal network of defects described by two and three real scalar fields [11, 12, 13, 14].

Also the three fields solutions in the Einstein equations for describing black holes with the cosmic string discussed by [15]. In general, these give us motivation to study three scalar fields. However, to make the present investigation as general as possible.

In most of the natural problems and modelings we encounter the nonlinear equations which could be solved by using different methods, such as Variational Iteration Method (VIM) [16], Modified Variational Iteration Method (MVIM) [17, 18], Homotopy Perturbation Method (HPM) [19, 20, 21, 22, 23], Adomian Decomposition Method (ADM) [24, 25], Modified Adomian

Decomposition Method (MADM) [26] and so on. Here we investigate the three last ones with comparing the results to show the accuracy of these methods and obtain a reliable solution for this physical problem.

The present work are organized as follows:
In the section 2 we will study topological defect by three scalar fields. Afterward, we will present the general form of ADM and MADM methods in section 3. Then, we will apply the ADM and MADM methods to three coupled scalar fields and obtain the corresponding solutions for the fields $\phi(x)$, $\chi(x)$ and $\rho(x)$ in section 4. In section 5, the result is showed in a table and the fields $\phi(x)$, $\chi(x)$ and $\rho(x)$ are drew in terms of position for three different values of parameter $r$. Finally, we conclude in section 6 that this solutions is capable to describe the topological defect.

## 2. Three scalar fields system

We start with the following Lagrangian density [1],

$$\mathcal{L} = \frac{1}{2}\frac{\partial\phi}{\partial x_\alpha}\frac{\partial\phi}{\partial x^\alpha} + \frac{1}{2}\frac{\partial\chi}{\partial x_\alpha}\frac{\partial\chi}{\partial x^\alpha} + \frac{1}{2}\frac{\partial\rho}{\partial x_\alpha}\frac{\partial\rho}{\partial x^\alpha} + U(\phi,\chi,\rho), \tag{1}$$

where $x^\alpha$ and $x_\alpha$ are as $x^0 = x_0 = t$, $x^1 = -x_1 = x$ and $U(\phi,\chi,\rho)$ is the potential which is a linear function of the three fields. The variation of Lagrangian density with respect to fields lead to equations of Euler-Lagrange in the following form

$$\frac{\partial}{\partial t}\left(\frac{\partial\mathcal{L}}{\partial\dot{\phi}}\right) = \frac{\partial\mathcal{L}}{\partial\phi}, \quad \frac{\partial}{\partial t}\left(\frac{\partial\mathcal{L}}{\partial\dot{\chi}}\right) = \frac{\partial\mathcal{L}}{\partial\chi}, \quad \frac{\partial}{\partial t}\left(\frac{\partial\mathcal{L}}{\partial\dot{\rho}}\right) = \frac{\partial\mathcal{L}}{\partial\rho} \tag{2}$$

where point symbol demonstrates derivative with respect to time. So, the above model we have

$$\frac{\partial^2\phi}{\partial t^2} - \frac{\partial^2\phi}{\partial x^2} + \frac{\partial U}{\partial\phi} = 0, \quad \frac{\partial^2\chi}{\partial t^2} - \frac{\partial^2\chi}{\partial x^2} + \frac{\partial U}{\partial\chi} = 0, \quad \frac{\partial^2\rho}{\partial t^2} - \frac{\partial^2\rho}{\partial x^2} + \frac{\partial U}{\partial\rho} = 0. \tag{3}$$

If the field is constant we get $\frac{\partial U}{\partial\phi} = \frac{\partial U}{\partial\chi} = \frac{\partial U}{\partial\rho} = 0$. However, for static fields configurations (i.e. $\phi = \phi(x)$, $\chi = \chi(x)$, $\rho = \rho(x)$) we have

$$\frac{\partial^2\phi}{\partial x^2} = \frac{\partial U}{\partial\phi}, \quad \frac{\partial^2\chi}{\partial x^2} = \frac{\partial U}{\partial\chi}, \quad \frac{\partial^2\rho}{\partial x^2} = \frac{\partial U}{\partial\rho}. \tag{4}$$

So the above equations of motion are a system of three non-linear differential equations. The energy density associated with these configurations could be written as

$$\epsilon = \frac{1}{2}\left(\frac{d\phi}{dx}\right)^2 + \frac{1}{2}\left(\frac{d\chi}{dx}\right)^2 + \frac{1}{2}\left(\frac{d\rho}{dx}\right)^2 + U(\phi,\chi,\rho) \tag{5}$$

We note, for existence of soliton solutions the following restrictions are require

$$U(\phi,\chi,\rho) = U(-\phi,\chi,\rho), \quad U(\phi,\chi,\rho) = U(\phi,-\chi,\rho), \quad U(\phi,\chi,\rho) = U(\phi,\chi,-\rho).$$

Let us now consider models described by three scalar fields, given by [6, 7, 8, 9]

$$U(\phi,\chi,\rho) = \frac{1}{2}(1 - \phi^2 - r(\chi^2+\rho^2))^2 + 2r^2\phi^2(\chi^2+\rho^2) \tag{6}$$

by inserting (6) into (4), the equations of motion for above system be in the form

$$\begin{aligned}
\frac{d^2\phi}{dx^2} &= -2\phi(1-\phi^2-r(\chi^2+\rho^2)) + 4r^2\phi(\chi^2+\rho^2), \\
\frac{d^2\chi}{dx^2} &= -2r\chi(1-\phi^2-r(\chi^2+\rho^2)) + 4r^2\phi^2\chi, \\
\frac{d^2\rho}{dx^2} &= -2r\rho(1-\phi^2-r(\chi^2+\rho^2)) + 4r^2\phi^2\rho.
\end{aligned} \tag{7}$$

In next section, we will study a brief review on ADM and MADM methods.

## 3. A brief review of ADM, and MADM methods

In here we explain ADM and MADM methods in below separately.

### *3.1. Fundamentals of Adomian decomposition method*

We start with a general nonlinear differential equation in the following form

$$Lu + Ru + Nu = g(t), \tag{8}$$

where linear term is represented by $Lu$ and $L$ is a linear operator and easily invertible. We choose $L$ as the highest-ordered derivative, $R$ is the reminder of the linear operator, namely a term consists only $u$ with a coefficient (constant or variable). The nonlinear term is represented by $Nu$, $L^{-1}$ is defined as $n$-fold integration for $L = \frac{d^n}{dt^n}$. For example, we can write $L = \frac{d^2}{dt^2}$ as

$$L^{-1}L = u - u(0) - tu\prime(0). \tag{9}$$

where $L^{-1} = \int_0^t \int_0^t [.] \, dtdt$. From equation Eq. (8) we get

$$Lu = g(t) - Ru - Nu, \quad L^{-1}Lu = L^{-1}g(t) - L^{-1}Ru - L^{-1}Nu, \tag{10}$$

in that case one can obtain $u$ as follows,

$$u = u(0) + tu\prime(0) + L^{-1}g(t) - L^{-1}Ru - L^{-1}Nu. \tag{11}$$

The first three terms are identified as in the assumed decomposition $u = \sum_{n=0}^{\infty} u_n$,

$$\begin{gathered} u_0 = u(0) + tu\prime(0) + L^{-1}g(t), \quad u_1 = -L^{-1}Ru_0 - L^{-1}Nu_0, \\ u_2 = -L^{-1}Ru_1 - L^{-1}Nu_1, \quad ..., \quad u_n = -L^{-1}Ru_{n-1} - L^{-1}Nu_{n-1}, \end{gathered} \tag{12}$$

also we write

$$Nu = \sum_{n=0}^{\infty} A_n(u_0, u_1, ..., u_n),$$

where $A_n$ are called Adomian polynomials and depend only on u components and make a rapidly convergent series (any nonlinearity are written in terms of $A_n$ and Nu need not even be analytic). The Adomian polynomials for one variable function are generated by the following formula :

$$A_n(u_0, u_1, ..., u_n) = \frac{1}{n!}[\frac{d^n}{d\lambda^n} N(\sum_{i=0}^{n} u_i \lambda^i)]_{\lambda=0}, \tag{13}$$

we write here the first five Adomian polynomials for convenience,

$$\begin{gathered} A_0 = f\left(u_0\right), \quad A_1 = u_1 f^{(1)}\left(u_0\right), \quad A_2 = u_2 f^{(1)}\left(u_0\right) + \left(\tfrac{1}{2!}\right) u_1^2 f^{(2)}\left(u_0\right), \\ A_3 = u_3 f^{(1)}\left(u_0\right) + u_1 u_2 f^{(2)}\left(u_0\right) + \left(\tfrac{1}{3!}\right) u_1^3 f^{(3)}\left(u_0\right), \\ A_4 = u_4 f^{(1)}\left(u_0\right) + \left[\left(\tfrac{1}{2!}\right) u_2^2 + u_1 u_3\right] f^{(2)}\left(u_0\right) + \left(\tfrac{1}{2!}\right) u_1^2 u_2 f^{(3)}\left(u_0\right) + \left(\tfrac{1}{4!}\right) u_1^4 f^{(4)}\left(u_0\right), \ ... \end{gathered} \tag{14}$$

The Adomian polynomials for two variables function $f(u(x), v(x))$ are as follow:

$$\begin{gathered} A_0 = f(u_0, v_0), \quad A_1 = u_1 f_{1,0} + v_1 f_{0,1}, \\ A_2 = u_2 f_{1,0} + v_2 f_{0,1} + (\tfrac{u_1^2}{2!}) f_{2,0} + u_1 v_1 f_{1,1} + (\tfrac{v_1^2}{2!}) f_{0,2} \\ A_3 = u_3 f_{1,0} + v_3 f_{0,1} + u_1 u_2 f_{2,0} + [u_1 v_2 + u_2 v_1] f_{1,1} + v_1 v_2 f_{0,2} + (\tfrac{u_1^3}{3!}) f_{3,0} + (\tfrac{u_1^2}{2!}) v_1 f_{2,1} \\ + (\tfrac{u_1 v_1^2}{2!}) f_{1,2} + (\tfrac{v_1^3}{3!}) f_{0,3}, \ ... \end{gathered} \tag{15}$$

where

$$f_{\mu,\nu}(u_0, v_0) = \frac{\partial^{\mu+\nu}}{\partial u^\mu \partial v^\nu} f(u(x), v(x))|_{x=0},$$

We can write the practical solution in n-term approximation [24]

$$\varphi_n = \sum_{i=0}^{n-1} u_i, \quad u = \lim_{n\to\infty} \varphi_n = \sum_{i=0}^{\infty} u_i. \tag{16}$$

where $u_0, u_1, u_2, \ldots$ are determined by the following recursive relation as mentioned above,

$$u_0(x) = f(x), \quad u_n(x) = -L^{-1}Ru_{n-1} - L^{-1}Nu_{n-1}, \quad n \geq 1 \tag{17}$$

*3.2. Fundamentals of modified Adomian decomposition method*

In ADM method the zeroth component $u_0(x)$ is usually identified by the function $f(x)$ defined in Eq. (17). It is obvious that Adomian method changes the differential equations to obtain an easily computable components. The closed form for the solution $u(x)$ if exists can be immediately obtained because of the rapid convergence presented by the method.

The modified decomposition method suggests that the function $f(x)$ defined above in Eq. (17) be decomposed into two parts, namely $f_0(x)$ and $f_1(x)$ such that $f(x) = f_0(x) + f_1(x)$. The proper choice of the parts $f_0(x)$ and $f_1(x)$ depends mainly on trial basis. In view of this decomposition of f(x), a slight variation only on the components $u_0(x)$ and $u_1(x)$ should be introduced. The proposed variation is that only the part $f_0(x)$ be assigned to the zeroth component $u_0(x)$, whereas the remaining part $f_1(x)$ be combined with the other terms given in $u_1(x)$ to define it. In view of this assumption, we formulate the following recursive relation for the modified decomposition method [26],

$$\begin{aligned} u_0(x) = f_0(x), \quad u_1(x) &= f_1(x) - L^{-1}(Ru_0(x)) - L^{-1}(Nu_0(x)), \\ u_n(x) &= -L^{-1}(Ru_{n-1}(x)) - L^{-1}(Nu_{n-1}(x)), \quad n \geq 2. \end{aligned} \tag{18}$$

The success of the modified method depends mainly on the proper choice of the parts $f_0(x)$ and $f_1(x)$. We have been unable to establish any criterion to judge what forms of $f_0(x)$ and $f_1(x)$ can be used to yield the acceleration demanded. It appears that trials are the only criteria that can be applied so far.

## 4. Solving of the three scaler fields

In this section, we are going to solve the equations system of Eqs. (7) by three methods of MADM and ADM as follows:

Now we expand the the system of Eqs. (7) and apply MADM formulism to them. So that we have,

$$\begin{aligned} L\phi &= -2\phi + 2\phi^3 + (4r^2+2r)\phi\chi^2 + (4r^2+2r)\phi\rho^2 \\ L\chi &= -2r\chi + 2r^2\chi^3 + 2r^2\chi\rho^2 + (4r^2+2r)\chi\phi^2 \\ L\rho &= -2r\rho + 2r^2\rho^3 + 2r^2\rho\chi^2 + (4r^2+2r)\rho\phi^2, \end{aligned} \tag{19}$$

assuming $\beta = 4r^2 + 2r$, and with the integration of the operator $L$ we can obtain the following expressions

$$\begin{aligned} \phi &= a_0 + a_1x - 2L^{-1}\phi + 2L^{-1}\phi^3 + \beta L^{-1}\phi\chi^2 + \beta L^{-1}\phi\rho^2 \\ \chi &= b_0 + b_1x - 2rL^{-1}\chi + 2r^2L^{-1}\chi^3 + 2r^2L^{-1}\chi\rho^2 + \beta L^{-1}\chi\phi^2 \\ \rho &= c_0 + c_1x - 2rL^{-1}\rho + 2r^2L^{-1}\rho^3 + 2r^2L^{-1}\rho\chi^2 + \beta L^{-1}\rho\phi^2, \end{aligned} \tag{20}$$

where, we let (base on modified decomposition)

$$\phi_0 = a_0, \quad \chi_0 = b_0, \quad \rho_0 = c_0. \tag{21}$$

$$\begin{aligned}
\phi_1 &= a_1 x - 2\int_0^x\int_0^x \phi_0(x)dxdx + 2\int_0^x\int_0^x A_0[\phi(x)^3]dxdx + \\
&+ \beta\int_0^x\int_0^x A_0[\phi(x)\chi(x)^2]dxdx + \beta\int_0^x\int_0^x A_0[\phi(x)\rho(x)^2]dxdx \\
\chi_1 &= b_1 x - 2r\int_0^x\int_0^x \chi_0(x)dxdx + 2r^2\int_0^x\int_0^x A_0[\chi(x)^3]dxdx + \\
&+ 2r^2\int_0^x\int_0^x A_0[\chi(x)\rho(x)^2]dxdx + \beta\int_0^x\int_0^x A_0[\chi(x)\phi(x)^2]dxdx \\
\rho_1 &= c_1 x - 2r\int_0^x\int_0^x \rho_0(x)dxdx + 2r^2\int_0^x\int_0^x A_0[\rho(x)^3]dxdx + \\
&+ 2r^2\int_0^x\int_0^x A_0[\rho(x)\chi(x)^2]dxdx + \beta\int_0^x\int_0^x A_0[\rho(x)\phi(x)^2]dxdx
\end{aligned} \tag{22}$$

Finally from equation (9) one can obtain $\phi_n(x)$, $\chi_n(x)$ and $\rho_n(x)$ as follows:

$$\begin{aligned}
\phi_n &= -2\int_0^x\int_0^x \phi_{n-1}(x)dxdx + 2\int_0^x\int_0^x A_{n-1}[\phi(x)^3]dxdx + \\
&+ \beta\int_0^x\int_0^x A_{n-1}[\phi(x)\chi(x)^2]dxdx + \beta\int_0^x\int_0^x A_{n-1}[\phi(x)\rho(x)^2]dxdx \\
\chi_n &= -2r\int_0^x\int_0^x \chi_{n-1}(x)dxdx + 2r^2\int_0^x\int_0^x A_{n-1}[\chi(x)^3]dxdx + \\
&+ 2r^2\int_0^x\int_0^x A_{n-1}[\chi(x)\rho(x)^2]dxdx + \beta\int_0^x\int_0^x A_{n-1}[\chi(x)\phi(x)^2]dxdx \\
\rho_n &= -2r\int_0^x\int_0^x \rho_{n-1}(x)dxdx + 2r^2\int_0^x\int_0^x A_{n-1}[\rho(x)^3]dxdx + \\
&+ 2r^2\int_0^x\int_0^x A_{n-1}[\rho(x)\chi(x)^2]dxdx + \beta\int_0^x\int_0^x A_{n-1}[\rho(x)\phi(x)^2]dxdx
\end{aligned} \tag{23}$$

for $n = 2, 3, 4, \ldots$
where $A_{n-1}[.]$ are $(n-1)$th Adomian polynomials. The exact solution will be

$$\phi = \lim_{n\to\infty}\sum_{i=0}^{n}\phi_i, \quad \chi = \lim_{n\to\infty}\sum_{i=0}^{n}\chi_i, \quad \rho = \lim_{n\to\infty}\sum_{i=0}^{n}\rho_i, \tag{24}$$

with applying initial conditions $\phi(0) = \frac{1}{2}$, $\phi\prime(0) = \frac{1}{2}$, $\chi(0) = \frac{\sqrt{2}}{4}$, $\chi\prime(0) = -\frac{\sqrt{2}}{4}$, $\rho(0) = \frac{\sqrt{2}}{4}$ and $\rho\prime(0) = -\frac{\sqrt{2}}{4}$ the constant coefficient from Eq. (3) will be

$$a_0 = \frac{1}{2}, \qquad a_1 = \frac{1}{2}, \qquad b_0 = \frac{\sqrt{2}}{4}, \qquad b_1 = -\frac{\sqrt{2}}{4}, \qquad c_0 = \frac{\sqrt{2}}{4}, \qquad c_1 = -\frac{\sqrt{2}}{4},$$

so the above information into equation (6) give us,

$$\begin{aligned}
\phi(x) &= \frac{1}{2} + \frac{1}{2}x + (\frac{1}{4}r^2 + \frac{1}{8}r - \frac{3}{8})x^2 + (-\frac{1}{12}r^2 - \frac{1}{24}r - \frac{1}{24})x^3 \\
&+(\frac{1}{12}r^4 - \frac{1}{96}r^3 - \frac{13}{192}r^2 - \frac{1}{148}r + \frac{1}{64})x^4 + \ldots + \Re(x),
\end{aligned} \tag{25}$$

$$\chi(x) = \rho(x) = \frac{\sqrt{2}}{4} - \frac{\sqrt{2}}{4}x + (\frac{3\sqrt{2}}{16}r^2 - \frac{3\sqrt{2}}{16}r)x^2 + (-\frac{\sqrt{2}}{48}r^2 + \frac{5\sqrt{2}}{48}r)x^3$$
$$+(\frac{23\sqrt{2}}{384}r^4 - \frac{\sqrt{2}}{24}r^3 + \frac{\sqrt{2}}{128}r^2 + \frac{5\sqrt{2}}{192}r)x^4 + ... + \Re(x), \quad (26)$$

where $\Re(x)$ is the reminder.

By using ADM method we obtain the following results

$$\phi(x) = (-\frac{19}{9600}r^5 - \frac{73}{1920} + \frac{13}{76800}r^3 + \frac{121}{19200}r^4 + \frac{403}{6400}r^2 + \frac{31}{960}r - \frac{19}{9600}r^6)x^5$$
$$+(\frac{61}{640} - \frac{203}{3840}r^2 - \frac{23}{15360}r^3 - \frac{5}{192}r + \frac{1}{1920}r^5 + \frac{1}{1920}r^6 + \frac{1}{3840}r^4)x^4$$
$$+(-\frac{1}{38400}r^3 - \frac{13}{320} - \frac{1}{19200}r^4 - 1/12\,r^2 - 1/24\,r)x^3 + (1/8\,r + \frac{1}{12800}r^3 - \frac{121}{320}$$
$$+\frac{1}{6400}r^4 + 1/4\,r^2)x^2 + \frac{67}{120}x + \frac{43}{84} + \ldots + \Re(x). \quad (27)$$

$$\chi(x) = \rho(x) = (-\frac{1169}{38400}r^2\sqrt{2} - \frac{871}{614400}r^6\sqrt{2} - \frac{1}{160}r\sqrt{2} - \frac{931}{76800}r^4\sqrt{2} + \frac{7}{76800}r^5\sqrt{2}$$
$$+\frac{197}{10240}r^3\sqrt{2})x^5 + (\frac{43}{3072}r^4\sqrt{2} + \frac{317}{7680}r^2\sqrt{2} - \frac{1}{96}r\sqrt{2} + \frac{21}{40960}r^6\sqrt{2}$$
$$-\frac{1343}{30720}r^3\sqrt{2} + \frac{1}{15360}r^5\sqrt{2})x^4 + (-\frac{13}{4800}r^4\sqrt{2} + \frac{5}{48}r\sqrt{2} - \frac{11}{307200}r^6\sqrt{2}$$
$$+\frac{401}{76800}r^3\sqrt{2} - 1/48\,r^2\sqrt{2})x^3 + (3/16\,r^2\sqrt{2} + \frac{1}{800}r^4\sqrt{2} - \frac{79}{25600}r^3\sqrt{2} - 3/16\,r\sqrt{2}$$
$$+\frac{1}{102400}r^6\sqrt{2})x^2 + (-\frac{1}{64}r^2\sqrt{2} - 1/4\,\sqrt{2} + \frac{1}{960}r^3\sqrt{2})x + 1/4\,\sqrt{2} - \frac{1}{6720}r^3\sqrt{2}$$
$$+\frac{1}{320}r^2\sqrt{2} + \ldots + \Re(x). \quad (28)$$

## 5. Discussions

In this section, we are going to discuss on obtained solutions. As the solutions of nonlinear equations are very sensitive to initial conditions and parameters thus we should choose them carefully, otherwise the result will be very different from our interest. For this purpose, we first solve system of (7) by Runge-Kutta method and the obtained values of $\phi(x)$, $\chi(x)$ and $\rho(x)$ are written in table 1 and 2 for interval $0 < x < 1$ in $r = 0$. Also, the obtained values of $\phi(x)$, $\chi(x)$ and $\rho(x)$ are calculated with MADM and ADM methods by Eqs. (25), (26), (27) and (28) for interval $0 < x < 1$ in $r = 0$ in table 1 and 2. Then, we compare the obtained results of MADM and ADM methods with numerical solution (Runge-Kutta method), and we specify them accuracy with column of error in table 1 and 2.

In what follows, we draw the $\phi(x)$, $\chi(x)$ and $\rho(x)$ with the choice of different values for real parameter $r = 0, \frac{1}{6}, 1$. Also, we add solution of HPM method in figures of 1 and 2 that it has come of Ref. [27]. We can see the diverse results in figures of 1 and 2 in interval $-1 < x < 1$. It is worth noting that the figures 1 and 2 show kink-like topological solutions.

## 6. Conclusion

In this paper, we studied on topological defect structure by a system of three scalar fields. For this purpose, we wrote the Lagrangian density in terms of three scalar fields and a potential functional of three fields. We note that the fields are functional of position, and by using of equations of Euler-Lagrange we obtained equations of field as a system of three scalar fields. In

**Table 1.** The results of field $\phi(x)$ for $r = 0$.

| $x$ | $Numerical$ | $MADM$ | $Error$ | $ADM$ | $Error$ |
|---|---|---|---|---|---|
| 0 | 0.5000000000 | 0.5000000000 | 0 | 0.5119047619 | $1.190 * 10^{-2}$ |
| 0.2 | 0.5847777033 | 0.5846916667 | $8.604 * 10^{-5}$ | 0.6082602709 | $2.348 * 10^{-2}$ |
| 0.4 | 0.6388794848 | 0.6377333333 | $1.146 * 10^{-3}$ | 0.6740968653 | $3.522 * 10^{-2}$ |
| 0.6 | 0.66272761418 | 0.6580250000 | $4.703 * 10^{-3}$ | 0.7104041009 | $4.768 * 10^{-2}$ |
| 0.8 | 0.65678514162 | 0.6450666667 | $1.172 * 10^{-2}$ | 0.7170933419 | $6.031 * 10^{-2}$ |
| 1 | 0.62092003909 | 0.5989583333 | $2.196 * 10^{-2}$ | 0.6902529763 | $6.933 * 10^{-2}$ |

**Table 2.** The results of fields $\chi(x) = \rho(x)$ for $r = 0$.

| $x$ | $Numerical$ | $MADM$ | $Error$ | $ADM$ | $Error$ |
|---|---|---|---|---|---|
| 0 | 0.3535533905 | 0.3535533905 | 0 | 0.3535533905 | 0 |
| 0.2 | 0.2828427124 | 0.2828427124 | 0 | 0.2828427124 | 0 |
| 0.4 | 0.2121320343 | 0.2121320343 | 0 | 0.2121320343 | 0 |
| 0.6 | 0.1414213562 | 0.1414213562 | 0 | 0.1414213562 | 0 |
| 0.8 | 0.0707106781 | 0.0707106780 | 0 | 0.0707106780 | 0 |
| 1 | $0.984455^{-16}$ | 0 | $9.84 * 10^{-17}$ | 0 | $9.84 * 10^{-17}$ |

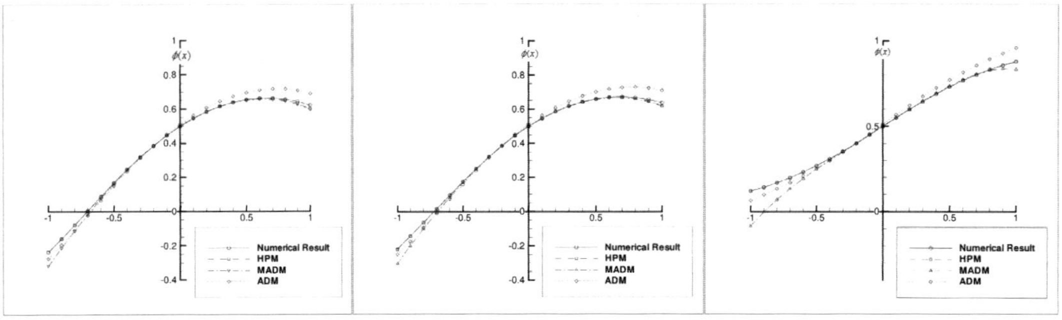


**Figure 1.** The graph of $\phi(x)$ for cases $r = 0$ (left hand), $r = \frac{1}{6}$ (middle) and $r = 1$ (right hand). The HPM calculated in Ref. [27].

order to solve this system, we used the ADM and MADM methods. In these methods, choice of initial condition is very sensitive, so that we chose them due to find kink-like solutions. Also the parameter $r$ plays an important role in the solution of this system of differential equations. Therefore, we considered its value with $r = 0, \frac{1}{6}, 1$. Afterward, we solve this system as numerical solution by Runge-Kutta method, so that tables 1 and 2 are included by three solutions of numerical, ADM and MADM in interval $0 < x < 1$ for $r = 0$. Finally, we plotted three scalar fields in terms of position, and compared them together. The figures 1 and 2 depicted the variation of three fields against position. Also we entered solution of HPM method (result of Ref. [27]) in figures 1 and 2.

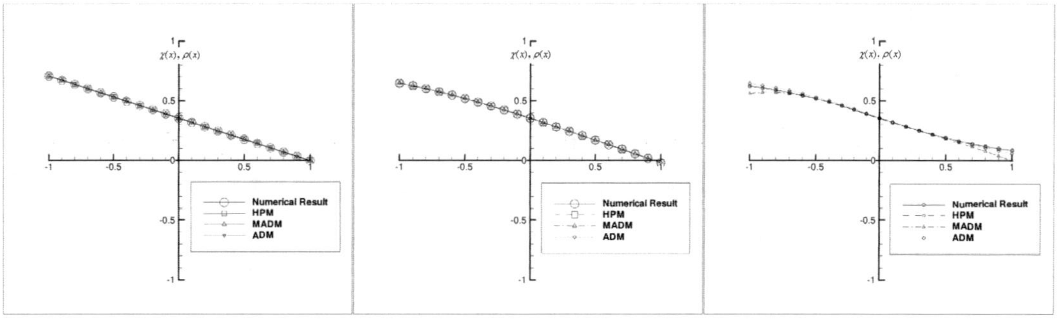


**Figure 2.** The graph of $\chi(x) = \rho(x)$ for cases $r = 0$ (left hand), $r = \frac{1}{6}$ (middle) and $r = 1$ (right hand). The HPM calculated in Ref. [27]

[1] J. Sadeghi, A. R. Amani, and A. Pourdarvish, Canadian Journal of Physics 86.7 (2008): 943-946. 1, 2
[2] A. de Souza Dutra, and P. E. D. Goulart, Physical Review D 84.10 (2011): 105001. 1
[3] J. Sadeghi and A. R. Amani, Acta Physica Polonica-Series A General Physics 113.2 (2008): 677-690. 1
[4] R. H. Brandenberger, International Journal of Modern Physics A 9.13 (1994): 2117-2189. 1
[5] A. Vilenkin and E. P. S. Shellard, Cosmic strings and other topological defects, Cambridge University Press, 2000. 1
[6] D. Bazeia, R. F. Ribeiro and M. J. dos Santos, Phys. Rev. D **54** 1852 (1996). 1, 2
[7] D. Bazeia, R. F. Ribeiro and M. J. dos Santos, Phys. Rev. E **54** 2943 (1996). 1, 2
[8] E. B. Bogomol'nyi, Sov. J. Nucl. Phys. **24**, 449 (1967). 1, 2
[9] R. Jackiw, Rev. Mod. Phys. **49** 681 (1977). 1, 2
[10] R. Rajaraman, Phys. Rev. Lett. **142** 200 (1979). 1
[11] J. R. Morris and D. Bazeia, Phy. Rev. D **54**, 5217 (1996). 1
[12] S. M. Carroll, S. Hellerman, and M. Trodden, Phy. Rev. D **61**, 065001 (2000). 1
[13] D. Bazeia and F. A. Brito, Phys. Rev. D **62**, 101701(R) (2000). 1
[14] D. Bazeia, L. Losano, and C. Wotzasek, Phys. Rev. D **66** (2002) 105025. 1
[15] V. P. Frolov and D. V. Fursaev, Class. Quant. Grav. **18** 1535 (2002). 1
[16] D. Altintan and Ö. Uğur, Journal of Computational and Applied Mathematics 259 (2014): 790-797. 1
[17] H. Tari, Physics Letters A 369.4 (2007): 290-293. 1
[18] T. A. Abassy, M. A. El-Tawil, H. El Zoheiry, Journal of Computational and Applied Mathematics 207.1 (2007): 137-147. 1
[19] C. Dong, Z. Chen and W. Jiang, Journal of Computational and Applied Mathematics 239 (2013): 359-366. 1
[20] J. H. He, Chaos, Solitons and Fractals 26.3 (2005): 695-700. 1
[21] G. Sliva, A. Brezillon, J. M. Cadou and L. Duigou, Journal of computational and applied mathematics 234.7 (2010): 2297-2302. 1
[22] X. Feng, Y. He and J. Meng, Journal of computational and applied mathematics 231.1 (2009): 288-301. 1
[23] F. Geng and M. Cui, Journal of computational and applied mathematics 235.8 (2011): 2405-2411. 1
[24] G. Adomian, Solving Frontier Problems of Physics: The Decomposition Method, Kluwer Academic, Dordrecht, 1994. 1, 4
[25] J. L. Li, Journal of Computational and Applied Mathematics 228.1 (2009): 168-173. 1
[26] Abdul-Majid Wazwaz, Applied Mathematics and Computation 173 (2006) 165176 2, 4
[27] B. Khabiri, A. R. Amani and M. Kazeminavaei, Proceddings on 3rd Con. Nonlinear Science and Complexity (Ankara-Turkey), July 28-31, 2010. 6, 7, 8

# Production of nanocrystalline cellulose from an empty fruit bunches using sulfuric acid hydrolysis: Effect of reaction time on the molecular characteristics.

**Ahmed A Al-Dulaimi[1], Rohaizu R[1] and Wanrosli W D[1]**

[1]Bioresources, Paper and Coating Division, School of Industrial Technology, Universiti Sains Malaysia, 11800 USM, Penang, Malaysia

Corresponding author: ahmedulaimi@gmail.com

**Abstract.** Nanocrystalline cellulose (NCC) was isolated from OPEFB pulp via sulfuric acid hydrolysis. The influence of reaction time to the molecular weight and surface charge of the NCC was investigated. Characterization of the product was carried out using zeta potential measurement and gel permeation chromatography test. Zeta potential measurement showed that the surface negative charge significantly increases with increasing reaction time. Gel permeation chromatography test indicates that molecular weight of NCC change variably with increasing of hydrolysis time. (Keywords: Nanocrystalline cellulose; acid hydrolysis; sulfate content; molecular weight)

## 1. Introduction

Nanocrystalline cellulose (NCC) is derived from the crystalline region of cellulose, which is having been isolated by strong acid hydrolysis. Nanocrystalline cellulose has attracted considerable academic interest due to its lack of toxicity, relative abundance, the negative surface charge and high surface area with zero-dimensional nanostructure, and it is applicable potency such as drug delivery [1], improve conductive polymer processability [2] and preparation of optically transparent film [3].

Acid hydrolysis is a diverse method which contains the dispersion of acid into the cellulose amorphous regions, then followed by cleavage of glycosidic bonds [4]. The reaction breaks the long chains into short fractions while the backbone structure leftovers [5]. The isolation of the NCC was first reported by [6], who used sulfuric acid to hydrolyze a lignocellulosic material to obtain a stable suspension of rod-like NCC structures. This method involves the diffusion of acid into the amorphous region of the lignocellulosic material and the subsequent cleavage of the glycosidic bonds [4] by the rapid protonation of glycosidic oxygen or cyclic oxygen, followed by the addition of water to break down the glycosidic bonds [5].

The strong acid hydrolysis, e.g. sulfuric acid and hydrochloric acid, is the common procedure applied for the isolation of nanocrystalline cellulose from pure cellulose at a controlled reaction time and temperature. For the NCC isolated by sulfuric acid hydrolysis, the suspension is generally shown higher acidic compared to the hydrochloric acid hydrolysis because of the sulfate groups attached to the NCC surface [7].

In this hydrolysis reaction, sulfuric acid reacts with the hydroxyl groups on the cellulose by esterification process when sulfuric acid was applied. This esterification process creates acid half-ester or so-called cellulose sulfate, as illustrated in Figure 1. The sulfate group that cloud surrounded the NCC particles induces negative charge and servicing the suspension stability in the aqueous medium by electrostatic interaction. The repulsive forces of the double layer electric effectively avoid the NCC agglomeration [8].

Production of cellulose from agricultural by-products is becoming an important agenda due to the availability of these materials in huge quantity that can be supplied at reasonable costs. In this regards, oil palm biomass residue has attracted attention for its potential to be utilized as an alternative for wood replacement in producing cellulose [9,10,11]. Previous work, we successfully isolate nanocrystalline cellulose from totally chlorine free (TCF) pulp for oil palm empty fruit bunches (OPEFB) fibers by sulfuric acid hydrolysis [9]. In this work, the varieties of molecular properties accompanied by increasing acid hydrolysis reaction time in terms of the sulfate content and molecular weight for NCC were investigated using zeta potential and gel permeation chromatography, respectively.

## 2. Material and methods

The OPEFB that used in this report were obtained from a local palm oil mill in Perak, Malaysia. OPEFB pulp was prepared using an environmentally friendly process as described in our previous works [12,13]. This method involves water pre-hydrolysis, soda pulping and then a totally chlorine-free (TCF) bleaching sequence. Nanocrystalline cellulose was prepared from the OPEFB pulp by treating the pulp with 58wt% sulfuric acid concentrations of sulfuric at a four period of time (40, 60, 80 and 100 min), at a fixed temperature of 45°C and at an acid/cellulose ratio of 1:12. Then, followed by centrifuging, washing with water, and re-centrifuging; with the process being repeated until the supernatant became turbid [14]. Then, the samples were sonicated using a rod-type ultrasonic sonifier (Branson Sonifier 450) for 5 min with an 800W constant power output. Dialysis against deionised water is an important step to remove the remaining acid molecules. Solid OPEFB-NCC was obtained by freeze-drying the sample.

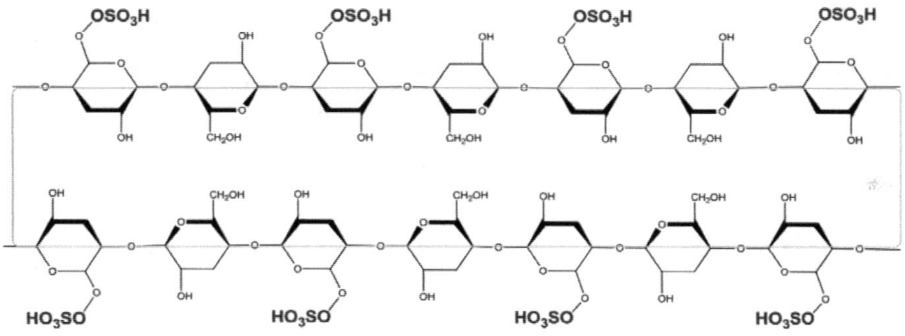


**Figure 1.**Structure of nanocrystalline cellulose (nanorod particles) isolated by sulphuric acid hydrolysis

### *2.1 Characterization methods*

After having undergone sonication for 5 min, the OPEFB-NCC suspension (0.05wt%) was analyzed for its Zeta potential using a Zetasizer Nano Z model ZEN 2600.

Gel Permeation Chromatography (GPC) measurements were carried out with waters GPC 510 pump equipped with UV and RI detectors using THF as the eluent at a flow rate of 0.7 mL/min at room temperature. Two Ultrastyragel linear columns linked in series (Styragel HR 1 and Styragel HR 5E) were used for the measurements. Standard polystyrenes of narrow disparity with molecular weight ranges from 0.82 to 1860 kg/mol were used for calibration. The number- and weight-average molecular weights were calculated using the Millenium software of Waters.

## 3. Results and discussion

NCC prepared by hydrochloric acid hydrolysis tends to aggregate; however, NCC prepared by sulfuric acid hydrolysis has higher dispersion stability because of the electrostatic repulsive forces among the crystalline particle effect of the present of sulfate groups on the cellulose surface [16].

Moreover, the influence of sulfuric acid hydrolysis reduces the possibility of the agglomeration of starch nanocrystals and limits their flocculation in aqueous medium has also been observed [17].
In the sulfuric acid hydrolysis, sulfate groups were induced to the surface of the cellulose material effect of the hydrolysis process. The introductions of the sulfate group occur at among of the three hydroxyl groups of cellulose surface, which then the cellulose will possess certain polyelectrolyte properties [5,18,19].

**Table 1.** Hydrolysis condition and NCC molecular properties

| Sample | NCC.40 | NCC.60 | NCC.80 | NCC.100 |
|---|---|---|---|---|
| *Hydrolysis condition* | | | | |
| Hydrolysis time (min) | 40 | 60 | 80 | 100 |
| $H_2SO_4$ concentration (wt. %) | 58 | 58 | 58 | 58 |
| Cellulose/acid ratio (g/mL) | 1:12 | 1:12 | 1:12 | 1:12 |
| Reaction Temperature (°C) | 45 | 45 | 45 | 45 |
| *NCC molecular properties* | | | | |
| Zeta potential (mV) | -36.2 | -38.8 | -46.2 | -69.8 |
| Molecular weight ($M_w$) | 16049 | 15198 | 15288 | 15403 |
| Molecular weight distribution ($M_n$) | 15997 | 15155 | 15238 | 15348 |

The examination of the stability of the NCC dispersion in an aqueous medium by zeta potential measurement (Figure 2) is an important parameter to study the polyelectrolyte properties for a sequence of samples prepared under different hydrolysis times. All of four of NCC suspensions showed a negative zeta potential in neutral water (Table 1). Figure 3 illustrates the interaction between the hydrolysis times versus zeta potential measurement. The hydrolysis time of 40 min showed the lowest negative value of -36.2 mV and this value was found to be increased relative with increasing of the hydrolysis time, with 100 min of the NCC isolation was the highest value (-69.8 mV).

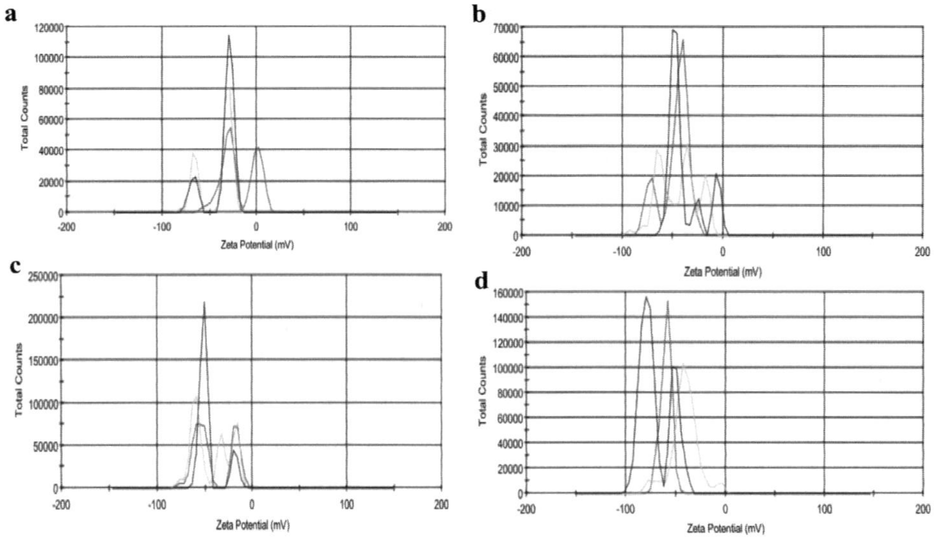


**Figure 2.** Zeta potential measurement of NCC for (a) 40 min, (b) 60 min, (c) 80 min, and (d) 100 min of sulphuric acid hydrolysis.

The presence of a larger amount of negatively charged of sulfate groups on the surface of the nanocrystalline cellulose effect of reaction time was ascribed in these findings [15]. Thus, the result shows that it is possible to increase the polyelectrolyte properties (surface charges) of nanoparticles by increase the reaction time of the sulphuric acid hydrolysis, which would lead to added electrostatic stabilization of the NCC suspensions against flocculation.

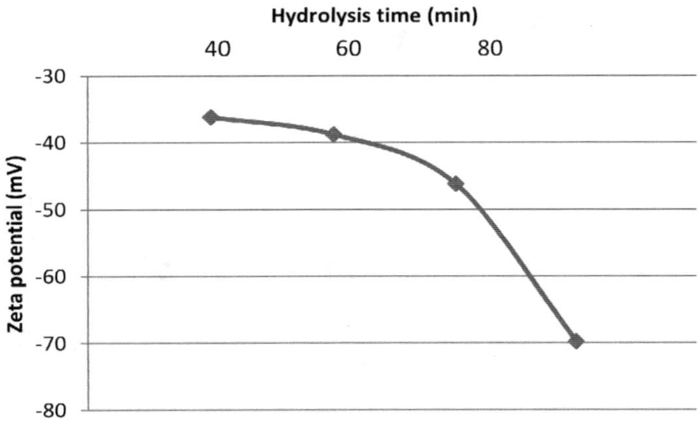


**Figure 3.** Effect of hydrolysis time to zeta potential of NCC.

The average number of molecular weights ($M_w$), molecular weight distribution ($M_n$) and the values of $M_w/M_n$ were determined by gel permeation chromatography (GPC) (Figure 4). The data listed in Table 1 illustrated irregular estimation about NCC molecular weight that accompanied the increasing of hydrolysis time.

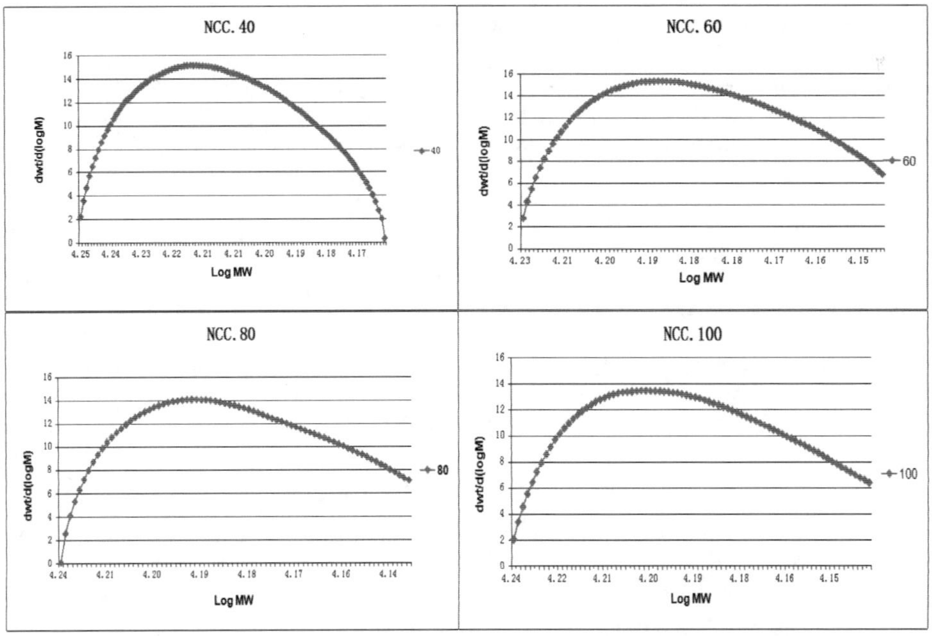


**Figure 4.** Gel permeation chromatography of NCC for (a) 40 min, (b) 60 min, (c) 80 min, and (d) 100 min of sulphuric acid hydrolysis.

Figure 5 shows that the molecular weight of NCC initially decreases when the reaction time go forward, however, after 60 min of reaction time, the NCC's molecular weight start to increase, which is in contrast to the hydrolysis concept of a cellulose fractioned by strong acid hydrolysis.

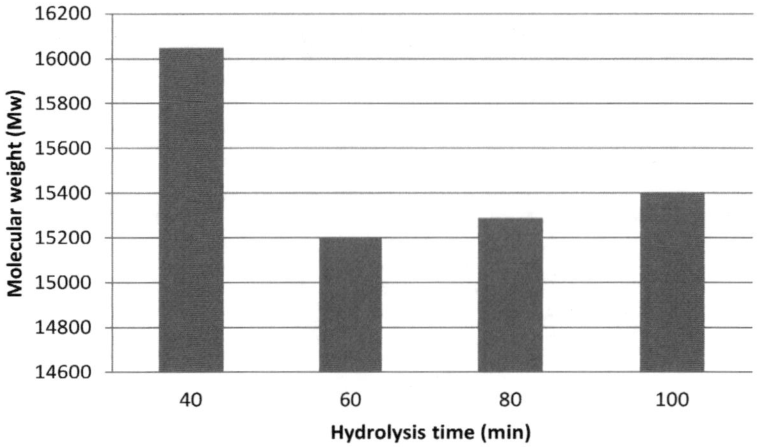


**Figure 5.** Effect of hydrolysis time to molecular weight of NCC.

These results elucidate that cellulose has to convert into sugars solubilization if further increased in the degradation process occurs, which could be observed by decreasing of the total yield of NCC. Moreover, TEM images support this observation, by did not show any significant changes in the morphological structure of the NCC, in terms of length and diameter distribution, after 60 min of reaction time [9].

However, an observation of increasing of the molecular weight after 60 min of reaction time most probably due to enhancement of the sulfate group on the surface of the NCC particle, which is supported by the zeta potential measurement in the Figure 3.

## 4. Conclusion

Nanocrystalline cellulose (NCC) was successfully isolated from OPEFB pulp, which is an abundant biomass residue in Malaysia by sulfuric acid hydrolysis. The characterizations of NCC by the zeta potential show that by increase the reaction time of acid hydrolysis, amount of surface charge that induces to the NCC surface was increased. The results of gel permeation chromatography indicate that in the process of cellulose hydrolysis lower molecular weight of the NCC was obtained, however, if further increased in the degradation process occurs, cellulose have to convert into sugars solubilization after reach of a certain amount of molecular weight.

**Acknowledgement**

The financial support from Universiti Sains Malaysia through Research University Grant No. 1001/PTEKIND/8140151 and 1001/PTEKIND/845025 is gratefully acknowledged.

Journal of Physics: Conference Series **622** (2015) 012047 doi:10.1088/1742-6596/622/1/012047

**References**

[1] ROMANY M,SHUPING D, Hirani A and YONG,WL (2009) *ACS symposium series*, Oxford Uni. Press,**1017**: 81-91
[2] Al-Dulaimi A A and Wanrosli W D (2014) *Advanced Materials Research* **1043** 105-8
[3] Rohaizu R and Wanrosli W D (2014) *International Journal of Chemical Engineering and Applications* **5** 468-73
[4] Dong X M, Revol J-f. and Gray D G(1998) *Cellulose* **5**19-32
[5] Lu P and Hsieh Y-L (2010) *Carbohydrate Polymers* **82** 329-36
[6] Ranby B G (1951) *Discussions of the Faraday Society* **11** 158-64
[7] Dong S and Roman M (2007) *Journal of the American Chemical Society* **129** 13810-1
[8] Y Habibi, L A Lucia, O J Rojas (2010) Chemical reviews **110** 3479-3500.
[9] Al-Dulaimi A A and Wanrosli W D (2015) *Industrial Crops and Products* Accepted
[10] Wanrosli W D, Rohaizu R and Ghazali A 2011 *Carbohydrate Polymers* **84** (1) 262-7
[11] Wanrosli W D, et al 2013 *Industrial Crops and Products* **50** 611-7
[12] Leh C P,WanRosli W D, Zainuddin Z and Tanaka R (2008) *Industrial Crops and Products* **28** 260-7
[13] Wanrosli W D, Zainuddin Z, and Lee L (2004) *Wood science and technology* **38** 191-205
[14] Bondeson D, Mathew A and Oksman K (2006) *Cellulose* **13** 171-80
[15] Roman M and Winter W T (2004) *Biomacromolecules* **5** 1671-7
[16] Araki J, Wada M, Kuga S and Okano T (1998) *Colloids and Surfaces A: Physicochemical and Engineering Aspects* **142** 75-82
[17] Angellier H, Putaux J L, Molina-Boisseau S, Dupeyre D and Dufresne A (2005) *Macromolecular Symposia* **221** 95-104.
[18] Dong X M, Kimura T, Revol J-F and Gray D G (1996) *Langmuir* **12** 2076-82
[19] Lin N and Dufresne A (2014) *Nanoscale* **6** : 5384-93

Journal of Physics: Conference Series **622** (2015) 012048 doi:10.1088/1742-6596/622/1/012048

# Analysis and Evaluation of principal climatic factors of NDVI in the Yarlung Zangbo River Basin

**ZHANG weil[1*] FU Xinfeng[2]**

1. School of energy and environment, Hebei University of Technology, tianjin,300401 China

2 .China Yellow River Institute of Hydraulic Research, Zhengzhou 450003, China;

The corresponding author's e-mail: zhzhangwei@yeah.net

**Abstract**. The normalized difference vegetation index (NDVI) of vegetation covers is influenced by many climatic factors such as precipitation and average temperature. In this paper, NDVI is calculated on the basis of waveband characteristic of NOAA/AVHRR data for the YarlungZangbo River Basin. Then, the spatial-temporal change of NDVI is analyzed. The basin's NDVI is in consistency with season from 2011 to 2013.Large value zone of NDVI is mainly distributed in the lower reaches and part of the middle reaches. Meanwhile, point-data of principal climatic factors are interpolated to spatial grid data with Kriging method consistent with NDVI based on DEM of the YarlungZangbo River Basin. Then, the relation ofprecipitation and normalized difference vegetation indexes or average air temperature is analyzed with F checking and two-sample variance test in the basin. The result is P=0, which shows that correlation analysis has larger confidence. The linear and logarithmic correlation ofprecipitation and normalized difference vegetation indexes or average air temperature is analyzed at confidence level of 0.05.The conclusion is thatprecipitation and normalized difference vegetation indexes and average air temperature in the basin vary with seasons and are closely correlated. The linear correlation coefficient ofprecipitation and normalized difference vegetation indexes is 0.8, and that of NDVI and average air temperature is about 0.77.And the logarithmic correlation coefficient ofprecipitation and normalized difference vegetation indexes is 0.71, and that of NDVI and average air temperature is 0.7.

*Keywords: average air temperature, vegetation cover, Yarlung Zangbo River basin, normalized difference vegetation index, precipitation,*

## 1. Introduction

Vegetation dynamics, especially over large scales, can be monitored using remote sensing Of the pectral indices derived from remote sensing which identify vegetated areas and their condition, the Normalised Difference Vegetation Index (NDVI) is still the most well-known and frequently used [12-

zhzhangwei@yeah.net

13]. NDVI is based on the differential reactance that plants exhibit for different parts of the solar radiation spectrum. Healthy green leaves strongly absorb photo synthetically active radiation for energy in photosynthesis, whereas internal mesophyll structures in the leaf scatter radiation in the near-infrared region to prevent overheating of the plant[12,14]. Calculated by obtaining the difference between the remotely sensed visible (red) and near-infrared bands and normalising it over thesum of the two, NDVI is a good index of the capacity of vegetation to absorb photosynthetically vigorous energy and therefore of land-cover which comprises unstressed vegetation[14,15].

In studying NDVI of basins, NOAA/AVHRR (National Oceanic Atmospheric Administration/ Advanced Very High Resolution Radiometer) has been on the spot for its unique characteristic. For its available information in studying vegetation cover, and its long series observations with low price, the NOAA/AVHRR data has been proved to be valuable for studying vegetation at regional, continental and global scale. At the same time the NOAA/AVHRR data has outstanding advantage in studying large or middle-scale vegetation distribution and dynamic variation [1-4].

Recently, the relationships between the normalized difference vegetation index and some primary climatic factors (such as average air temperature and precipitation) are important for researches of global changes. Weiss et al. researched relationships between climatic factors and regular and inter-annual vegetation in central New Mexcico in USA [5]. The linear relationship between vegetation NDVI and antecedent rainfall in arid to semi-arid regions has been well documented[16-18]. In china, some scholars also have reported the relationships between main climatic factors and NDVI in different regions. The study of the spatial features of spring vegetation NDVI coupling with temperature over Northern Hemisphere presented the high relationship between the Normalised Difference Vegetation Index and air temperature [6]. Climatic factors that drove NDVI variation in china from 1983 to1992 was reported [7]. The research of the relationship between NDVI of several typical vegetation types and seasonal and inter - annual precipitation change in northern China from 1983 to 1992, and impact of spatial distribution of precipitation on vegetation were reported [8]. The relationship of NDVI of Northeast China sampled area and climatic factors such as air temperature was reported[9]. Also some researches is in advance in relationship of the Normalised Difference Vegetation Index and precipitation in the Yellow River Basin, of which the relationship of vegetation cover and precipitation etc. is reported[10]. SunRui et al researched the relationship between the incomplete vegetation coverage changes and rainfall with 8km resolution multi-temporal NOAA AVHRR-NDVI databases in the Yellow River Basin from 1982 to 1999 [11]. NDVI data derived from satellite imagery to measure the different types of land use . AS a particularly high and cold region. However, there was little study on the relationship of normalized difference vegetation indexes and some primary climatic factors as a particularly high and cold region in the Yarlung Zangbo River Basin. In this paper, it included the NDVI changed with temporal and spatial variation and the relationships was indicated in the Yarlung Zangbo River Basin, where the NDVI was changing corresponding with primary climatic factors.

## 2. The geographical position

The Yarlung Zangbo River Basin is at 27°9′~31°9′north latitude, 81°9′~97°1′east longitude. The basin's area is more than 2.4048 $\times 10^5 km^2$. The altitude of most of the basin is above 3000m. From the lower reaches to the riverhead, named JieMaYangZong glacier, the basin is composed of humid、semi-dry and semi-humid regions and arid areas.

The hydrological and meteorological stations, which lie in humid and semi-dry and semi-humid regions, are located at middle and lower reaches of the Basin (see figure 1, red points are stations). The precipitation of the Basin is specially uniform in space, and the maximum precipitation of humid regions is more than 4000mm, while, only 10mm in dry region. Similarly, the air temperature is special uniform in spatial distribution, which changes with terrain regularly. The temperature of the basin changes in the region about -20℃ at headstream and approximate 40℃ at the mouth of the Yarlung Zangbo River during the whole year. The great difference of main climatic factors (precipitation and average air temperature) results in different ecological and environmental conditions. In the Basin, as an index of vegetation cover, NDVI can reflect ecological and environmental conditions and be influenced by main climatic factors. So far, there is no paper for the relationship between NDVI and main climatic factors in the Yarlung Zangbo River Basin, so it is in great significance to research the relationship between NDVI and main climatic factors in the basin.

Figure 1 Geographic position of Yarlung Tsangpo River Basin and its hydrographic and meteorological stations

## 3. The source of data and its disposal method

### 3.1 The source of data

The NOAA/AVHRR data are from Earth Resources Observation System of USA, which are mainly data of daytime from Jan. 1 2001 to Dec. 31 2003. the spatial resolution of the data is 1.1km×1.1km. table 1 is the band characteristics of NOAA. The DEM(Digital Elevation Model) of the Basin is from 1:250000 digital data of National Fundamental of GIS, and the projection is longitude and latitude one, and is transferred to ALBERS projection with Arc/info. The data DEM is provided by the Institute of Chinese Academy of Sciences (CAS), the monthly data of precipitation and average air temperature and Geographical Sciences and Natural Resources Research (IGSNRR) in the Basin is provided by Hydrological information center, Hydrology and Water Resources Survey Bureau of Tibet Autonomous Region.

### 3.2 The disposal method of data

The NOAA data is disposed of with radiometric calibration、geometric rectification、mosaic data and ALBERS projection. The definition of NDVI is got by reflectivity of the first channel of AVHRR (visible light) and second channel (near- Infrared light):

$$NDVI=(CH2-CH1)/(CH2+CH1) \quad (1)$$

Where, CH1and CH2 is reflectivity of visible light(0.58～0.68um) and near- Infrared light(0.73～1.10um), respectively.

Based on the monthly average air temperature of 2011～2013, which is got from hydrological station and meteorological station, Grid spatial data of the monthly average air temperature is obtained with method of Kriging interpolation. The average air temperature is revised with elevation when interpolating, that is, the average air temperature of meteorological stations is revised to sea-level height with elevation, then interpolated, the result is revised to actual height with DEM of 1.1 km×1.1km. When revising, the rate of air temperature decreases with elevation is 0.0065°C/m. Using ENVI, monthly average air temperatures on basin surface is obtained, then, its relationship with NDVI of the basin is analyzed.

Table.1 Band characteristics of AVHRR

| Band | Spectrum scope | Resolution /km | Representative application |
|---|---|---|---|
| B1 | 0.58～0.68 | 1.1 | Cloud layer, Snow, ice and vegetation at daytime |
| B2 | 0.73～1.10 | 1.1 | Cloud layer, vegetation, water and Water vapor at daytime |
| B3 | 3.55～3.93 | 1.1 | Cloud layer, Snow, ice, vegetation, wate, fire examination at daytime |
| B4 | 10.30 ～ 11.30 | 1.1 | Cloud layer, vegetation and water at daytime |
| B5 | 11.50 ～ 12.50 | 1.1 | Cloud layer, Snow, ice, vegetation, Sea-level temperature at daytime |

## 4. Temporal and spatial changes of NDVI in the basin

### 4.1 spatial changes of NDVI in the basin

One monthly composite NDVI figure of 2011—2013 is analyzed. As NDVI value is the largest in July and August, and its contrast in space is more obvious than that of other month in the basin, NDVI figure of August in 2013 is chosen to show vegetation cover distribution of the basin( see Figure 2). The larger value of NDVI is largely scattered in the inferior reaches and part of the middle reaches, while, the value of NDVI in the Middle-Upper reaches and headwater of the basin is smaller from Figure 2. Seeing the spatial distribution of NDVI, there is a relationship between the large value of NDVI and main branches, that is, the value of NDVI in main branches area is comparative large, correspondingly, the value of NDVI is smaller in the area far from mainstream and main branches.

### 4.2 Temporal dynamic changes of NDVI in the basin

In order to show temporal dynamic changes of NDVI in the basin, monthly composite NDVI figure is got from daily NDVI data of 2011—2013, and The average value of NDVI WAS calculated for the whole basin. as a result, the figure of monthly NDVI changes with time from 2001 to 2003 is obtained (seeing Figure.3). From the figure.3, it can be aberseved that NDVI of the basin changes greatly with season. The NDVI value changes smoothly before May; Between MAY and June,NDVI of the basin had a rapid creaseing ,and then,the NDVI was in higher value and keeped stablely beteen June and August;From August, NDVI was decreaseing slowly, and the minimum of NDVI was observed beteen February and March in second year. The monthly NDVI changes tendency equation of 2011-2013 is got with the method of trend analysis (seeing equation.2). Another result was abtained that NDVI value increaseed slowly from 2001 to 2003,and the general tendency of NDVI between August and December was becoming reducing from 2001 to 2003.( According to Figure 2, from 2011 to 2013, the value of NDVI was decreasing gradually. Furthermore, in every year, from August to December, the tendency of reduction was increased for the value of NDVI.)

$$NDVI_{TREND}=0.0016T+0.1636 \qquad (2)$$

Where, $NDVI_{TREND}$ is corresponding NDVI value with trend line, and T is month(0～36). NDVI of the basin was changed greatly with season,where the mainly reason was the climatic factors.(In Yarlung Tsangpo River, climatic factors such as raining, atmospheric temperature and sunlight resulted in the value of NDVI followed commutative seasons.)

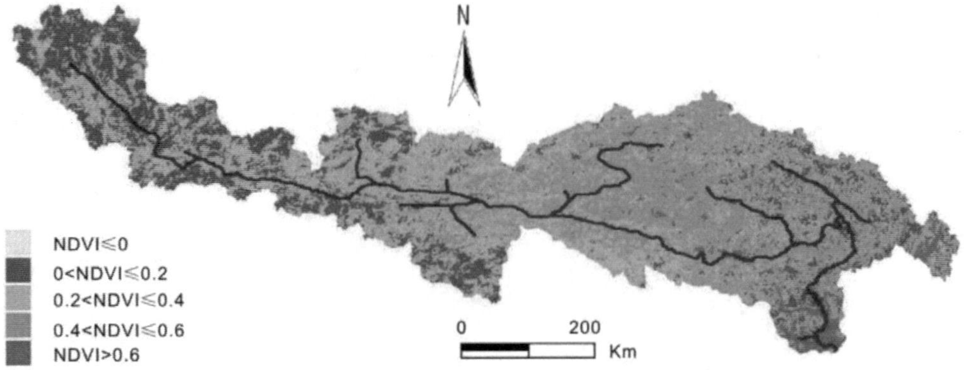


Figure 2 Spatial distribution of normalized difference vegetation index in Yarlung Tsangpo River Basin

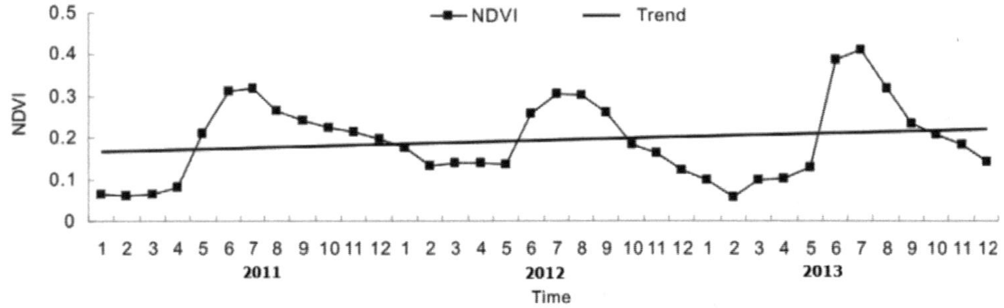


Figure 3 change of monthly normalized difference vegetation index in the Yarlung Tsangpo River Basin

## 5. The relationship between normalized difference vegetation index and main climatic factors

### 5.1 The relationship between normalized difference vegetation index and precipitation

To report the relationship between precipitation and normalized difference vegetation indexes of the basin, monthly precipitation data got from hydrological and meteorogical stations is transformed to monthly precipitation spatial Grid data with the method of Kriging interpolation in the basin. Then, in the every month, the average precipitation of the basin is received with calculation. And then, the monthly average precipitation and normalized difference vegetation indexes is analyzed. It is obvious that both of the average precipitation and NDVI change with time in the every month (see Fig.4)

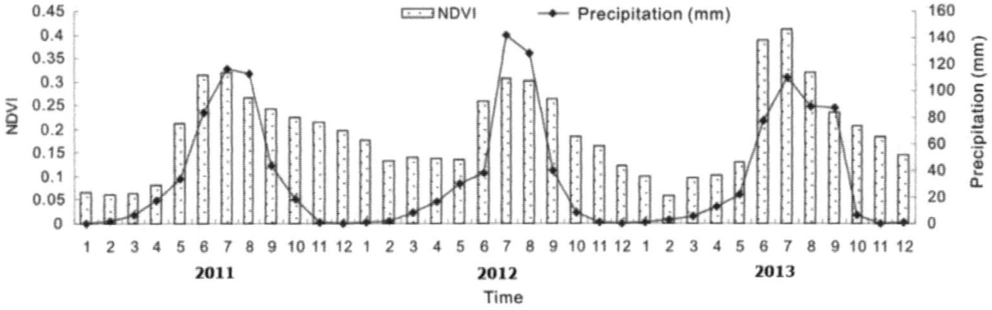

Figure 4 relation of precipitation and normalized difference vegetation indexes in the Yarlung Tsangpo River Basin

F checking and two-sample variance test were applied to s analyze the relation of precipitation and normalized difference vegetation indexes. At confidence level of 0.05, P=0, the linear and logarithmic correlation equations is got with correlation analysis:

$$NDVI=0.0017P+0.1334 \quad (3)$$

$$NDVI=0.0289\ln(P)+0.1299 \quad (4)$$

Where, P is precipitation.

From 2011 to 2013, linear correlation coefficients of precipitation and normalized difference vegetation indexes is about 0.8, and the logarithmic correlation coefficient is 0.71, which indicates that precipitation is one of the most important factors that influences NDVI in the basin.

**5.2 The relationship between NDVI and average air temperature**

To report the relationship between NDVI of the basin and average air temperature, the monthly average air temperature data got from hydrological and meteorological stations is transformed to monthly average air temperature spatial grid data accompanied with DEM by using the method of Kriging interpolation, wich is analyzed with NDVI. It is can be seen clearly that both NDVI and the monthly average air temperature change greatly with time (see Fig.5).

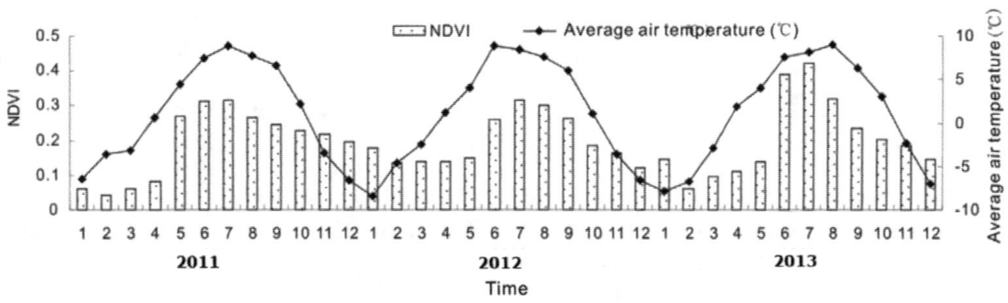


Figure 5.the relation of average air temperatures and NDVI in the Yarlung Tsangpo River Basin

From Figure 5, the average air-temperature increases three months ahead of NDVI, F checking and two-sample variance test were used to analyze the relation of NDVI and the monthly average air temperature, at confidence level of 0.05, P=0, The linear and logarithmic correlation equations of the average air temperature and normalized difference vegetation indexes is obtained with correlation analysis:

$$NDVI=0.0123Ta+0.1334 \quad (5)$$

$$Ta=7.2301\ln(NDVI)+13.842 \quad (6)$$

Where, Ta is average air temperature of the basin.

From 2011 to 2013, the logarithmic correlation coefficient is 0.7 and the linear correlation coefficient of NDVI and average air temperature in the every month is about 0.76, which indicates that precipitation is one of the main factors that influence the NDVI in the basin.

## 6. Conclusion

As an index of denoting vegetation cover, the change of NDVI reflects the condition of the vegetation cover in the basin. The air temperature and precipitation are the main factors which influenced vegetation cover of the basin. The paper analyzes temporal and spatial distribution of the normalized difference vegetation indexes in the basin, and then, analyzes the relationship between the normalized difference vegetation indexes and average air temperature or the relationship between the normalized difference vegetation indexes and precipitation. The following conclusions are obtained:

(1).Larger value of the relationship between the normalized difference vegetation indexes is distributed in the inferior reaches and part of the central-point reaches, while the value of NDVI in the Upper reaches and headwater of the basin is smaller. Generally speaking, the NDVI value increases gradually from upstream to the downstream of the basin,. The value of NDVI is comparatively large in mainstream or main branches and nearby area.

(2) NDVI value changes small from February to April, and NDVI value increases rapidly after May, and NDVI value is largest and keeps stable from June to August, and then, the NDVI value reduces slowly until February of the next year. The NDVI values from 2011 to 2013 is analyzed that annual mean NDVI value increases slowly.

(3). Average air temperature and precipitation in the basin are main climatic factors that influenced the change of NDVI. The change of NDVI is generally consistent with the changes of average air temperature and precipitation in time, which changes greatly with season. The change of precipitation is consistent with the change of NDVI in the basin in time. While average air temperature increases ahead of NDVI three months. The relation of precipitation and normalized difference vegetation indexes or average air temperature and normalized difference vegetation indexes is analyzed with F checking at confidence level of 0.05, which indicate both of correlation analysis has larger confidence. The linear correlation coefficient of precipitation and normalized difference vegetation indexes is 0.8, and the logarithmic correlation coefficient of precipitation and normalized difference vegetation indexes is 0.71. And the linear correlation coefficient of average air temperature and normalized difference vegetation indexes is about 0.77, and the logarithmic correlation coefficient of average air temperature and normalized difference vegetation indexes is 0.7.

(4). The NOAA/AVHRR data is obtained only from 2011 to 2013, which needs to be got more in the following research. Moreover, vegetation cover interpreted from NOAA data should be proved with underlying surface vegetation type in the basin.

In the future work, it will be integratedly analyzed that the vegetation coverage in river basins and underlying surface vegetation type in the basin with NOAA data.

**Acknowledgement**

This project was supported partially by The National Natural Science Foundation of China(40561002) and Hebei Science and Technology Plan Project(12223606)

**References**

[1] Justice C O,Hoben B N,Gwynne M D（1986） Monitoring East African vegetation using AVHRR data•InternationalJournal of Remote Sensing, 7,453~1474.

[2] Turker C J, Townshend J R G, Goff T E(1985) African land-cover classification using satellite data•Science, 227. 369~375.

[3] Tucker J（1979）Red and photographic infrared linear combinations for monitoring vegetation•Remote Sensing of the En-vironment，8，127~150•

[4] Defries R S, Townshond J R G（1994） NDVI-derived land cover classification at a global scale•International Journal of Remote Sensing，15(17)，3567~3586•

[5] Weiss J L, Gutzler D S, Allred Coonrod J E,etal（2004）Seasonal and inter-annual relationships between vegetation and climate in central New Mexico•USA• Journal of Arid Environments，57，507~534•

[6] GONG Daoyi; SHI Peijun; HE Xuezhao（2002）Spatial features of the coupling between spring NDVI and temperature over northern hemisphere•Journal of Geographical Science, 57(5)，505~514

[7] CHEN Yun Hao LI Xiao Bing and SHI Pei Jun（2001）•Variation in NDVI driven by climate factors across CHINA, 1983~1992•Journal of Plant Ecology，25(6)，716~720•

[8] LI Xiao-bing; WANG Ying; LI Ke-rang（2000） NDVI sensitivity to seasonal and Inter-annual rainfall variations in northern China•journal of geographical science， 55，82~89

[9] Tang Haiping, Chen Yufu(2003)intra-annual variability of NDVI and its relation to climate in northeast China transect•Quaternary Sciences,23(3),318~325

[10] YANG Shengtian; LIU Changming2; SUN Rui1(2003)The Vegetation Cover over Last 20 Years in Yellow River Basin•Journal of Geographical Sciences,57(1),679~692
[11] SUN Rui, LIU Chang-ming, ZHU Qi-jiang(2001) Relationship Between the Fractional Vegetation Cover Change and Rainfall in the Yellow River Basin Journal of Geographical Sciences, 56(6),667~672
[12] Bulcock, H.H., Jewitt, G.P., 2010. Spatial mapping of leaf area index using hyper-spectral remote sensing for hydrological applications with a particular focus on canopy interception. Hydrol. Earth Syst. Sci. 14, 383-392.
[13] Sims, N.C., Colloff, M.J., 2012. Remote sensing of vegetation responses to ooding of a semi-arid oodplain: Implications for monitoring ecological effects of environmental ows. Ecol. Indic. 18, 387-391.
[14] Wang, J., Rich, P.M., Price, K.P., 2003. Temporal responses of NDVI to precipitation and temperature in the central Great Plains, USA. Int. J. Remote Sens. 24,2345-2364.
[15] Otto, M., Scherer, D., Richters, J., 2011. Hydrological differentiation and spatial distribution of high altitude wetlands in a semi-arid Andean region derived from satellite data. Hydrol. Earth Syst. Sci. 15, 1713-1727.
[16] Groenvald, D.P., Baugh, W.M., 2007. Correcting satellite data to detect vegetation signal for eco-hydrologic analyses. J. Hydrol. 344, 135-145.
[17] Ji, L., Peters, A.J., 2003. Assessing vegetation response to drought in the northernGreat Plains using vegetation and drought indices. Remote Sens. Environ. 87,85-98.
[18] Peng, J., Dong, W., Yuan, W., Z, Y., 2012. Responses of grassland and forest to temperature and precipitation changes in Northeast China. Adv. Atmos. Sci. 29,1063-1077.

# The Study of Band Structure of Graphite Intercalation Compound Containing Sodium Calculated Using Density Functional Theory

**Ahmad Nazrul Rosli[1], Izzati Fatimah Wahab[1], Noriza Ahmad Zabidi[2] and Hasan Abu Kassim[3].**

[1] Faculty of Sciences and Technology, Universiti Sains Islam Malaysia (USIM) Nilai, 71800 Negeri Sembilan, Malaysia

[2] Department of Physics, Center for Defence Foundation Studies, Universiti Pertahanan Nasional Malaysia, 59200 Kuala Lumpur, Malaysia.

[3] Department of Physics, Faculty of Science, University of Malaya, 50603 Kuala Lumpur, Malaysia

Email: anazrul84@yahoo.com

**Abstract**. Sodium intercalation in graphite (GIC-Na) was investigated by the first principle calculation. The structure of GIC-Na was calculated using density functional theory (DFT) with the aid of CASTEP module of Material Studio. The exchange correlation functional has been treat by local density approximation (LDA) and generalized gradient approximation (GGA). It was shown that, unlike potassium GIC and lithium GIC, the band gap of GIC-Na was not induced and has same value of band gap with bulk graphite.

**Keyword: density functional theory, graphite intercalation compound, Na.**

## 1. Introduction

Since 1960s, graphite intercalation compounds were been developed by various method to achieve novel features in optical, structural, electronic and magnetic terms. Many atoms or molecules were introduced to fit important applications. Intercalation-induced doping altering the Fermi energy $E_F$, band structure, and density of states (DOS) at $E_F$ on graphite is of interest [1]. GICs, over graphite are well known due to the dramatic change of electrical conductivity. Their properties can be varied by choosing appropriate intercalants. Alkali metals had been capturing many attention toward GICs implementation because of the simplicity of their crystal structures. They obtained the simplest electronic configuration among other elements, therefore easily compared to the graphite band structure [2].

In bulk graphite, two dimensional carbon layers are held together via weak van der Waals forces. It is this feature which makes it possible to react chemically with various atoms or molecules [3]. Graphite behaves as a semiconductor with a vanishing energy gap, and in order to account for the magnetic and electrical properties, it is necessary to introduce impurities to it [4]. Carbon containing in the graphite would have four valence electrons. Three of them involves in $\sigma$ bonds, and the last one participates in $\pi$ bonding. This last electron contributes to most of transport properties in graphite. Upper and lower $\pi$ bands overlap by 30meV along the Brillouin zone boundaries, creating semi-metal signature.

Several attempts have been made to explain the electrical properties of graphite intercalation compound by a band structure. Alkali metals are able to intercalate graphite and modify its structure and properties [5]. Many researches have been done to study the graphite intercalation compound consisting alkali metals like potassium and lithium, but several only been investigated onto the structure of sodium-GIC. In contrast to lithium and potassium, the sodium intercalates are said to be difficult to be synthesized.

Sodium intercalated into GIC could be prepared only in the presence of impurities (such as oxygen, water, hydrogen or metal alloys) or with imperfectly graphitized carbon [6]. The liquid phase interaction of molten sodium with graphites provide only high stage compositions [7]. Metrot and co-workers have shown that intercalation from sodium vapour at high temperature yields the pure eight- stage $NaC_{64}$ compound and that lower stage compounds, seventh-stage ($NaC_{48}$) and sixth-stage ($NaC_{56}$) can be obtained from reactions performed at low temperature [2]. The problem of sodium intercalation into graphite is still one of the most interesting and extraordinary in the solid-state physics of intercalation compounds. The graphite intercalation compounds containing sodium are mostly present in high stages under high temperature and pressure [7]. The study of band structure of sodium-GIC is proposed to see the suitability of this material in semiconductor applications. Although graphite is said to be zero-bandgap semiconductor, its band widening is still an important art. Nevertheless, the absence of band gap in graphite limits its practical applications [8]. However, to date, no first principle study has been performed to address how electronic properties of graphite are modified by layered of sodium intercalants.

## 2. Calculation Method

In this research, CASTEP module of Materials Studio was used to develop the structure of graphite intercalation compound with sodium (GIC-Na). The plane wave basis set has been used. Two possible structures of GIC-Na were made up through parameters mentioned in Table 1 with space group hexagonal P63/mmc. The band structure has been calculated using Kohn-Sham equation with different approximation to treat exchange correlation potential. To identify the quantitative difference between bulk graphite and GIC-Na, graphite structure was also developed and set to the ground state so that comparison can be made between the bulk graphite and also the sodium GIC structures. Graphite acts as the reference structure along this study with same lattice parameter and carbon atoms positions. The lattice parameter, space group and basic coordinate of GIC intercalate with Li, K and Na has been discussed by Wang et al. [12].

**Table 1**. Lattice constant and atomic coordinates of graphite, GIC-Na (a type) and GIC – Na (b type)

| Type of structure | Lattice parameter (A, deg) | Atomic coordinates (fractional) |
|---|---|---|
| 2 layer of graphene | $a = 2.456, b = 2.456, c = 6.696$<br>$\alpha = 90.000, \beta = 90.000, \gamma = 120.000$ | $C_1$ 0.000, 0.000, 0.250<br>$C_2$ 0.667, 0.333, 0.750 |
| GIC – Na (a type) | | $C_1$ 0.000, 0.000, 0.250<br>$C_2$ 0.667, 0.333, 0.750<br>Na 0.333, 0.667, 0.750 |
| GIC – Na (b type) | | $C_1$ 0.000, 0.000, 0.250<br>$C_2$ 0.667, 0.333, 0.750<br>Na 0.333, 0.667, 0.500 |

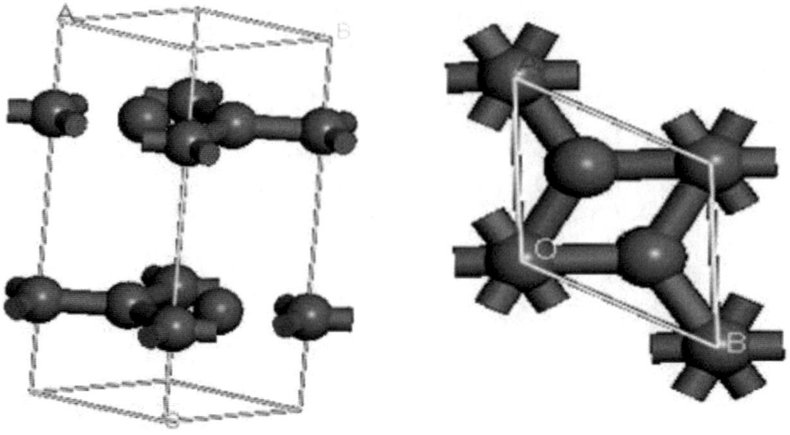

**Figure 1.** A structure of GIC-Na (a type) from two different angle of view. Na atom (purple) at the same level with graphene layer in the center of carbon (grey) ring

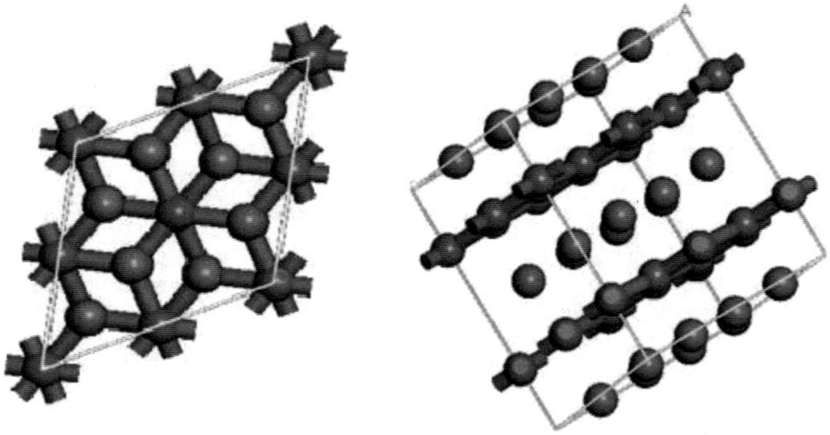

**Figure 2.** A structure of GIC-Na (b type) from two different angle of view. The Na atoms (purple) sit in between 2 graphene layers (carbon in green colour).

## 3. Band structure of GIC-Na

The intercalation of sodium atoms takes place into graphite structure with atomic coordinates such in Table 1. Two structures of GIC-Na were possible to be generated as the first stage intercalation compound. As stated before, low stages of GIC-Na are known to undergo difficulties to be synthesized experimentally compared to other alkali metal GICs. Figure 1 and figure 2 show the structure of sodium intercalation into graphite. After optimization with local density approximation (LDA) and general gradient approximations (GGA), the lattice parameters inserted were optimized as indicated in Table 2. Lattice parameters inserted changed after optimized with LDA and GGA-PBE.

**Table 2.** The lattice parameters of GIC-Na (a type) and GIC-Na (b type) after optimized with LDA and GGA.

| **Atomic Coordinates (fractional)** | **Approximation** | **Lattice parameter (Å, deg)** |
|---|---|---|
| GIC-Na (a type) | LDA | $a = 3.835349, b = 3.835349, c = 4.849585$ |
| | GGA - PBE | $a = 3.794112, b = 3.794112, c = 8.190601$ |
| GIC-Na (b type) | LDA | $a = 3.492933, b = 3.492933, c = 8.253563$ |
| | GGA - PBE | $a = 2.682171, b = 2.682171, c = 15.455735$ |

The band structure was calculated along high symmetry k point in the Brillouin zone of the GIC-Na model. From the results in Figure 3, 4 and 5, the band structures of GIC-Na for both atomic arrangements show zero band gap after energy optimization using LDA and GGA-PBE approximations. These were indicated by the interception of valence band and conduction band near the Fermi level. Fermi level in graph is labelled with dashed line. The value of band gap for GIC-Na was proven had not induced from the graphite band gap. The intercalation of sodium seems did not alter the conductivity behaviour of graphite compound. Thus the electronic properties of graphite intercalation compound when intercalated with sodium does not change the electronic properties of graphite.

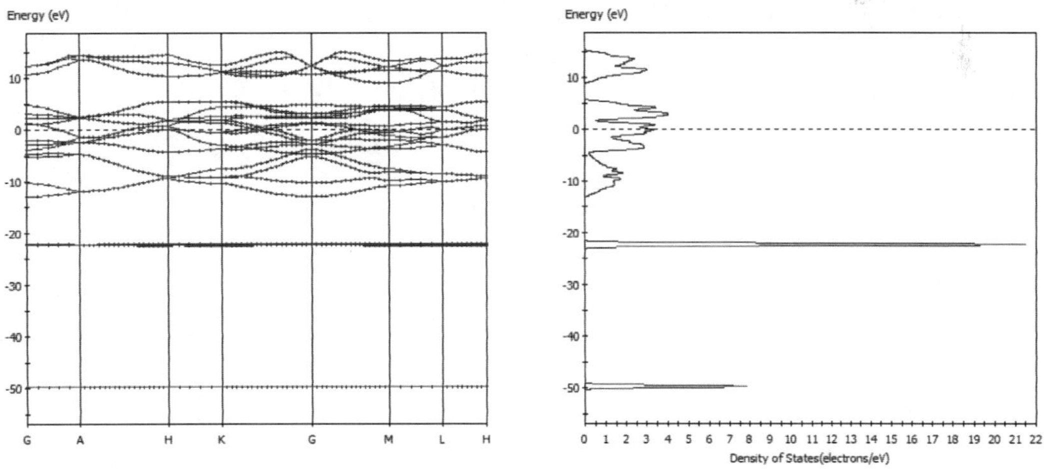


**Figure 3.** Band structure and DOS for GIC-Na (a type) optimized with LDA.

Journal of Physics: Conference Series **622** (2015) 012049 doi:10.1088/1742-6596/622/1/012049

Density of states graph over energy revealed that carriers were probable to mostly occupy near Fermi level and it was clear that at energy equals to zero, no band gap appeared. The density of state near Fermi energy were enhanced in both cases. In the electronic transition view, for donor GIC, electrons transfer from donor to graphite atoms occupy anti-bonding $\pi*$ band. While for acceptor GIC, charge transfer from graphite to acceptor empties the top of the $\pi$ bonding band. Charge transfer between graphite and intercalates increases number of electrons or holes around the Fermi energy and makes the GIC behaving metallic as in graphite.

Graphite structure optimization resulted band structure and density of state (DOS) are shown in Figure 6. Both approximations produced out almost same band structures and DOSs. High state density was shown in conduction band indicates that more states were available to be occupied by electron. This proved the metallic behavior of graphite, which is high electrical conductivity. Clear zero band gap is seen at the H point, indicating the position of corner point in Brillouin zone. In the band structure of graphite, the density of states near Fermi level (labeled dashed line in graphs) were entirely came from the p band around the Brillouin zone corner and there is no band near Fermi level around the zone center [11]. Our calculated band structure of two graphene layers shown similar result with the calculated band structure done by Noriza. A. Z. et al. [9-10]. We are using this layer as a base when we try to intercalate the sodium inside graphene layer.

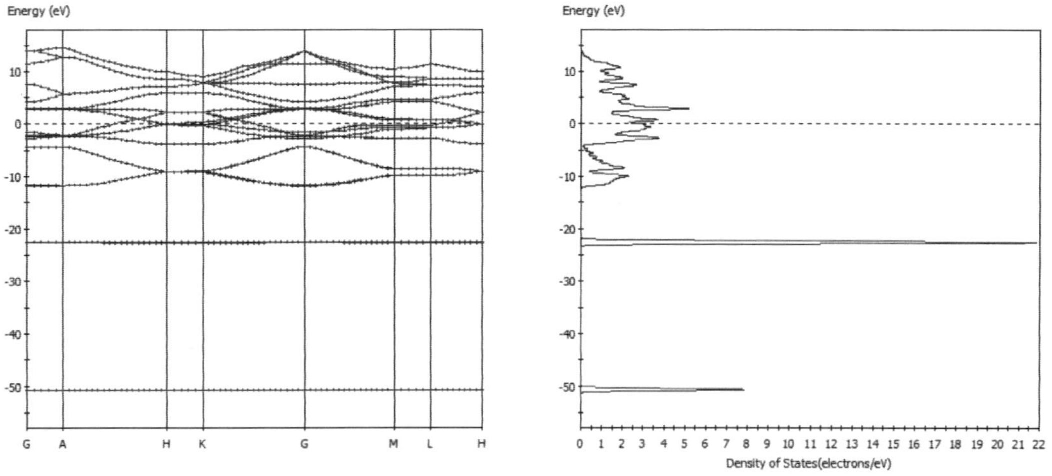


**Figure 4.** Band structure and DOS for GIC-Na (a type) optimized with GGA - PBE.

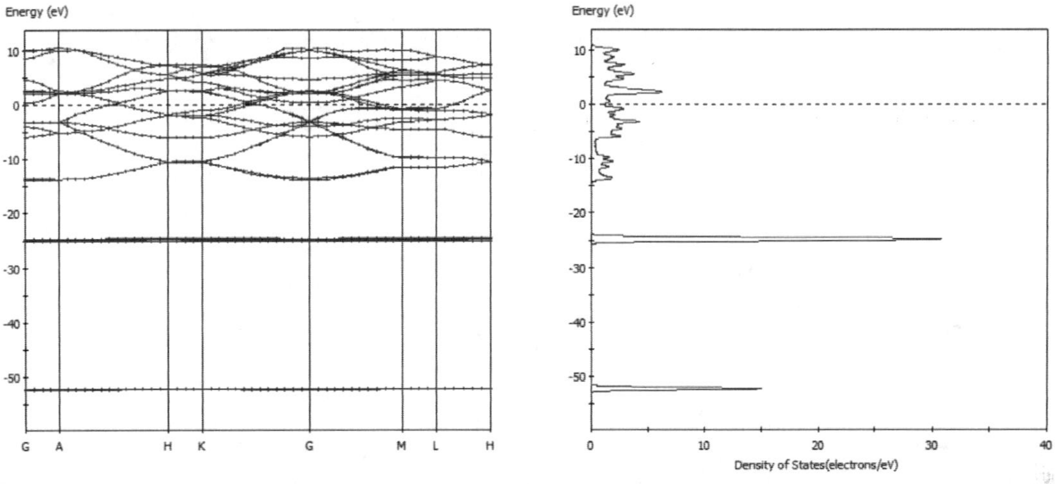


(a)

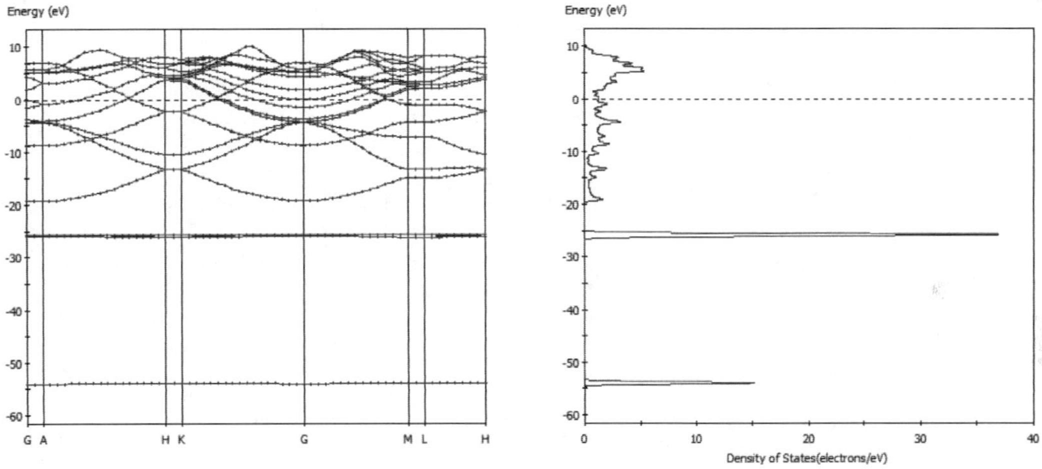


(b)

**Figure 5.** Band structure and DOS for GIC-Na (b type) optimized with (a) LDA and (b) GGA.

## 4. Comparison with the Literature Review

Through reviewing the literatures, the value of bandgaps of graphite intercalation compound containing lithium (GIC-Li) and potassium (GIC-K) were known. GIC-Li and GIC-K possessed bandgaps of 0.6eV and 0.5eV respectively. It was estimated that GIC-Na must have band gap induced at certain value in between zero to 0.6eV since it should not differ too far from GIC-Li and GIC-K when they are in same group of periodic table, having same valence electron. Despite of that, intercalation of sodium seemed does not behave the same way as these two alkali metal intercalants.

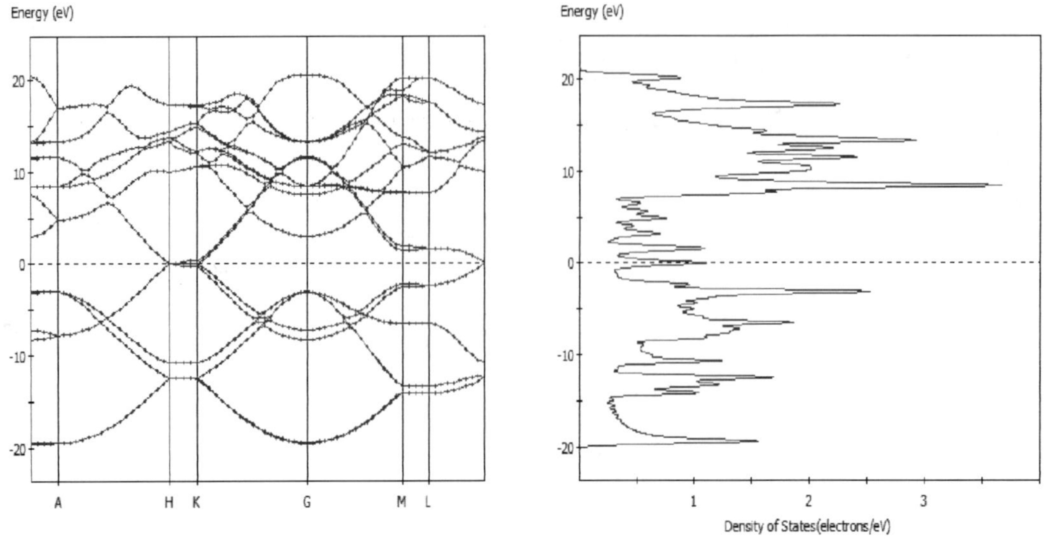


(a)

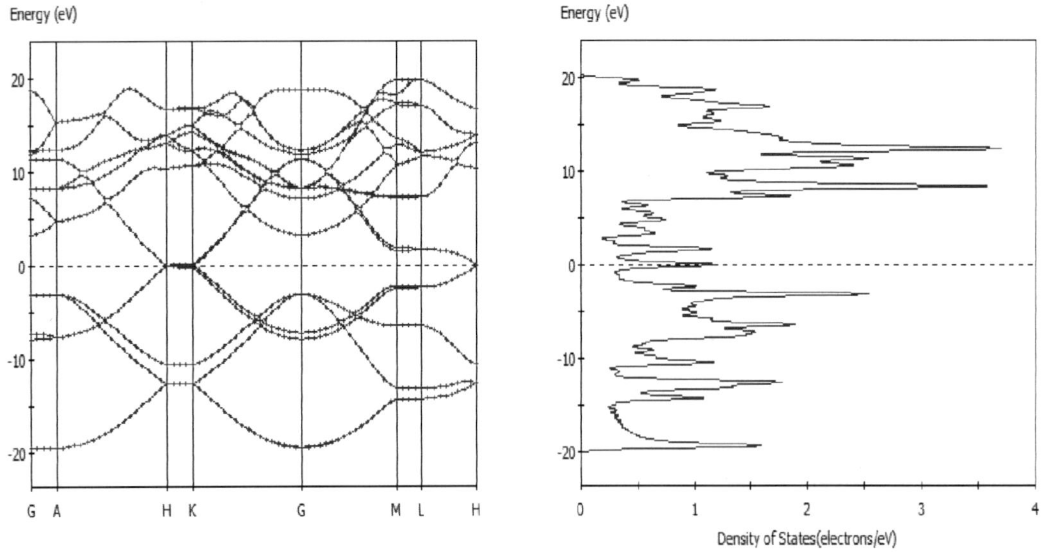


(b)

**Figure 6**. Graphite undergo (a) LDA and (B) GGA approximations.

## 5. Conclusion

Bandgaps are supposedly very important in the way that they allow control of electrical conduction and energy storage. This study proved that graphite intercalation compound containing sodium (GIC-Na) is not capable of these applications and does not behave such that of lithium GIC and potassium GIC. Although sodium ions are smaller than potassium ions, they cannot be easily intercalated into graphite and always form unstable compounds at low stages.

## Acknowledgments

We would like to thanks Ministry of Education Malaysia for providing the financial through Fundamental Research Grant (FRGS) code USIM/FRGS/FST/32/50413. We also want to thanks Universiti Sains Islam Malaysia (USIM) for Short Term Grant, PPP/USG-0113/FST/30/10513.

## References

[1] Oelhafen P, Pfluger P, Hauser E and Guntherodt J 1980. Evidence for an alkali-like conduction band in alkali graphite intercalation compounds *Solid State Communications* ***33*** 241-244

[2] Metrot A, Guerard D, Billaud D and Herold A 1979 New results about the sodium-graphite system. *Journal of Synthetic Metals* **1** 363- 369

[3] Weller T E, Ellerby M, Saxena S S, Smith R P and Skipper N T 2005. Superconductivity in the intercalated graphite compounds $C_6Yb$ and $C_6Ca$. *Journal of Nature Physics* **1** 39-41

[4] Slonczewskit J C and Weiss P R 1957 Band structure of graphite *Journal of Physical Review*. **109** (2) 272- 279

[5] Akuzawa N, Yoshioka J, Ozaki C, Tokuda M, Ohkura K and Soneda Y 2002 Preparation and characterization of sodium- graphite intercalation compounds *Journal of Mol. Cryst. Liq. Cryst.* **388** 415-421

[6] Pascal GE and Fouletier M 1988 Electrochemical intercalation of sodium in graphite *Journal of Solid State Ionics* **28-30** 1172-1175

[7] Udod I A 1997 Sodium-graphite intercalation compound of the first stage: two-dimensional structure and stability *Journal of Synthetic Metals* **88** 127-131

[8] Guo X and Zhou Y G 2013 Band gaps of graphene on layered ZnO substrate: a first principle study *Journal of Applied Physics* ***113*** 054307

[9] Zabidi N A, Kassim H A and Shrivastava K N 2010 Dft calculation of band structure of carbon chain pulled from graphene *AIP Conf. Proc.* **1250** (1) 241-244

[10] Zabidi N A, Kassim H A and Shrivastava K N 2008 Energy band crossing points in multilayers of graphene *AIP Conf. Proc.* **1017** 326

[11] Kamakura N, Kubote M and Uno K 2008 Band dispersion and bonding character of potassium on graphite *Surface Science* **602** 95-101

[12] Wang Z, Selbach S M and Grande T 2014 Van der Waals density functional study of the energetics of alkali metal intercalation in graphite *RSC Adv.* **4** 4069

# RFID - based Staff Control System (SCS) in Kazakhstan

**N Saparkhojayev**
PhD, Senior Lecturer, Almaty Management University, 227 Rozybakiev str, Almaty, Kazakhstan

E-mail: nursp81@gmail.com

**Abstract.** RFID - based Staff Control System (SCS) will allow complete hands-free access control, monitoring the whereabouts of employee and record the attendance of the employee as well. Moreover, with a help of this system, it is possible to have a nice report at the end of the month and based on the total number of worked hours, the salary will be allocated to each personnel. The access tag can be read up to 10 centimeters from the RFID reader. The proposed system is based on UHF RFID readers, supported with antennas at gate and transaction sections, and employee identification cards containing RFID-transponders which are able to electronically store information that can be read / written even without the physical contact with the help of radio medium. This system is an innovative system, which describes the benefits of applying RFID- technology in the Education System process of Republic of Kazakhstan. This paper presents the experiments conducted to set up RFID based SCS.

## 1. Introduction

RFID-based equipment monitoring and tracking system is a complex, integrated system that offers an effective solution of managing items especially for large scale environment [1]. It combines the RFID technology and security devices to ensure the items are always been monitored and secured remotely. The system enable the organization to track and monitor selected individual to access locations inside the university, permit movement, record the data of arrival/departure and also enable the viewing of record via this system. This paper proposes the different components of RFID technology and focuses on its main core competencies, scalability and security. It will then provide detailed description of RFID – based SCS in Kazakhstan. RFID is wireless automatic identification technology that is gaining attention and is considered by some to emerge as one of the pervasive computing technologies in history. As the technology grows very precipitately, RFID has received considerable attention worldwide and widely used in controlling and tracking objects ranging from human identification to product identification. Previous research has successfully showed that RFID has been increasingly expanded in various fields such as retail supply chain, animal identification, metro pass cards and e-passports, asset tracking, postal and courier services, construction industry, education, medical etc. Developments in RFID technology continue to relinquish capacities of the memory, wider reading ranges, and faster processing. It is not possible that the technology will ultimately replace barcode but however, RFID will continue to grow in its established niches where barcode or other optical and wireless technologies are not effective, such as in the chemical container industry and livestock industry. RFID enables tracking and controlling of items over distances that range from about a centimeter to hundreds of meters. RFID can track any items starting from inventories, mobile handy equipment, and moreover, it can track people in real time as the tagged item travels around the organization. Tracking of mobile equipment can include wheelchairs, infusion pumps, and blood supplies. RFID systems have been widely used in many different application areas, such as: product tracking via manufacturing and assembly, inventory control, access to parking lot and its control, tracking of container, ID badges and access control, equipment tracking in hospitals, etc. [2]. Compared to other automatic identification technologies, and especially, compared to optical barcode systems, RFID-technology has important advantages, and among of them, the most important one is the following: tag data can be read automatically beyond the line of sight, through certain materials, and from a range of several meters [3].

## 2. Comparison of RFID technology with barcode technology

Both RFID and barcode are automatic identification technologies that enable automated processes and improve overall operational control and management. Barcode technology relies on – line of sight between the barcode and the reader, meaning that the barcode must be visible to the reader. On the other hand, RFID does not rely on – line of sight, meaning that the RF Tag can be invisible to the reader. This is because RFID is a radio frequency based identification system, which transmits information from a RF Tag on demand to a reader using radio waves that can penetrate packaging materials.

RFID technology is more complex than the bar code, and it is because of the following facts:

- RFID technology is embedded and read with no requirement for line of sight;
- Tags can be reprogrammed easily, and it is capable of working in suitable and harsh environments;
- Ready to carry 96 bits of information – compare this with 16 bits for bar code;
- Fraud controlling increases, and moreover, cloning become non-existence;
- Improves antitheft protection;
- Better efficiency and cost saving;
- Simultaneous multi-tag reading;
- High speed data capture;
- Possibility of having a unique ID and portable database;
- Profit enhancement;
- Better supply chain and management of inventory;
- Reducing counterfeiting;
- Tracking work-in progress;
- Reducing human errors and rework.

## 3. RFID System Components

Radio-frequency identification (RFID) is a technology that uses radio waves to transfer data from an electronic tag – called an RFID tag or label, which is attached to an object – through a reader for the purpose of identifying and tracking the object.

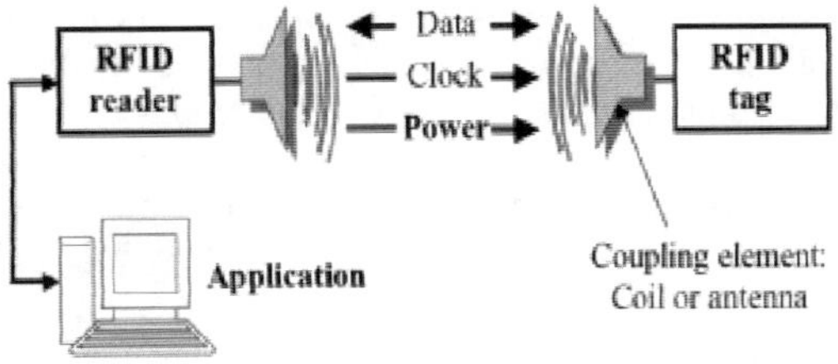


**Figure 1.** The workflow of RFID technology.

A conventional RFID system is made up of four components mainly,

1. RFID Reader;
2. RFID Antenna;
3. RFID transponder (or tag) electronically programmed with unique data;

4. Host computer with appropriate application software.
RFID systems consists of an antenna and a transceiver, which read the radio frequency and transfer the information to a processing device, and a transponder, or tag, which is an integrated circuit containing the RF circuitry and information to be transmitted. The system requires the usage of RF active readers able to communicate with the main station using 3 different communication mediums: serial cable, LAN cable, and WLAN antenna.
Our system relied on a MIFARE RFID-tag, specifically, the MIFARE MF1ICS50 typed RFID-tag. This type of tag was developed by the company NXP to be used in a contactless smart card according to ISO/IEC 14443 Type A. The MIFARE MF1ICS50 IC is used in such applications as public transportation ticketing, which major cities of the world have adopted as their e-ticketing solution. The MF1ICS50 chip consists of a 1 K-byte EEPROM, a RF-Interface and a Digital Control Unit. Energy and data are transferred via an antenna comprising a coil with a few turns directly connected to the MF1ICS50 [4].
As RFID-reader, EHUOYAN's YHU638 was used due to its cheapness and ease of use. This reader enables the contact-free reading and writing of operations and works on a 13.56 MHz frequency [5].

## 4. The description of the SCS

This system is able to check entrance/leaving times for any staff, control the location of any person who is registered in the system, and at the end of the month, this system prints reports about all staff based on some pre-assigned requests. The most important point of the use of this system is that it teaches staff be responsible and come to the work earlier, without being late or even absent. People know that this system is managed and controlled by the administration of the University, so they try to come to work at the time, because the system is able to get their time. The idea of this system is to supply every person with an ID-card, which is composed of unique ID. This system timestamps every swipe of this ID-card so that the system controls entrance and leave times of the person. This system is capable of the following:

- Accurate monthly report of staff attendance;
- No paperwork for HR department, less problem;
- Reports are saved in secured .PDF format, so that at the end of each month administration will have a nice documented attendance.

This system maintains a daily record of a person's arrival time to work and departure time from work to home. Besides this, the system features the name, position and the assigned number (so-called ID) of each personnel. The system also protects the employees by providing the exact number of hours they worked, making it much harder for employers to cheat out of their wages. Staff Control System is an application built on Java that checks every staff's arrival/departure time and provides print/save report-paper that much more simplifies the staff's attendance check-in. The application uses Postgre SQL as Database to store and exports reports, whose are stored in .PDF-format, so that, anything is handled on Windows - desktop, with no handwritten procedures. The application has additional plug-ins that directly refer to database. There are additional plug-ins that consist of RFID reader, Mifare Cards and tiny Java-based module that operates with this equipment. The application was built on NetBeans IDE, so nothing is derived from external source. The ID-cards of the employees are embedded with RFID-tag, which read by a reader. This RFID-reader is connected to a database through a computer. This method is more effective for preventing problems encountered when checking attendance manually.
To maintain perfect data flow and compact view of structure, it was decided to perform only one staff's table (also room) and temporary table of arrival, which can be cleaned up by end of the year. Those two tables perform and maintain all necessary information to prepare monthly attendance report.

Journal of Physics: Conference Series **622** (2015) 012050 doi:10.1088/1742-6596/622/1/012050

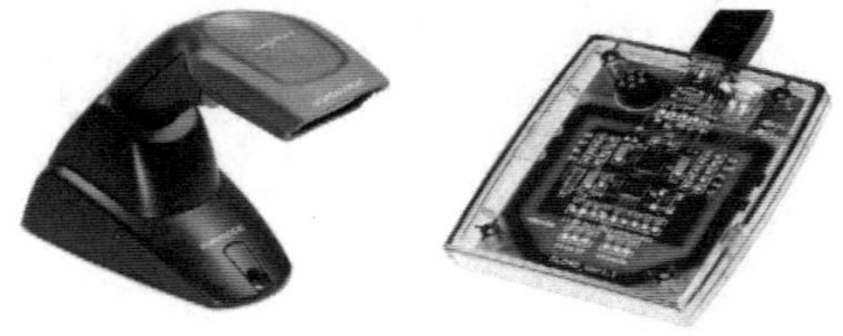

**Figure 2.** Two different types of reader: left one-barcode reader, and the right one is RFID-reader.

SCS reader requires to be separated from desktop computer, with Java and PostgreSQL installed in this desktop. Furthermore, connection to server's database is required to be established; however, it is not strict condition, as SCS reader module stores all data in text file.

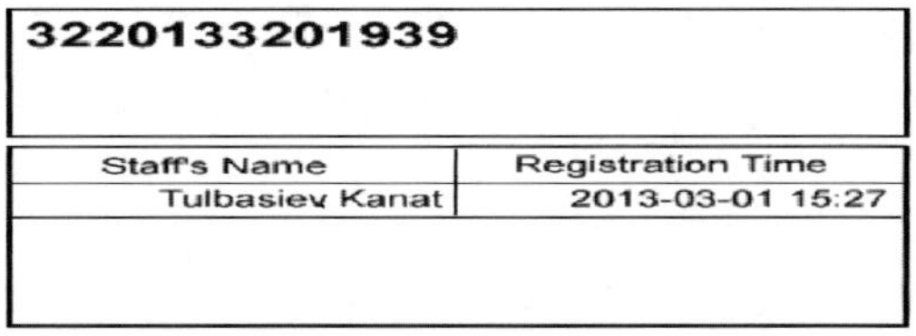


**Figure 3.** SCS reader module GUI.

Readers can be one or many, but each of them requires specific desktop computer. Reader part has its own launcher (launcher.exe), the shortcut of which should be placed to Windows Startup Folder: X:\Users\Iuserl\AppData\Roaming\Microsoft\Windows\StartMenu\Programs\Startup.

It is highly recommended to shut down desktop machine, to prevent any damage and to perform more lifelong service time: Windows Task Scheduler – at 22:00, operation "shutdown –s";

Database setup: PostgreSQL to be installed with archived database structure, which comes with installation folder (postgres.backup). For any login, password and destination database value changes should be also modified in configuration file (conf.txt) in the given format as "login password destination".

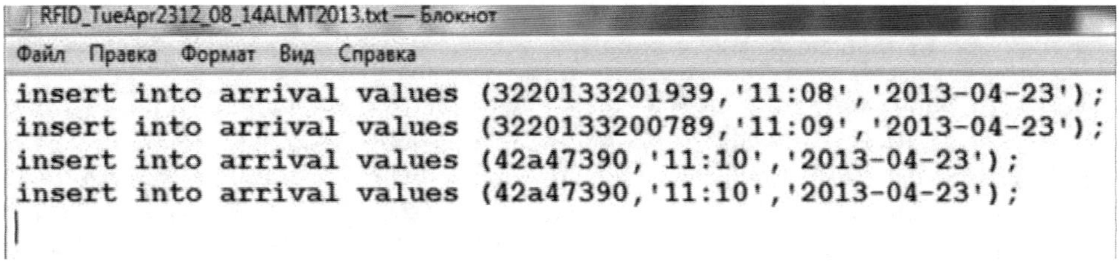
RFID_TueApr2312_08_14ALMT2013.txt — Блокнот
Файл Правка Формат Вид Справка
```
insert into arrival values (3220133201939,'11:08','2013-04-23');
insert into arrival values (3220133200789,'11:09','2013-04-23');
insert into arrival values (42a47390,'11:10','2013-04-23');
insert into arrival values (42a47390,'11:10','2013-04-23');
```

**Figure 4.** Registration log-text file.

The best way to secure data can be gained by preventing any access to the data, so this way, nothing can be changed, so in this application, even HR-Department is restricted from an access to this application. SCS application (Figure 5) provides simple user-friendly GUI, with static and efficient functions, which operate with database, without modifying it.

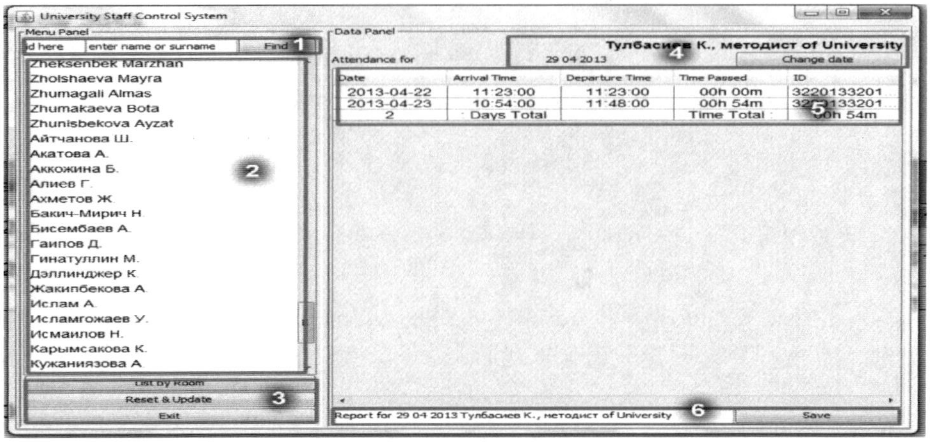

**Figure 5.** SCS main page.

SCS application panel explanation

1. Search by ID and Name criteria: ID line should be consistent and correct; otherwise, no result will be acquired; and Name criteria search is more relevant for inputting word. Search input may include from one to many characters. Found staff will be highlighted in query.

2. Staff List: The whole list of staff that appeared in data – table "person". The personnel is ordered by person.first_name criteria. Double – click for future procedure.

3. Menu Buttons: List by Room/List by Name – changes query type from staff to rooms. Reset & Update – refreshes list appearance, cleans – up data panel. Exit – closes the application.

4. Date and Information panel: Shows the staff's degree and faculty or just working type. Change date button refers to day – month specification. Chosen date appears in report.

5. Data panel: Shows arrival/departure log for specified staff and date. Total time and days are labeled at the bottom of the table.

6. Report label: This line input refers to name of the report, by default it is set to current date and staff's criteria. Save buttons exports .pdf report.

## 5. Conclusion

In this research paper, the idea of implementing the system controlling the staff based on RFID-technology is discussed. The authors we have consulted in our research have shown how a system relying on RFID- technology may be developed. This system is flexible, which means that it may be extended by adding more modules. The cards that have been employed for this specific system are RFID-cards, and the algorithm used has shown stable and reliable results; moreover, this algorithm has secured important data that we have stored on these cards. Regarding the algorithm, there is research paper [6] that discusses the structure of the algorithm, which have been implemented for this system. These cards can be put to use at the university and may replace personnel ID cards. As demonstrated, personnel and students, alike, can use these cards for many purposes; additional functions can always be incorporated into the system and greater security provided to the cards. RFID-technology continues to develop, and the time has come for us to avail ourselves of its promise and convenience. In the research [7], the authors have shown the system for checking the students' attendance in Kazakhstan. The idea is similar, however, there were some differences, and this implemented system have shown an importance of using and developing such systems for educational purposes in Kazakhstan. The main aim of this research has been to demonstrate potential uses of RFID-technology and build a system reliant on it. For the future work, this research should be extended by adding more modules and making some updates or changes. We are planning to extend this system by adding some new modules, specifically "Library module", "Doors Access Control module", "Payment module", "Parking lot module", and so on. There was research done in [8], which showed how to build and implement Library Management system based on RFID. Simultaneously, other cards should be checked and be replaced, because cards which were used for this research seemed to be secure less, and new cards should have enough memory size so that we can keep more data inside of them. Furthermore, the possibility of using some additional tools like GPS, GSM and so on is considered, and the project for implementing such a system is started. We plan to use GPS and GSM technologies in educational system, and the work that was done in [9] is an impulse for this project implementation.

## 6. References

[1] S. Sharma, S.L. Shimi, and S. Chatterji, "Radio Frequency Identification (RFID) based Employee Monitoring System (EMS)", International Journal of Current Engineering and Technology, Vol.4, No.5, October 2014.

[2] K. Domdouzis, B. Kumar, and C. Anumba, "Radio Frequency Identification (RFID) applications: A brief introduction", Advanced Engineering Informatics, Vol. 21, 2007, pp 350-355.

[3] S. A. Weis, S. E. Sarma, R. L. Rivest, and D. W. Engels, "Security and Privacy Aspects of Low-Cost Radio Frequency Identification Systems." Security in Pervasive Computing, 2003, pp 201-212.

[4] NXP official web-site. For additional information, visit http://www.nxp.com.

[5] Ehuoyan Technology Co, Ltd. Official website: http://www.ehuoyan.com.

[6] Saparkhojayev, Nurbek, and Olzhas Shaiken. "An Implementation of MD5 Hash Algorithm for RFID Tags." International Journal of Computer Applications 72 (2013).

[7] Nurbek Saparkhojayev and Selim Guvercin, "Attendance Control System based on RFID-technology", IJCSI International Journal of Computer Science Issues, Vol. 9, Issue 3, No 1, May 2012.

[8] M. Dhanalakshmi and U. Mamatha, "RFID based library management system", Proceedings of ASCNT, pp.227- 234, 2009.

[9] S.B. Patil and R.M. Walli, "Design and Development of fully automatic AT89C52 based low cost embedded system for rail tracking", International Journal of electronic communication and soft computing science and engineering", Vol. 1, Issue 1.

Journal of Physics: Conference Series **622** (2015) 012051 doi:10.1088/1742-6596/622/1/012051

# The mode competition in a two-mode semiconductor laser

**G Baimenshina**
*Kazakh National Technical University, Department of Applied Physics, Kazakhstan*
E-mail: gulnaz.baimenshina@sdu.edu.kz

**Abstract.** In this paper the dynamics of a two-mode semiconductor laser model is considered. The dynamic evolution of a two –mode semiconductor laser can be described by nonlinear rate equations, which take in account the gain saturation. Changes in parameters of the system can lead to changes of the character of the solutions. Qualitative analysis of the system is used for investigating the typical bifurcations and allowed conditions for the existence of different types of mode oscillations in the system. The phase portraits of steady state solution of system are plotted using numerical methods. The competition phenomenon between modes is observed.

## 1. Introduction

Over the past decade instabilities in laser have become a widely discussed topic in quantum optics. Semiconductor laser is an excellent example of nonlinear dynamic systems. In current paper we consider a model of a two-mode semiconductor laser. The study of multimode lasers is by far one of the most actual problems of nonlinear dynamics and laser physics. Notable in this system is the phenomenon of mode competition, arising from connectivity modes in the laser.

The phenomenon of mode competition can also be found in the systems with entirely different kind of nature, such as in environmental, biological, chemical, economy systems. For example, in the economy system the competition between producers under capitalism plays a major role in the development of the proportionality between the different areas of production [1]. The competition phenomenon is also observed in a well-known biological model of Voltaire, which describes the coexistence of several populations [2].

In recent years, a considerable amount of theoretical and experimental studies have been devoted to the practical application of these mechanisms in the creation of new types of lasers and other optical quantum devices.

Most laser generators tend to work in the multimode regime. Such as ring [3, 4], semiconductor lasers and also lasers with feedback.

We will analyze the competition events occurring in the two-mode semiconductor laser. To determine the conditions of stable and unstable states of lasing modes, we will use a qualitative and numerical analysis of the system. Qualitative analysis of a two-mode laser model, gives a clear explanation of the mode competition in the laser.

## 2. The Model

In this section we will discuss a two-mode laser model. The dynamical behaviour of such laser can be described by a set of coupled first order differential equations. The model is applicable to all solid-state lasers, including semiconductor lasers as well.

Rate equations for the laser photon density ($S_i$) in the two-mode approximation, that take into account the saturation of the medium gain has the following dimensionless form[5, 6]:

$$\begin{cases} \dot{S}_1 = (\alpha_1 - \theta_{11}S_1 - \theta_{12}S_2)S_1 \\ \dot{S}_2 = (\alpha_2 - \theta_{22}S_2 - \theta_{21}S_1)S_2 \end{cases}, \tag{1}$$

where, the net amplification rate $\alpha_i$, $i = 1, 2$ equals to the difference of gain $G_i = a(N - N_{th})$ and loss $1/\tau_{ph}$ in resonator $\alpha_i = G_i - 1/\tau_{ph}$,

$N$ - carrier concentration,
$N_{th}$ - carrier density at threshold,
$\tau_{ph}$ - the photon lifetime of $i_{th}$ - mode,
$\theta_{ii}$ - self-saturation coefficient,
$\theta_{ij}$ - cross-saturation coefficient, ($i$=1, 2, $j$=2, 1).

Differential equations for the concentration of carriers charge are omitted, since we are interested in time scales of the order of $\tau_{ph} \approx 1$ ps, and typical carrier lifetime of $\tau_e \approx 1$ns.These parameters correspond to semiconductor lasers.

Thus, the concentration of charge carriers can be considered constant and equal to the initial value $N_0$. In this case $G_i = a(N_0 - N_{th}) = aN_0$.

## 3. Discussions and Results

Let us now examine the stability of equilibrium points of the system using qualitative analysis. Solutions of nonlinear differential equations depend on their parameters. There may be critical parameter values, at which the character of the solution changes completely. For this reason, it is important to know the qualitative behavior of the solutions and examine stability boundaries of equilibria.

Differential equations for the concentration of carriers charge are omitted, since we are interested in time scales of the order of $\tau_{ph} \approx 1$ ps, and typical carrier lifetime of $\tau_e \approx 1$ns.

Setting the differential terms of the equations (1) to zero,

$\dot{S}_i = 0$, $i$=1, 2..,

we obtain four possible steady- state solutions for the photon densities:

$$(S_1^*, S_2^*) = (0,0), \tag{2}$$

$$(S_1^*, S_2^*) = (\frac{\alpha_1}{\theta_{11}}, 0), \tag{3}$$

$$(S_1^*, S_2^*) = (0, \frac{\alpha_2}{\theta_{22}}), \tag{4}$$

$$(S_1^*, S_2^*) = (\frac{\alpha_1\theta_{22} - \alpha_2\theta_{12}}{\theta_{11}\theta_{22} - \theta_{12}\theta_{21}}, \frac{\alpha_2\theta_{11} - \alpha_1\theta_{21}}{\theta_{11}\theta_{22} - \theta_{12}\theta_{21}}). \tag{5}$$

Let us consider the stability of steady states solutions (2 ) - ( 5) using linear approximation . Jacobian matrix of the system ( 1 ) has the following form:

$$\mathbf{J}=\begin{bmatrix}\frac{\partial\dot{S}_1}{\partial S_1} & \frac{\partial\dot{S}_1}{\partial S_2}\\ \frac{\partial\dot{S}_2}{\partial S_1} & \frac{\partial\dot{S}_2}{\partial S_2}\end{bmatrix}=\begin{bmatrix}\alpha_1-2S_1\theta_{11}-\theta_{12}S_2 & -\theta_{12}S_1\\ -\theta_{21}S_2 & \alpha_2-2S_2\theta_{22}-\theta_{21}S_1\end{bmatrix}$$

The first equilibrium point (2) is at the origin and in all parameters of the system is an unstable node ,since $\mathrm{tr}(\mathbf{J})=\alpha_1+\alpha_2>0$ and $\det(\mathbf{J})=\alpha_1\alpha_2>0$.

The solutions ( 3) and ( 4) are stable nodes when determinant $\det(\mathbf{J})=-\alpha_i(\alpha_j-\theta_{ji}\frac{\alpha_i}{\theta_{ii}})$ has a positive and trace $\mathrm{tr}(\mathbf{J})=\alpha_j-\alpha_i-\theta_{ji}\frac{\alpha_i}{\theta_{ii}}$ has a negative values , that is, under appropriate conditions $\alpha_j<\theta_{ji}\frac{\alpha_i}{\theta_{ii}}$.

The solution of the system is transformed into the saddle, when a determinant has a negative value $\det(\mathbf{J})=-\alpha_i(\alpha_j-\theta_{ji}\frac{\alpha_i}{\theta_{ii}})$ , which is reached at $\alpha_j>\theta_{ji}\frac{\alpha_i}{\theta_{ii}}$ ,where $i$, $j$ = 1,2; i, j = 2,1 for the solutions of ( 3 ) and ( 4 ) respectively.

The nontrivial solution ( 5 ) is a stable node in case of performance ratio $\frac{\theta_{12}\alpha_2}{\theta_{22}}<\alpha_1<\frac{\theta_{11}\alpha_2}{\theta_{21}}$, that implies the inequality : $\theta_{12}\theta_{21}<\theta_{11}\theta_{22}$. (6)

If the solution doesn't satisfy above condition the steady state solution is an unstable saddle. Thus, the stability of solution 5 depends on the relation between the products of the coefficients of saturations.

Let us now explain the physical meaning of the conditions of stability of the steady states solutions . Absolute instability of trivial solutions ( 2) means that the laser oscillates always at least in one mode .

Solution (3) shows that the laser generates in the first mode, when $\alpha_2<\frac{\theta_{21}\alpha_1}{\theta_{11}}$, that corresponds to $G_2<G_1$ . A similar pattern can be observed for the condition (4), in this case the laser generates in a second mode when $\alpha_1<\theta_{12}\frac{\alpha_2}{\theta_{22}}$, that corresponds to the inequality $G_2>G_1$ .

Thus, we can conclude that two-mode laser prefer to oscillate in a mode that has the largest gain.

Using numerical methods, such as Runge-Kutta of fourth order, it is possible to integrate the nonlinear equations of the system in different parameters and get the phase portraits of system states. The phase portraits of steady states solutions (3) and ( 4) are illustrated in Figure 1,a,b. respectively.

Solution (3) (see Figure 1, a) shows that the laser generates in the first mode. For the solution 4 (see Figure 2, b), the laser generates in a second mode since. We can see that these numerical results confirm the results of qualitative analysis of the system presented in previous section.

For the solution 5, there are two modes that oscillate simultaneously, since all integral curves tend to one intersection point of these modes. The type of the phase portrait in this case is the stable node (see Figure 2, a).

According to expression (6) for the solution (5), we can formulate a rule of coexistence of oscillation of two modes simultaneously (see Figure 2, a) in the laser: the product of the coefficients of cross- saturation must be less than the product of the coefficients of its own saturation.

The outcome of the competition for solution (5) in unstable state, that has the form of a saddle, depends on the initial conditions. The mode 1 or mode 2 will be oscillated according to initial values of $S_1$ and $S_2$. These laser states are plotted in Figure 2, b.

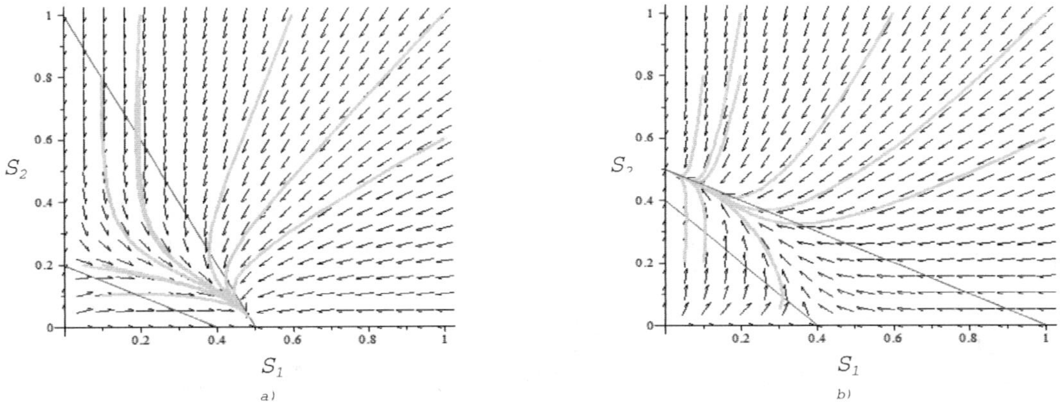


**Figure1.** The phase portraits of steady states solutions ( 3 ) and ( 4)

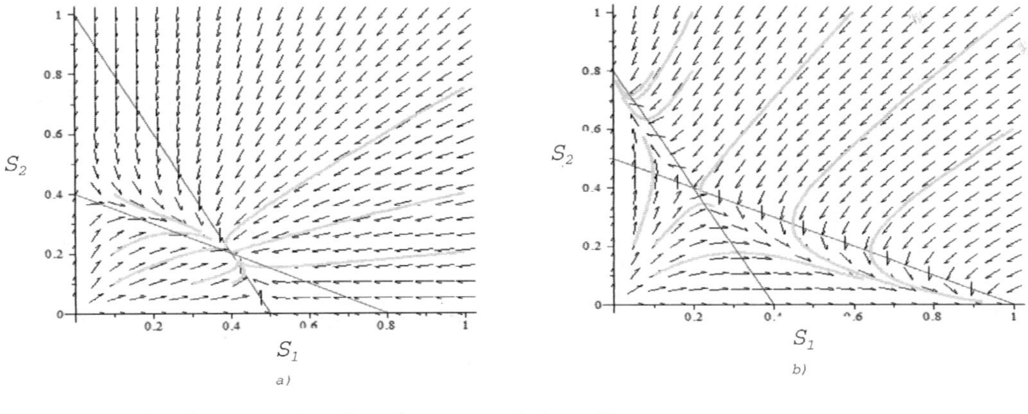


**Figure 2.** The phase portraits of steady states solutions (5)

**Conclusion**

Thus, the two-mode laser has 3 types of mode generation : in the first case - one of the modes will suppress the other , in the second case - two modes will oscillate simultaneously, in the third case – the laser works as a trigger , where the oscillation in one of two possible modes depends on its initial conditions .

**References**

Journal of Physics: Conference Series **622** (2015) 012051 doi:10.1088/1742-6596/622/1/012051

[1] Clifton J A 1977 Cambridge Journal of Economics 1 137–151
[2] Volterra V 1931 Chapman R.N. (ed) (McGraw–Hill))
[3] Menegozzi L N Lamb W E 1971 Phys. Rev. 8 2103-2125
[4] Tehrani M. Mandel L. 1977 Optics Letters 1 196-198
[5]Siegman A E 1986 Lasers (Mill Valley: University science books)
[6]Numai T 2004 Fundamentals of semiconductor lasers(New York:Springer-Verlag)

Journal of Physics: Conference Series **622** (2015) 012052 doi:10.1088/1742-6596/622/1/012052

# Dependence in Classification of Aluminium Waste

**Y Resti**[1]
Jurusan Matematika FMIPA Universitas Sriwijaya,
Jl. Raya Palembang-Prabumulih Km.32 Inderalaya 30662, Ogan Ilir, Sumatera Selatan, Indonesia

E-mail: fyresti@yahoo.com

**Abstract.** Based on the dependence between edge and colour intensity of aluminium waste image, the aim of this paper is to classify the aluminium waste into three types; pure aluminium, not pure aluminium type-1 (mixed iron/lead) and not pure aluminium type 2 (un-recycle). Principal Component Analysis (PCA) was employed to reduction the dimension of image data, while Bayes' theorem with the Gaussian copula was applied to classification. The copula was employed to handle dependence between edge and colour intensity of aluminium waste image. The results showed that the classifier has been correctly classifiable by 88.33%.

## 1. Introduction

In an automatic sorting system, a classification of objects into a few groups mostly base on the images. An image usually represented by a vector in an n.m dimensional space. In practice this spaces are too large to allow robust and fast classification so dimensionality reduction of data is needed at first. For classification, an object image allocated to a group base on a rule, for example a discriminant score or a posterior probability. Linear Discriminant Analysis (LDA) and Quadratic Discriminant Analysis (QDA) are methods of classification was employed a grouping rule base on discriminant score. These methods assume that the joint probability density function is multivariate Gaussian distributed, also the marginal. However, this discriminant score cannot accurately classify object whose joint probability density function is different from its marginal distribution. A grouping rule using posterior probability base on Bayes' theorem is more flexible. In the theorem, the likelihood function can be nonparametric. If we have information about the image characteristics of each group, then we can use this information and the posterior probability based on Bayes' theorem in order to classify the objects. Dependence between the image characteristics can be also involved in this theorem by copula function.

PCA is a statistically techniques which is widely used for dimensionality reduction and recorded a great performance in many application both social and natural science. [1]-[7] applied Principal Component Analysis (PCA) and LDA in face recognition, while [15] applied PCA and QDA in classification of aluminium waste. This paper propose combination of PCA and Bayes' theorem with copula function to classify aluminium waste base on the images into a three groups; pure aluminium, not pure aluminium type-1 (mixed iron/lead) and not pure aluminium type-2 (un-recycle) involving edge and colour intensity of image. PCA as a technique to reduce the data dimension at first and Bayes' theorem with copula function as a technique to classification of aluminium waste image.

[1] Y Resti.

Copula was employed to handling the dependence between edge and colour intensity of the images. Using copula, joint distributions can be modelled curiosities of marginal distributions. Copula models have been applied in several areas such as finance [10]-[11], insurance [12]-[15], environmental [16] and classification studies [9], [17]-[18].

## 2. Materials and methods

The aluminum waste image is taken using a webcam by placing it on a conveyor belt with a particular pose (eg, front, top, back, down or side view), and then saved in jpeg format with a resolution of 320 x 240 pixels. For each of 85 different aluminum wastes, the images were taken in five pose; front view, top view, back view, down view and side view, so we collected 425 different images. Each image pose has been taken edge and colors component of RGB (Red, Green and Blue). The average of red, green and blue colors component is defined as color intensity.

Aluminum waste classification based on the dependence between edge and color intensity combine PCA and Bayes' theorem with copula function. Principal Component Analysis (PCA) was employed to reduction the dimension of image data, while Bayes' theorem with the Gaussian copula was applied to classification. The copula was employed to handle dependence between edge and color intensity of aluminum waste image.

### *1.1. Principal Component Analysis (PCA)*

The objective of PCA approach is to get principal components which explain as much of the total variation in the data as possible with as few of these principle components as possible. The principal component is linear combination of observed variables.

For image data of aluminum waste, let observed variables are image poses, $X_1, X_2, \cdots, X_p$. The linear combination of $p$ image poses variables principal component is defined as,

$$Y_r = a_{r1}X_1 + a_{r2}X_2 + \cdots + a_{rp}X_p = \mathbf{a}_r^T\mathbf{X}_k \tag{1}$$

where $\mathbf{a}^T$ is eigenvector of covariance matrix **S** which is orthogonal with each other. The eigenvector can be found by maximizing $\mathrm{var}(Y_r) = \mathbf{a}_r^T\mathbf{S}\mathbf{a}_r$ with constraints $\mathbf{a}^T\mathbf{a} = 1$.

### *1.2. Copula in Bayes's theorem*

For the purpose of classification, the Bayes' theorem stated in terms of density the following,

$$P(A = a|C = c) = \frac{P(C = c|A = a)P(A = a)}{P(C = c)} \tag{2}$$

where $P(A = a|C = c)$ is the posterior probability, $P(C = c|A = a)$ is the likelihood function, $P(A = a)$ is the prior probability and $P(C = c)$ is the data probability. An object image allocated to group $g$ that have the image characteristic $C$ given its attributes $A$ if $g$ has the largest posterior probability $P(A = a|C = c)$ between other group. For continuous attributes, a copula function can be used for modeling the dependence structure in the likelihood function.

Let $f_1(c_1|a)$ and $f_2(c_2|a)$ are the marginal density of image edge and colour intensity, respectively, while $F_1(c_1)$ and $F_2(c_2)$ are the marginal distribution successively. Let copula density function of the variables is $h(F_1(c_1), F_2(c_2)|a)$ and $P(A = a)$ is the prior probability based on the uniform distribution. Thus for this case, the copula in Bayes' theorem such as [8] can be rewritten as,

$$P(A = a|C_1 = c_1, C_2 = c_2) = \frac{h(F_1(c_1), F_2(c_2)|a) f_1(c_1|a) f_2(c_2|a) P(A = a)}{f(c_1, c_2)} \tag{3}$$

For Gaussian bivariate copula, the density function $h(F_1(c_1), F_2(c_2)|a)$ is defined as,

$$h(\Phi(z_1), \Phi(z_2)) = \frac{1}{2\pi\sqrt{1-\rho^2}} \exp\left[ -\frac{1}{2}\left( \frac{z_1^2 + 2\rho z_1 z_2 - z_2^2}{\sqrt{1-\rho^2}} \right) \right] \quad (4)$$

where $\Phi(z_1)$ and $\Phi(z_2)$ are the marginal standard Gaussian distribution of image edge and colour intensity, and $\rho$ is parameter of Gaussian copula can be found by maximum likelihood estimation.

## 3. Results and Discussion

For each image edge, red, green and blue (RGB) colour component of image, the explanatory variables consists pixel size of down view ($x_1$), top view ($x_2$), side view ($x_3$), front view ($x_4$) and back view ($x_5$). We employed PCA in this research is to find as few of significant image poses variable as possible represented by principle components.

Table (1) shows eigenvalue, proportion of variance and cumulative proportion of principal component for image edge. The small eigenvalue shown the information of the image poses are useless so can be reduction without information loss. The cumulative proportion of the variance of the first component can explain 71% of the total variance and the second component when added to 82% means that if only taking one image pose alone is sufficient to represent the other four variables of image pose.

Table (1) Eigenvalue, proportion of variance and cumulative proportion of principal component for edge

| | Component | | | | |
|---|---|---|---|---|---|
| Edge | 1 | 2 | 3 | 4 | 5 |
| eigenvalue | 4734.72 | 744.60 | 506.62 | 389.22 | 286.03 |
| proportion of variance | 0.71 | 0.11 | 0.08 | 0.06 | 0.00 |
| cumulative proportion | 0.71 | 0.82 | 0.90 | 0.96 | 1.00 |

Table (2) shows eigenvalue, proportion of variance and cumulative proportion of principal component for red, green and blue colour respectively. For red colour, the cumulative proportion of the variance of the first component explained 95% of the total variance. It means that if only taking one image pose alone is sufficient to represent the other four variables of image pose. For each of green and blue colour component, the cumulative proportion of the variance of the first component explained 96% and 94% respectively and it also means that if only taking one image pose alone is sufficient to represent the other four variables of image pose.

Table (2) Eigenvalue, proportion of variance and cumulative proportion of principal component for red, green and blue colour

| | Component | | | | |
|---|---|---|---|---|---|
| Red | 1 | 2 | 3 | 4 | 5 |
| eigenvalue | 4747.22 | 170.80 | 60.72 | 26.12 | 21.30 |
| proportion of variance | 0.95 | 0.03 | 0.01 | 0.01 | 0.00 |
| cumulative proportion | 0.95 | 0.98 | 0.99 | 1.00 | 1.00 |
| | Component | | | | |
| Green | 1 | 2 | 3 | 4 | 5 |
| eigenvalue | 6348.84 | 149.40 | 54.44 | 25.80 | 11.90 |
| proportion of variance | 0.96 | 0.02 | 0.01 | 0.00 | 0.00 |
| cumulative proportion | 0.96 | 0.98 | 0.99 | 1.00 | 1.00 |
| | Component | | | | |

| Blue | 1 | 2 | 3 | 4 | 5 |
|---|---|---|---|---|---|
| eigenvalue | 4904.84 | 172.90 | 68.61 | 28.50 | 18.71 |
| proportion of variance | 0.94 | 0.03 | 0.01 | 0.01 | 0.00 |
| cumulative proportion | 0.94 | 0.97 | 0.98 | 0.99 | 1.00 |

The linear combination of image edge for the first principal component is

$$Y_{edge} = 0.456x_1 0 + .440x_2 + 0.492x_3 + 0.432x_4 + 0.412x_5$$

It means that the side view image pose has the most significant contribution to increasing of image edge comparing to the other image poses. Any increasing in one unit pixel of side view image pose will increase image edge at 0.492. Taking the side view image pose only has been representing the image edge. The linear combination of the first principal component for red, green and blue colour are given successively,

$$Y_{red} = 0.480x_1 0 + .493x_2 + 0.420x_3 + 0.421x_4 + 0.415x_5$$
$$Y_{green} = 0.481x_1 + 0.483x_2 + 0.422x_3 + 0.423x_4 + 0.422x_5$$
$$Y_{blue} = 0.486x_1 + 0.494x_2 + 0.419x_3 + 0.416x_4 + 0.415x_5.$$

It means that the top view image pose has the most significant contribution to increasing of red color composition comparing to the other image poses. Any increasing in one unit pixel of top view image pose will increase red color composition at 0.493. Taking the top view image pose only has been representing the red color composition. Similar as red colour, for both green and blue color component, the top view image poses of the components have the most significant contribution comparing to the other image poses. Any increasing in one unit pixel of each top view image pose will increase green and blue color composition successively at 0.483 and 0.494. Taking the top view image pose only of each color has been representing the green and blue color composition respectively. The PCA shown edge and red, green and blue colour has one significant image pose only, then the colour intensity is average of red, green and blue pixel in one pose only.

Testing 55 different images of aluminum waste to classify into 3 groups was employed. Table 3 and 4 shown results classification aluminum waste based on the image. Specially, the results in table 3 based on posterior probability as a classification rule, while in table 4 based on discriminant score such as done in [15] that employed QDA (Quadratic Discriminant Analysis). Employing Bayes' theorem with Gaussian copula such as given in (3) and (4) to classification 55 different images of aluminum wastes based on dependence between edge and colour intensity, we have been correctly classifiable by 88.33%. It is a performance improving of the method comparing with QDA that has been correctly classifiable by 80% as shown in Table 4.

Table 3. Result of Bayes Classification based on Gaussian Copula

| Observed group | Result of Classification 1 | 2 | 3 | % correct |
|---|---|---|---|---|
| 1 | 25 | 0 | 0 | 100% |
| 2 | 5 | 15 | 0 | 75% |
| 3 | 0 | 1 | 9 | 90% |
| | | | Total | 88.33% |

Journal of Physics: Conference Series **622** (2015) 012052 doi:10.1088/1742-6596/622/1/012052

Table 4. Result of QDA Classification

| Observed group | Result of Classification 1 | 2 | 3 | % correct |
|---|---|---|---|---|
| 1 | 25 | 0 | 0 | 100% |
| 2 | 10 | 10 | 0 | 50% |
| 3 | 0 | 1 | 9 | 90% |
| | | | Total | 80% |

## 4. Conclusion

In this paper, we employed the combination of Principal Discriminant Analysis (PCA) and Bayes' theorem with Gaussian copula to classify the aluminium waste represented by the images, based on the dependence between edge and colour intensity of aluminium waste images. The combination of techniques has been correctly classifiable by 88.33% of 55 different images of aluminium wastes. It is a performance improving comparing with QDA (Quadratic Discriminant Analysis) that has been correctly classifiable by 80%. A better performance of classification can be applied to get a faster automatic sorting system.

## 5. References


[1] W.S. Zheng, J.H. Lai and P.C. Yuen, "GA-Fisher: A New LDA-Based Face Recognition Algorithm with Selection of Principal Components", *IEEE Transactions on Systems, Man, and Cybernetics - Part B: Cybernetics*, Vol. 35, No. 5, 2005, pp. 1065-1078.

[2] A. Eleyan and H. Demirel, "PCA and LDA Based Neural Network for Human Face Recognition" in *Face Reognition*, edited by K. Delac and M. Grgic, ISBN 978-3-902613-03-5, Vienna: I-Tech Education and Publishing, 2007, pp. 95 – 106.

[3] N. Nain, P. Gour, N. Agarwal, R. P. Talawar and S. Chandra, "Face Recognition using PCA and LDA with Singular Value Decomposition(SVD) using 2DLDA", *Proceedings of the World Congress on Engineering,* Vol. I, London, U.K., July 2 - 4, 2008.

[4] S. Singh, M. Sharma and N. S. Rao, "Accurate Face Recognition Using PCA and LDA", *International Conference on Emerging Trends in Computer and Image Processing*, Bangkok, 2011, pp. 62-68.

[5] J. A. Dargham, A. Chekima, E. G. Moung, "Fusion of PCA and LDA Based Face Recognition System", *International Conference on Software and Computer Applications,* Singapore: IACSIT Press, vol. 41, 2012, pp. 111-116.

[6] S.K. Hese and M.R. Banwaskar, "Performance of PCA and LDA for Face Recognition", *International Journal of Advanced Research in Electronics and Communication Engineering*, Vo. 2, No.2, 2013, pp. 149-154.

[7] S. Fernandes and J. Bala, "Performance Analysis of PCA-based and LDA-based Algorithms for Face Recognition", *International Journal of Signal Processing Systems,* Vol.1 No.1, 2013, pp.1-6.

[8] R.Salinas-Gutierrez, A.Hern´andez-Aguirre, M.J.J.Rivera-Meraz, and E.R.Villa-Diharce, Supervised Probabilistic Classification Based on Gaussian Copulas, Advances in Soft Computing, *Lecture Notes in Computer Science*, 2010, vol. 6438, pp 104-115.

[9] C.M. Bishop, *Pattern Recognition and Machine Learning*. Springer, 2007.

[10] D.E. Li, On default correlation: a copula function approach. *The Risk Metrics Group, Working Paper,* 2000, no. 99-07.

[11] A. Shamiri., N.A. Hamzah and A. Pirmoradian, Tail Dependence Estimate in Financial Market Risk Management: Clayton-Gumbel Copula Approach, *Sains Malaysiana*, 2011, vol. 40 no. 8, pp. 927–935.

[12] E.W. Frees, and P. Wang, Copula credibility for aggregate loss models. *Insurance: Mathematics and Economics*, 2006, vol. 38, pp. 360-373.

[13] Y. Resti, N. Ismail and S.H. Jaaman, Handling the dependence of claim severitieswith copula models, *Journal of Mathematics and Statistics*, 2010, vol. 6 no. 2, pp. 136-142.

[14] Y. Resti, N. Ismail and S.H. Jaaman, Mathematical modeling for claim severities using normal and t copulas. *International Journal of Applied Mathematics and Statistics* 2012, vol. 27 no.3, pp. 8-19

[15] Y. Resti, Statistically Classification of Aluminum Waste, *Proceeding of the 1st International Conference on Education, Technology and Science*, Jambi, Indonesia, November 19-20, 2014, pp. 414-417.

[16] G. Wong, M.F. Lambert and A.V. Metcalfe, Trivariate copulas for characterization of droughts, *Australian and New Zealand Industrial and Applied Mathematics Journal*, 2008, vol. 49, pp 306-323.

[17] Hastie, T., Tibshirani, R., Friedman, J.: The Elements of Statistical Learning: Data Mining, Inference, and Prediction. Springer (2009)

[18] Mackay, D.J.C.: Information Theory, Inference, and Learning Algorithms. Cambridge (2008)

Journal of Physics: Conference Series **622** (2015) 012053 doi:10.1088/1742-6596/622/1/012053

# Development of Identification System of cans And Bottle

**Irsyadi Yani[1], Ihsan Budiman[2]**
Mechanical Engineering Department, Faculty of Engineering
Sriwijaya University, Indonesia

E-mail: yani_irs@yahoo.com

**Abstract.** The objectives of this research was developed an integrated simulation model of an intelligent sorting system based on machine vision, which is focused on object detection, identification and automatic sorting system of cans and bottle. In this research we used 60 cans and 40 bottles for database with five direction of each sample (upper, bottom, diagonal left and right side). Performance of the identification system for correct cans and bottles are 91.33% with estimated amount of 21,600 items per hour. The success of the identification depends on the size and type of the objects.

## 1. Introduction

Recycling of solid waste has a big challenge in sorting areas around the world. Without solid-waste sorting system effective and efficient, recyclable solid waste from various human activities have a negative impact on the recycling business. Recycling of solid waste problem in developing countries, such as Indonesia, has a number of aspects related to them, such as technological, official, financial, environmental and communal aspects.

The recycling of solid-waste cost billions each year, for example recycling a ton of aluminium saves the equal in energy of 2,350 gallons of petrol. This is equal to the total of electricity used by the basic home over a period of 10 years (www.aluminum.org, 2009). In 2004, whereas only 51.2 percent of beverage cans were recycled in USA, which is equivalent of energy about 15 million barrels of oil [Aluminium Association, 2009].

Sorting of reusable waste is a main stage in recycling industry. In developed countries, such as USA, Europe and Japan, traditional sorting systems are still usual methods when used to recyclable items. However, nowadays, the development and improvements of automated sorting methods are actively carried out using more scientific ways. In Indonesia, recycling plan is still at its beginning step. Public understanding on the importance of recycling waste is relatively too low. Hence, the goal of the future research is to develop an automated identification and sorting system of recyclable containers (cans and bottles) that utilizes an intelligent machine vision system. The future system is significantly

[1] Irsyadi Yani.
[2] Irsyadi Yani.

contributing towards providing an efficient and cost effective method for sorting recyclable containers, saving of resources, reduction of ecological pollution and build-up of landfills.

Cans and bottles are one type of non-degradable materials in the solid waste. In fact, recycling cans and bottles have a number of significant environmental benefits, including reducing waste, reducing landfill and reducing greenhouse gas emissions. Table 1 shows number of waste production in 2008.

Table 1. Estimation of Waste Generation by Composition

| Type of Waste | Amount (million | Percentage (%) |
|---|---|---|
| Garbage | 22.4 | 58% |
| Plastics | 5.4 | 14% |
| Papers | 3.6 | 9% |
| Others | 2.3 | 6% |
| Woods | 1.4 | 4% |
| Glasses | 0.7 | 2% |
| Rubbers / Leathers | 0.7 | 2% |
| Fabrics | 0.7 | 2% |
| Metals | 0.7 | 2% |
| Sands / Inert Material | 0.5 | 1% |
| ALL | 38.5 | 100% |

Waste quantity collected from household is greater than other non-household sources. Waste quantity collected from each sources is estimated (Figure 1). The question asked to fill out the data from last 5 years at least.

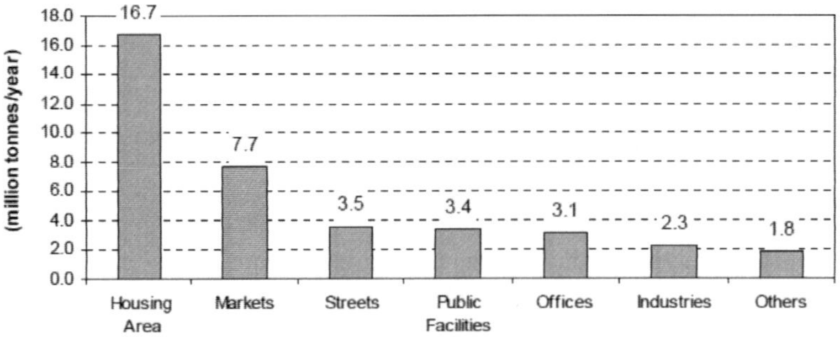


**Figure. 1** Estimation of Waste Quantity Collected by Local Government

## 2. Sorting System

In any recycling activity, sorting of reusable waste is the first and most important step. Many developed country such as USA, Japan, and so countries in Europe are common still use manual sorting system as method of sorting recyclable items. However, work on the development and improvements of automated sorting are actively carried out (M. Ebner 2009; G. Pingju et al. 2008; D. A. Wahab et al. 2003). Out of many technologies, imaging technologies such as infrared, near infrared, sensors, X-rays and optics have been deployed in the development of automated sorting system. While the research and development of containing sorting based on the shape, size and colour characteristics generally is still ongoing.

Many approaches have been proposed for development of an intelligent sorting system which involves object detection, classification and position based on image and signal processing (Vetter, Poggio, et al., 1992; Sung and Poggio, 1994; Sinha, 1994; Moghaddam et al., 1995; Rowley et al., 1998; Aparna et al., 1999; Yang, 2002;). However, lighting condition, image variation, false detection rate, data acquisition in complex background, real time implementation and capturing the entire area are still posing a number of challenges for recyclable and non-recyclable object detection and classification (Y. Irsyadi et al., 2009; E. Scavino et al., 2009).

*1.1. Manual Sorting System*

Many countries including Indonesia (www.aluminum.org, 2009), waste cans are sorted into different types using manual sorting system. In the manual sorting systems, sorting and separation of alumina cans and non-alumina currently requires extensive manual process used to remove contaminants such as glass, plastic bottles, dirt and other non-metallic materials. Furthermore, the waste can be washed by the merchant waste cans and then sent to the recycled industry, where waste can be separated again according to the type of waste alumina cans, non-alumina cans and non- recycled cans.

The manual sorting faces some major problems such as labour intensive industry which incurs relatively high processing cost, Inconsistent quality of end products, skill And commitment essential for efficient sorting. As such, There is a need to maintain a relatively stable and skilled work team. The high processing cost does not only involve wages, pensions, and insurance payable for crew members working on the production line, but also the expenses on health and safety facilities. Labourers working at the manual sorting facilities are exposed to microorganisms, organic dust, and fungi, which can cause severe infections (Waste Cap. 2008). Hence there is an important need to automate the sorting process for efficient safe and clean recycling.

*1.2. Automated Sorting System*

The Automated sorting systems are classified into mechanical and optical. Mechanical sorting is based on the difference of physical features between fibre materials and other contaminations (non-fiber e.g. plastic and glass bottles, metals, cardboard). Gravity, size, shape, magnetization, and conductivity are considered as physical features. Recovered cans from a commingled waste stream are generally composed of a range of object sizes, shapes, thickness and stiffness. Finally it seems that the mechanical sorting is inadequate for achieving commercially viable throughputs and accuracy.

The greatest advantages of an optical sorting systems include: consistent and reliable efficiency of production with a relatively high hit rate and purity; and low operational cost due to reduced number of manual workers on the production line example.

## 3. Procedure For Identification System

The procedure for the identification of cans and bottle consist of a series of computational steps which are conventionally identified as image acquisition, image pre-processing, feature extraction and identification. Each of these steps is described in the following; an example of pattern recognition of an actual object is presented.

The first step in the pre-processing block is to take the image from inspection zone after trimming unnecessary boundary portion of the image. After that, the background noise is eliminated from the image using combined operation of threshold and morphological operation erosion with 7×7 minimum convolution filter.

The features extraction stage consists of quantization of image, calculating co-occurrence matrix and calculates energy in different quantization levels. In feature extraction phase, both colour and grey scale images are considered. For colour image each of the three colour components - red, green and blue - are considered separately. For grey scale image, standard grayscale transformation is obtained from the original RGB image

The image is then normalized; the edges are extracted and stored as a logical image. Sometimes, if the edges are not closed due to lack of contrast in the image, the system closes the missing points in order to get a better picture. To store the image in normalized form, the normalization conditions are as follow:

- The object placed right in the middle on the conveyor belt.
- Background of the conveyor belt is black.
- Lighting for each position tried to be kept the same.
- Distance between the camera and objects is 60 cm.
- Webcam resolution used is 100 x 100 pixels.

Matching process is undertaken by comparing the value of the object with the values in the database to find the closest values

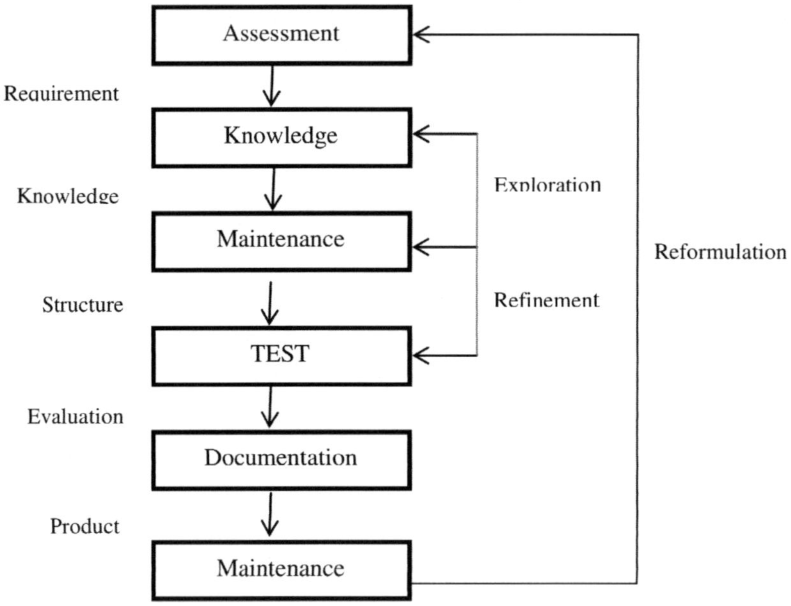


**Figure. 2** Stages of Development of Expert System (Durkin, 1994, p. 40)

## 4. Development of The Automatic Sorting System

The sorting system configuration has two main modules, namely, identification module and the sorting module. In this section, the two modules are presented and described by giving details on the related theory and functional methods.

The procedure ("Pattern recognition") for the identification of a beverage cans consists of a series of computational steps which are conventionally identified as:

- Image acquisition

- Image pre-processing
- Feature extraction
- Pattern recognition

Each of these steps is described in the following; an example of pattern recognition of an actual beverage cans is presented. This procedure is meant to operate in real time while cans and bottle samples are travelling on the conveyor belt; thus, all the stages have been featured to be as fast as possible and other reliable but slower techniques (one example is the Fourier analysis of the image) are not included.

Once an object on the conveyor is detected by a webcam, the webcam automatically capture a picture and send it to the PC. The procedure is:

- Connect webcam to the computer through the USB.
- Show window to view Image at runtime
- This will open a small window Webcam interface
- Automatic Capture image
- Capture the image of that direct into the variable data.

Some factors, such as lighting conditions, speed of the conveyor belt, camera gain, contrast and brightness, height of the camera above the object, may strongly influence the quality and sharpness of the final image; experiments are carried on in order to find an optimum configuration. The image of a cans or bottle on the conveyor belt is shown with a resolution of 320*240 pixels and low lighting conditions. The white circular objects appearing at the top and bottom of the image are the supports of auxiliary lights, not used in this image

Image with background noise, due to the irregularities in the image scanning and capturing, should be processed for good performance of the system. Threshold technique is used for background noise elimination

The edge detection is performed on each colour ("layer") R, G and B in order to be independent of the colour and intensity of the background. An edge being in general a sharp change either of colour or intensity or both, also internal edge are found in the first part of this procedure, but subsequently removed. In the beginning, each layer is filtered with an averaging mask in order to avoid the edge detection of single spikes in the brightness. The edges in the filtered layer are detected by means of a modified Sobel mask of size 7x7 which proves to be more reliable than the standard 3x3 mask when the edges are not sharp and the brightness of the layer grows over a distance of many pixels. The edges obtained for each of the 3 layers are combined in order to obtain the final raw edge image.

An object in the binary image is identified by a complete closed edge made of a sufficient number of pixels. For sake of simplicity and speed the minimum number is taken as a fraction of the minimum perimeter among the cans or bottle in the database. However, the analysis of very small edges would just be time consuming without producing false positives in the identification of cans or bottles.

The edges of an object are discarded if they touch the boundary of the image, as the correct feature extraction would be prevented by the lack of a part of the object in the binary image.

In Sobel operator method, a mask is used to compute $G_x$ at the center point of a $3 \times 3$ image region. Whereas, a similar mask is used to compute Gy. These masks are often referred to as Sobel operators. Figure 3 shows a grid of $3 \times 3$ image and its associated $G_x$ and $G_y$ operators.

| | | |
|---|---|---|
| $z_1$ | $z_2$ | $z_3$ |
| $z_4$ | $z_5$ | $z_6$ |
| $z_7$ | $z_8$ | $z_9$ |

**(a)**

| | | |
|---|---|---|
| –1 | –2 | –1 |
| 0 | 0 | 0 |
| 1 | 2 | 1 |

**(b)**

| | | |
|---|---|---|
| –1 | 0 | 1 |
| –2 | 0 | 2 |
| –1 | 0 | 1 |

**(c)**

**Figure. 3**(a) 3 × 3 image region; (b) mask used to compute $G_x$ at centre point of the 3 × 3 region; (c) mask used to compute $G_y$ at that point. These masks are often referred to as the Sobel operators.

In the experiment, it is observed that the performance of the vision system is extremely influenced by the lighting arrangement. For calibration and adjusting the lighting, three different lighting techniques namely front lighting-directional-bright field illumination, front lighting-directional- dark field illumination, and diffuse front lighting are considered in this study. In both front lighting-directional-bright field illumination and diffuse front lighting, the images from the inspection zone show some reflection problems such that the beverage cans object on the conveyor belt seems to be. Moreover, the reflection from the surface of the object is not uniform. It is important that the texture information of the objects is analysed. Even one object of the same colour in whole body showed different colour combination in histogram analysis of the segmented portions of the image due to non-uniform lighting. In front lighting-directional-dark field illumination, image from inspection zone is distinctive for texture analysis and the object surface is illuminated uniformly. Moreover, front lighting-directional-dark field illumination is widely used in surface scratches or texture analysis (Pham D.T., & Alcock, R. J., 2003; Burke M.W., 1996), thus, this illumination technique is adopted for this experiment.

Since we have proposed a method of cans and bottle identifications system there is no standard data bank on which we can verify our method. So, the first task was to create a database of the object. We have collected 5 photographs for each of 60 cans and 40 bottle type. In order to store the feature values, a database with two tables has been created using Excel. One of the tables is used to store raw sample data images and another is used to store the template data.

In this section a relative comparison is made based on the outcomes of our proposed method for cans and bottle with different resolutions. Cans and bottle must be identified considering only one component out of Red, Green and. When we consider other grades of cans and bottle then it is essential to combine three components to make robust decision in cans and bottle type selection.

The correct identification rate is calculated based on the percentage of the number of objects are classified into their respective type. When the window size is 3×3 then achieved classification success rates are 88% for cans and bottle. When the template size is 7 × 7 then the achieved average classification success rate is 91.33%. In order to get the optimum performance in terms of classification success rate and computation time, the suggested window size is not greater than 8×8.

**Table 2** Identification Succsess Rate

| Object | Object | Success Rate (%) |
|---|---|---|
| 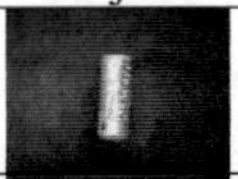 | Cans | 93 |
| 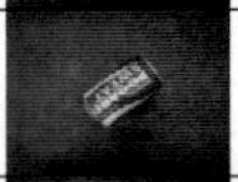 | Cans | 80 |
| 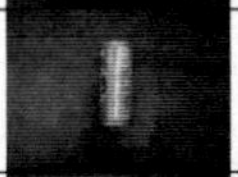 | Cans | 100 |
|  | Cans | 100 |
|  | Cans | 86 |
|  | Bottle | 100 |
|  | Bottle | 100 |
|  | Bottle | 86 |
| 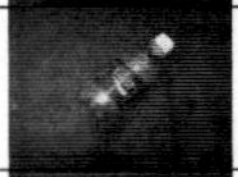 | Bottle | 86 |
|  | Bottle | 80 |

## 5. Conclusion

The major importance of this study is on the development of cans and bottle identification method for automated sorting system. The identification system performance for correct cans identification is 91.33% with estimated throughput of 21,600 things per hour. The success of the identification depends on the size and type of the objects. The most important point addressed in this work is that the method, which uses computer vision, can be implemented easily to sort multiple types of solid waste, (cans and bottles). Further work should focus on to other solid waste sorting, such as paper, plastic, and glass. Finally, identification system with edge detection method is recommended to use for automated sorting system of the solid waste.

## 6. References

[1] J. Petek and P. Glavic, An integral approach to waste minimization in process industries, *Resource, Conservation and Recycling,* Elesevier, 17: 169–188, 1996.

[2] R. K. Pati, P. Vrat, and P. Kumar, Economic analysis of paper recycling visa–vis wood as raw material. Int J Prod Econ; 103:489–508, 2006.

[3] S. Faibish, H. Bacakoglu, and A. A. Goldenberg, An Eye-Hand System for Automated Paper Recycling, in *Proceedings of the IEEE International Conference on Robotics and Automation*, Albuquerque, New Mexico, pp 9-14, 1997.

[4] M. K. Ramasubramanian, R. A.V enditti, C. M. Ammineni and M. Mallapragada, Optical Sensor for Noncontact Measurement of Lignin Content in High-Speed Moving Paper Surfaces, *IEEE Sensors Journal,* 5(5), pp. 1132-1139, 2005.

[5] F. A. Hottenstein, G. R. Kenny, T. Friberg, and M. Jackson, High-Speed automated optical sorting of recovered paper, in *Proc. TAPPI Recycling Symposium*, Vol. 1, Atlanta, GA, pp. 149–158, 2000.

[6] R. A. Venditti, M. K. Ramasubramanian, and C. K. Kalyan, A Noncontact Sensor for the Identification of Paper and Board Samples on a High Speed Sorting Conveyor, *Appita Journal: Journal of the Technical Association of the Australian and New Zealand Pulp and Paper Industry*, 60(5), pp. 366-371, 2007.

[7] Gschweitl, and K. Heinz, Method for sorting waste paper, *European Patent*, EP0873797, (1998).

[8] C. M. Chandini, Design of lignin sensor for identification of waste paper grades for an automatic waste paper sorting system, *Master's thesis, North Carolina State University*, Raleigh, NC, USA, 2001.

[9] Z. Khalfan, and S. Greenspan, Optical Paper Sorting Method Device and Apparatus, *US Patent* No. 2006/0124511, (2006).

[10] A. G. Doak, M. G. Roe, and G. R. Kenny, Multi-Grade Object sorting system and method, *US Patent* No. US2007/0002326, (2007).

[11] K. K. Chakravarthi, Development of online stiffness sensor for high speed sorting of recovered paper, *Master's thesis, North Carolina State University*, Raleigh, NC, USA, 2006

[12] M. K. Ramasubramanian, R. A. Venditti, and P. K. Gillella, Sensor systems for high speed intelligent sorting of waste paper in recycling, available at: [http://www.osti.gov/bridge/servlets/purl/919471-VGwxA0/919471.PDF] (Accessed 20 November 2008).

Journal of Physics: Conference Series **622** (2015) 012054 doi:10.1088/1742-6596/622/1/012054

# Characterizations of Bio-char from Fast Pyrolysis of Meranti Wood Sawdust

**M A F Mazlan[1], Y Uemura[1], N B Osman[1], S Yusup[2]**

[1] Centre for Biofuel and Biochemical Research, Universiti Teknologi PETRONAS 31750 Tronoh, Perak, Malaysia

[2] Biomass Processing Lab, Universiti Teknologi PETRONAS, 31750 Tronoh, Perak, Malaysia

E-mail: amir11@hotmail.com

**Abstract.** In this research, Meranti wood sawdust (MWS) was pyrolyzed in a fixed bed drop-type pyrolyzer under an inert condition. The first part of the study is to determine the influence of pyrolysis temperature (450, 500 and 550 °C) on the yield of pyrolysis products. Pyrolysis of the waste MWS material generated the highest amount of bio-char with approximately 38 wt.% at pyrolysis temperature of 450 °C. Next, the char product (from pyrolysis at 450 °C) was analyzed to compare its characteristics with the raw MWS feedstock. The major component of the char is carbon element, significantly contributed to its high calorific value. TGA profile shows the MWS char could withstand high temperature of up to 400 °C. Under extensive heating, particle size of the bio-char from SEM images decreased due to breakage and shrinkage processes.

## 1. Introduction

Depleting fossil fuel resources calls for alternative energy assets. Economic and environmental concerns are the main driving forces for exploiting new sustainable sources of fuel. Biomass is one of the key potential energy resources in Malaysia to overcome increasing energy needs while being environmentally friendly. Malaysia produces many agricultural, plantation, and forestry residues which could be converted into valuable fuels and chemical products. These waste materials have no or little economic values and often present as a disposal problem. In 2010, Malaysia's annual log production was approximately 24 million $m^3$ [1]. Multiplying by the ratio of wood waste of 0.78, about 18 million $m^3$/year of wood waste was generated [2]. Among hardwood waste, Meranti wood sawdust (MWS) is a significant residue from logging and deforestation in Malaysia [3]. Meranti is a local and common name or otherwise scientifically known as S*horea sp.* [3].

Thus, there is a need to convert these by-products into useful and profitable bioenergy and biomaterials products. One of the promising techniques to accomplish that is through a fast pyrolysis process. Fast pyrolysis is the heating of an organic material, for instance wood, at a high temperature (450 to 550 °C) and short retention time (a few seconds of vapor product), in the absence of oxygen or air. So, the material does not combust as no oxygen is present. However, the three main biomass pseudo-components of cellulose, hemi-cellulose, and lignin that form the material are thermo-chemically decomposed into charcoal and volatiles. Some of these volatiles could be condensed into liquid, called bio-oil. All that is left is the carbon latticework of the charred material, called bio-char [4,5]. So, pyrolysis of lignocellulosic biomass yields three products: one liquid, bio-oil; one solid, bio-char; and one gaseous, non-condensable gas. Therefore, pyrolysis of waste materials into pyrolysis products could add a substantial economic value and promote a greener environment [5,6].

Pyrolysis of different biomasses produces different yields and characteristics of products. Meranti wood has been chosen for this study because for the meantime, pyrolysis of Meranti wood research has not been reported in any literature. The aims of this study are to determine the effect of pyrolysis temperature on the products yield and to analyze the MWS feedstock and its bio-char product.

## 2. Materials and methods

### *2.1. Preparation of raw materials*

Meranti wood sawdust was collected from sawmills in Perak, Malaysia. The biomass was dried in a drying oven (Memmert, 100-800 model) for 24 hours at 105 °C to minimize the free moisture that could affect the performance of pyrolysis reaction [4]. Then, the biomass was sieved to the size range of 0.15-0.5 mm. By referring on the author's previous research review, smaller particle size favours higher rate of fast pyrolysis reaction [6]. However, very small particles are difficult to handle. The raw feedstock is stored in an air-tight container.

### *2.2. Experimental setup*

Figure 1 shows the schematic diagram of a fixed bed drop-type pyrolyzer used to carry out fast pyrolysis of the feedstock for the present study. The experimental setup was connected to vacuum and nitrogen gas lines used to create an inert condition. A biomass holder zone was created by closing two ball valves. The design would allow the biomass to drop inside the heated zone only after the desired temperature reached up. The pyrolyzer consisted of a cylindrical reactor made of stainless steel with 53 mm internal diameter and 155 mm height. The reactor was installed inside an electrical heater and was insulated to minimize heat loss. A thermocouple was inserted inside the reactor to record the pyrolysis temperature. The reactor was connected to two condensers to improve the condensation efficiency and immersed in a cooled ethylene glycol bath to condense the pyrolysis vapors into bio-oil at below -10 °C. The outlet of the vapor-trap was connected with a gas sampling bag to collect the non-condensable gases (NCG). The bio-char was accumulated in the reactor.

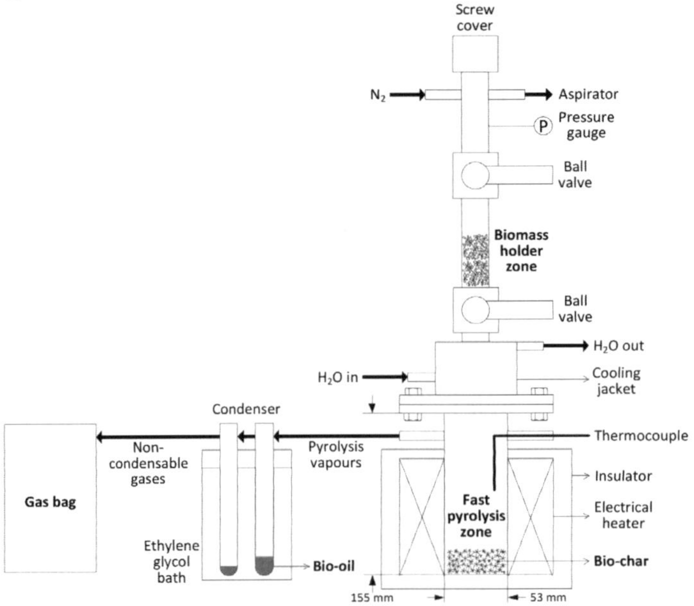


**Figure 1.** Schematic diagram of the fixed bed drop-type pyrolyzer [7]

### *2.3. Pyrolysis procedure*

In all experiments, 10 g of feedstock was put into the reactor. The first part of the study is to determine the impact of pyrolysis temperature on the distribution of pyrolysis products. The biomass was dropped to the fast pyrolysis zone under several temperatures (450, 500 and 550 °C). Duration of the experiment was held at the selected temperature for 10 minutes. The pyrolysis reaction produced bio-char, bio-oil, and non-condensable gases. The char and bio-oil yields were determined based on the weight changes of the reactor and condensers, respectively, before and after the experiment. The mass of the char and bio-oil are given within the experimental error of < ±3 wt.%, as shown in Figure 2. The yield of non-condensable gases was calculated using a mass balance equation [7]:

$$\text{Bio-char (wt.\%)} + \text{bio-oil (wt.\%)} + \text{NCG (wt.\%)} = 100 \text{ (wt.\%)} \qquad (1)$$

### *2.4. Analysis of feedstock and pyrolysis product*

The second part involves the analyses of raw MWS and MWS char product from the pyrolysis reaction at 450 °C. The proximate analysis of moisture, ash, and volatile contents was performed according to ASTM E871-82, D1102-84, and E872-82 standards, respectively [8,9,10]. Fixed carbon content was determined by the difference of the three analyses. The elemental analysis was carried out using a Perkin Elmer 2400 CHN Analyzer. The O content was determined by the difference of the C, H, and N. The calorific value was measured by an IKA-WERKE C2000 bomb calorimeter. All the aforementioned analyses were repeated at least twice to verify the reproducibility and then averaged. Thermogravimetric analysis (TGA) was performed using a TA Instruments model QA50 with 60 mL/min of $N_2$ flowrate at a heating rate of 10 °C/min. Morphology analysis was done via a TM3030 Scanning Electron Microscopy (SEM).

## 3. Results and discussion

### *3.1. Impact of pyrolysis temperature on products yield*

After each pyrolysis experiment, the products namely bio-char, bio-oil, and non-condensable gases were obtained. Calculation of pyrolysis products yield was based on mass percentage. It is well known that the key parameter among operational factors of pyrolysis reaction is the pyrolysis temperature [4,5,11]. For the present study, pyrolysis experiments of MWS were performed at various temperatures (450, 500 and 550 °C). The effect of pyrolysis temperature on products yield is illustrated in Figure 2. Pyrolysis of MWS generated the highest amount of bio-char with 38.7 wt.% at pyrolysis temperature of 450 °C. As pyrolysis temperature increased from 450 to 550 °C, the bio-oil and non-condensable gases yields increased. As contrast, the yield of char decreased because of the continuous volatilization reaction became more preferred. Similar pattern of pyrolysis products distributions were observed by other researchers [11,12,13].

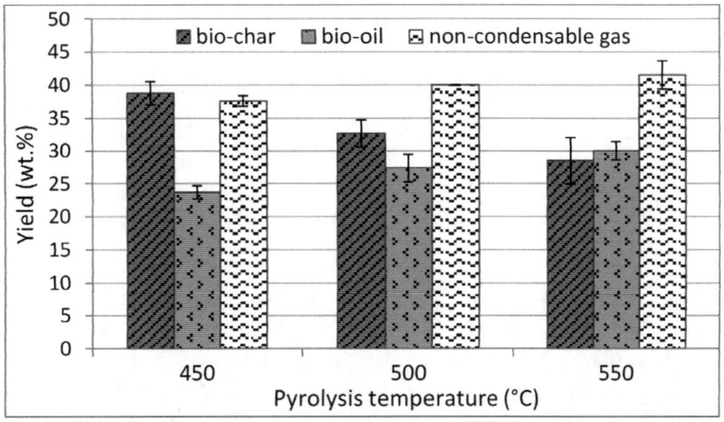


**Figure 2.** Pyrolysis of MWS

The experimental setup of this study is a complete closed system. It is possible, therefore that a high percentage of product recovery is anticipated. Based on the carbon balance calculation between input (feedstock) and output (pyrolysis products), the recoveries for the pyrolysis of MWS in this research was 98.6 %, by using the equation [12]:

$$\text{Product recovery} = (m_{char} + m_{oil} + m_{gas}) / m_{feedstock} \quad (2)$$

*3.2. Characterization of raw MWS and MWS bio-char*

The main characteristics of the feedstock and char product samples are summarized in Table 1. As expected from the proximate analysis, moisture content of the bio-char has decreased when compared to the initial feedstock due to a dehydration reaction from the heating process. In addition, the moisture content for the raw MWS feedstock is low when comparing to other lignocellulosic biomass from colder countries, mostly contributed by the hot climate in Malaysia [11,13,14]. For ash content, it is easy to assume that the ash content is higher after the pyrolysis process. However, this data must be interpreted with caution because after multiplying the percentage of ash in char with the yield of char, the calculated ash content of the char remained constant. During pyrolysis process, some of volatile matter decomposed into vapors, hence showing a reduced amount of volatile matter in the char. In term of fixed carbon, more than half of the MWS char contained fixed carbon substance. This finding, while preliminary, suggests that the MWS char could withstand higher temperature.

**Table 1.** Proximate and ultimate analysis

| Properties | Raw MWS | MWS char |
|---|---|---|
| Proximate analysis (wt.%) | | |
| moisture | 6.2 ± 0.3 | 2.9 ± 0.2 |
| ash | 1.2 ± 0.1 | 3.0 ± 0.2 |
| volatile matter | 69.9 ± 1.7 | 42.5 ± 1.0 |
| fixed carbon | 22.7 | 51.6 |
| Elemental analysis (wt.%) | | |
| C | 41.7 ± 0.1 | 84.9 ± 0.3 |
| H | 5.7 ± 0.1 | 2.3 ± 0.1 |
| N | 0.1 ± 0.1 | 0.4 ± 0.1 |
| O | 52.6 | 12.4 |
| Calorific value (kJ/g) | 19.6 ± 0.8 | 28.5 ± 1.2 |

The major component of the char was carbon element, which had a composition of approximately double than the original feedstock. The calorific value of the char was significantly higher than the feedstock. These results might explain a good correlation between the carbon content and calorific value. There are, however, other possible factors such as the structural or chemical bond of the char that could affect the calorific value [4,11,13]. Some of the suggested applications for bio-char are as carbon sequestration agents, soil improvers, activated carbons or catalysts [13,14,15]. On the other hands, it is apparent that the calorific value of bio-char is very high, indicating a potential usage as a solid fuel. The calculated energy consumption of the experiment was approximately 7200 J.

Figures 3 and 4 display the thermogravimetric, TG (in wt.%) and differential thermogravimetric, DTG (in wt.%/°C) curves for the raw MWS and MWS bio-char. High rate of decomposition for raw MWS started approximately at 250 °C whereas for MWS char took place after 400 °C. The highest rate of decomposition for MWS char happened at approximately 650 °C. These findings further corroborate the previous statement that the MWS char could withstand to a higher temperature. It has been established that pyrolysis of lignocellulosic biomass could be divided into four stages which are dehydration (removed at temperature above 105 °C), followed by decomposition of hemicellulose which mainly occurs between 210 to 310 °C. It decomposes easier than cellulose or lignin as it comprises of amorphous structure and rich in branches. Next, cellulose is broken down at around 300 to 400 °C. Thermal stability of cellulose is higher than hemicellulose as it is made up of a long polymer of glucose without branches. The activities of chemical bonds in lignin cover a widespread range which leads to the decomposition of lignin occurring in a broader range of temperature between 150 and 900 °C [4,16,17]. Based on the guidelines and TGA data, the lignocellulosic compound of a biomass could be estimated. Hence, Figure 3 shows the estimated lignocellulosic composition of the raw MWS.

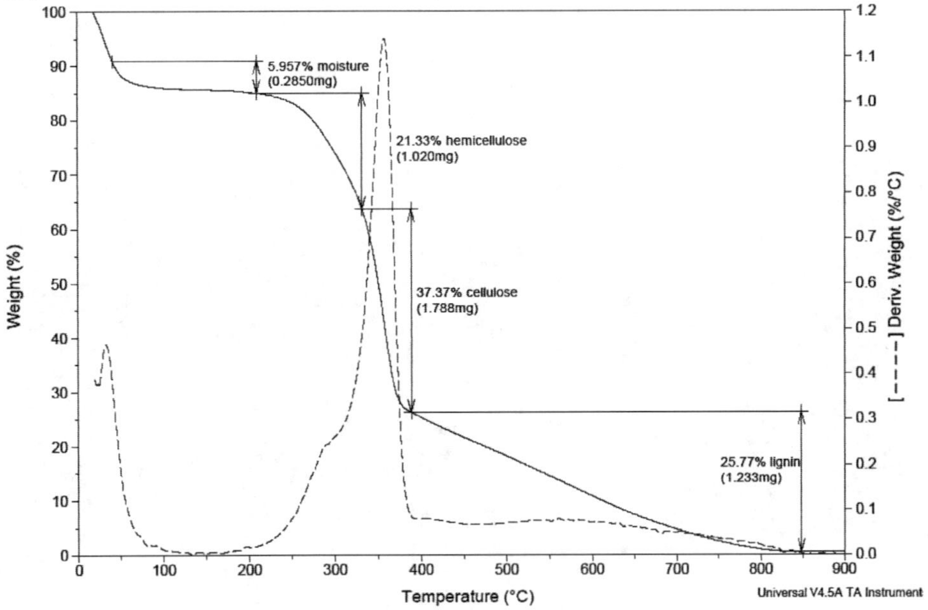


**Figure 3.** TGA and DTG of raw MWS

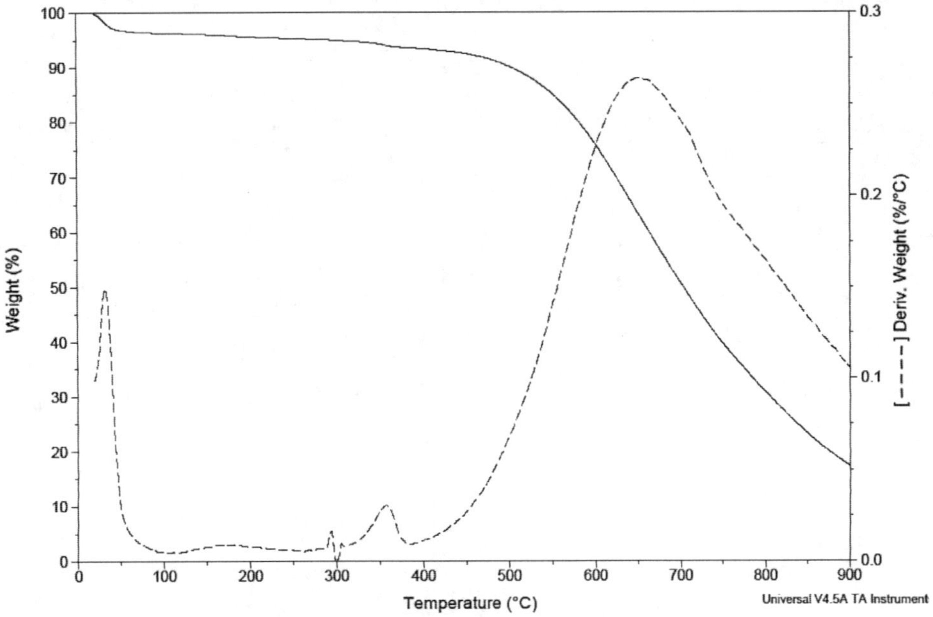


**Figure 4.** TGA and DTG of MWS char

Figure 5 and 6 exhibit the morphology of the raw MWS and MWS bio-char. In this analysis, five measurements from five particle samples were performed. Focusing closely through the SEM instrument at 180x magnification, the particle size of the char decreased when compared to the raw feedstock, as shown in Figure 5. Further magnifying to microporous level, as shown in Figure 6, the structures of the feedstock undergo breaking and shrinking mechanisms from extensive heating, to form charred materials. The average vessel sizes decreased from 0.018-0.027 mm (MWS raw feedstock) to 0.010-0.013 mm (MWS char). These observations were consistent with the report by Alvarez et al. [14]. It is projected that the degree of structure breakage will increase with the higher pyrolysis temperature.

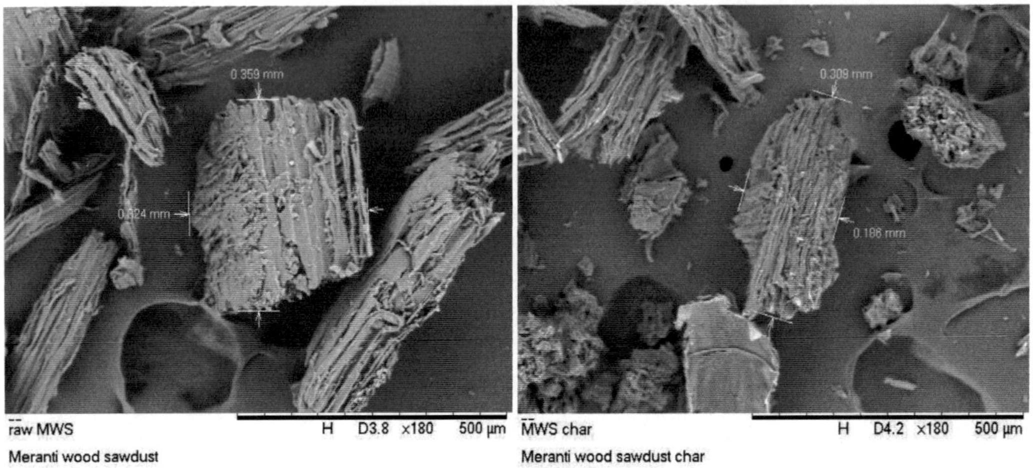


**Figure 5.** Comparison of raw MWS with MWS char (magnification: 180X)

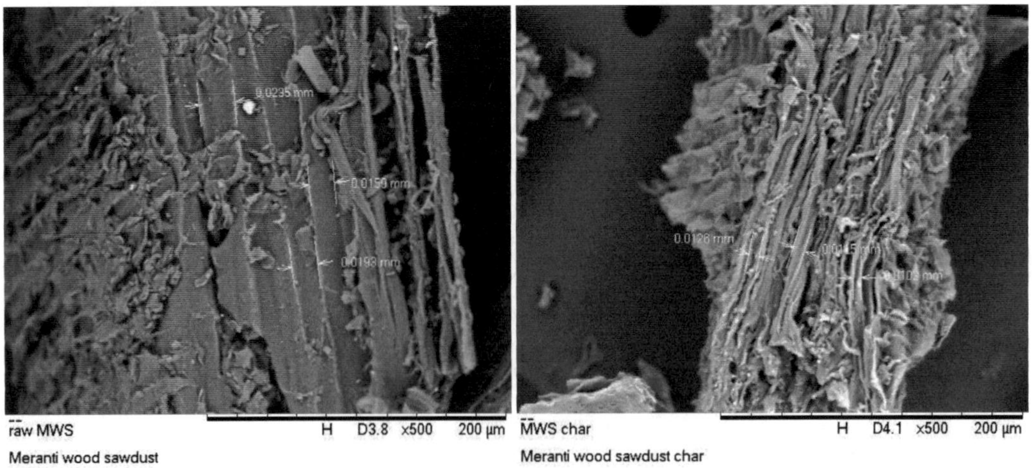


**Figure 6.** Comparison of raw MWS with MWS char (magnification: 500X)

## 4. Conclusions

The effect of pyrolysis temperature on the products yield and the characterization of the bio-char were discussed. Pyrolysis of MWS generated the highest amount of bio-char with 38.7 wt.% at 450 °C. Major component of the MWS char is carbon element, with 84.9 wt.% and its calorific value is high, which is 28.5 kJ/g. The particle size of the char decreased, due to the breakage of the structure from high heating of pyrolysis reaction. High decomposition rate for MWS char started approximately at 400 °C, indicating the char could withstand to a high temperature. A further study, which takes more varieties of wood feedstock into account, needs to be undertaken before the association between pyrolysis of different woody types and its pyrolysis products yield and characterizations are more clearly understood.

**Acknowledgment**
This research was carried out with the financial support of the MOE LRGS (Future Biorefineries: Unlocking the Potential of Fine Chemicals and Value-Added Byproducts) and the Mitsubishi Corporation Education Trust Fund.

**References**
[1] Malaysian Timber Statistics 2009-2011, Malaysian Timber Industry Board (2012).
[2] The Asian Biomass Handbook, The Japan Institute of Energy (2008).
[3] N.B. Osman, H.T. Othman, R.A. Karim, M.A.F. Mazlan, Biomass In Malaysia: Forestry-based Residues, *International Journal of Biomass and Renewables* 3 (2014) 7-14.
[4] D. Mohan, C.U. Pittman, P.H. Steele, Pyrolysis of wood/biomass for bio-oil: A critical review, *Energy Fuels* 20 (2006) 848-890.
[5] S. Yaman, Pyrolysis of biomass to produce fuels and chemical feedstocks, *Energy Conversion and Management* 45 (2006) 651-671.
[6] M.A.F. Mazlan, Y. Uemura, N. Osman, S. Yusup, Review on Pyrolysis of Hardwood Residue to Biofuel, *Applied Mechanics and Materials* 625 (2014) 714-717.
[7] S.R. Naqvi, Y. Uemura, S. Yusup, Catalytic pyrolysis of paddy husk in a drop type pyrolyzer for bio-oil production: The role of temperature and catalyst, *Journal of Analytical and Applied Pyrolysis* 106 (2014) 57-62.
[8] ASTM, E871-82, Standard test method for moisture analysis of particulate wood fuels, Annual Book of ASTM Standard, American Society for Testing and Materials (2013).
[9] ASTM, E871-82, Standard test method for ash in wood, Annual Book of ASTM Standard, American Society for Testing and Materials (2013).
[10] ASTM, E872-82, Standard test method for volatile matter in the analysis of particulate wood fuels, Annual Book of ASTM Standard, American Society for Testing and Materials (2013).
[11] L. Burhenne, M. Damiani, T. Aicher, Effect of feedstock water content and pyrolysis temperature on the structure and reactivity of spruce wood char produced in fixed bed pyrolysis, *Fuel* 107 (2013) 836–847.
[12] N. Ali, Y. Uemura, H. Afif, N. Osman, W.N. Omar, T. Tsutsui, Effect of operating conditions and fractional condensation on pyrolytic products, *Journal of the Japan Institute of Energy* 92 (2013) 1014-1020.
[13] F.A. López, T.A. Centeno, I. García-Díaz, F.J. Alguacil, Textural and fuel characteristics of the chars produced by the pyrolysis of waste wood, and the properties of activated carbons prepared from them, *Journal of Analytical and Applied Pyrolysis* 104 (2013) 551–558.
[14] J. Alvarez, M. Amutio, G. Lopez, J. Nunes, M. Olazar, J. Bilbao, Upgrading the rice husk char obtained by flash pyrolysis for the production of amorphous silica and high quality activated carbon, *Bioresource Technology* 170 (2014) 132–137.
[15] P. Basu, Biomass gasification, pyrolysis and torrefaction: Practical design and theory, 2nd ed., Academic Press (2013).
[16] H. Yang, R. Yan, H. Chen, C. Zheng, D.H. Lee, Characteristics of hemicellulose, cellulose and lignin pyrolysis, *Fuel* 86 (2007) 1781-1788.
[17] A.V. Bridgwater, Review of fast pyrolysis of biomass and product upgrading, *Biomass Bioenergy* 38 (2012) 68-94.

# Broadening of the Spectral Atomic Lines Analysis in High Density Argon Corona Plasma by Using Voigt Profile

**M.Nur[1], N. Bonifaci[2], A. Denat[2] and V.M. Atrazhev[3]**

[1]Department of Physics, Faculty of Science and Mathematics, Diponegoro University, Semarang, Indonesia
[2]G2E.Laboratory, CNRS and Joseph Fourier University 25 rue des Martyrs, 38042 Grenoble, France
[3]Joint Institute for High Temperatures (JIHT) and Russian Academy of Sciences, Izhorskaya St. 13, 2125412, Moscow, Russia
E-mail: m.nur@undip.ac.id

**Abstract.** Studies of spectrum emission from high density argon plasma corona has been done. The analysis of the boardening of spectral atomic lines of Ar-I profile has been curried out by using an empirical approximation based on a Voigt profile. Full-width at half-maximum (FWHM) of the spectral-lines of 763.5 nm has been determined from atmospheric pressure until liquid state. The study liquid argon was curried out in a variation of temperature from 87.5 K to 151.2 K and hydrostatics pressure from 2.1 MPa to 6.4 MPa. These pressure gives the densities $N_\infty$ (i.e. density very far from ionization zone) a variation from 1.08 $10^{22}$ to 2.11 $10^{22}$ $cm^{-3}$. FWHM of Voigt approximation ($W_v$) of the line 763,5 nm of 'Ar I for: the emission lamp very low pressure ($W_V$ = 0,160 nm) and our corona discharge at a pressure of 3,5 MPa ( $W_V$ = 0,67 nm) and at a pressure of 9,5 MPa ($W_V$ = 1,16 nm). In gas, corona plasma has been generated from 0.1 MPa to 9.5 MPa. We found that the broadening spectral line increase by increasing densities both for . the spectral-lines of 763.5 nm and 696.5 nm. We concluded that broadening of spectrum cause of *Van der Waals* force.

## 1. Introduction

The applications of corona plasma technology have increased in different scientific and technological fields (gas purification, detoxification, medical, chemical analysis, ozone production, etc.) [1]. Nur et. al. detected ions wind phenomenon and measured mobilities of charges carrier in argon corona discharge from atmospheric pressure until liquid state [2]. In liquid and gaseous cryogenic helium experimental data on spectral shape of the line 706 nm emitted by corona discharge are published by Bonifaci et. al. [3]. The convolution of a Lorentz and a Gauss function, commonly known as the Voigt function, is important in many branches of physics, e.g., atomic and molecular spectroscopy, atmospheric radiated transfer, plasma physics, astrophysics, etc.[1] Plasma parameters can be studied by using atomic and moleclar emission spectroscopy. In case of plasma corona Argon high density broadening of emission spectrum line are due to the Lorentzian (due to the pressure broadening) distributions, and the Gaussian (due to the Doppler broadening) [4]. In this case, its is necessery to lead to the Voigt profile function, which turns out to be the convolution of Lorentzian and

Journal of Physics: Conference Series **622** (2015) 012055 doi:10.1088/1742-6596/622/1/012055

Gaussian distribustions. The van der Waals broadening is produced by the dipolar interaction between an excited atom (the emitter) and the dipole induced by it over a neutral perturbing atom in the ground state. Christova et al [5] analyzed of the profiles of the argon 696.5 nm spectral line excited in non-stationary wave-guided discharges. This group used 696.5 nm line to "measure" discharge themperature

*Voigt Profile*

In general boradening line of atomic emission or ratational molecular emission can be identified by Full Width Half Maximum (FWHM). In 1968, Whiting has proposed an empirical analytical approximation Voigt profile until the second order [6]. This approximation has been used by Aeschliman et al [6], it gives satisfactory results and precise. It is well known that the voigt profile, which corresponds to the composition of two types of Lorentzian and Gaussian enlargement, would provide a distribution of intensity, observed experimentally, an atomic line or rotational line molecular. This profile is written as follows [7]:

$$I_\lambda = \frac{2}{\pi}\frac{\lambda_0}{W_l} I^g{}_{\lambda_0} \int_{-\infty}^{\infty} \frac{\exp\left\{-\frac{2,772\lambda_0^2}{W_g^2}\left(\frac{v}{c}\right)^2\right\} d\left(\frac{v}{c}\right)}{1+\frac{4}{W_l^2}\left\{(\lambda-\lambda_0)-\lambda_0\left(\frac{v}{c}\right)\right\}^2} \quad (1)$$

Or most of the symbols are defined in Figure 1 $I^g{}_{\lambda_0}$ is a fictitious maximum intensity of a line corresponding to a pure thermic enlargement, c is the velocity of light, v is the velocity integrated over all molecules that contribute to intensity of $\lambda$.

The Voigt profile consists of two types of profile, given by the equations:

Gaussian profile

$$I_\lambda = I_{\lambda_0} \exp\left[-2,772\frac{(\lambda-\lambda_0)}{W_g}\right] \quad (2)$$

The Lorentzian-type profile

$$I_\lambda = I_{\lambda_0} \frac{1}{1+4\left[\frac{(\lambda-\lambda_0)}{W_l}\right]^2} \quad (3)$$

In 1968, Whiting [6] has proposed an empirical analytical approximation Voigt profile until the second order. This approximation has been used by Aeschliman et al [7], it gives satisfactory results and precise. The empirical formula is written for voigt profile after Whiting [6]:

$$\frac{I_\lambda}{I_{\lambda_0}} = \left(1-\frac{W_1}{W_v}\right)\exp\left[-2{,}772\frac{(\lambda-\lambda_0)}{W_V}\right]+\left(\frac{W_1}{W_v}\right)\frac{1}{1+4\left[\frac{(\lambda-\lambda_0)}{W_V}\right]^2}$$

$$+0{,}016\left(1-\frac{W_1}{W_v}\right)\left(\frac{W_1}{W_v}\right)\left\{\exp\left[-0{,}4\left(\frac{\lambda-\lambda_0}{W_V}\right)^{2{,}25}\right]-\frac{10}{10+\left[\frac{(\lambda-\lambda_0)}{W_V}\right]^{2{,}25}}\right\} \quad (4)$$

$\frac{I_{\lambda_0}}{I_\lambda}$ Is the normalized intensity of line $\left(\frac{\lambda_0-\lambda}{W_v}\right)$ is the wavelength normalized and $W_g$, $W_l$, $W_v$ are respectively the widths at mid-height of Gauss, Lorentz and Voigt.

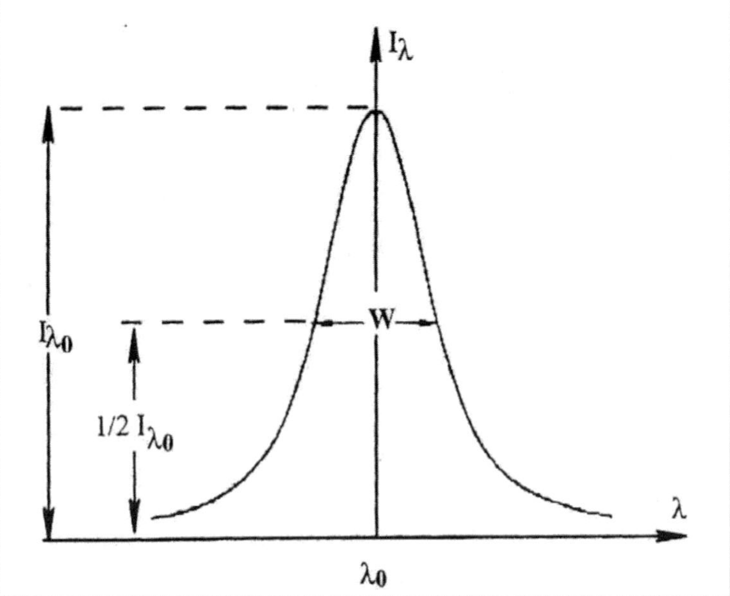


**Figure 1.** Diagram of profile for Voight. W=Wg for enlargement natural pressure and stark effect

In this paper, the broadening of the spectrum will be discussed of the corona plasma emission wavelength of argon for 763.5 . nm and 696.5 nm. Gas pressure ranging from 1 bar to 100 bars. We present also the broadening of argon micro-plasma spectrum that was formed in the argon liquid. Measurements of the line profiles atomic transitions were analyzed assuming the combined effects of Lorentzian (pressure) and Gaussian (kinetic) contributions which can be most conveniently superimposed using the Voigt profile.

## 2. Method

Experimental system for this research is shown in Figure 1. Spectroscopy equipment consisted of the lens Spectrosil B which serves to focus the light emission from the plasma into the slit (25 mm) of a spectrograph with medium resolution HRS Jobin-Yvon. The corona discharge reactor can be moved towards the vertical and horizontal position. By this movement, the slit can be placed at specific locations of the emissions. Spectrograph with a focal length of 600 mm equipped with a number of gratings 1200 grooves per mm, and connected with a photodiode detector or CCD (model LN/CCD-512 SF & SB, Princeton Instruments, Inc.). The detector is connected to an EG & G Optical Multichannel Analyzer (OMA) which has a spectral range between 200-850 nm (model 1460 EG & G Princeton Applied Research). To reduce the parasitic beam, detector was cooled at a temperature of -40°C.

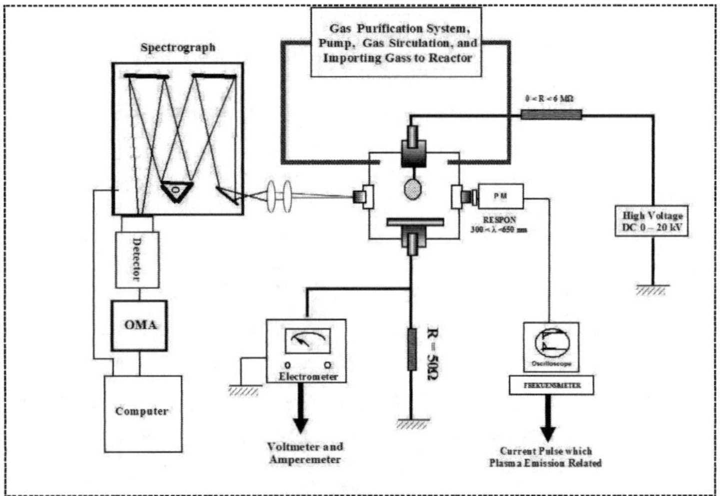


**Figure 2.** Scheme series of experiments

Simultaneously performed well gas discharge current measurements in the reactor with an electrometer Keithley (model 610C), and voltage can be read directly from a DC generator RHSR/20PN60 Spellman. In addition to the above, the light emission was also detected through the help photo-multiplicator (Dario 56AVP model) which has a spectral response between 300 and 650 nm. Current pulse was detected with a Tektronix oscilloscope (model 7633)

## 3. Results and Discussion

### *3.1. Characterization of Current-Voltage and Ozone concentration*

The analysis of the atomic lines of Ar-I profile has been done by using an empirical approximation based on a Voigt profile, proposed by Whiting [5]. This approximation is given by equation (4). Full width half maximum of the spectral-lines has been determined from atmospheric pressure until liquid state. The study argon liquid was curried out in a variation of temperature from 87.5 K to 151.2 K and hydrostatics pressure from 2.1 MPa to 6.4 MPa. These pressure gives the density $N_\infty$ (i.e. density very far from ionization zone) a variation from 1.08 $10^{22}$ to 2.11 $10^{22}$ $cm^{-3}$. The figure 3 shows the experimental profiles and Voigt approximation of the line 763.5 nm of 'Ar I for : the emission lamp very low pressure ($W_l$ = 0.157 et $W_v$ = 0.160 nm) and our corona discharge at a pressure of 3.5 MPa ($W_l$ = 0.32 et $W_v$ = 0.67 nm) and at a pressure of 9.5 MPa ($W_l$ = 1.06 et $W_v$ = 1.16 nm).

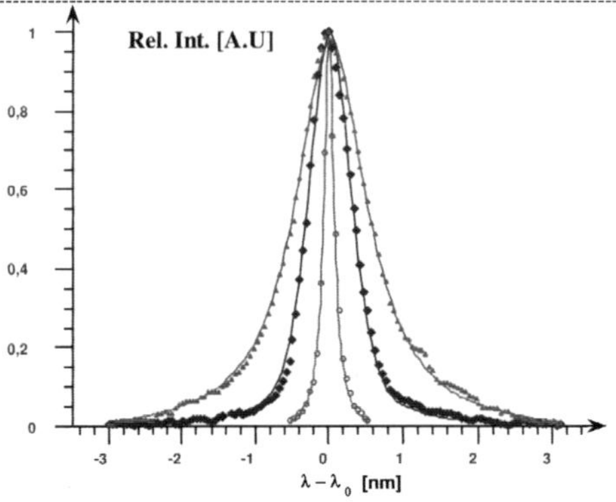


**Figure 3**. Line 763,5 nm of argon. Experimental Points: ○ Radiated by lamp very low pressure, (♦)corona discharge in the pressure of $P_\infty$ = 3,5 MPa et,(▲)in the pressure of $P_\infty$ = 9,5 MPa. Full line shows Voigt profile.

There are different causes of spectral-line broadening. In our case, we suppose that the plasma take in place in the surrounding of the active electrode, dense enough and its temperature is not excessively high. Broadening due to the thermic agitation effect (Doppler effect) can be ignored. It rest now broadening collisions (pressure effect) that are several phenomena: Stark effect, energy transfer by resonance, and, broadening by *Van der Waals* force. According to Griem [8], Stark effect is more important that the level of atom emitter is high. In this study, the emitter levels of the spectral-lines analyzed are not enough high cause its represent of the 4p and 4p' forward 4s and 4s' transitions. In addition, in our case, the current of the discharge is not very high (several 100 μA) and in consequent, the glow discharge can be considered as a plasma with weak ionization degree where the electron density $N_e$ is weak enough. This leads us to consider that the line broadening 4s-4p by the Stark effect is negligible compared to other effects (resonance and/or Van der Waals). Muñoz, et al, [9] used the Van der Waals broadening of spectral atomic lines to measure the gas temperature of an argon–helium microwave plasma at atmospheric pressure. In our case, we work in the high density argon gas and argon liquid, we certain that broadening of spectral lines wrere due to Van der Waals force.

*3.2 Broadening of the line 696.5 nm and line 763.5 nm of Ar-I*

In the figure 4 we plotted the broadening depending on the density of the fluid ($N_\infty$) for line spectrum of 763.5 nm and figure 5 for line spectrum od 696.3 nm. However, the density must be considered is the actual density ($N_p$) of the medium plasma (ie the ionization zone which emits light). Effect in determining the temperature $T_p$ (plasma temperature) which is certainly higher than the temperature $T_\infty$ the middle distance of the tip and the pressure $P_p$ in the plasma produced by a glow discharge is equal to either continue the test pressure $P_\infty$ gas or hydrostatic pressure applied to the liquid. Our work was to estimate the temperature Tp of the bright area to calculate Np. To strengthen our hypothesis, we discussed the work of Aeschliman et al [7]. In this work on the broadening and displacement of argon spectral line showed that the Stark effect is completely negligible. Konjevic (1999) explained that Van der Waals broadening results from the dipole interaction of an excited atom with the induced dipole of a neutral ground-state atom of density $N_\infty$ it is really our condition of experiment in high density gas and liquid state [10]. According to [11] in microplasma van der Waals contributions were the most important effects in the spectral line broadening. The Voigt full-widths at half-maximum (FWHM) normally combination between Gaussian and Lorentzian, and we used Whiting's [6] empirical analytical approximation Voigt profile until the second order (equation 4). We found that this formula gives satisfactory results and precise. An observed line generally has a Voigt profile and its FWHM can be measured directly.

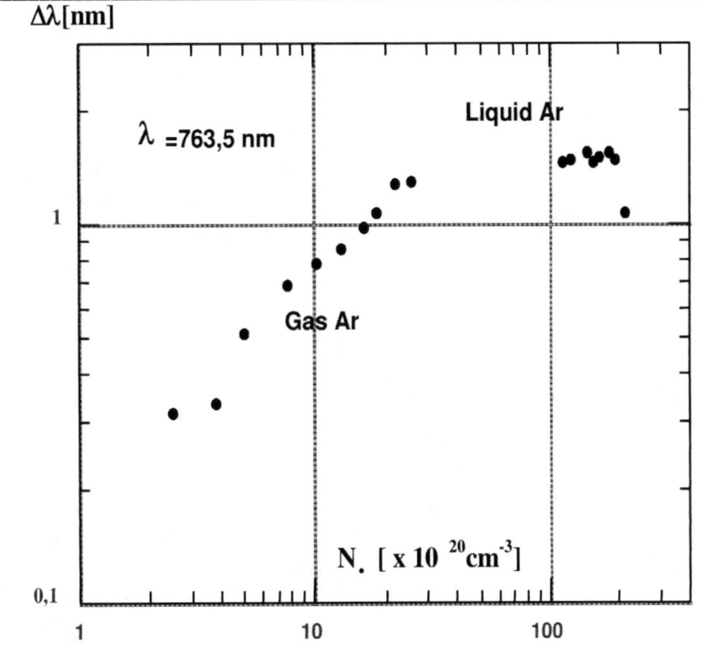


**Figure 4**.Broadening of the line 763,5 nm of Ar-I as function of density $N_\infty$ produced by argon corona plasma in gaseous and liquid Ar.

In liquid argon, there appears a correlation between enlargement and the hydrostatic pressure applied to the liquid. Unfortunately, we have not made sufficient steps by varying the hydrostatic pressure at different temperatures of tests to quantify this result. More systematic studies of this phenomenon should be undertaken. For now, we can only say that the atoms emit light are not in the condensed phase for the following reasons. There are atomic emission lines of Ar-I. Hydrostatic pressure plays an important role on the line broadening that this expansion does not correspond to that seen with trials in the gas at the same pressure $P_\infty$. For example in the liquid, $P_\infty$ = 2.1 MPa and $T_\infty$ = 87 K, the expansion observed for the line at 696.5 nm is equivalent to that measured in the tests in the gas at $P_\infty$ = 89 MPa and $T_\infty$ = 300 K

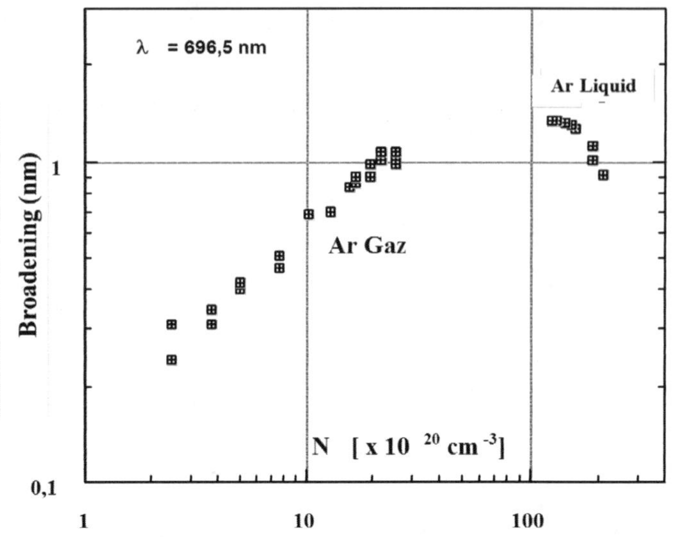


**Figure 5**. Broadening of the line 696.5 nm of Ar-I as function of density N∞ produced by argon corona plasma in gaseous and liquid Ar.

## 4. Conclusion

In this work, we have shown that Voigt approximation can be used for measuring broadening of the spctrum line emissio by plasma corona generated from gas state to liquid state in Agron. We found that the broadening spectral line increase by increasing densities both for . the spectral-lines of 763.5 nm and 696.5 nm. We concluded that broadening of spectrum cause of Van der Waals force. Plasma corona cenbe generated in the liquid state in a variation of temperature from 87.5 K to 151.2 K and hydrostatics pressure from 2.1 MPa to 6.4 MPa or densities $N_\infty$ (i.e. density very far from ionization zone) a variation from 1.08 $10^{22}$ to 2.11 $10^{22}$ $cm^{-3}$. In liquid argon, there appears a correlation between enlargement and the hydrostatic pressure applied to the liquid. Unfortunately, we have not made sufficient steps by varying the hydrostatic pressure at different temperatures of tests to quantify this result. More systematic studies of this phenomenon should be undertaken.

## Acknowledgments

M.Nur thanks to Center for Plasma Research Team of Faculty of Science and Mathematics for some kinds of assistance

## References

[1] Franz Schreier, 2011, Optimized implementations of rational approximations for the Voigt and complex error function, *Journal of Quantitative Spectroscopy & Radiative Transfer* 112 pp 1010–1025
[2] M. Nur,1997 Etude des décharges couronne dans l'argon et l'azpte très purs: transport des charges, spectroscopie et influence de la densité", *PhD Thesis*, Universite Joseph Fourier, Grenoble, France

[3] N Bonifaci, F Aitken, Hai Van Nguyen, V M Atrazhev, K von Haeften and R. Rincon 2012 Shape of atomic lines emitted by cryoplasma in Helium, *Journal of Physics: Conference Series* 397 (2012) 012066

[4] M Nur, N Bonifaci, A Denat 2014 Ionic Wind Phenomenon and Charge Carrier Mobility in Very High Density Argon Corona Discharge Plasma, *Journal of Physics: Conference Series* 495 (495), 012041

[5] Christova, M., Gagov, V.,Koleva, I. 2000 Analysis of the profiles of the argon 696.5 nm spectral line excited in non-stationary wave-guided discharges", *Spectrochim. Acta* Part B 55, 815–822.

[6] Whiting, E.E.,1968,An Empirical Approximation to The Voigt Profile", *J. Quant. Spectros. Radiant. Transfer.* Vol. 8. Pp. 1379 - 1384.

[7] Aeschliman, D.P., Hill, R.A., and Evans, D.L.1976, Collisional Broadening and Shift of Neutral Argon Spectral Lines, *Phys. Rev.* A Vol. 14 No.4.

[8] Griem H. R., 1974 *Spectral line broadening by plasmas*, (Academic Press, New York ; London)

[9]Muñoz, J., Dimitrijevic, M.S, Yubero, C., Calzada, M.D. 2009, Using the van der Waals broadening of spectral atomic lines to measure the gas temperature of an argon–helium microwave plasma at atmospheric pressure, *Spectrochimica Acta*, Part B 64, 167–172

[10] N. Konjevic, 1999, Plasma broadening and shifting of non-hydrogenic spectral lines: present status and applications, *Physics Reports* 316, pp 339- 401

[11] B.N. Sismanoglu, J. Amorim, J.A. Souza-Corrêa, C. Oliveira and M.P. Gomes 2009, Optical emission spectroscopy diagnostics of an atmospheric pressure direct current microplasma jet, *Spectrochimica Acta Part B* 64 pp1287–1293

# Statistical Analysis to Select Evacuation Route

**Z Musyarof, DY Sutarto, DR Atika and RB Fajriya Hakim**

Department of Statistics, Faculty of Mathematics and Natural Science, Islamic University of Indonesia, Jalan Kaliurang KM 14.5, Sleman, Indonesia

E-mail: zakymusyarof@yahoo.com, hakimf@fmipa.uii.ac.id

**Abstract.** Each country should be responsible for the safety of people, especially responsible for the safety of people living in disaster-prone areas. One of those services is provides evacuation route for them. But all this time, the selection of evacuation route is seem does not well organized, it could be seen that when a disaster happen, there will be many accumulation of people on the steps of evacuation route. That condition is dangerous to people because hampers evacuation process. By some methods in Statistical analysis, author tries to give a suggestion how to prepare evacuation route which is organized and based on people habit. Those methods are association rules, sequential pattern mining, hierarchical cluster analysis and fuzzy logic.

## 1. Introduction

We have anxiety panic attacks because we are creatures of habit [10]. We can say that habit is like a routine work. Habit makes us behave in a similar that habit when we were panic. One of the panic moment that may happens is when a disaster happen, because disaster is threaten lives.

Humans often experience anxiety attacks because human beings are accustomed to doing routine things. The habit causes us to behave similarly when we experience a panic. One moment of panic that may occur is when disaster strikes. Based on common practice, then in the face of disaster, people will behave the same as the disaster threatens their safety.

These things happened when Kelud Mountain, Kediri, erupted. Traffic jam occur in main evacuation route, Wates, Kediri, Province of East Java. Officers had difficulty to untie the traffic jam because of the large number of vehicles, doesn't comparable to existing roads. The traffic jam made people scared. They screamed to call out for help from officers, while hail is happened. Sand and volcanic ash is also fell. On the sky, bright and red storm is happened. Thunder is also heard by all the people [3].

All this time, the selection of evacuation route doesn't well organized. People is forced as soon as possible to goes to main road, and unfortunately main road is limited especially in developing countries. With that condition, many accumulation of people on the steps of evacuation route when a disaster happen. That will cause occurrence of traffic jam and hamper evacuation process. And of course will endangers people lives.

There are many ways to describe emergency management and the importance purpose is saving lives and property in disasters. Science and technology were applied in planning and managing the extreme events that can injure or kill large numbers of people, do extensive damage to property and disrupt community life [4]. Statistics as a science had been applied in disaster management [9]. This method will use a statistical method to make simulation method in managing the evacuation road for mitigation when Merapi Volcano erupt. One of many Statistics methods is association analysis, which is useful for discovering interesting relationships which is hidden in large data sets. This relationships can be displayed in the form of association rules or sets of frequent items (frequent road passed through by local residents). For illustrative purposes, our discussion of association rules will be represented in the term of market transaction.

Let $I = \{i_1, i_2, ..., i_d\}$ be the set of all items in a market transaction data and $T = \{t_1, t_2, ..., t_N\}$ be the set of all transactions recorded in database. Each transaction $t_i$ contains a subset of items which is chosen from $I$. In association analysis method, a set of zero or more items is termed an itemset. The transaction width is defined as the number of items present in a transaction. A transaction $t_j$ is said to contain an itemset X if X is a subset of $t_j$. An important property of an itemset is its support count, which refers to the number of transactions that contain a particular itemset. Mathematically, the support count, $s$(X), for an itemset X can be stated as follows:

$$s(X) = |\{t_i|X \subseteq t_i,\ t_i \in T\}|,$$

where the symbol |.| denote the number of elements in a set. An association rule is an implication expression of the form X → Y, where X and Y are disjoint itemsets. The strength of an association rule can be measured in terms of its support and confidence. Support determines how often a rule is applicable to a given data set, while confidence determines how frequently items in Y appear in transactions that contain X. The formal definition of these metrics are:

$$\text{Support, } S(X\rightarrow Y) = s(X\cup Y)/N$$
$$\text{Confidence, } C(X\rightarrow Y) = s(X\cup Y)/s(X)$$

Support is an important measure because a rule that has very low support may occur simply by chance. A low support rule is also likely to be uninteresting from a business perspective because customers seldom buy together. Support is often used to eliminate uninteresting rules. Support also has a desirable property that can be exploited for the efficient discovery of association rules. Confidence, on the other hand, measures the reliability of the inference made by a rule. For a given rule X → Y, the higher the confidence, the more likely it is for Y to be present in transactions that contain X. Confidence also provides an estimate of the conditional probability of Y given X. In the selection of an evacuation route, this method will be applied to see which road in that route frequently traveled together.

In this case, support determines how often a rule is applicable to the road taken by the locals while to save themselves. Confidence determines how frequently road Y traveled after road X taken during evacuation process. A rule that has very low support may occur simply by chance or we could say that the attempt to save themselves, they are just guessing which road can be passed safely without having been familiar with the road conditions as best as possible which evacuation route to be taken.

Association rules says that when someone do A, and he or she do B too will get a support and confidence value. The bigger support and confidence, then more often the occurrence of A → B happen from all. In selecting for evacuation route, this method will use to see association rules of activities and driving social habit. This method will sees is it true that if someone drive on A road will drive on B road too or will drive on other roads too. The rule is different in each village.

This method is used to select the evacuation route, therefore, should pay attention to the patterns formed road sequence. Evacuation route to be sequenced from a starting location to the point closest to the epicenter of the safest distance. Method pattern formed associations combined with the sequential pattern.

Roads conditions in each other is different, there are good roads and there are damaged roads too. Each road will be grouped using hierarchical cluster analysis. Grouping will produce some graded group, very prone to very safe. Variables that will be used is road width, road condition and ease of access road. These variables doesn't have an indicator to be evaluated, so these variables evaluated using fuzzy logic.

The selection of evacuation road is done by activities and driving social habit by considering roads conditions in each village. Besides those, population in each village is also considered. The selection of evacuation road was directed away from disaster-prone areas of last disaster happen.

## 2. Related Research

Information from IKONOS images can be to create a map of evacuation routes and the impact area of cold lava floods. The preparation of evacuation route consider to eight parameters. Those parameters are disaster-prone area, slope, length and width of roads, material of roads, the existence of bridges and roads direction [2].

Association rules method can also used in education database. This method is used to seek for unique rule of Teacher Certification in data of Nomor Unik Pendidik dan Tenaga Kependidikan (NUPTK), codes of teachers in Indonesia. They found there are 184 rules in the data [1].

Fuzzy logic has been widely applied in real life, one of all is using fuzzy logic to find best supplier for a clothing store. Researcher use Fuzzy Principal Component Analysis and Analytic Hierarchy Process (AHP). Those two method are methods of decision-making that use qualitative and quantitative measurement. Qualitative measurement define issue and quantitative measurement provide an assessment. This methods use primary data from questionnaires and observation [7].

A specific application of "automated geo-spatial data acquisition and mapping" offers the transformation of verbally given geo-information into a situation map. For a successful transformation it is essential to develop a domain specific knowledge base. The presented Disaster Management Data Model is based on the Command and Control Information Exchange Data Model. Although the C2IEDM is a sophisticated standard, the DM points out important considerations that have to be taken into account for disaster management ontologies in general [8]. Another application is used in health. Epidemics, chronic diseases which are the major social disasters follow strategic-virulent disasters that affect the ecosystem of a spatial zone probabilistic study is made on the health demographic data. Using the Collocation rule, the effected area of dengue is found and results are obtained. Using the Participation index the symptoms that influence spreading the dengue were found and results are obtained. i.e., spatial knowledge for the health campaign. Identifying dengue effected area using spatial maps as spatial knowledge i.e. to identify spreading of dengue was proposed and proved [6].

Discovering association rules in sequences is an important data mining problem that is useful in many scientific and commercial domains. Predicting events and identifying sequential rules that are inherent in the data help domain experts to learn from past data and make informed decisions for the future. Several different approaches have been investigated for sequential data mining. Algorithms for discovering associations in sequential data, and episodal associations, use all frequent episodes. The entire set of association rules is produced and significance criterion such as J-measure for rule ranking are used to determine the valuable rules. An approach that uses temporal constraints on transactional sequences was presented in. We present a new approach that uses Minimal Occurrences With Constraints And Time Lags (MOWCATL), to find relationships between sequences in the multiple data sets. In addition to the traditional frequency and support constraints in sequential data mining, MOWCATL uses separate antecedent and consequent inclusion constraints, along with separate antecedent and consequent maximum window widths, to specify the antecedent and consequent patterns that are separated by a maximum time lag. The MINEPI algorithm was the first approach to find minimal occurrences of episodes [5].

## 3. Research Methodology

In the process of selecting evacuation route, three methods of Statistics will used be used. First method is association rules, one of many methods of data mining. This method is used to see the rules of people when driving and activity. Selecting evacuation route process will be based on those rules that will be obtained. Those rules can called as local people habit. Those rules were later obtained will be different for each village, so also will decrease accumulation of people on the evacuation route sections risk. Each rule have support and confidence in percentage. Support is got from:

$$Supp(X) = \frac{X\ incident\ (transaction)}{All\ incidents\ (transactions)}$$

And confidence is got from:

$$Conf(X \rightarrow Y) = \frac{Supp(X\ U\ Y)}{Supp(X)}$$

With sequential mining, sequence of lines considered. The route that obtained is not distorted from the nearest lane to the lane farthest point of disaster and back to the nearest road again. The route begins to form on the road closest to the disaster epicenter.

The next method is two methods that are combined, those two methods are hierarchical clustering analysis and fuzzy logic. Fuzzy logic is used to assess the assessment variables. Those variables are road condition, road width and ease of access road. Those three variables doesn't have indicator to rate the assessment variables, so the assessment is conducted using simulation of fuzzy logic. Fuzzy set will be divided to seven stratified groups, those groups are:

1. very unsafe,
2. unsafe,
3. bit unsafe,
4. normal,
5. bit safe,
6. safe, and
7. very safe.

Fuzzy set for the three variables is different. Road condition variable is divided into:

1. very bad,
2. bad,
3. bit bad,
4. normal,
5. bit good,
6. good, and
7. very good.

Road width variable is divided into:

1. very narrow,
2. narrow,
3. bit narrow,
4. normal,
5. bit wide,
6. wide, and
7. very wide.

Ease of access road variable is divided into:
1. very hard,
2. hard,
3. bit hard,
4. normal,
5. bit easy,
6. easy, and
7. very easy

Disaster affected area from the last disaster is also used as a consideration to select evacuation route. The first selecting evacuation route process is decide some driving local people habit. The rule (habit) with the largest support becomes the main recommendation to be the main evacuation route. Smaller support rules used as routes that local people is used to reach the main evacuation route. Those all rules directed away from disaster affected area from the last disaster. All rules routes are grouped into seven determined groups. Evacuation route selection consider to which group they are grouped. Population of villages is also considered to select evacuation route.

For example, one of some main evacuation route of Merapi Volcano Evacuation Route, Sleman, Yogyakarta, is used. All the evacuation route is as shown in **Figure 1**. Merapi Volcano evacuation route is used only to make it easier, as authors live near Merapi Volcano.

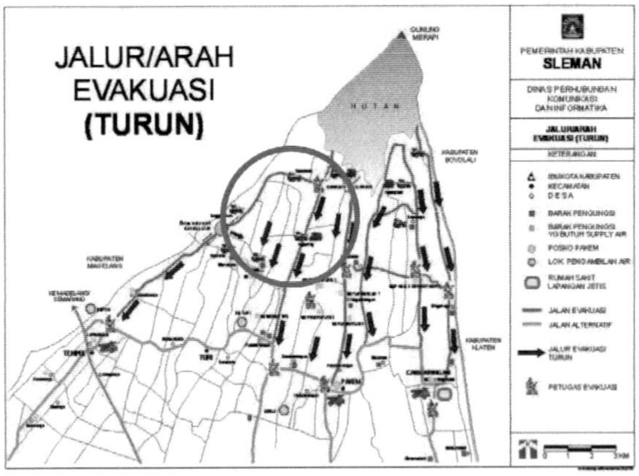


**Figure 1.** Merapi Volcano evacuation route. **Source** : Sleman District Government (slemankab.go.id)

Evacuation route that used as a simulation is route circled in red, the routes is as shown in **Figure 2**.

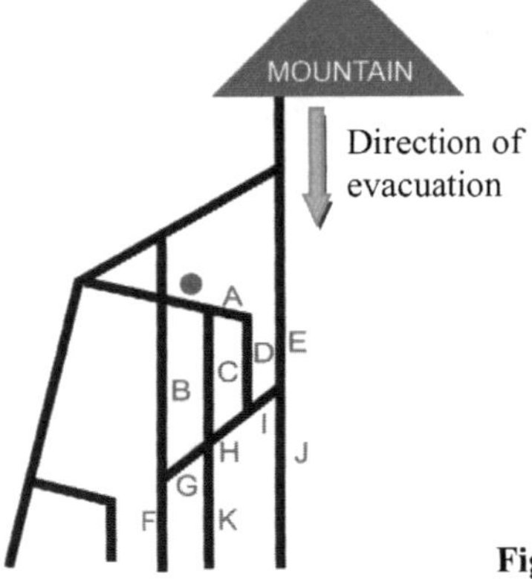


**Figure 2.** One of Merapi Volcano evacuation routes.

● : starting point of Merapi volcano evacuation route simulation

For example, the road condition of B is shown **Figure 3**, road condition of C is shown **Figure 4** and road condition of D is shown **Figure 5**.

**Figure 3.** Road B condition.

**Figure 4.** Road C condition.

**Figure 5.** Road D condition.

And the affected area of the last Merapi Volcano eruption in 2010 is as shown in **Figure 6**.

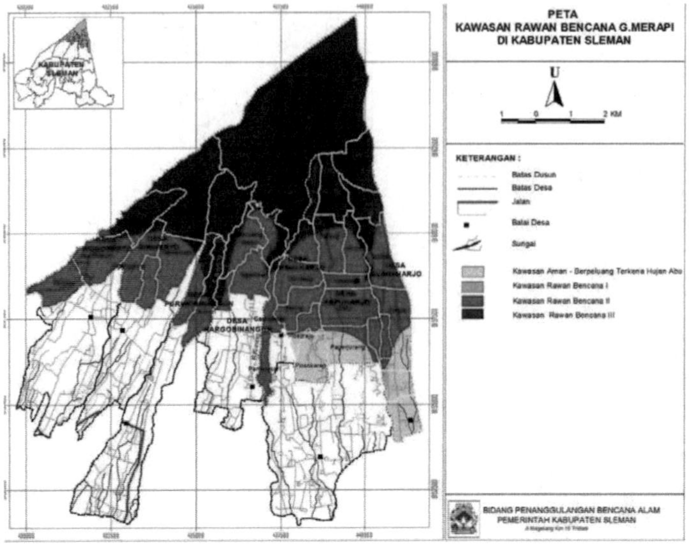


**Figure 6.** Affected area of Merapi Volcano eruption (based on 2010 eruption).
**Source** : Sleman District Government (slemankab.go.id)

### 4. Simulation Study

This study will use a simulation with local people living in the disaster-prone areas. A thousand people were then placed in one area (in Fig. 1 colored red dots). Then circumstances made as if it is similar to when Merapi Volcano will reach the highest point of warning. After being given a sign to immediately evacuate, then the thousand people hasten to save themselves down the road that they consider the best way to save themselves. After being in a safe place that is determined, then every person who follows the simulation, telling which line he passes to reach a safe point. The result is then recorded as follows.

**Table 1.** Example of the route that taken by local residents
(This data is not real, used only to show how the data, these data is not exist.)

| Travel (person) | Route |
|---|---|
| **1** | C, H, I, J |
| **2** | C, K |
| **3** | C, G, F |
| **4** | A, D, I, J |
| **5** | B, G, K |
| **...** | ... |
| **997** | B, F |
| **998** | A, D, H, K |
| **999** | C, H, I, J |
| **1,000** | C, K |

Then we get the following measurements of rules for two routes,

$A \rightarrow I$ : 0.17,

$C \rightarrow I$ : 0.1,

$C \rightarrow F$ : 0.32,

...
B → F : 0.24,

Those support values is not real, those four support value is just a made-up and there are no support value for another route.
From those rules, seems that rule of "C → F" get the highest value of support. So, rule "C → F" is the first recommendation to become main evacuation route. However, to reach route F from C, people have to go through route G. Thus the existing rules instead of C → F but became C → G → F. Therefore, this rules recommends that the main evacuation route is C → G → F.

The second biggest support is B → F rule route, so that rule become second recommendation to become main evacuation route or ordinary evacuation route to reach main evacuation route for a while. The third is A → I rule route and become third recommendation, and so on until the smallest support. Road condition of those all rule route are also considered. It should be noted that this method using sequential mining, so the results that obtained is started from the closest from volcano to farthest.

From direct observation of the existing roads, it appears that the road B (**Figure 3**) has poor road conditions and can harm people when they pass through in large numbers and panic. The condition of road C (**Figure 4**) rather bad, while the condition of the road D (**Figure 5**) look good, but after the barrel located in the middle of the road, the left side of the road was severely damaged. Those three roads have same width. By considering the real roads conditions, then the best choice is road C. Direct observation was also conducted to assess the other roads.

From observation, seems that route B (**Figure 3**) is damaged and endanger local people if they pass through that route. Route C (**Figure 4**) is bit damaged and route D (**Figure 5**) is in very good condition but after the drum, there is landslide on the left side of the road. Those three route is also have same width.

Authors saw directly roads then assessed in the opinion of authors (in fuzzy set). In simulation, roads condition for each route is below:

**Table 2.** Roads condition for each route.

| Route | Road Condition | Width | Ease of Access |
|---|---|---|---|
| **A (1)** | Bad | Normal | Very easy |
| **B (2)** | Bad | Normal | Easy |
| **C (3)** | Bit bad | Normal | Easy |
| **D (4)** | Very bad | Normal | Easy |
| **E (5)** | Good | Wide | Hard |
| **F (6)** | Bit bad | Bit wide | Bit hard |
| **G (7)** | Normal | Bit narrow | Bit hard |
| **H (8)** | Bad | Bit narrow | Bit hard |
| **I (9)** | Normal | Bit narrow | Easy |
| **J (10)** | Very good | Very wide | Hard |
| **K (11)** | Good | Wide | Hard |

Using hierarchical clustering (single linkage, manhattan), got dendrogram below:

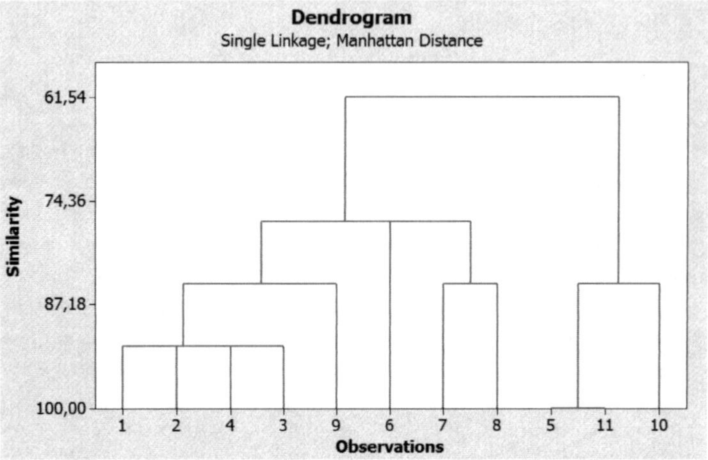


**Figure 6.** Dendrogram of roads condition

From dendrogram, choosed object is grouped into 4 groups, got groups:

**Table 3.** Groups.

| Group | Ease of Access |
|---|---|
| **1** | 1, 2, 4, 3, 9 |
| **2** | 6 |
| **3** | 7, 8 |
| **4** | 5, 11, 10 |

From the result of association rules, got rules C → G → F become the first recommendation to become main evacuation route. From the hierarchical clustering, road C is grouped into Cluster 1, G is grouped into Cluster 3 and F is grouped into Cluster 2. There's no road from road C, G and F that is grouped into worst cluster, so route C → G → F still can become the first recommendation to become main evacuation route.

The second evacuation route and so on can be determined in the same manner. These results can then be provided to the competent authorities in the evacuation problem. Thus dividing routes could consider road capacity and the number of actual residents in order to avoid excess capacity at one point causing congestion and excessive panic.

## 5. Result and Discussion

Results and conclusions are as follows:

1. in simulation, from the result of association rules, got rules C → G → F become the first recommendation to become main evacuation route. From the hierarchical clustering, road C is grouped into Cluster 1, G is grouped into Cluster 3 and F is grouped into Cluster 2. There's no road from road C, G and F that is grouped into worst cluster, so route C → G → F still can become the first recommendation to become main evacuation route. The second evacuation route and so on can be determined in the same manner.
2. selection of evacuation route with this method can decrease accumulation of people on the steps of the evacuation route risk,

3. if this method is used in real life, the hardest part is to record activity and driving people data. Same person must recorded from first place, then their travel road, until their destination. And because data mining analysis need big data to obtain relevant result, and
4. for the development of this method, small area estimation method can be used to estimate the disaster-prone area and many simulation possibility of evacuation process, so probability of people safe on every part of evacuation route can be calculated.

## 6. References

Results and conclusions are as follows:

[1] Amiruddin P et. al. 2010 *Penerapan Association Rules Mining Pada Data Nomor Unik Pendidik dan Tenaga Kependidikan Untuk Menemukan Pola Sertifikasi Guru* (*Association Rules Mining Aplication To Find Teacher Sertification Rules In Nomor Unik Pendidik dan Tenaga Kependidikan Data*) 15-16

[2] Ardana D M S and Purwanto T H. 2012 *Penentuan Jalur Evakuasi dan Dampak Banjir Lahar Dingin Gunung Merapi* (*Select An Evacuation Route and The Impact Of Cold Lava Flood From Merapi Mountain*) 149

[3] BeritaJatim.com April 24th 2014 *Warga histeris, jalur evakuasi macet total,* http://m.beritajatim.com/peristiwa/198365/warga_histeris,_jalur_evakuasi_mmace_total.html#.U1jF5_mSxmw

[4] Drabek 1991 *Introduction to emergency management* ch 1

[5] Harms S K. et al *Discovering Sequential Association Rules with Constraints and Time Lags in Multiple Sequences* 1-2

[6] Kameswara R et. Al 2013 *Spatial mining system for disaster management* 6

[7] Muhajir M 2011 *Pendekatan Metode Fuzzy Principal Component Analysis dan Analytic Hierarchy Process dalam Pemilihan Supplier Terbaik Produk Bahan Baku Baju*" (*Fuzzy Principal Component Analysis and Analytic Hierarchy Process Methods Approach to Select Best Raw Material Dress Products Supplier*) 17

[8] Lucas et. al 2007 *Information Mining for Disaster Management* 6

[9] Savoiu G *The Holistic Concepts of Disaster Management and Social Cohesion* Statistics and Method, Scientific Bulletin – Economic Sciences Vol. No. 10/Issue 1

[10] The Secret of Mindpower and NLP April 22th 2014 *Anxiety Panic Attacks* http://www.the-secret-of-mindpowerandnlp .com/an.xiety-panic-attacks.html

# Food powder analysis by using transversely excited atmospheric $CO_2$ laser-induced plasma spectroscopy

**Ali Khumaeni,[1*] Zener Sukra Lie,[2] Wahyu Setiabudi,[1] Koo Hendrik Kurniawan,[2] and Kiichiro Kagawa[3]**

[1]Department of Physics, Diponegoro University, Semarang, Indonesia
[2]Maju Makmur Mandiri Research Center, Kembangan, Jakarta, Indonesia
[3]Research Institute of Nuclear Engineering, University of Fukui, Fukui, Japan

*Email: khumaeni@fisika.undip.ac.id

**Abstract**. A direct and sensitive analysis of food powder sample has successfully been carried out by utilizing the special characteristics of pulsed transversely excited atmospheric (TEA) $CO_2$ laser. In this study, a food powder was placed in a container made of copper plate and covered by a metal mesh. The container was perpendicularly attached on a metal surface. A high-temperature luminous plasma was induced on a metal surface 5 mm above the mesh. Once the plasma was produced, a strong shock wave was induced, blowing-off of the powder from the container to enter into the plasma to be dissociated and excited. By using this method, a semi-quantitative analysis of food powder was made. The detection limits of Cr in the powdered agar and Cd in the powdered rice were 9 mg/kg and 50 mg/kg, respectively.

## 1. Introduction

Analysis of food powder sample has become an interesting subject. This is due to the fact that the food materials contain the important nutrition for human being such as calcium, sodium, and potassium. Thus, the development of the technique for rapid quantitative analysis with high precision to measure the concentration of essential elements for nutritional purpose is urgently necessary [1]. Beside that, it is also necessary to examine the existence of heavy elements such as chromium (Cr) and cadmium (Cd), which are hazardous for human health and might be accidentally contained in the powder food material during the growth of the food plant and industrial process.

Generally, the method applied for food powder analysis is Inductively Coupled Plasma Atomic Emission Spectrometry (ICP-AES) or Inductively Coupled Plasma Mass Spectrometry (ICP-MS) [2]. Those methods perform high precision and sensitivity in elemental analysis. However, in those methods, the samples must first be liquid often being prepared by either wet digestion or dry ashing. Thus, the analysis requires delicate sample preparation, labor intensive and expensive equipment.

Recently, laser-induced breakdown spectroscopy (LIBS) has become a tremendous method for rapid qualitative and quantitative analysis [3-4]. In conventional LIBS technique, a pulsed laser with high peak power and short pulse duration such as Nd: YAG laser is mostly employed to induce laser plasma at atmospheric pressure. In this method, the vaporization, atomization, and excitation processes take place at the same time due to a single treatment of laser irradiation. However, due to the short pulse duration (8 ns) and high power density (about $10^{10}$-$10^{11}$ W/cm$^2$) of the Nd: YAG laser, the blowing-off of powder occurs when the laser is directly irradiated onto the powder sample. Therefore, in this technique, analysis is mostly made after pressing the powder sample into a pellet [5].

We found that a direct analysis of a powder sample can be made by utilizing the specific characteristics of a TEA $CO_2$ laser [6]. The TEA $CO_2$ laser is longer in wavelength (10.64 μm) and in pulse duration (200 ns) compared to the Nd:YAG laser. Thus, the TEA $CO_2$ laser is suitable to the analysis of powder samples. In the study, a powder sample was placed into a small hole in a metal plate and covered with a metal mesh. The TEA $CO_2$ laser was then focused on the powder surfaces through the metal mesh. However, the sensitivity of the elemental result was rather low. Therefore, this technique cannot be employed for a quantitative analysis of minerals and hazard materials at low concentration in the organic powder samples.

To solve this problem, we propose a unique method by utilizing the specific characteristics of TEA $CO_2$ laser. In this method, a powder sample was placed into a container and covered by a metal mesh. The container was then perpendicularly attached on an iron metal surface. The fine powder particles are dissociated and excited in the gas plasma induced by a TEA $CO_2$ laser on metal surface. Compared to our previous method [6], the present method is much superior. Namely, the luminous plasma has higher temperature and larger diameter because the plasma was induced on metal surface, while in the previous method, the plasma was induced by assisting a metal mesh, resulting in lower temperature. Therefore, the dissociation and excitation processes can effectively takes place in the plasma induced by using our present method.

## 2. Experimental setup

The basic experimental setup applied in this work is the same as the one used in previous research [5]. A TEA $CO_2$ laser (Shibuya SQ 2000, pulse energy of 3 J, wavelength of 10.6 μm, pulse duration of 200 ns in full width of half maximum, beam cross section of 3 mm x 3 mm), which was constructed by Shibuya Company for laser marking, was employed in this study. The laser beam was focused by a ZnSe lens (f = 200 mm) onto a target sample through a ZnSe window. The laser energy was set at 1.5 J by inserting an aperture in the front of the focusing lens. The spot size of the laser beam on the sample surface was 2 x 2 $mm^2$, which results in an irradiance of 0.19 $GW/cm^2$ for an energy of 1.5 J.

The sample was placed in a metal chamber with dimensions of 12 cm x 12 cm x 12 cm. During the experiment, nitrogen gas was flowed into the metal chamber, in which the pressure of the surrounding gas was kept constant at 1 atmosphere. The flow rate of the gas was 4 liters per minute (Lpm).

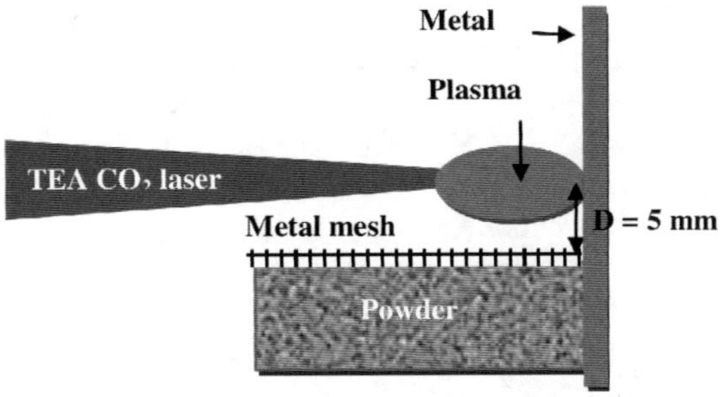


Figure 1. Illustration of the sample holder used in this study.

The samples used in this study were a powdered agar, a herb medicine (Tsumura Kackontou) and a standard polished rice (NIES certified reference material, produced by Environment Agency National Institute for Environmental Studies, Japan) containing mineral and hazardous impurity. Prior to the analysis, 0.5 g powder sample was placed into a container made of copper plate with a dimension of 15 x 22 x 6 $mm^3$ and covered by a metal mesh made of stainless steel wires (lattice constant of 0.2 mm and wire diameter of 0.1 mm). The mesh functions to control the blowing-off of the powder particles caused by the plasma shock wave. The container was then perpendicularly attached on an iron metal surface as shown in Fig. 1. The sample preparation and data acquisition were made at the University of Fukui, while the data processing and analytical study were carried out at Diponegoro University.

The plasma emission spectrum was obtained by detecting laser plasma radiation using optical multi channel analyzer (OMA) system (ATAGO Macs-320) consisting of a 0.32 m focal-length spectrograph with a grating of 1200 groves/mm, a 1024-channel photodiode detector array, and a micro-channel plate image intensifier. The spectral resolution of the OMA system is 0.2 nm. The light emitted from the laser plasma was collected by an optical fiber ($\theta = 27^0$ in solid angle) and fed into the OMA system. One end of the fiber was placed at 15 cm distance from the focusing point of the laser light and set perpendicularly to the path of laser beam. Another OMA system with an intensified charge-coupled device (CCD) (Lambda vision LVICCD 1012), was also used in this experiment. This OMA has higher spectral resolution of 0.02 nm. We use the OMA system with an ICCD having higher resolution because we will improve the sensitivity of analytical detection for impurity analysis. By using the higher resolution spectrometer, the background emission and the noise can be suppressed and the analytical lines can clearly be identified with an adjacent lines. Thus, the detection sensitivity of analyte can be improved. This OMA will be used to detect poisonous heavy metals contained in the food powder at low concentration.

## 3. Results and discussion

Initially, the study was addressed to search the optimum experimental condition. First, we examined the distance (D) between the focusing point of laser beam on a metal surface and the metal mesh used to cover the powder in the container (Fig. 1). Figure 2 shows the emission spectra taken from the herb medicine powder using the present technique with the D of (a) 10 mm (red line), and (b) 5 mm (blue line). The herb medicine contains Ca of approximately 400 mg/kg. In this experiment, the TEA $CO_2$ laser energy was set at 1500 mJ. The lattice constant and the wire diameter of mesh used to control the blowing-off of the powder were 0.2 mm and 0.1 mm, respectively. During the experiment, the nitrogen gas was flowed in the chamber in order to avoid gas breakdown taking place in front of the metal due to the interaction between dust and TEA $CO_2$ laser beam. It is seen in the figure that the emission intensities of ionic Ca lines at 393.3 nm and 396.8 nm for the case of D of 5 mm is much higher than that of 10 mm. The Ca has a specific doublet ionic lines at 393.3 nm and 396.8 nm with the excitation energies of approximately 3 eV from the ground state of ionic energy level. It is assumed that the amount of powder particles entering into the gas plasma region is much larger for the case of 5 mm of D. Therefore, we chose the D of 5 mm for further practical applications.

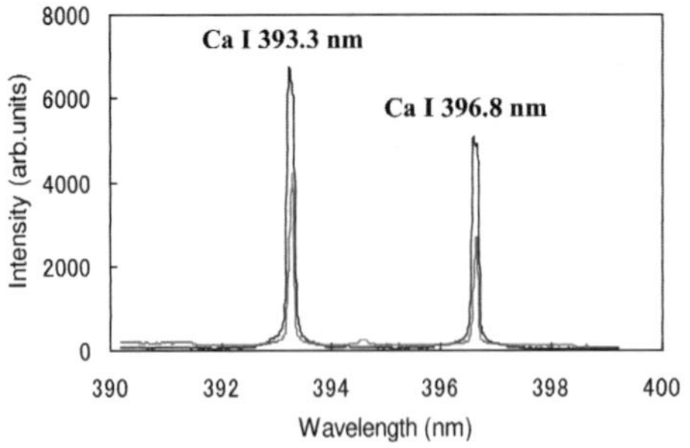


Figure 2. Emission spectra taken from the herb medicine sample using the metal-induced plasma technique with the D of 10 mm (red line) and 5 mm (blue line).

Another experiment was also made in order to get the best experimental circumstances. To this end, the same technique as used in Fig. 2 (D = 5 mm) with a different lattice constant of mesh, namely 0.2 mm and 0.4 mm, was employed. The sample used in this experiment is the same as in Fig. 2. The results certified that the lattice constant of 0.2 mm is much more effective to result in higher intensity than that of 0.4 mm. This is probably that for the case of mesh with the lattice constant of 0.2 mm, the powder particles entering into the gas plasma can be suppressed and controlled, so that the amount of evaporated powder particles entering into the gas plasma region is suitable. While, in case of lattice constant of 0.4 mm, the amount of evaporated powder particles are too much, lowering the temperature of gas plasma. Detail mechanism of plasma generation and dissociation and excitation processes of atoms in the powder sample was shown in the previous work [7]. From this point, we assumed that the control of the blowing-off of the powder being sent into the gas breakdown plasma is crucial issue.

To estimate the temperature of plasma induced by using the present technique, a Boltzmann method was employed. To this end, the emission spectrum of copper (Cu) taken from the $CuSO_4.5H_2O$ powder was used. Figure 3(a) shows how the emission intensities of Cu I 510.5 nm and Cu I 521.8 nm change with time. It is seen in the figure that the emission intensity of Cu I 510.5 nm and Cu I 521.8 nm (blue and red curves, respectively) is rather high at the initial stage. With increasing the time, the Cu intensities slowly decrease and last very long time up to around 60 μs. Figure 3(a) also presents how the plasma temperature changes with time (green curve); the plasma temperature can readily be derived by the ratio of the emission intensity of Cu I 521.8 nm (excitation energy of 6.2 eV) to that of Cu I 510.5 nm (excitation energy of 3.8 eV), with the assumption of Boltzmann's distribution [10]. It is seen that at the initial stage, the plasma temperature is approximately 5750 K. The plasma temperature keeps almost constant up to 20 μs and finally slowly decreases up to 60 μs.

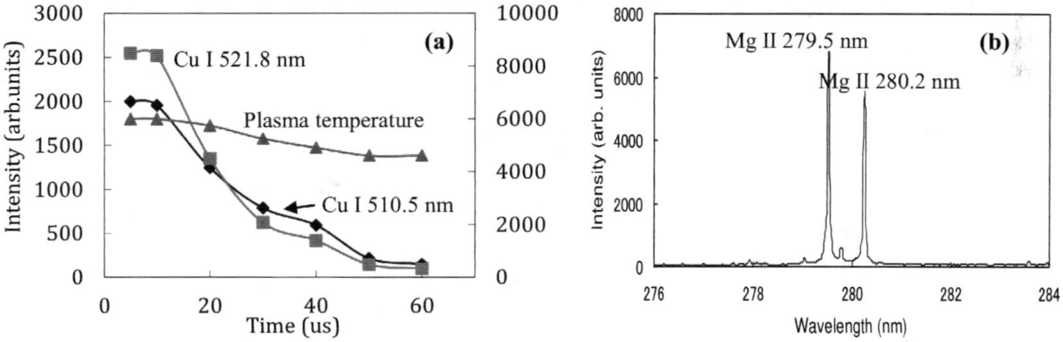


Figure 3. (a) Time profiles of emission intensities of Cu lines and plasma temperature taken from the $CuSO_4.5H_2O$ powder sample using the metal-induced plasma technique.
(b) Emission spectrum of magnesium taken from the powdered rice.

At initial, the present technique was employed to perform the direct analyses of food powder samples. It is well known that food powders contain the primary source of nutrition for human being such as magnesium (Mg). Figure 3(b) shows the emission spectrum of magnesium taken from the powdered rice. Two specific ionic emission lines of Mg II at 279.5 nm and 280.2 nm were strongly detected. The ordinary powdered rice sample contains magnesium of around 400 mg/kg. This result certified that the present method can be used to detect the nutrition elements such as Mg in the food powder.

For further exploration of the capability of this technique, an analysis of poisonous-heavy-metals in the powdered agar and the powdered rice was made. This analysis is necessary because the heavy metals are hazardous elements for human health. Figure 4(a) shows the emission spectrum of Cr taken from the powdered agar. The powdered agar contains Cr of approximately 1000 mg/kg. It can be seen that the strong emission lines of Cr I 425.4 nm and Cr I 427.4 nm clearly appear. It should be mentioned that Cr I 428.9 nm cannot clearly be seen in this spectrum because it overlaps with Fe I 428.8 nm. Other strong emission lines of Fe also occur in the spectrum. By using Cr I 425.4 nm line, the signal to background (S/B) ratio and the detection limit were estimated. The S/B ratio of Cr line was approximately 10 and the detection limit of Cr in powdered agar was approximately 9 mg/kg; the detection limit was derived by calculating what concentration of the signal yielded 3 times of the noise level is because the 3 times of the noise is clearly identified as a signal that can be distinguished from the noise. It should be noted that the detection limit of Cr produced by using the present method is much lower compared to the case of X-ray fluorescence spectroscopy (XRF) method (detection limit of Cr in the powdered agar is approximately 170 mg/kg). The other heavy metal was also detected from the powdered rice containing 1000 mg/kg of Cd. The Cd is poisonous heavy metals usually contained in the rice during the growing process due to the environment pollution. Figure 4(b) shows the emission spectrum of Cd taken from the powdered rice. It is seen that strong emission lines of Cd neutral at 346.6 nm and 361.0 nm can clearly be observed. Compared to the Cr lines, the efficiency of the Cd atomic excitation is much lower. Thus, total emission intensity of Cd lines is much lower than that of Cr lines. By using Cd I 361.0 nm line, the S/B ratio and the detection limit of Cd in the powdered rice were approximately 7 and 50 ppm, respectively. The results certified that this present technique has high possibility to be applied to the practical quantitative analysis of powders.

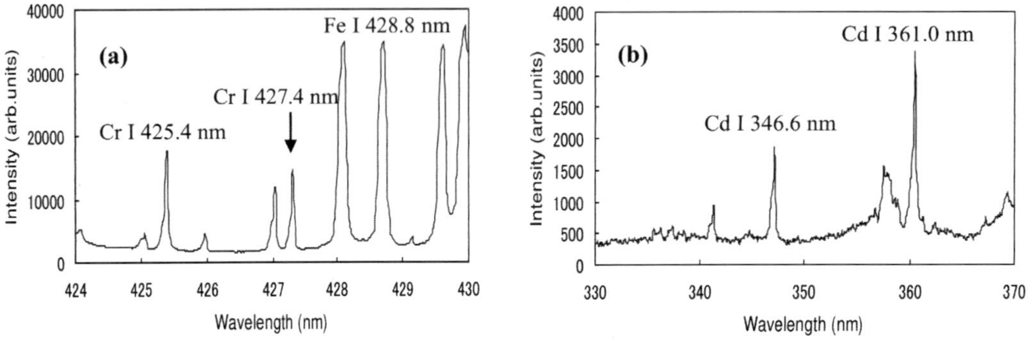


Figure 4. Emission spectra of the heavy metals taken from (a) the powdered agar, (b) the powdered rice.

## 4. Conclusion

The direct analysis of powder sample has been demonstrated by using the specific characteristics of a TEA $CO_2$ laser. In this study, a powder was placed in a container, equipped with a metal mesh, perpendicularly attached on a metal surface. High-temperature and large-size luminous plasma was induced when a TEA $CO_2$ laser was focused on the metal surface just 5 mm above the mesh. Due to strong shock wave induced by the plasma, a lot of fine powder particles were blown-off and finally entered into the gas plasma region to be dissociated and excited. By using this present technique, a semi quantitative analysis of food powder sample can successfully be carried out. The detection limit of Cr in the powdered agar and Cd in the powdered rice was approximately 9 mg/kg and 50 mg/kg, respectively. This technique can potentially be used to be applied to checking the powder containing low concentration elements such as heavy metals in the food powder.

### References

[1] A. Ryan, ICP-AES Instrument at Work Varian (1997).
[2] A. Tame, and D. Hoobin, ICP-AES Instrument at Work Varian (1999).
[3] Miziolek W, Palleschi V, Schechter I 2006 *Laser-induced Breakdown Spectroscopy (LIBS), Fundamentals and Applications*. Cambridge University Press, New York.
[4] Quarles C D Jr, Gonzales J J, East L J, Yoo J H, Morey M, Russo R E 2014 J. Anal. At. Spectrom **29** 1238
[5] Gondal M A, Hussain T, Yamani Z H, and Baig M A 2007 *Talanta* **72**, 642
[6] Khumaeni A, Ramli M, Deguchi Y, Lee Y I, Idris N, Kurniawan K H, Lie T J, Kagawa K 2008 *Appl Spectrosc* **62** 1344

[7] Khumaeni A, Lie Z S, Niki H, Kurniawan K H, Tjoeng E, Lee Y I, Kurihara K, Deguchi Y, Kagawa K, 2011 Anal. Bioanal. Chem **400** 3279

# Power System Harmonic Elimination to Improve Power Quality

K.Chandrasekaran,
Department of Electronic Engineering,
University Tunku Abdul Rahman,Perak, Malaysia.
E-mail: kchandra@utar.edu.my

V.K. Ramachandaramurthy
Power Quality Research Group,
University Tenaga Nasional, Malaysia.

**Abstract-** An improvised RLC interface filter for a Dynamic Voltage Restorer (DVR) is proposed. The RLC filter is connected in the front end between the IGBT based Voltage Source Inverter (VSI) and the injection transformer and is able to eliminate voltage harmonics in the system and also switching harmonics generated from VSI. The voltage at the sensitive load end is pure sinusoidal. In this method, the DVR produced Pulse Width Modulation (PWM) voltage with voltage harmonic canceling the voltage harmonic generated from the supply main. The VSI handles harmonic power. The low order switching harmonics generated by the IGBT based VSI is suppressed. The DVR has greater voltage injection capability. Good dynamic and transient results recorded and Total Harmonic Distortion (THD) at the sensitive load end is minimized. The voltage at the sensitive load is sinusoidal and at 1.0 pu. PSCAD/EMTDC is used to validate the performance of the interface filter and the DVR. Simulated results are presented.

**Keywords**: Series Active Power Filter (SAPF), Voltage harmonics, Passive filters, H-Bridge inverter

## 1. INTRODUCTION

Devices such as arc furnaces, adjustable speed drives, computer power supplies, switch-mode power supply etc. are some typical nonlinear characteristic loads used in most of the industrial applications and are increasing rapidly due to technical improvements of semiconductor devices, digital controller and flexibility in controlling the power usage. The alternate switching actions of these power devices and its non-linearity characteristics result in a distorted input current. The power converter will behave like a current source, injecting harmonic and the interaction with the main system impedance will result in voltage distortion. Voltage THD above 5 % is undesirable above limit IEEE Std. 519. Unwanted heat in generated in equipment connected to harmonic source. High level of voltage distortion will cause equipment mal-function and damage.

Power system rich in harmonics have poor operating power factor and low efficiency. Harmonics can damage power factor correcting capacitors at plant facility and at utility distribution points. Harmonics can initiate system resonance that can severely disrupt system operation. A method is proposed to reduce the voltage THD. Rectifier or dc supply connected with a large filter capacitor on its dc side is a voltage type harmonic source [1][2][3]. One of the problems of power system harmonics is the supply voltage distortion at the Point of Common Coupling (PCC). Generally, passive LC filter is used to attenuate the voltage ripple from inverter switching [4][5]. For specific harmonic elimination, banks of passive filters are connected parallel with the load at the load end [6][7][8]. This will suppress the specific harmonics such as $5^{th}$, $7^{th}$ and $9^{th}$ orders. This method is not cost effective. The output voltage on the LC filter capacitor is controlled by switching operation of the PWM inverter, where the LC filters will introduce a time delay and cause resonance in the output AC voltage. Publication [9][10] discusses the regulating technique done to control the output voltage across the LC filter capacitor. Various control techniques has been introduced to control harmonics in inverter such as digital control strategies based on repetitive control, dead-beat control, and discrete-time sliding-mode control [11]. The techniques have some drawbacks, such as complexity, sensitivity to parameter variations and loading conditions, and steady-state errors in distorting condition.

Current ripple of PWM inverter is reduced with large inductor however, the voltage drop across it is increased [12]. The disadvantages of large inductor are the cost, weight and also increase voltage stress on the inverter switches.

Several controller schemes were proposed to control the DVR such as PI controller and park's transformation [13], DVR with Sliding Mode Control [14], Indirect Control of Capacitor Supported DVR [15], instantaneous p-q theory [16], synchronous d-q theory [17] and fuzzy logic controller [18]. Majority of the proposed controller algorithms were based on the Synchronous Rotating Frame Theory (SRFT) or d-q theory while the remaining controllers were too complex to implement as significant computing time was required. It is observed that many researchers addressed the problems of power quality and special attention is given to the voltage sag/swell but few research papers are published on multilevel converters based DVR. The developed controllers are also complex to implement and required more number of components. The total harmonic distortion (THD) of the load side voltage obtained by some research papers is above IEEE-519 Standards. Also the results obtained for the RMS of the load side voltage of some researchers is shown to be lower than the desired level and as there is not much published literature on multilevel based DVR, it is necessary to establish the compensating performance of seven-level cascade multilevel based DVR by proposing new controlling techniques.

This paper proposes an improvised smoothing filter design method with minimum transient current overshoot problem without decreasing the active damping performance of inverter systems. The VSI model is a 7-level H-Bridge as shown in Figure 1. The number of H-bridge cells in a Cascaded H-Bridge (CHB inverter) is determined by the inverter operating voltage, harmonic requirements, and cost. A phase-shifted carrier-based modulation scheme is used for the multilevel inverter. The main advantage of PWM control based DVR control is including fast switching speed of the power switches beside PWM technique offers simplicity and good response and high switching frequencies can be used to improve on the efficiency of the converter without incurring significant switching losses [19]. The basic idea of PWM based multicarrier phase shift modulation technique is varying the ON or OFF switching periods of the IGBTs switches of multilevel converter at a constant frequency so that the ON periods are longest at the peak of the wave. The control of switch duty ratio adjusts the output voltage level. The multilevel inverter with $s$ voltage levels requires $(s - 1)$ triangular carriers. In the phase-shifted multicarrier modulation, all the triangular carriers have the same frequency and the same peak-to-peak amplitude, but there is a phase shift between any two adjacent carrier waves, given by $\theta cr = 360°/(s - 1)$.

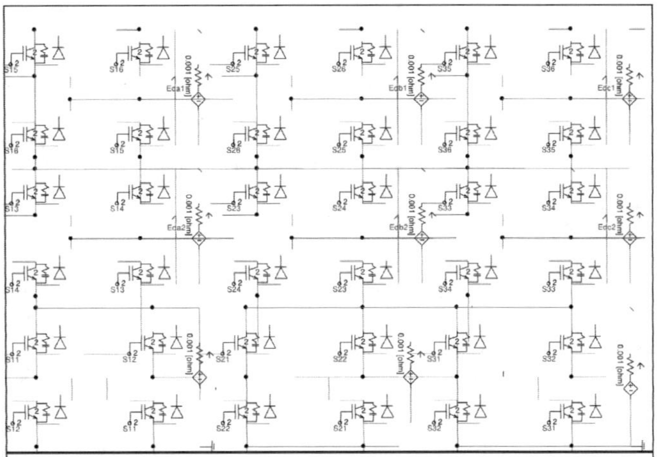

Figure. 7-level H-Bridge inverter model in PSCAD

Phase shifted multicarrier modulation model in PSCAD is shown in Figure 2. The three phase structure of seven level cascaded inverter is illustrated in Figure 1. Each dc source is connected to a three phase inverter. Each inverter level can generate three different voltage outputs, +Vdc, 0, and –Vdc by connecting the dc source to the ac output by different combinations of the four switches. The ac outputs of each of the different full bridge inverter levels are connected in series such that the synthesized voltage waveform is the sum of the inverter outputs.

From Figure 2, when switch S15 is at $240^0$, the phase shift of S16 is at $60^0$. Similarly S13 at $120^0$, phase shift of S14 is at $300^0$, S11 is $0^0$ and S12 is $180^0$.

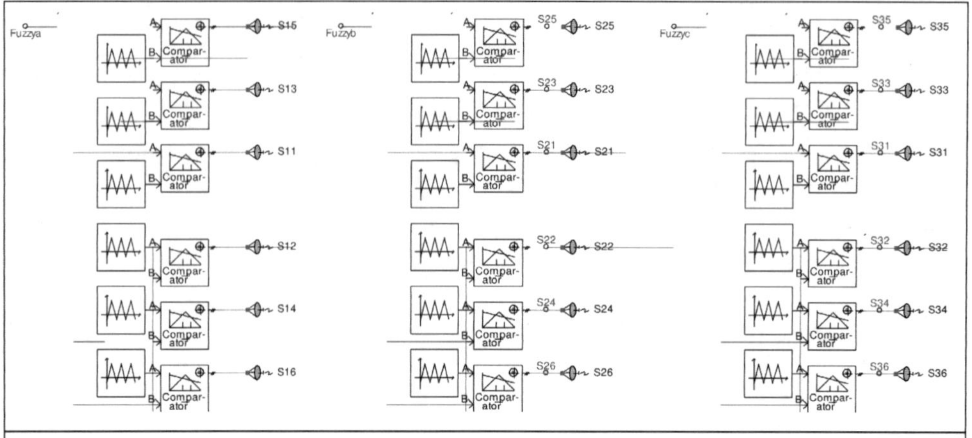


Figure 2. Phase-shifted multicarrier modulation model in PSCAD

## 2. RLC FILTER SYSTEM

The single-phase one leg equivalent circuit of an inverter system is shown in Figure 3. From the Figure,

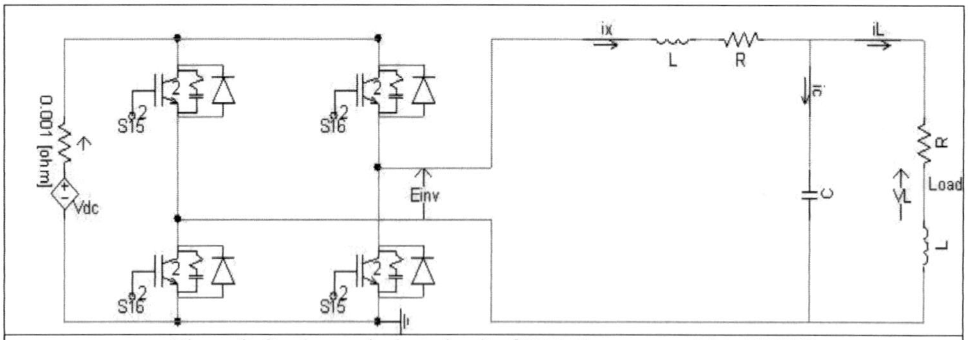


Figure 3. One leg equivalent circuit of PWM inverter model in PSCAD

*E*inv is the VSI output voltage, supply current is *i*x, current through capacitor is *ic*, load current is *i*L and *VL* is voltage across the load.

The dc supply voltage *Vdc* is constant and the gate switch of VSI is ideal. The output voltage of the VSI is written as;

$$Einv = VL + R_f\, ix + L_f \frac{dix}{dt} \quad (1)$$

Separating both the capacitor voltage and the current through the inductor into average over a switching cycle and harmonics ripple component,

$$Einv = \overline{VL} + \tilde{V}L \quad (2)$$

$$ic = \bar{i}x + \tilde{i}x \quad (3)$$

Substituting eq. 2 and 3 in eq. 1 gives

$$Einv = \overline{VL} + \widetilde{VL} + R_f\,(\bar{i}x + \tilde{i}x) +$$

$$L_f \frac{d}{dt}(\overline{ix} + \tilde{ix}) \quad (4)$$

Considering $\overline{VL}$ and $R_f\,\tilde{ix}$ is small as compared to $L\frac{d}{dt}\tilde{ix}$, then the ripple component of the filter inductor current is

$$\tilde{ix} = \frac{1}{L_f}\int (Einv - \overline{Einv})\,dt \quad (5)$$

and the average value of output voltage is

$$Einv = \overline{VL} + R_f\, ix + L_f \frac{dix}{dt} \quad (6)$$

The capacitor voltage harmonic, with reference to Figure 3;

$$ic = ix - iL \quad (7)$$

Splitting the capacitor and inverter output currents into the average and harmonic components,

$$ic = \overline{ic} + \tilde{ic} \quad (8)$$

$$ix = \overline{ix} + \tilde{ix} \quad (9)$$

Substituting eq. (3) and eq. (8)-(9) into eqn. (7), the following equation is obtained

$$\overline{ic} + \tilde{ic} = \overline{ix} + \tilde{ix} - \overline{iL} - \tilde{iL} \quad (10)$$

Thus,

The average and harmonic components of the filter capacitor current can be obtained as

$$\overline{ic} = \overline{ix} - \overline{iL} \quad (11)$$

$$\tilde{ic} = \tilde{ix} - \tilde{iL} \quad (12)$$

Considering, the filter capacitor is large, and all the inductor harmonic current flows through the capacitor. Thus,

$$\tilde{ic} = \tilde{ix}$$

The ripple component of the filter capacitor voltage is

$$\tilde{V} = \frac{1}{C_f}\int \tilde{ic}\,dt = \frac{1}{C_f}\overline{ix}\,dt \quad (13)$$

## 3. LC FILTER VALUES

The reactive power of the LC filter is

$$P_{rec} = \omega_s\, L_f\, (\bar{I}_x^2 + \bar{I}_{pav}^2) + \omega_s\, C_f\, (\bar{V}L^2 + \bar{V}L_{av}) \quad (14)$$

where,

$\omega_s = 2\pi f_s$ and $f_s$ is fundamental frequency, $\bar{I}_X$ and $\overline{VL}$ are rms value of the fundamental component of the inductor current and load voltage. The harmonic components is much smaller than the fundamental components and eq. 14 is simplified as

$$P_{rec} = \omega_s\, L_f\, (\bar{I}_X^2) + \omega_s\, C_f\, (\overline{VL}^2) \quad (15)$$

The rms value of the current through the inductor is

Splitting the current $I$x in terms of real and imaginary, eq. 15 is rewritten as

$$P_{rec} = \omega_s\, L_f\, [\bar{I}_{re}^2 + (\bar{I}_{im} - \omega_s\, C_f\, \bar{V}L)^2] + \omega_s\, C_f\, \overline{VL}^2 \quad (16)$$

Capacitance is given as

$$C_f = k\frac{Vdc}{L_f\, f_{sw}^2 \bar{V}_{av}} \quad (17)$$

Inductance is given as

$$L_f = \frac{\bar{V}}{\overline{iL}\, f_s}\{k\frac{\bar{V}_{dc}}{\bar{v}_{AV}}[1 + 4\pi^2 (\frac{fs}{f_{sw}})^2 k\frac{\bar{V}_{dc}}{\bar{V}_{ac}}]\}^{1/2} \quad (18)$$

where,

$k$ is the modulation index.

From extensive simulation, the value of $R_f$ is selected at 2.0 ohm giving good dynamic performance, actively damps the resonance oscillations, during voltage disturbances. The value of L and C component is determined to minimize the reactive power in these components. From eq. 17 and 18, the capacitance and inductance values are 75 μF and 0.09 mH, respectively for this model.

## 4. PROPOSED CONTROL SCHEME

Journal of Physics: Conference Series **622** (2015) 012058 doi:10.1088/1742-6596/622/1/012058

In this proposed control strategy as shown in Figure 4, the synchronous angular frequency of the supply voltage is detected by using phase locked loop (PLL). The generated reference angular frequency is used to generate the zero sequence component of the system voltage based on eq. 19.

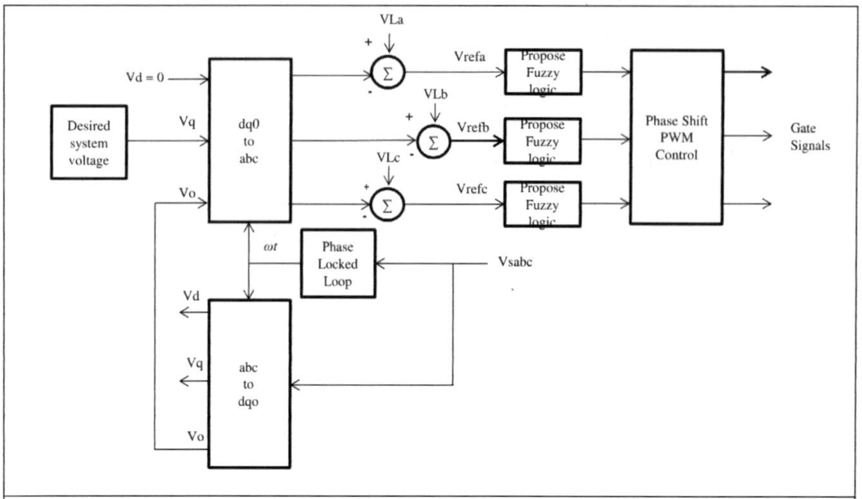


Figure 4. Proposed controller of DVR system multi-carrier

$$\begin{bmatrix} v_{zero\ sequence} \\ v_d \\ v_q \end{bmatrix} = \sqrt{\tfrac{2}{3}} \begin{bmatrix} 1 & 0 & 0 \\ 0 & \cos(\omega t) & \sin(\omega t) \\ 0 & -\sin(\omega t) & \cos(\omega t) \end{bmatrix} \begin{bmatrix} 1/\sqrt{2} & 1/\sqrt{2} & 1/\sqrt{2} \\ 1 & -1/2 & -1/2 \\ 0 & \sqrt{3}/2 & -\sqrt{3}/2 \end{bmatrix} \begin{bmatrix} v_a \\ v_b \\ v_c \end{bmatrix} \tag{19}$$

By applying the inverse of the park transformation under consideration of ($V_d = 0$, $V_q = V_{Smax}$, $V_0 = V_{zero\ sequence}$), the desired system voltages are generated as shown in (20). Position of d-q vector is shown in Figure 5.

$$\begin{bmatrix} v_a \\ v_b \\ v_c \end{bmatrix} = \sqrt{\tfrac{2}{3}} \begin{bmatrix} 1 & 0 & 0 \\ 0 & \cos(\omega t) & -\sin(\omega t) \\ 0 & \sin(\omega t) & \cos(\omega t) \end{bmatrix} \begin{bmatrix} 1/\sqrt{2} & 1 & 0 \\ 1/\sqrt{2} & -1/2 & \sqrt{3}/2 \\ 1/\sqrt{2} & -1/2 & -\sqrt{3}/2 \end{bmatrix} \begin{bmatrix} v_{zero\ sequence} \\ 0 \\ V_{S\max} \end{bmatrix} \tag{20}$$

The reference compensation voltages to control DVR are generated by comparing the calculated desired system voltages in (20) and the measured load side voltages, and then the error signals are subjected to a phase shift multicarrier PWM controller in order to determine the desired switching pattern for the IGBTs of DVR.

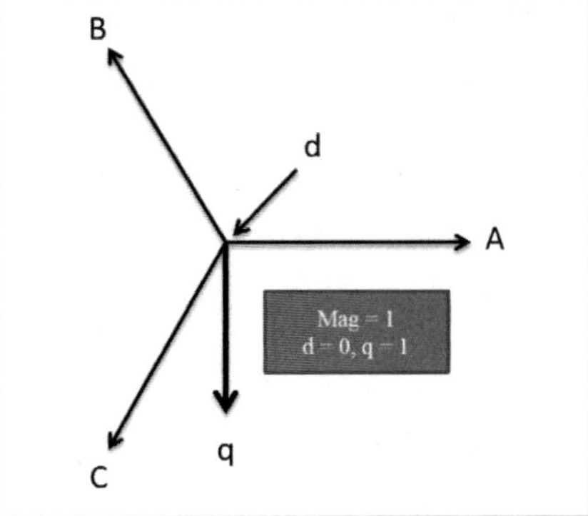


Figure 5. D-Q vector representation

## 5. RESULTS AND DISCUSSION

The model is connected to three-phase at 415 V, 50 Hz, power source, as shown in Figure 6. The sensitive load is three-phase R-L circuit rated at (R = 8.0 ohms, L = 0.05 mH) with a parallel connected 3-phase diode bridge rectifier series with a 0.9 mH reactance, as nonlinear load. Voltage harmonics distortion is introduced when a three-phase diode Bridge rectifier is connected at the main supply.

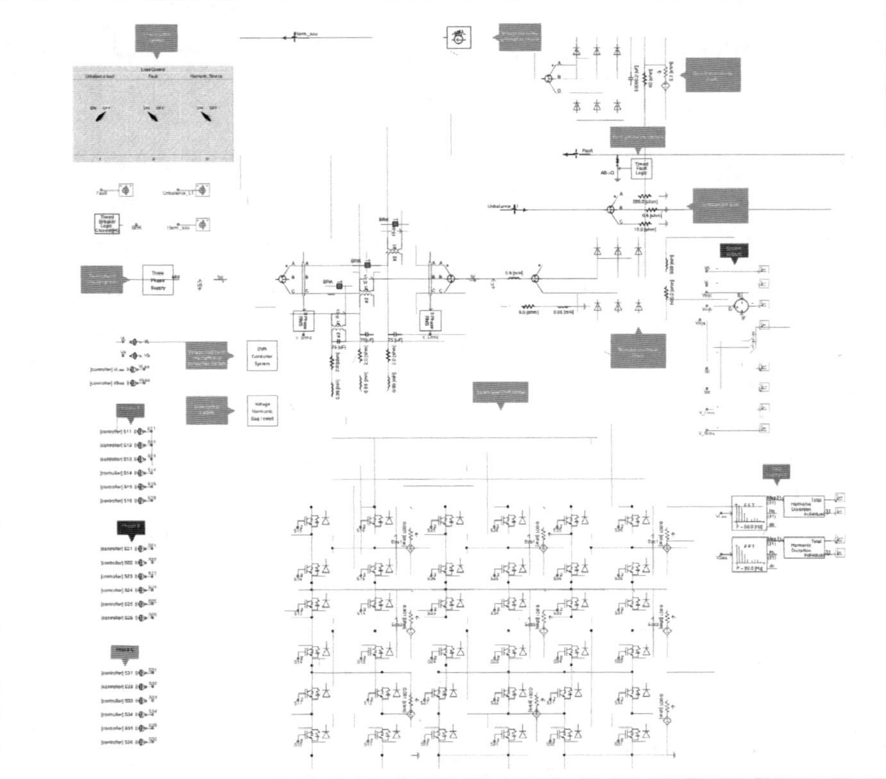

Figure 6. Power circuit of DVR model in PSCAD

### 5.1 Compensation of voltage harmonics

Voltage harmonics is initiated at the source side between ($t$ = 0ms to 300ms), distorting both source and voltage at sensitive load as presented in Figure 7. THD level recorded at source side is 12.56 %. Figure 7 also shows voltage at the sensitive load and the injected voltage by DVR.

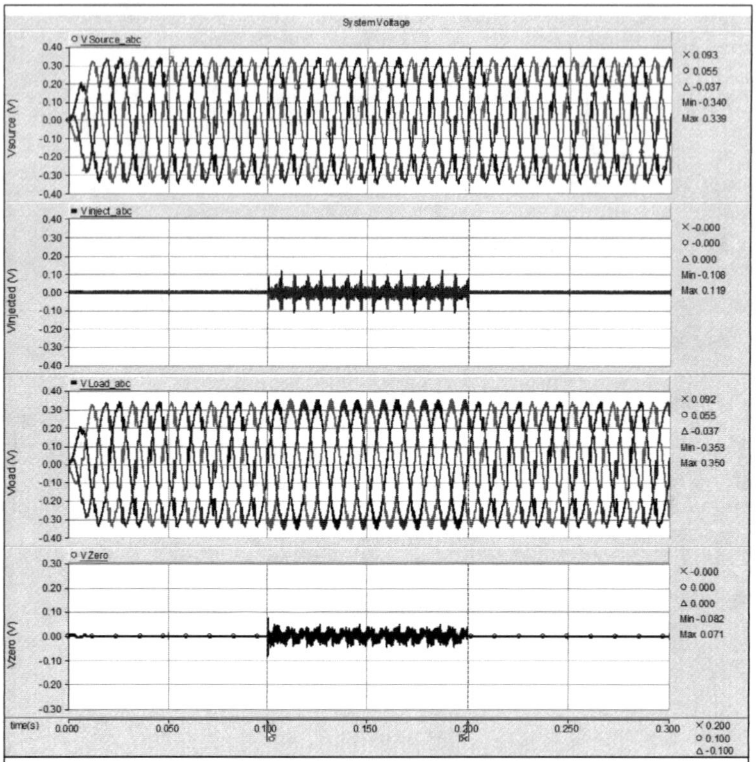


Figure 7. Load voltage compensated

As can be seen from Figure 7, the proposed DVR is able to inject the appropriate three-phase voltage component to correct the supply voltage anomalies due to three-phase harmonic distortion. Under normal operation the DVR is doing nothing. It quickly injects the necessary voltage components to smoothen the load voltage upon detecting voltage harmonics. The load voltage waveform is sinusoidal maintained at 1.0 pu. The source and load current waveform is shown in Figure 8. As seen from the figure, the load current is compensated and is sinusoidal.

Journal of Physics: Conference Series **622** (2015) 012058 doi:10.1088/1742-6596/622/1/012058

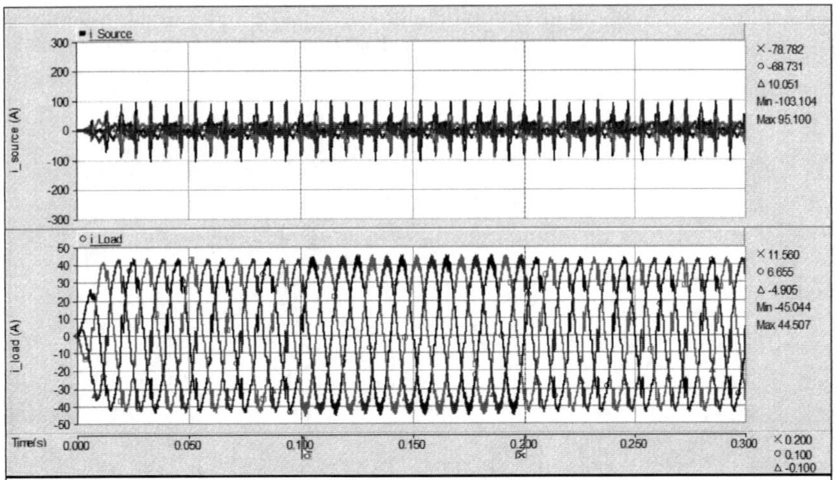


Figure 8. Source and load current waveform

When the DVR is connected, the voltage THD is considerably reduced from 12.56 % to 1.793 % at sensitive load end as shown in Figure 9. This is a significant improvement. The system parameters used in the model are shown in Table 1.

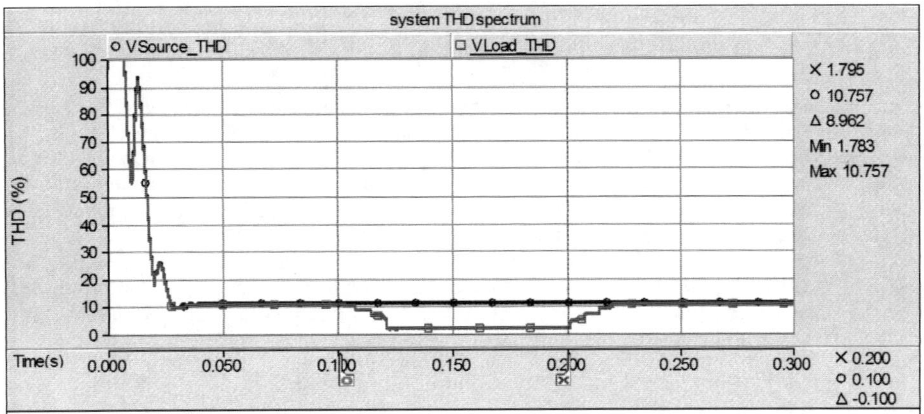


Figure 9. Comparison of THD level.

Table 1

| | |
|---|---|
| Supply voltage | 415 V |
| *F*s | 50 Hz |
| *F*sw | 10 kHz |
| *f*cut-off | 650 Hz |

To evaluate performance of the proposed DVR based Seven-level cascaded multilevel inverter against other published literature, a comparison matrix is shown in Table 2. The Table shows that the voltage THD level of the current model developed in PSCAD has the lowest value. The main contributions of this study are as follows: seven level cascade multilevel converter based DVR for power quality improvement of the distribution systems has been developed and simulated in PSACD/EMTDC. Improvement in system voltage transient. New simple controller schemes based d-q theory to overcome the disadvantages of the existing controller schemes by reducing the complexity, number of signal measurements and computing time has been developed.

Table 2

| Description | [20] | [21] | [22] | [23] | [8] | Current Study |
|---|---|---|---|---|---|---|
| Year | 1995 | 2008 | 2009 | 2012 | 2012 | 2014 |
| Operating voltage and frequency | 480 V, 60 Hz | 2300 V, 50 Hz | 346 V, 50Hz | 400 V, 50Hz | 400 V, 50Hz | 415 V, 50Hz |
| Inverter type | Passive Clamped Resonant DC Link Inverter | Diode Clamped | 3-level | 3-level | 3-level | Cascade Multilevel |
| RMS of the load side voltage after compensation | 283 V | Not given | 210 V | 240 V | 240 V | 245 V |
| THD of the load side voltage after compensation | 3.39 % | 2.23 % | 3.31 % | 3.9 % | 4.35 % | 1.793 % |

## 6. CONCLUSION

This project presents a new model of DVR based Seven-level cased multilevel converter and, the controller. The simulation results show that voltage imperfection mitigation, and the enhancement of voltage stability has been achieved. The performance comparison of the proposed DVR and its controller with some previously reported DVR modules shows that it is more efficient in terms of cost, component, power consumption and losses. It is also found that the absolute beneficial of DVR can be achieved with connection of the distributed generation sources such as solar cells, wind generation system etc in the DC side of the proposed DVR model. The simulation results have demonstrated excellent suppression of voltage harmonic capabilities of DVR using the proposed new RLC filter carrier based PWM technique. The DVR is an attractive custom power device for power quality improvement specifically to suppress voltage harmonics and also to mitigate voltage sag and combination both as a hybrid unit.

## 7. REFERENCES

[1] Dayi Li, Jun Tian, "A Novel Active Power Filter for the Voltage-Source Type Harmonic Source", Electrical Machines and Systems, International Conference, pp. 2077-2080, 17-20 Oct. 2008.
[2] R. B. Gonzatti, "Hybrid active power filter applied to harmonic compensation of current-source type and voltage-source type nonlinear loads", Power Electronics Conference, pp. 1257-1262, 27-31 Oct, 2013.
[3] R. Chan Bahaudin I.Daut, "Investigation of Influence of Voltage Source Type on Harmonic Characteristics", The 4th International Power Engineering and Optimization Conf. (PEOCO2010), Shah Alam, Selangor, MALAYSIA: pp. 391-395, 23-24 June 2010
[4] B. Swathi, "Minimization of Resonant Frequency Oscillations in Terminal Voltages of AC Motors Using Active Damping Technique", International Journal of Engineering Research and Development, Volume 5, Issue 6 (December 2012), PP. 20-31.
[5] Kamalesh Hatua, "Active Damping of Output LC Filter Resonance for Vector-Controlled VSI-Fed AC Motor Drives", IEEE Transactions on Industrial Electronics, Vol. 59, NO. 1, pp. 334-342, January 2012.

[6] Nisha G. K, "Harmonic Elimination of Space Vector Modulated Three Phase Inverter", IMECS, March 14-16, 2012.
[7] Ab. Hamadi, "A New Hybrid Series Active Filter Configuration to Compensate Voltage Sag,
Swell, Voltage and Current Harmonics and Reactive Power", IEEE International Symposium on Industrial Electronics, pp.286-291, 2009.
[8] M. A. Mulla, "A Simplified Control Strategy for Series Hybrid Active Power Filter that Compensat Voltage Sag, Swell, Unbalance and Harmonics" IEEE, pp. 1-6, 6-8 Dec. 2012.
[9] Patricio Salmer, "A Control Strategy for Hybrid Power Filter to Compensate Four-Wires
Three-phase Systems", IEEE Transactions on Power Electronics, Vol. 25, No. 7, pp.1923-1931, July 2010.
[10] A.Z.M. Shahriar Muttalib, "Design and Simulation of an Inverter with High Frequency Sinusoidal PWM Switching Technique for Harmonic Reduction in a Standalone/ Utility Grid Synchronized Photovoltaic System", International Conference on Informatics, Electronics & Vision, pp. 1168-1173, 2012.
[11] Mohammad Monfared, "Analysis, Design, and Experimental Verification of a Synchronous Reference Frame Voltage Control for Single-Phase Inverters", IEEE Transactions on Industrial Electronics, Vol. 61, No. 1, pp. 258-269, January 2014.
[12] Khaled H. Ahmed, "Passive Filter Design for Three-Phase Inverter Interfacing in Distributed Generation", Electrical Power Quality and Utilisation, Journal Vol. XIII, No. 2, pp. 49-58, 2007.
[13] Mohammad Faisal, "PI Controller and Park's Transformation Based Control of Dynamic Voltage Restorer for Voltage Sag Minimization", The 9th International Forum on Strategic Technology (IFOST), pp.276-279, October 21-23, 2014, Cox's Bazar, Bangladesh.
[14] G.V.Nagesh Kumar, "DVR with Sliding Mode Control to alleviate Voltage Sags on a Distribution System for Three Phase Short Circuit Fault", IEEE Region 10 Colloquium and the Third International Conference on Industrial and Information Systems, Kharagpur, INDIA , pp.1-4, December 8 -10, 2008.
[15] Bhim Singh, Indirect Control of Capacitor Supported DVR for Power Quality Improvement in Distribution System" Power and Energy Society General Meeting - Conversion and Delivery of Electrical Energy in the 21st Century, 2008 IEEE, pp.1-7, 20-24 July 2008.
[16] K.Chandrasekaran, "Dynamic Voltage Restorer as Sag mitigator and Harmonic restore", pp. 904-909, IEEE Singapore, 5 Sept. 2008.
[17] D. Rajasekaran, "Mitigation of Voltage Sags and Voltage swells by Dynamic Voltage Restorer", Froc. of Int. Conf on Advances in Recent Technologies in Communication and Computing, pp. 36-40, 20 11.
[18] D. Rajasekaran, "Dynamic Voltage Restorer Based on Fuzzy Logic Control for Voltage Sag Restoration", Second International Conference on Sustainable Energy and Intelligent System (SEISCON 2011),pp26-30,July. 20-22, 2011.
[19] A. L. Fawzi "Modeling and Simulation of Different System Topologies for Dynamic Voltage Restorer", Electric Power and Energy Conversion Systems International Conference, EPECS '09. pp. 1-6, 2009.
[20] Subhashish Bhattacharya, "Synchronous frame based controller implementation for a hybrid series active filter system", Industry Application Conference, pp. 2531 - 2540 vol.3, 8-12 Oct 1995.
[21] P. Boonchiam, "Diode-clamped Multilevel Voltage Source Converter Based on Medium Voltage DVR", pp. 62-67, International Journal of Electrical Power Energy Systems Engineering, Waset, Spring 2008.
[22] Ab. Hamadi, A New Hybrid Series Active Filter Configuration to Compensate Voltage Sag, Swell, Voltage and Current Harmonics and Reactive Power", IEEE International Symposium on Industrial Electronics, pp. 286-289,Seoul Olympic Parktel, Seoul, Korea July 5-8, 2009.
[23] Alireza Javadi, "A Novel Transformerless Hybrid Series Active Filter", IEEE Industrial Electronics Society, pp.5312- 317, 25-28 Oct. 2012.

# Development of Landslide Early Warning System Using Macro-bending Loss Based Optical Fibre Sensor

**Ahmad Marzuki*, Muhammad Heriyanto, Ika Dedy Setiyadi, Sorja Koesuma**
Optics & Photonics Lab., Physics Department, Sebelas Maret University

amarzuki@mipa.uns.ac.id

**Abstract**. This paper presents the design of a simple and cheap landslide early warning system which mainly consists of a displacement fibre sensor, mechanical displacement converter, and Short Messaging Service (SMS) gateway equipped with a siren. Displacement fibre optic sensors were made by wrapping a polymer optical fibre (POF) around a holey elastic cylinder connected to a mechanical displacement converter that converts a real land displacement in centimetres order of magnitude into millimetres order that fibre optic sensor can detect. From the experimental results we suggest an optical fibre sensor that has ability to monitor land displacement in the range of 40 cm, sensitivity of (5.9 ± 0.2) dB/cm and linearity 99.5% as well as the way of improving sensor performance to meet the real need. A whole system has been tested making use of a slider attached to the mechanical displacement converter. Once a nonzero continuous displacement for 5 seconds or a downward land displacement of 10.0 cm occurs, the system will activate the siren and spread an alert via SMS automatically.

## 1. Introduction

Geologically, a landslide can be described as downward movement processes of materials forming an inclined ground. Some movements are rapid, take seconds to occur, while others may take years. There are many different causes of landslides, they are usually related to slope instability caused by the failure of earth materials to overcome the force of earth gravity that pulls everything downward. Steep slope and water content are two important factors that can serve as an alert to a potential landslide problem. Ideally, the areas affected by landslide are free from any occupation. Unfortunately, many areas with high landslide potential to occur are versatile and over decades have been densely populated. In many places, evacuating people from their founding fathers' lands is not easy. Therefore, landslide can cause not only property damages but also injuries and deaths.

Landslide early warning systems are of great benefit since they can reduce loss of life and property. Fibre optic sensor is one of sensors currently used for this purpose. Compared with other types of sensors, fibre optic sensor shows a number of advantages, two of them are due to their high sensitivity and the possibility to measure in hostile environments. There exist a great variety of fibre optic sensors developed for landslide monitoring, such as Fibre Bragg Grating (FBG) [1, 2, 3], Optical Time Domain Reflectometer (OTDR) [4, 5], and Brillouin optical time domain reflectometry (BOTDR) [6]. These sensors are known to have particularly high sensitivity, accuracy and reliability. Despite these outstanding properties, these types of sensors are very expensive and for most region in the developing and under developing countries their price are unaffordable. In this work, a landslide early warning system works based on land movement data collected by an intensity modulated fibre optic sensor was made. The system was designed to provide an early warning via SMS and siren simultaneously to the local residents on the impending landslide occurrence nearby. Due to its simplicity and low cost, this sensor can be an ideal landslide monitoring technology for use in developing countries.

## 2. Experiment

Figure 1 is schematic arrangement model of landslide monitoring system used in this experiment. It mainly consists of three parts: mechanical displacement converter, displacement fibre sensor, and Short Messaging Services (SMS) gateway. Early landslide movement is usually slow; it might be millimetres or centimetres per year. Significant incremental displacement sometimes occurs only in years of higher than normal rainfall. In such a case, continuous displacement of millimetres or even centimetres per day has to be regarded as the initial criteria for early warning. Nevertheless, the magnitude scale of displacement fibre sensor developed in this experiment is in millimetres. In order that landslide movement which is typically measured in centimetres order can be monitored by the sensor, a displacement converter is required. In this experiment, this is done by a pulley system composes of two pairs of identical wheels. Wheel A and C have radius of 10.0 cm, while wheel B and D have radius of 1.0 cm. As shown, wheel A and B are on the same axis and so are the wheel C and D.

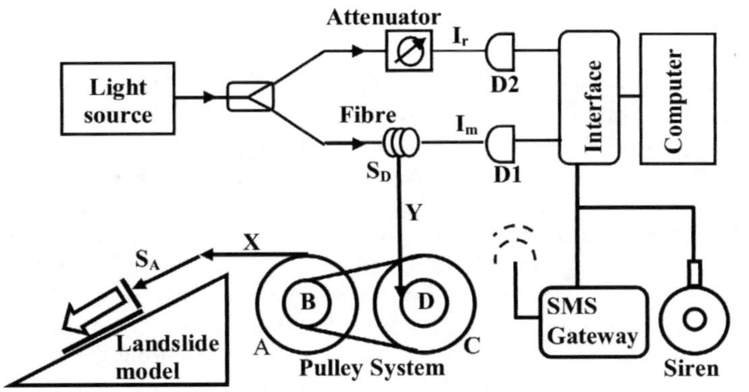


Figue 1. Schematic arrangement model of landslide monitoring

As shown, wheel A of radius $r_A$ and B of radius $r_B$ are on the same axis and therefore have the same angular speed (ω), i.e.:

$$\omega_A = \omega_B$$

Taking the angular speed and linear speed ($v$) relationship as $v = \omega r$, we have:

$$v_B = \frac{r_B v_A}{r_A} \tag{1}$$

Similarly, wheel C and D are on the same axis and therefore

$$v_C = \frac{r_C v_D}{r_D} \tag{2}$$

Further from Figure 1, wheel B is coupled by belt E to wheel C and hence $v_B = v_C$. From equation (1) and (2), therefore we have

$$\frac{v_D}{v_A} = \frac{r_D r_B}{r_C r_A}$$

Remembering that linear displacement ($S$) of any point on the wheel can be expressed as $S = vt$, we can calculate the relationship between linear displacement of wheel A and D. If X rope is displaced by landslide movement by $S_A$ then the Y rope that goes to the fibre sensor will be displaced by $S_D$ as expressed by

$$S_D = \frac{r_D r_B}{r_C r_A} S_A = \frac{1}{100} S_A \qquad (3)$$

where $r_A$, $r_B$, $r_C$, and $r_D$ are the wheel radius of A, B, C and D, respectively. The multiplication factor as given by equation (3) can be varied by varying the wheels radius or adding more wheels into the pulley system.

Further from Figure 1, light coming out from the light source was coupled into a 50/50 fibre splitter. One arm of the fibre splitter was connected to the modulated fibre and the other one to the reference fibre. Light detected by both light detectors (D1 & D2) were recorded and processed by a computer program to calculate the transmission coefficient ($T$) defined by

$$T = \frac{I_m}{I_r} \qquad (4)$$

Light intensity detected from the modulated fibre arm ($I_m$) was designed to be less than that detected from the reference one ($I_r$).

To set the transmission coefficient prior to modulation equals to 1, an attenuator was added to the reference arm. The way in which the fibre coil responds any landslide movements is shown in Figure 2. Fibre coil, which is a hollow mandrel rubber wrapped with polymer optical fibre, was positioned between two walls: fixed and moveable wall. As the suspected land is displaced by $S_A$ (Figure 1), the moveable wall as shown in Figure 2 is pulled by the wire connected to the pulley system by $S_D$ and thereby compressing the coil system and the mandrel circular cross-section changes to an ellipse. As described in the next section, this change will result in decreasing light intensity emerging from the coil ($I_m$). Applying equation (4), the transmission coefficient ($T$) decreases accordingly. To obtain a sensitive fibre sensor, a suitable cylindrical rubber diameter had to be chosen. In this experiment, the most suitable diameter was obtained by comparing the gradient of the optical loss-displacement curve as function of the fibre coil diameter (1.7 cm, 1.4 cm, 1.0 cm) as well as the number of turns of the fibre coil (1, 2, 3, 4 turns). The optimum configuration (coil diameter and number of turns) are then judged by the magnitude of their gradient and the width of their linear part.

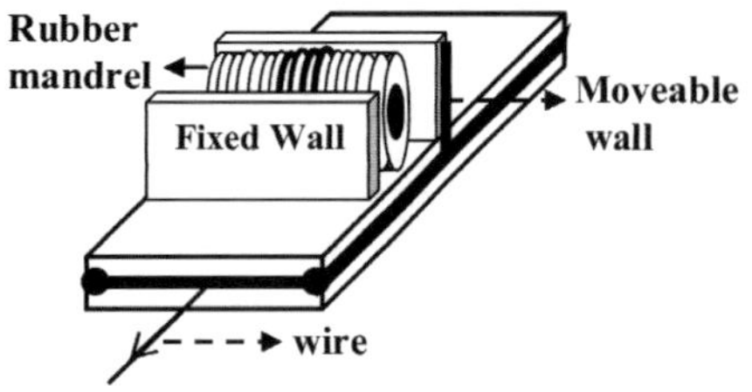


Figure 2. Modulation mechanism in fibre sensor.

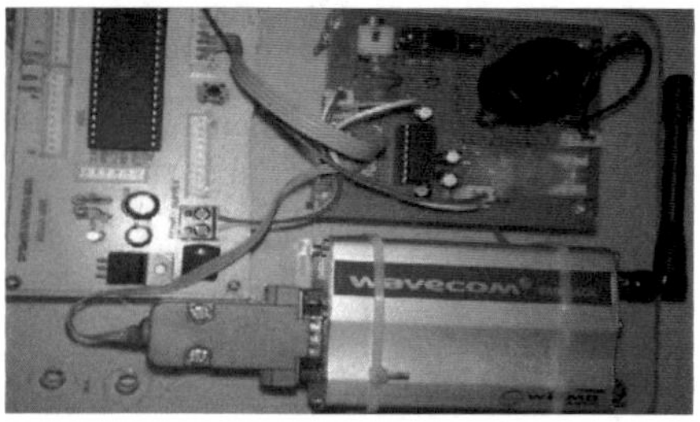

Figure 3. SMS gateway used to send an alert.

In order to examine the sensor system performance, a model of a land movement and an SMS gateway device were made. Figure 3 is an SMS gateway device used in this experiment. This device consists of three main components: micro-controller, TTL to RS232 converter, and modem. Looking back to Figure 1, triggered by a land movement light coming out of the modulated fibre arm ($I_m$) decreases. This intensity change is detected by detector in the form of voltage change. For this purpose, detector is connected to a port of ADC ATMega8535 microcontroller. This microcontroller is functioned as a control system where voltage data variation taken from detector is processed. To make the system be able to spread a landslide alarm (siren sounds and SMS), the microcontroller is connected to a siren and a Wavecom modem. A computer program containing a defined critical condition at which landslide warning has to be spread out was given to microcontroller. For convenience, the critical condition set in this experiment is the occurrence a nonzero continuous displacement for 5 seconds or a downward land displacement of 10.0 cm.

## 3. Results and Discussion

Figure 4 presents how bend losses vary as function of the land model displacement measured at different coil mandrel diameters and number of turns of each coil. Since fibre attenuator was added in our experimental setup (Figure 1), the initial transmission coefficient can be managed always equals to 1. It is well known that a bend loss can be expressed in term of a transmission coefficient as

$$Loss(dB) = 10.^{10}Log\left(\frac{1}{T}\right) = 10.^{10}\log\left(\frac{I_r}{I_m}\right) \quad (5)$$

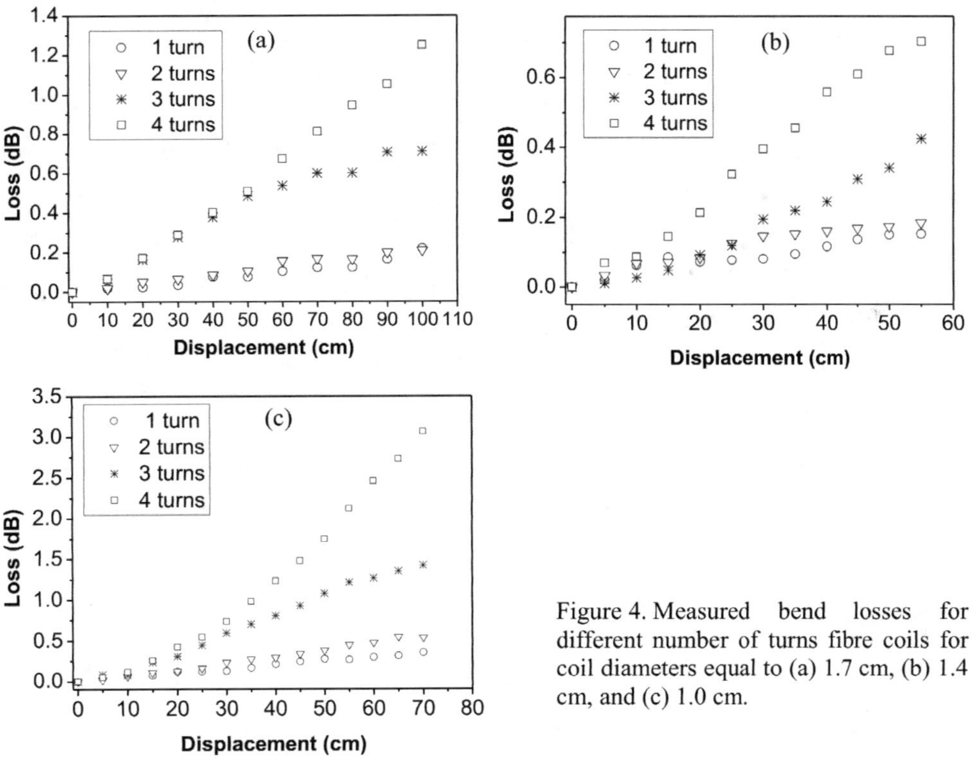


Figure 4. Measured bend losses for different number of turns fibre coils for coil diameters equal to (a) 1.7 cm, (b) 1.4 cm, and (c) 1.0 cm.

Here, the attenuator manages the initial intensity of light passing through fibre reference ($I_r$) such that before displacement was given $I_r/I_m=1$. In other word, the initial loss of all curves can always be set equal to 0. The attenuator added in this setup can thus be functioned as a zero button, which is required for calibration purposes. In general, fibre loss at a given radius is higher for a coil with more turns. The change is more obvious for coils with a diameter of 1.0 cm (Figure 4.c). These typical results show good agreement with that obtained by other researchers investigating similar sensor designed for microdisplacement [7].

It is clear from Figure 4 that the bend loss increases as the displacement of the land model increases. As the wire in Figure 2 is displaced by $S_D$, the circular shape of the coil changes to the elliptical shape (Figure 5). The vertical axis and horizontal axis of the circular coil become minor and major axis of the ellipse, respectively. From geometric optics point of view, this increase is caused by light radiated out of the fibre at the high curvature point. As shown in Figure 4.b, light is coupled into a multimode fibre within the acceptance angle $\beta$ and incident on the core-cladding interface at the angle of $\theta$. For a given angle $\beta$, there will be a certain number of allowed $\theta \geq \theta_{CR}$ for the light to travel through the fibre. Around the maximum curvature point $X$, many angles $\alpha$'s decrease down to bellow their corresponding critical angles such that total internal reflection cannot occurs. Consequently, part of the light is radiated out of the fibre.

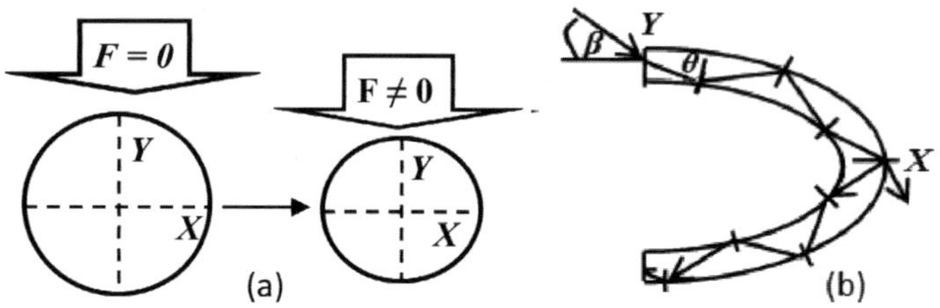


Figure 5. (a) Coil cross section change due to an applied force. (b) Light leaks out of the bent fibre edge.

There are three groups of coils used in this experiment, i.e., circular coil with the diameter of 1.7 cm, 1.4 cm, and 1.0 cm. For a given displacement, $S_D$, they are transformed into ellipse with different curvature. To understand how this is different, let's see Figure 4a. As no external force is applied ($F=0$), the coil is in circular shape. This shape turns to an elliptical shape as an external force ($F \neq 0$) is applied. Both shapes therefore have the same perimeter $p$. Taking the ellipse perimeter ($p_{el}$) as

$$p_{el} = 2\pi\left(\frac{a^2+b^2}{2}\right)^{1/2} \tag{6}$$

where $a$, $b$ are semi-major and semi-minor of the ellipse, respectively; and noting that the ellipse has the same perimeter as its corresponding circular coil of radius $r$, i.e., $p_{cir} = 2\pi r$, we can write

$$2\pi r = 2\pi\left(\frac{a^2+b^2}{2}\right)^{1/2}$$

and therefore

$$a = \left(2r^2 + b^2\right)^{1/2} \tag{7}$$

Looking back to Figure 2, if the moveable wall is displaced by $S_D$, an ellipse is formed (Figure 4). Accordingly, its semi-minor is $b = r - \frac{S_D}{2}$ and from equation (7) its semi-major is

$$a = \left(r^2 + rS_D - \frac{S_D{}^2}{4}\right)^{1/2} \tag{8}$$

Knowing that the curvature of an ellipse at any point $(x,y)$ is given by

$$\kappa = \frac{1}{a^2 b^2}\left(\frac{x^2}{a^4}+\frac{y^2}{b^4}\right)^{-3/2} \tag{9}$$

Its maximum curvature, which is the curvature at points ($-a$,0) and ($a$,0), can be found as expressed by

$$\kappa_{\max} = \frac{a}{b^2} = \left(r^2 + rS_D - \frac{S_D^{\,2}}{4}\right)^{1/2}\left(r^2 - rS_D + \frac{S_D^{\,2}}{4}\right)^{-1} \tag{10}$$

Figure 5 shows how $\kappa_{\max}$ of an ellipse formed from a circular coil with diameter $r$ changes as it is squashed by $S_D$ by a moveable wall shown in Figure 1. It is clear from Figure 6 that for a given displacement $S_D$, the maximum curvature of a small ellipse is much higher than the bigger one. As the moveable wall is displaced farther, the curvature of the small size ellipse changes more rapidly than the bigger one. Since the bending loss ($\alpha$) of a fibre is related to the radius of curvature ($R=1/\kappa$) as given by [8]

$$\alpha = C_1 \exp(-C_2 R) \tag{11}$$

where $C_1$ and $C_2$ are constants, then by substituting equation (10) into equation (11) we have

$$\alpha = C_1 \exp\left(-C_2\left(r^2 - rS_D + \frac{S_D^{\,2}}{4}\right)\left(r^2 + rS_D - \frac{S_D^{\,2}}{4}\right)^{-1/2}\right) \tag{12}$$

As a consequence, higher loss is achieved by a fibre coiled onto a smaller radius of rubber mandrel than the larger one (Figure 4). The loss change resulted from the successive displacements is more rapid in small size coil than the bigger one. The higher the curvature, the more rapid the loss will be [8]. This effect is much more significant if the number of turn increases.

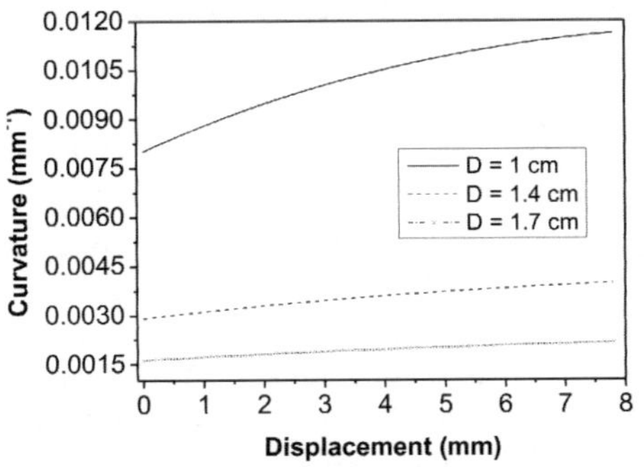


Figure 6. Maximum curvature variation of ellipses formed from different size of circular coils (diameter of 1.0 cm, 1.4 cm, and 1.7 cm).

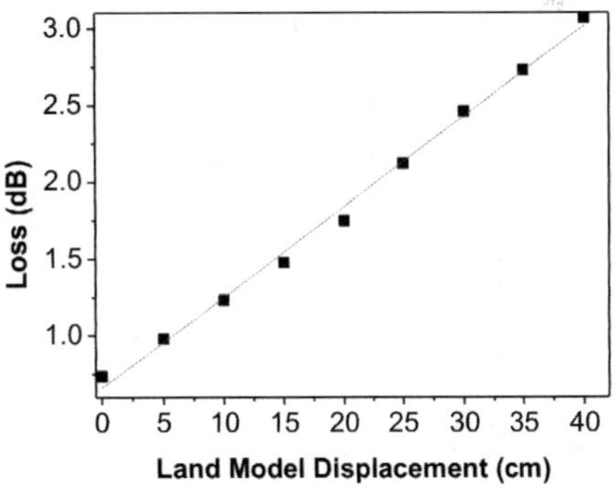


Figure 7. Plot showing the linear part of loss Vs displacement curve of a four turns fibre coil formed on mandrel with a diameter of 1.0 cm

Since sensitivity of a sensor is defined as the change in output of the sensor per unit change in the parameter being measured, fibre sensor made by wrapping four polymer optical fibres onto rubber mandrel with the diameter of 1 cm and with the number of turns is 4 is favourable among those presented in this paper. Taking the linear part of the loss-displacement curve belongs to this favourable fibre sensor shown in Figure 4.c; we can modify this curve as shown in Figure 7. Sensor configured in this way has sensitivity of (5.9 ± 0.2) dB/cm and displacement range 40 cm. The sensitivity and displacement range of this sensor can still be improved by playing with pulley system and/or number of coil turns. Adding a pair of two coaxial wheels exactly the same as that given in Figure 1, for example, will make a tenfold increase in measurable displacement range, i.e., from displacement range of 40 cm to 4 m.

Fig. 8. (a). A slider used as landslide model. (b). Mechanical displacement converter used to convert land displacement in centimeters order into displacement order within which fibre sensor can detect.

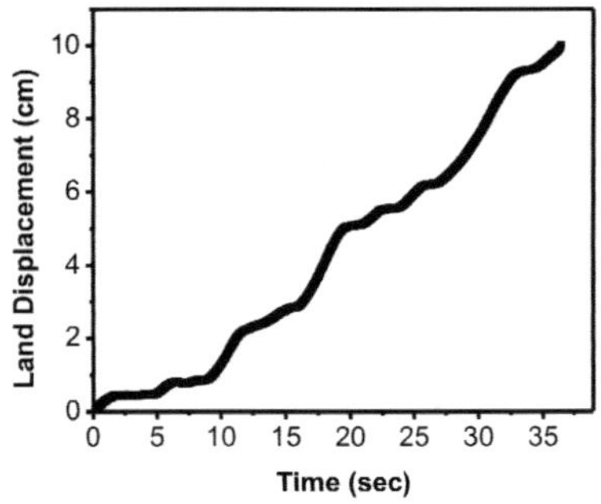


Fig. 9. Land displacement obiained from landslide model

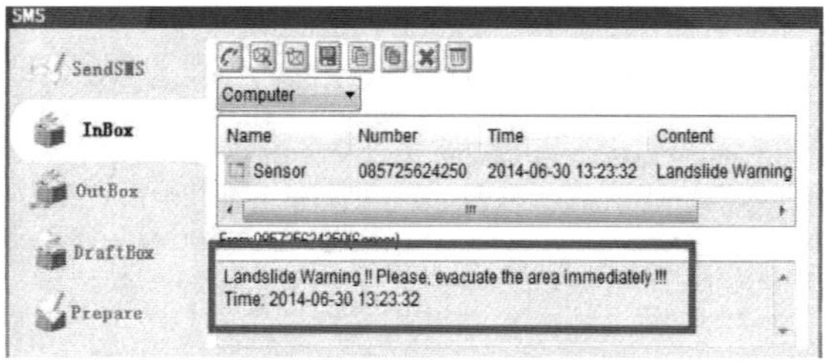


Fig. 10. Example of an SMS alert sent when a landslide is detected by sensor.

A model has been made to test the system performance. A slider (Figure 8.a) attached to a mechanical displacement converter (Figure 8.b) is set to move downward, resulting in displacement of the wall pushing the fibre coil so that its shape turns to ellipse (Figure 2). As the slider displaces continuously downward, the coil curvature will also continuously increase and the fibre loss increases accordingly. Figure. 9 is a typical land displacement history taken from the model made in this research. The actual time required for the land to make a notable displacement varies from site to site [9, 10, 11]. One should note that landslide is an accelerated motion with the initial velocity equals to zero. Alert should therefore be spread from the early stage of motion to provide people to have enough time to evacuate from the danger zone. An early warning system designed in this research is aimed to alert people in a landslide danger zone whenever the sensor detects a continuous nonzero land displacement for few seconds or the magnitude of the land displacement reaches a certain value. People should know that an early step of land movement occurs. Since movement can occur at anytime, alert in the form of SMS is not enough. In order to provide people living in the danger zone with the most effective warning system, a siren has been added to the EWS system developed in this experiment. Alert is sent once sensor detects a nonzero continuous displacement for 5 seconds or the land displaces downward a distance of 10.0 cm. Example of the SMS alert is shown in Figure 10.

## 4. Conclusions and Future Work

A simple landslide early warning system built based on macro-bending loss in polymer optical fibre has been developed and successfully tested. Fibre optic sensor system presented in this paper allows detection of landslide at the early stage of land movement using only a simple and cheap fibre sensor (polymer optical fibre). The system consists of three main parts: displacement fibre sensor system, pulley system and SMS gateway device equipped with a siren. Typical fibre sensor that measures displacement in micrometers or milimeters scale has been combined with a pulley system to form a new system enabling the scale of measurement which can be scaled up to centimeters. Three groups of fibre coil each has different radius and number of turns have been made and compared. It has been shown that sensor system built by wrapping 4 turns of fibre polymer around a rubber mandrel with diameter of 1.0 cm has better sensor performance than others investigated in this paper. This sensor has ability to monitor land displacement in the range of 40 cm and sensitivity of $(5.9 \pm 0.2)$ dB/cm. The key principle to improve the fibre sensor performance to meet the real need has also been presented, i.e., by managing the pulley system as well as the radius and the number of a used coil. Furthermore, we suggest that alert should be spread whenever sensor detects a nonzero land movement.

Continuing this research, our future work will focus on developing this simple and cheap landslide early warning system such that it is ready for field test and provide an alternative landslide early warning system having performance comparable to that of other already established systems.

**Acknowledgments**

This work is supported by the Indonesian Ministry of National Education and Culture for Hibah Unggulan Utama UNS (grant No. 165/UN27.11/PN/2013).

## References

[1] Chuan L, Yong-Gui Z, L Hao, Zhou W, Chen Z and Ning R 2008 *J. Automat Constr* **17** (5) 641-644.
[2] Tsair-Chun L and Yung-Li L 2012 J. *Opt. Commun.* **285** (9) 2363-2367.
[3] Tang T G, Wang Q Y and Liu H W 2009 J. *Opt. Laser. Eng.* **47** 156–160.
[4] Aulakh N S, Chhabra J K, Singh N and Jain S 2004 J. *Exp. Tech.* ***28*** *(3) 37-42.*
[5] Wang B J, Li K, Shi B and Wei G Q 2009 J. *Landslides* **6** 61–68.
[6] Wang A, Miller M S, Plante A J, Gunther M F, Murphy K A and Claus R O 1996 J. *Appl. Opt.* **35** (15) 2595-2601.
[7] Zendehnam A, Mirzaei M, Farashiani A and Farahadi H L 2010 *J. Phys.* **74** 591-603.
[8] Marcuse D 1976 *J. Opt. Soc. Am.* **66** (3) 216-220.
[9] Lowry B, Gomez F, Zhou W, Mooney M A, Held B and Grasmick J 2013 J. Engineering Geology **166** 160-169.
[10] Herrera G, Gutiérrez F, García-Davalillo J C, Guerrero J, Notti D, Galve J P, Fernández-Merodo J A and Cooksley G 2013 J. Remote Sensing of Environment **128** 31–43.
[11] Zhang Y, Li H, Sheng Q, Wu K and Chen G 2011 J. Environm. Sci **23** S100–S105.